# The Land Between

**Library of Congress Cataloging-in-Publication**
A record in the CIP catalog has been requested for this book
of the Library of Congress.

**Bibliographic information published by the Deutsche Nationalbibliothek**
The Deutsche Nationalbibliothek collects this publication in the Deutsche Nationalbibliografie; detailed bibliographic data are available on the Internet at Nationalbibliografie; detailed bibliographic data are
available on the Internet at http://dnb.d-nb.de.

Published with the financial support of the Research Centre of the Slovenian Academy of Sciences and Arts, and ARIS research program "Historical Interpretations of the 20th Century" (P6-0347).

Front cover ©: Giulio Quaglio (draft), a cover for Jurij Andrej Gladič's poem Vetus et Nova Carnioliae Ecclesiastica memoria, copper engraving, 1721–1723, photo courtesy Ana Lavrič, Franc Stele Institute of Art History, ZRC SAZU.

ISBN 978-3-631-91046-7 (Print)
E-ISBN 978-3-631-91047-4 (E-PDF)
E-ISBN 978-3-631-91048-1 (E-PUB)
DOI.10.3726/b21313

© 2024 Peter Lang Group AG, Lausanne
Published by Peter Lang GmbH, Berlin, Germany

info@peterlang.com - www.peterlang.com

All rights reserved.
All parts of this publication are protected by copyright.
Any utilization outside the strict limits of the copyright law, without the permission of the publisher, is forbidden and liable to prosecution. This applies in particular to reproductions, translations, microfilming, and storage and processing in electronic retrieval systems.

The Land Between

Oto Luthar / Marjeta Šašel Kos / Petra Svoljšak /
Martin Pogačar / Peter Štih / Dušan Kos /
Peter Kos / Igor Grdina / Alja Brglez

# The Land Between
A History of Slovenia

Berlin - Bruxelles - Chennai - Lausanne - New York - Oxford

**Library of Congress Cataloging-in-Publication**
A record in the CIP catalog has been requested for this book
of the Library of Congress.

**Bibliographic information published by the Deutsche Nationalbibliothek**
The Deutsche Nationalbibliothek collects this publication in the Deutsche Nationalbibliografie; detailed bibliographic data are available on the Internet at Nationalbibliografie; detailed bibliographic data are
available on the Internet at http://dnb.d-nb.de.

Published with the financial support of the Research Centre of the
Slovenian Academy of Sciences and Arts, and ARIS research program
"Historical Interpretations of the 20th Century" (P6-0347).

Front cover ©: Giulio Quaglio (draft), a cover for Jurij Andrej Gladič's poem
Vetus et Nova Carnioliae Ecclesiastica memoria, copper engraving,
1721–1723, photo courtesy Ana Lavrič, Franc Stele
Institute of Art History, ZRC SAZU.

ISBN 978-3-631-91046-7 (Print)
E-ISBN 978-3-631-91047-4 (E-PDF)
E-ISBN 978-3-631-91048-1 (E-PUB)
DOI.10.3726/b21313

© 2024 Peter Lang Group AG, Lausanne
Published by Peter Lang GmbH, Berlin, Germany

info@peterlang.com - www.peterlang.com

All rights reserved.
All parts of this publication are protected by copyright.
Any utilization outside the strict limits of the copyright law, without
the permission of the publisher, is forbidden and liable to prosecution.
This applies in particular to reproductions, translations, microfilming,
and storage and processing in electronic retrieval systems.

# Contents

Preface .................................................................................................. 9

Introduction ........................................................................................ 11

## From Prehistory to the End of the Ancient World

History Created by Archaeology ...................................................... 15

The Roman Empire: Conquest and Pax Romana ............................ 41

From the Marcomannic Wars to the Settlement of the Slavic Tribes ......... 67

## The Early Middle Ages

The Early Slavs .................................................................................. 89

The Carolingian Period of the 9th Century .................................... 103

## Feudalism

Reorganization of the Marches and a Shift of Ethnic and Language Borders ................................................................................. 121

From Autonomy to the Unification of the Alpine and Danube Basin Regions ........................................................................................ 129

"Tres Ordines Slovenorum": Society, Economy, and Culture ........ 147

The Stars of Celje ............................................................................ 171

The Bloody Fall of the Middle Ages ............................................... 189

## The Early Modern Period

From Humanism to Reformation .................................................................. 207

From Counter-Reformation Rigor to Baroque Exuberance ....................... 227

Scholars, Officials, and Patriots Changing the World ................................ 245

## Modernization and National Emancipation

French Rule ................................................................................................... 273

The Pre-March Era, the Time of Non-freedom .......................................... 283

"The Year of Freedom," the 1848 Revolution, and United Slovenia ......... 299

The Slovenes in the Constitutional Era ....................................................... 315

Unity and National Existence ...................................................................... 349

In the Shackles of Political Parties ............................................................... 355

The Other Side of History: Herstory ........................................................... 389

## From the Habsburg Monarchy to the Kingdom of Yugoslavia

Divided by the Great War ............................................................................. 397

The Making of the New State ...................................................................... 407

The Kingdom of Serbs, Croats and Slovenes .............................................. 413

Dictatorship and the Crisis ........................................................................... 429

"A Nation Torn Apart": World War II in Slovenia ..................................... 447

## Slovenia after the Liberation: The "People's Republic" and the Time of Socialism

The Establishment of the "New Order" .............................................................. 475

The First Five-Year Plan and Self-Management ................................................ 493

"Liberals" vs. "Conservatives" .............................................................................. 507

From Crisis to Conflict and Beyond ................................................................... 519

History Does (Not) Repeat Itself ......................................................................... 559

Epilogue ................................................................................................................. 563

A Note on Pronunciation and Regional Toponyms ........................................... 567

List of Figures ....................................................................................................... 569

List of Maps .......................................................................................................... 575

Selected Bibliography .......................................................................................... 577

Notes on Contributors ......................................................................................... 595

Index ...................................................................................................................... 597

# Preface

We were led to write this book not only because we have yet to see a concise history of Slovenia to date, but also because we have noticed that articles about Slovenia, written mostly by journalists and politicians, often turn out to be inadequate and incorrect.

We acknowledge that—in the wake of the collapse of socialism and after the outbreak of wars in former Yugoslavia—the lack of knowledge about the history of Central and Eastern Europe led to a stream of factual errors and misleading interpretations. Their repetition led to new misinformation and questionable analysis, which only added to the existing confusion. Our aim in this book is not to clarify and correct these interpretations, but to provide a detailed and accurate scholarly account of the history of our country. We merely wish to describe the events and processes that happened to those who resided here in the past and to those who reside here today.

In an effort to avoid a one-dimensional description of major events and leading personalities, we have discussed parts of this book with many colleagues from Slovenia, Croatia, Serbia, Bosnia and Herzegovina, Austria, Great Britain, and the United States. We will not attempt to list their names as any list would be incomplete. However, we hope that we have acknowledged the most important contributions through quotations in the text. Nevertheless, there are still a few people who cannot be omitted. Following the chronology of the book, we would like to thank Breda Luthar, Ivan Turk, Janez Dular, Rajko Bratož, Dejan Djokić, Emil Brix, and Peter Vodopivec.

The authors of this book would also like to thank Mateja Belak, Manca Volk Bahun, and Iztok Sajko for creating the maps, our translators Manca Gašperšič, Olga Vuković, Alan McConnell-Duff, and Paul Towend for their hard work, our copyeditors Catherine Baker, Damjan Popić, Hanna Szentpéteri, Miha Zemljič, and Tadej Turnšek for their invaluable contribution to improving the language of our manuscripts, Jean McCollister for her proofreading, and Hanno Hardt for his editorial comments. The only way to thank them properly is to paraphrase the words of Ivo Andrić: The work of a good translator sometimes borders on magic, and seems like real heroism.

The authors,
Ljubljana, 2024

# Introduction

In the summer of 2002, a large, interactive map of the world was installed in front of the Victoria and Albert Museum in London. The map was made up of several small pieces that resembled a giant jigsaw puzzle. Since it was part of a playground in front of a London museum, only children were allowed to access it. In fact, the map was designed for them. They were encouraged to learn, whether from it or on top of it. They were supposed to educate themselves about the countries of the world from the installation, such as their size, shape, and location. The idea itself was brilliant, but, unfortunately, the designers of the map made some minor mistakes. As it often happens, the curators of the museum divided the countries into more important and less important ones.

This mishap seemed to have happened due to technical reasons. Since the map also included pictures of the most famous cities, the mapmakers were simply running out of space. So, they just covered Slovenia with a photo of Venice, since Venice seemed much more important than a tiny country that no one really knew. And just like that, just like many times before, Slovenia's more influential neighbors spread across the country, this time metaphorically.

Having already had the idea of writing a book about the history of Slovenia, we quickly realized that our main task would be to make visible that tiny portion of the map where the average connoisseur of European history and geography would expect to find pieces of Italy or Austria. In order to do this, we decided to include certain Central European countries which have had a decisive influence on the form and shape of Central Europe in modern history. We concluded they could be used to present the history of our country as an integral part of that area.

Later on, once we started writing the book, we found ourselves encountering the same situation again and again. When working on our respective chapters, each of us would find Slovenia, or the Slovenian lands, as a representation of a region or space between two different worlds, an extension between Europe and its periphery. We also encountered this extension in the views of the foreigners who have described this area. These people have experienced the Slovenian lands as distinct from other surrounding areas. From their point of view, the landscape, temperament, history, tradition, and language were different from everything else in that same region. At first glance, they compared it to "Switzerland or the Tyrol," until their "closer inspections" showed them that the temperament was Slavic. The language was similar to Serbian or Croatian, but with noticeable differences. It was observed to be more archaic and complicated, and far more difficult to learn.

Nevertheless, the most puzzling concept for foreigners to understand is Slovenian history. Most people who are interested in the past find it hard to recognize the fact that Slovenians survived so many centuries of foreign rule. Therefore, we decided to depict the history of Slovenia as a history of a particular place, especially when dealing with the prehistoric periods and the time before the arrival of the Slavs. Well aware of the usual brief and superficial description of this period, we decided to present it in its full complexity as comprehensively as possible. For us, this is also a way to show that the Slavs did not settle in an empty space and simply replace the Celto-Roman inhabitants of the earlier times. They were first mentioned by Roman historians as early as the mid-6th century. However, it is only from the 9th century onwards that Medieval historians have been in a position to give these neighboring communities particular names (which they are still known by today), instead of lumping them all together into one generalized Slavic group. However, these writers envisaged them as ready-formed peoples, whereas the reality is more complex. Their formation was a matter of reciprocal acculturation.

We approached the later periods in a similar manner, which is why the Slavic territory is not depicted as a typical frontier march under various margraves. We also tried not to portray them as expendable defenders of the empire against the threat of the Hungarians, sometimes other Slavs, and later the Ottomans. We also do not believe that the Slovenians only made two important appearances in history throughout the entire feudal era. Namely, a brief period of glory under the Counts of Celje as well an early revival of national sentiment and language during the Reformation.

Similarly, in the later periods, especially in the 20th century, most Slovenians joined the resistance movement against the German, Italian, and Hungarian occupation during World War II. Later on, they gained independence without being dragged into the wars that accompanied the disintegration of Yugoslavia.

Our goal with this book is simple: instead of a briefly outlined overview, we offer a concise but complete history of the Slovenian people, thereby affording the reader a perspective that goes beyond the stereotypical depictions of the area. In short, we wish to explore the 2,000-year history of the different dominant ruling cultures in the modern Slovenian territory, and what was hidden beneath their administrative surface. In other words, we wish to present you with a history of the place beneath the aforementioned photo of Venice, for all the readers who wondered what was hiding underneath and for all those who are interested in the history of one of Europe's smallest countries.

# From Prehistory to the End of the Ancient World

# History Created by Archaeology

## The Ice Age and the Decline of the Hunter-Gatherer Community

The history of any country should reach back to the very first traces left by humankind. In the territory we now call Slovenia, this is most likely the end of the Lower Paleolithic Age (c. 250,000 years ago). Some rare stone artifacts from that period are known to originate from the caves Jama v Lozi and Risovec in the Pivka Basin. The Pivka Basin, rich in water sources, food for animals, and the raw material suitable for stone knapping, was an attractive living space for Paleolithic hunter-gatherers. It is therefore not a coincidence that this is an area with the highest frequency of Paleolithic sites in Slovenia.[1]

While no fossil remains of Pleistocene hominins have been confirmed to date in Slovenia, Neanderthals (*Homo sapiens neanderthalensis*) and anatomically modern humans (*Homo sapiens sapiens*) left many traces of their presence nevertheless.[2] The Neanderthals lived in Europe during the Middle Paleolithic (250,000 to 40,000 years ago), before they became extinct and were replaced by modern humans. The Middle Paleolithic sites in Slovenia are mainly found in caves and are between 120,000 and 40,000 years old. At that time, the Mousterian culture, attributed to the Neanderthals, was at its height. Interestingly, Neanderthals had visited those same caves from time to time without being aware of any predecessors, although they used the same routes and pursued similar goals throughout the millennia. This was especially true for the multilayered Paleolithic cave sites Betalov spodmol in the Pivka Basin and Divje babe I near Reka in the Idrijca river valley.

Caves were a natural shelter for both Paleolithic hunter-gatherers and cave bears. Neanderthal tools have often been found alongside cave bear remains which has in the past led to the conclusion that Neanderthals were specialized hunters of this mighty beast. However, newer findings and data obtained during systematic archaeological excavations of Divje babe I has allowed archeologists

---

1 Mitja Brodar, *Stara kamena doba v Sloveniji/Altsteinzeit in Slowenien*, Ljubljana, 2009. The text "The Ice Age and the Decline of the Hunter-Gatherer Community" was rewritten by Matija Turk.
2 The oldest human remains found in Slovenia are part of the cranium from the Ljubija stream on Ljubljansko barje, dated to the Mesolithic.

to at least partially understand the Neanderthals' way of life.[3] While many hearths have been preserved in the cave, only rare remains of hunted animals such as marmots, wolves, brown bears, chamois, and roe deer have been found around them. The Neanderthals visited Divje babe I mostly in search of limb bones and cave bear skulls. They crushed them in order to reach the nutritious bone marrow and brain. In the colder seasons of the Last Glacial Period, Neanderthals also sought shelter in the cave and might have then encountered cave bears that regularly used caves as hibernation dens. There may have been some competition for shelter between man and beast, although little is known about this. Interestingly, Neanderthals and cave bears, both highly adapted to Ice Age conditions, became extinct before the Last Glacial Period reached its peak, while anatomically modern humans and the brown bear—two far less specialized species—survived.

At Divje babe I, characteristic Neanderthal tools such as scrapers and denticulated tools were present for a very long time—from the Last Interglacial Period (c. 115,000 years ago) until their extinction—which suggests that tradition played an important role for the Neanderthals. Nonetheless, they were also innovative and were already working with bones to make a unique bone musical instrument. This approximately 60,000-year-old bone musical instrument (known as the Neanderthal flute) is the oldest wind instrument in the world to date (Fig. 1).[4] This outstanding Slovenian Paleolithic find utterly transformed deeply rooted conservative conceptions of Neanderthals' cognitive abilities, since until then they had not been credited with even the most rudimentary artistic skills up until that point. This Neanderthal musical instrument sheds a new light on both our extinct predecessors and on the origin of music. Divje babe I is also an invaluable cave site because it is the only Paleolithic site in Slovenia where cultural finds of the late Neanderthals and early anatomically modern humans have been found.

During the early Upper Paleolithic (c. 40,000 to 30,000 years ago), modern humans, bearers of the Aurignacian culture, settled the present-day Slovenian territory. Characteristic finds of the Aurignacian are bone points, which were hafted on wooden shafts and used as spears. An outstanding number of bone

---

3   Ivan Turk, *Divje babe I: Upper Pleistocene Palaeolithic Site in Slovenia*, pt. 2: *Archaeology*, Ljubljana: Založba ZRC, 2014.
4   Matija Turk et al., "The Mousterian Musical Instrument from the cave Divje babe I (Slovenia): Arguments on the Material Evidence for Neanderthal Music Behaviour," *L'Anthropologie* 122/4 (2018): 679–706.

points, more than 130, were found in the cave Potočka zijalka (1,675 m) on Mt. Olševa, the first Paleolithic site discovered in Slovenia and excavated by professor Srečko Brodar between 1928 and 1935. In addition, one of the world's oldest bone needles has been found at Potočka zijalka and several perforated cave bear mandibles, interpreted as flutes.

A large quantity of cave bear remains was found at Potočka zijalka, as well as the enigmatic teeth of a musk ox—a characteristically arctic animal—and, more recently, the teeth of another arctic species, the wolverine.[5] Another important Aurignacian Alpine site is the cave Mokriška jama (1,500 m) on Mt. Mokrica, which was a typical cave bear den. In comparison to Potočka zijalka, Mokriška jama yielded only a scarce number of bone points and stone tools.

The Aurignacian was followed by Gravettian culture (30,000 to 20,000 years ago), which saw great prosperity of artistic expression in Europe. Higher population density and increased mobility accelerated the cultural and technological development of anatomically modern humans.

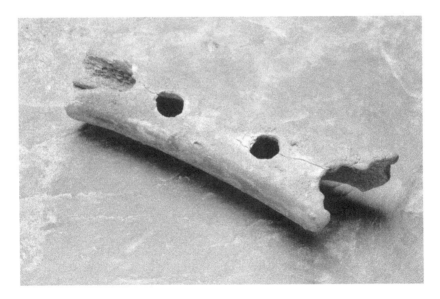

**Figure 1.** Neanderthal musical instrument from Divje babe I.

---

5 Srečko Brodar and Mitja Brodar, *Potočka zijalka. Visokoalpska postaja aurignacienskih lovcev/Potočka zijalka. Eine hochalpine Aurignacjägerstation*, Ljubljana: Slovenska akademija znanosti in umetnosti, 1983.

The Late Upper Paleolithic or the Epigravettian (20,000 to 11,500 years ago) was characterized by fast changes in the natural environment, such as glaciation, warming, extinction and the retreat of some animal species. Significant cooling approximately 20,000 years ago caused the last great advance of continental ice and Alpine glaciers, which forced the human population and animals from the affected area to move to the south. The Paleolithic hunting-gathering economy was at its height, and composite hunting weapons became widely used. Characteristic finds of this period are backed bladelets and points that served as barbs fastened to the tips of wooden spears. Pronounced microlithization of stone armatures indicates the invention and use of a bow. The richest Epigravettian cave site in Slovenia is the Ciganska jama, a cave near Kočevje, where the marmot and reindeer were the most haunted animals. Among the open-air sites it is worth mentioning Zemono near Vipava, where two engraved stones were found, a rare example of Paleolithic artistic expression found in Slovenia.

Some 11,500 years ago, the Last Glacial Period was followed by rapid warming and forest expansion. This period of time conventionally refers to the end of the Ice Age (Pleistocene) and the beginning of the Holocene, the current geological epoch. It also marks the beginning of the Mesolithic (Middle Stone Age). The transition from the Paleolithic to the Mesolithic is not well known in Slovenia, and Mesolithic sites are generally poorly researched. Breg pri Škofljici, Pod Črmukljo, Mala Triglavca, and Viktorjev spodmol are among the most significant ones. Much like in the Paleolithic, people still did not have permanent settlements, and subsisted largely by hunting forest animals, such as red deer and boar. The bow became the main hunting weapon. Among lithic tools, microlithic artifacts of geometric forms predominate, particularly scalene triangles and trapezes, which were attached to the tips of arrowheads. By the end of the Ice Age, people began to exploit the mountain environment yet again, as evidenced by numerous open-air sites recently discovered in the Julian Alps in the Upper Soča Valley. At this time, wolves began to be domesticated and the first dogs appeared. After the Mesolithic, people began to turn to farming and raising livestock.

## Herders and Agriculturalists of the Neolithic and Copper Ages

The end of the Ice Age brought about great change: the sea encroached on what had been the continental bay of Trieste up until that time, glacial lakes emerged in the Alps, and flora and fauna were altered. People began to establish permanent settlements, and with that, their population numbers increased greatly, since cultivation and animal husbandry provided extra food supplies. This development marked a major advance, perhaps even a revolution, in the

Neolithic period. People began cultivating specific types of grains, such as wheat and barley, and raising domesticated animals such as goats, sheep, pigs, and cattle.

We now know that the Middle East had already been taken over by Neolithic man by the 8th millennium BCE, even though early evidence places agriculturalists in Greece early in the 7th millennium. Inhabitants of the upper Danubian region cultivated the land during the 6th millennium and made ceramic dishes, which they decorated with uniform patterns. Around that time, a Neolithic population inhabited the central Balkans, characterized by the Vinča culture and named after the Serbian site of Vinča, near Belgrade. Influences from the Carpathian basin and the eastern Adriatic coast also reached Slovenian territory. The oldest open settlement discovered so far is located in western Slovenia, in the foothills of Sermin near Koper—more precisely, the region of Capris, present-day Koper—and dates from the 6th or early 5th millennium BCE. Thus far, the only other finds from the oldest Neolithic are from the caves of the Karst hinterlands of Trieste. This area was probably still settled by hunters, who came into contact with herders from the Gulf of Kvarner and Dalmatia, from where sheep and goat herders brought Danilo ceramics and Hvar pottery to the Karst region. The ceramics are named after Danilo, a site near Šibenik, while the pottery is named after the island of Hvar. The Sermin site suggests that the herders had settled permanently in the littoral. Interesting information about the Eneolithic (Copper Age) in western Slovenia comes from cave sites (at Mala Triglavca and Podmol near Kastelec), which were still the main shelters for people who mostly raised livestock and hunted. Zoological and botanical studies provide evidence that herdsmen spent some time in several of the caves. The remains of sheep and goat feces in caves, the presence of grass pollen in the Neolithic cultural strata at Podmol, as well as the presence of mixed oak forest and typical pasture vegetation all indicate that herds grazed near the cave. Bones provide further evidence that sheep, goats, pigs, and domestic cattle were raised.

Stone tools of the time were mainly fashioned from local stone;[6] the finds of crucibles (vessels used for smelting) indicate that Copper Age metal workers

---

6 They suggest that smaller stone quarries were in use in the 5th millennium BCE, although there is still no direct evidence. Stone tools, mainly axes, from this time have also been found on the slopes of Pohorje, in Slovenske Gorice, in the Prekmurje region, and elsewhere. The first copper artifacts were supposedly brought from the east by metallurgists searching for copper ore in these areas. Copper in its elementary state is rarely found in nature, and oxide carbonate and sulphide copper ores are important for its production. Slovenia does have underground sulphide deposits. See Anton Velušček,

probably smelted the ore in open fireplaces. By blowing, they could increase ventilation and thus accelerate the smelting process. Ventilation aids included elongations for bellows and blowpipes. The oldest copper finds at Slovenian sites are from the first half of the 4th millennium BCE and include axes and daggers, which were probably fabricated in the Middle East. Domestic crafts began to develop somewhat later, as evidenced by the remains of smelting vessels with traces of copper, which originated in the mid-4th millennium BCE and are from Hočevarica in the Ljubljansko barje (Ljubljana Marshes). Remnants of a clay mold from the late 4th millennium BCE were found at the Maharski prekop (Mahar Canal). Single and double-part clay molds for small hatchets and clay vessels for molten metal from Dežman pile dwellings (Dežmanova kolišča) date back to the 3rd millennium BCE, i.e., the Late Copper Age, a time when metalwork was becoming more firmly established in present-day Slovenia. Nozzles for blowpipes and air bellows have also been found here, specifically in the pile dwellings at Špica by the Ljubljanica river in Ljubljana. The pile dwellings near Ig in the Ljubljana Marshes (Fig. 2) were first researched in the late 19th century by Karl Dežman (Deschmann), the curator and later director of the Regional Carniolan Museum in Ljubljana (the modern day National Museum of Slovenia). This period is characterized by the settlement of people belonging to the Vučedol culture (named after a site in northern Croatia), whose presence has been documented mainly on the Ljubljana Marshes.

Herders and livestock breeders did not settle in the interior of Slovenia, where only hunter-gatherers had previously lived, until the 5th millennium BCE.[7] The new population settled on promontories above rivers (such as the town of Moverna Vas near Semič), along river bends, in caves, by lakesides and in marshlands, as well as on hills. They also had hillfort settlements (such as Gradec near Mirna), fortified by stone enclosures. A burial site, which marks the transition from the Late Neolithic to the Copper Age, was discovered at the cave Ajdovska jama, near Nemška Vas, and close to Krško, in the easternmost part of the Lower Carniola (Dolenjska) region. The cave has two entrances and a complex ground plan with several shafts. The dead were placed upon the cave floor according to a predetermined arrangement and usually surrounded by a circle of stones. Beside them various vessels, bracelets, pendants, and necklaces, as well as stone axes and arrowheads. Corn-filled vessels, various animal bones, and fireplaces suggest burial rites, which included eating at the site or offering

"Neolithic and Eneolithic Investigations in Slovenia (Neolitske in eneolitske raziskave v Sloveniji)," *Arheološki vestnik* 50/1 (1999): 59–79.
7 The Late Neolithic and Eneolithic periods. See Hermann Parzinger, *Studien zur Chronologie und Kulturgeschichte der Jungstein-, Kupfer- und Frühbronzezeit zwischen Karpaten und Mittleren Taurus*, Mainz am Rhein: Von Zabern, 1993.

food to the dead. Anthropologists have since identified the remains of 31 individuals: 7 men, 8 women, and 16 children.

**Figure 2.** Pile dwellings on the Ljubljana Marshes.

Most of the known sites from the 4th and 3rd millennia BCE are found in the Ljubljana Marshes, whose inhabitants had maintained contact with Central Europe, the Pannonian Plain, Caput Adriae, northern Italy, Dalmatia and Balkan regions. Some finds even indicate a connection with eastern Europe and the inhabitants of the steppes. In the early 3rd millennium BCE, members of the Vučedol culture also started to establish permanent settlements in the Ljubljana Marshes and elsewhere in the area of contemporary Slovenia (specifically Vinomer, above Metlika, and the Ptuj Castle), but not in the Trieste Karst, where artifacts have been discovered in caves. The pile dwellings in the Ljubljana Marshes were an exceptional discovery. Over 40 of them from various time periods have been unearthed in the past 145 years, and even the imperial court in Vienna was interested in Dežman's excavations. The reason for these pile dwellings was probably the desire for security.[8]

---

8 Research and excavations continue at the Ljubljana Marshes. The platforms' supporting posts, on which the rectangular houses with double-eave roofs rested, were made of water-resistant oak and ash wood. The oldest dwelling on the Resnik Canal (Resnikov prekop) near Ig dates from the 5th millennium BCE. Anton Velušček (ed.), *Resnikov prekop—The Oldest Pile-dwelling Settlement in the Ljubljansko barje*, Ljubljana: Založba ZRC, 2006; see also his *Stare gmajne Pile-dwelling Settlement and its Era: The Ljubljansko barje in the 2nd Half of the 4th Millennium BCE*, Ljubljana: Založba ZRC, 2009. According to analyses, inhabitants of these pile dwellings were mainly engaged in hunting, but domestic sheep and cattle bones also point to livestock, which had become

**Figure 3.** Anthropomorphic clay statuette from Ig on the Ljubljana Marshes.

A characteristic feature of the lake dwellers is their beautifully decorated earthenware of good quality, which is indicative of creative, artistic practices. Clay spindles, bone needles, and remnants of threads reveal that women wove and sewed. Wool must have also been in use, as is indicated by the remnants of sheep of a suitable age and size. Interesting patterns on the ceramics, particularly on anthropomorphic statuettes, may suggest the type of patterns on women's dresses. Statuettes, together with beautiful vases, are the most striking relics from that time (Fig. 3).

The transition from hunting-gathering to the cultivation of land was accompanied by a continuation of gathering as well as hunting and fishing,

---

more important in the 4th millennium BCE. Pile dwellings are also well documented in the Eneolithic Period; see Anton Velušček (ed.), *Hočevarica: An Eneolithic Pile Dwelling in the Ljubljansko Barje*, Ljubljana: Založba ZRC, 2004.

although grazing and dairy farming also became increasingly important.[9] Indeed, food remains indicate that people ate not only fish and shellfish, but also stag, deer, wild boar, and even bear. Buffalo and bison were eaten too, though their remains are rarer. People also gathered forest fruits and nuts, while grape pips were found at several pile dwellings in the Ljubljana Marshes. The remains of harpoons consisting of three parts (a wooden shaft, a bone extension, and a horn-tipped prong), used for hunting beavers, otters, and larger fish, were discovered near the village of Ig. Inhabitants undoubtedly cultivated the land, as indicated not just by the permanent settlement but also by stone hoes, millstones, and harvest knives. Charred wheat grains have also been uncovered, as have cultivated plants like wheat and barley, and the remains of large clay pots, in which they probably preserved their food.

The lake inhabitants of the time also used simple, hollowed-out canoes, known as logboats, which have been found in considerable numbers. Besides the logboat, remains of one of the oldest wooden carts (a wheel with an axle) in Europe was found on the site at Stare Gmajne, dating from the late 4th millennium BCE. According to radiocarbon dating, the largest number of logboats belongs to the 1st millennium BCE, yet logboats from Stare Gmajne were in use ever since the 4th millennium BCE. Pile dwellings disappeared in the mid-2nd millennium BCE, most likely because the lake gradually drained and became marshland.

## The Flourishing of Metallurgy in the Bronze Age and the First Hillforts

During the Early Bronze Age, between the 22nd and the 16th century BCE, the Ljubljana culture reached its height and spread into the littoral region and along the coast to southern Dalmatia. However, its decline at the end of this period has not yet been fully explained. The transition to the 2nd millennium BCE was certainly not as important a turning point in central Slovenia as it was in the Aegean regions, where society was already based on a more highly developed economy. People in the Aegean regions became socially stratified and lived in cities as this was the age of the blossoming Minoan and Mycenaean cultures. In the Danubian region, the population enjoyed a higher level of development as

---

9   Mihael Budja, "Neolithic Transition to Farming in Northern Adriatic. Lactose Tolerance, Dairying and Lipid Biomarkers on Pottery," *Archaeologia Adriatica* 7/1 (2014): 53–75.

well, due to ore deposits in the Carpathians and contacts with more developed cultures.

At that time, the Slovenian regions were on the sidelines of a changing cultural landscape. During the Middle Bronze Age, between late 16th to 14th century BCE, two characteristic cultures had settled in the region. A population that buried its dead in barrows lived in northeast Slovenia (Styria). Their settlements were located in the lowlands and on the hills. Their material culture (which has been poorly researched) was characteristic of Central Europe. To the west, in the Karst region and in Istria, the population known as the Castellieri (written as *kaštelirji* in Slovenian) cultural group lived in fortified hilltop settlements encircled by defensive stonewalls.

While the settlements in Slovenia have been poorly researched, those in the Trieste hinterland are far better known. It is interesting that the inhabitants of the Karst settlements lived according to the old ways as recently as the late 14th and 13th centuries BCE, and their flat cremation burials (known as the Urnfield Culture) were a significant novelty throughout Europe. The Karst inhabitants faced great changes. One of the most significant and best-preserved settlements is Debela Griža in the vicinity of Volčji Grad, near Komen. It had been reinforced with an imposing double defensive wall in low-lying places and with a single wall where karstic sinkholes protected the settlement. Caves still remained attractive occasional dwelling places for Bronze Age man, for example, at Podmol near Kastelec, and the cave below the Predjama Castle. The Karst dwellers began to bury their dead in flat cremation burial grounds as late as the 10th century BCE. The main finds from the Late Bronze Age and in the area of the Castellieri culture are in the vicinity of Škocjan.

At the end of the 14th century BCE, a different way of life began in Europe as new peoples arrived. The new population lived in a different kind of settlement and in diverse ways, which considerably changed the appearance of the settlement. This is especially visible when looking at the burial of the dead. The dead were incinerated, and their ashes preserved in urns, which were simply interred in the earth without additionally covering the barrow. Their culture was known as the Urnfield Culture.

These sudden changes are difficult to explain. Perhaps they were simply caused by great technological advances, which led to the stratification of society in various European regions. However, as has been said, it was probably large-scale migrations that brought these changes to Europe. These migrations are considered to have caused the decline of the Mycenaean culture in Greece, the downfall of the Hittite empire in Asia Minor, and the fall of various important eastern cities like Troy, Byblos, and Ugarit. Egypt was also threatened; the sources mention people from the sea. However, in 1189 BCE they were conquered by Pharaoh Ramses III.

These new peoples, who cremated their dead, also settled in present-day Slovenia, leaving their traces in Styria, in the Prekmurje region, and in central Slovenia. Among the two most significant settlements are Oloris near Doljni Lakoš, and Rabelčja Vas in Ptuj. These diverse Slovenian sites have yielded various types of bronze axes and bronze jewelry (sometimes intricately wrought), and different, simply shaped ceramic urns and other vessels. In some places, like the cemetery of the Rabelčja Vas settlement, these are the only objects to have been found in graves. The settlements in Slovenia belonged to a culture which was relatively uniform throughout the broad expanse from western Hungary to eastern Croatia and northern Bosnia. However, there were not many settlements in the Slovenian region during the Early Bronze Age.

The end of the 2nd millennium BCE brought about even more changes. New peoples most likely arrived in these parts again and altered the appearances of settlements, although their inhabitants were still recognizable by their flat cremation cemeteries. In the southeastern Alpine area, one may speak of a cultured region only in the Late Bronze Age (specifically, in the late 11th and 10th centuries BCE). Various groups were formed at that time, which did not differ greatly, but had local particularities indicated by their material culture. In Slovenia, along the Drava and Mura rivers, these groups are categorized as the Ruše group (Ruše, Maribor, Ormož), the Dobova group, which was the second largest and was represented by the inhabitants of Posavje along the Sava River, and the third group, in central Slovenia, which encompassed those who belonged to the Ljubljana cultural group (Ljubljana, Mokronog, Novo Mesto). Their burial goods did not essentially differ and indicate that the society of the time was only slightly stratified. However, long-term changes began to occur during the approach of the 1st millennium BCE. These would end in the 8th century BCE, with the onset of the Iron Age.[10]

---

10  Innovations penetrated the Bronze Age world from two directions, the Mediterranean area and the lower Danube region. Both regions had a higher level of development: the coastal world was constantly open to various influences and contacts, and the Danube region had rich ore deposits and advanced metalworking crafts. Innovations from the Mediterranean arrived mainly in the Karst and Inner Carniola (Notranjska) regions, while those from the lower Danube mostly reached eastern Slovenia. See Janez Dular, "Ältere, mittlere und jüngere Bronzezeit in Slowenien—Forschungsstand und Probleme (Starejša, srednja in mlajša bronasta doba v Sloveniji—stanje raziskav in problemi)," *Arheološki vestnik* 50/1 (1999): 81–96.

The majority of Bronze Age settlements originated only in the Late Bronze Age, although some had already been occupied throughout the entire Bronze Age, such as Brinjeva gora near Zreče. The houses were, for the most part, wooden, single-room huts: some rested on a stone foundation and featured a clay-covered hearth on the inside. Two important lowland settlements were Oloris (near Dolnji Lakoš) and Ormož.[11] Oloris was established in the foothills of the Lendava slopes. This is the first lowland settlement of the middle and Late Bronze Age to have been discovered in Slovenia. The settlement was situated in the bend of a nearby stream, the Črnec, whose moving river-bed left a swamp in its wake. A wooden fence encircled the village. Remains of a wooden conduit for drinking water were found on the settlement's northern border, in a ditch of the original stream. The walls of the houses, which probably had double-eave roofs, were made of wooden posts, interlaced with branches and plastered with clay. The houses were built close together around a central courtyard, where stoves were placed and around which the life of the settlement revolved. The houses had fireplaces, and pits were dug under the floor to preserve produce. Paleobotanical research has shown that the settlers had cleared the forests to create arable surfaces and had cultivated pastures.

The settlement in Ormož was partly protected by the Drava River and partly by a natural gorge. The exposed sides of the settlement were protected by an earth embankment with palisades and a deep ditch in front of it. This was one of the most significant settlements of the Late Bronze Age in the southeastern Alpine region.[12] It was constructed according to a plan, as revealed by the remains of a road network and the remnants of houses alongside it. The houses were constructed similarly to those in Oloris, except that they were larger; one even had two rooms. According to analyses of animal bones, the inhabitants raised livestock, mainly cattle, but also pigs, sheep, goats, and horses. The adjacent burial grounds were also uncovered. The settlement was still inhabited in the Iron Age, but disintegrated after 600 BCE.

Burial customs elsewhere were different.[13] Little is known about the cults of those times, except for burial rituals, which are accessible to historians through

---

11   Janez Dular, Irena Šavel, and Sneža Tecco Hvala, *Bronzezeitliche Siedlung Oloris bei Dolnji Lakoš*, Ljubljana: Založba ZRC, 2002.
12   Janez Dular and Marjana Tomanič-Jevremov, *Ormož. Befestigte Siedlung aus der späten Bronze- und der älteren Eisenzeit*, Ljubljana: Založba ZRC, 2010.
13   At Mokronog in Lower Carniola (Dolenjska), two large, flat cremation cemeteries were discovered, dating from the Late Bronze Age and the beginning of the Iron Age (10th–8th century BCE), but many finds were destroyed due to sand excavation. The

an analysis of burials and burial objects: vessels are almost always found in graves, as well as jewelry and various other objects. Possibly, the anthropomorphic figurines from Ormož, and various amulets in symbolic shapes (wheel—sun, sickle—moon), which were used as pendants, could indicate the presence of religious ideas at that time.

Bronze Age hoards, especially of tools and weapons, are particularly remarkable.[14] They are thought to be the unrecovered property of traveling traders, buried during times of migration and danger. Although their significance has not yet been clearly explained, the prevailing opinion suggests that these were cult offerings by individuals or larger communities who dedicated them to deities and demons. The Karst region of Škocjan, with its renowned Škocjanske jame (Škocjan Caves), as well as other caves in the vicinity, are an expressive natural phenomenon, which itself evokes a religious atmosphere. The people undoubtedly felt that the region of Škocjan was a holy place, while the cave Mušja jama (Fliegenhöhle, Grotta delle Mosche) has since been proven to be a cult site by the discovery of material evidence. The cave, which measures 50 m in depth, was inaccessible to the people of those times, who threw precious objects into the cave, mainly bronze weapons and vessels which they had previously ritually burned. With these offerings, one guesses, warriors were commending themselves to the gods of the underworld. This must have been a well-frequented area, since the objects, which testify to its supra-regional importance, reflect influences not only from the Pannonian Plain but also from Italy and the Aegean world, as well as the western Balkans. The significance of this cult site disappeared almost completely after 500 years, in the 7th century BCE.[15] Nevertheless, it was not completely forgotten, as some objects in the cave date back to Roman times.

---

        burial ground at Mestne njive in Novo Mesto, where large covered urns were found, is important, though the related settlements have not been investigated.
14    They were buried at carefully chosen places, most often in isolated areas outside the settlements. See Biba Teržan (ed.), *Hoards and Individual Metal Finds from the Eneolithic and Bronze Ages in Slovenia*, Ljubljana: Narodni muzej Slovenije, 1995 and 1996.
15    Biba Teržan, Elisabetta Borgna, Peter Turk et al., *Il ripostiglio della Grotta delle Mosche presso San Canziano del Carso. Ripostigli delle età del bronzo e del ferro in Slovenia* III, Ljubljana: Narodni muzej Slovenije, 2016.

## The Hallstatt Princes and "Situla Art"

During this period (8th–4th century BCE), the tribes that settled in present-day Slovenia remain anonymous, while others in neighboring areas were already known by name: the Histri in Istria, the Iapodes in Lika and in the Una Valley in Bosnia, and the Liburni in northern Dalmatia. This was when the first city states emerged in Greece and when the Greek script developed on the basis of the Phoenician alphabet. It was also the time of Homer's epics and the decline of the Mycenaean Age, reflected in the Iliad and the Odyssey.

Under these remote influences, Central European society underwent a transformation as well: the leading class wielded economic and thus also military power, which promoted universal progress. At this time, ironwork had become one of the most significant branches of the economy. The earlier phase, which ended with the settlement of the Celts, is also known as the Hallstatt period (after the Austrian site of Hallstatt), and the latter is referred to as the La Tène period (after a Swiss site). With the arrival of the Celts, tribes in these parts became known by name for the first time in history.[16]

During the Early Iron Age, major changes occurred again in settlement patterns, as hitherto unoccupied regions were settled by their original inhabitants or because newcomers from the Danube areas were attracted to these parts by rich iron ore deposits.

Various tribes had settled the lands of what is now Slovenia, yet their material culture indicates that they differed among themselves in settlement structure, burial practices, and in objects for everyday use. The various Hallstatt groups are distinguished as: Lower Carniola (Dolenjska), Inner Carniola (Notranjska), Upper Carniola (Gorenjska), Carinthia (Koroška), Styria (Štajerska) and Posočje (formerly Sv. Lucija, along the Soča River) groups. Speaking of different groups does not necessarily mean that tribes living in a certain area were ethnically differentiated. In our opinion, however, the inhabitants of the Posočje group, who lived close to Italy, were under the Veneti's strong influence. Their community was among the most developed in modern day Slovenia. The Lower Carniola (Dolenjska) community was also highly developed, judging from the wealth of the finds and from the high degree of social stratification reflected in burial finds and "Situla Art." In particular, its swift development became possible as a result of its advanced metalworking craft, the most significant economic branch in this community aside from the all-important livestock-raising and cultivation

---

16 Boštjan Laharnar and Peter Turk, *Iron Age Stories from the Crossroads*, Ljubljana: Narodni muzej Slovenije, 2018.

of land. Iron had become so important in the 8th and early 7th centuries BCE that it even replaced bronze in ornament production, although bronze was far more attractive.

In Greece, the Iron Age had already begun in the mid-11th century BCE, while ironwork only began to flourish in modern day Slovenia in the 8th. Earlier, iron objects were imported, including the oldest iron sword (from the 10th century BCE) in the cave Mušja jama, which had come from the Aegean region. Limonite ore was plentiful, particularly in the Upper and Lower Carniola (Gorenjska and Dolenjska) regions.[17]

Most of the settlements from the Hallstatt period were situated on hills and minor elevations. They were encircled by sturdy defensive walls called hillforts. The fortifications used huge rocks for both frontal sides, with stone chippings and a mixture of clay poured between them. At the hillfort above Vir near Stična, excavations at the Cvinger site, whose perimeter measures 2.3 km, have revealed that houses stood close to the outer defenses, although a corridor allowed defenders free access to the wall. The houses were simple, rectangular wooden structures, erected either on piles driven into the ground, or else with upright supporting pillars placed on horizontal foundation beams (Fig. 4).

---

17   These deposits, nowadays exhausted, were accessible by surface mining. Evidence of iron production is found mainly in the slag, uncovered at most of the hillforts, and frequently outside the walls, indicating that ore was smelted close to the settlements and in shaft-like stoves with built-in fireplaces. To obtain higher temperatures for smelting, bellows were used to blow air through spout-shaped clay vents into the fireplaces. They produced ingots, which were suitable for transportation and used mainly to manufacture weapons and tools. See Janez Dular and Sneža Tecco Hvala, *South-Eastern Slovenia in the Early Iron Age. Settlement—Economy—Society/ Jugovzhodna Slovenija v starejši železni dobi. Poselitev—gospodarstvo—družba*, Ljubljana: Založba ZRC, 2007.

**Figure 4.** Iron Age house at Kučar near Podzemelj in the Lower Carniola (Dolenjska) region.

Most of the hillforts can be found in the Karst region as well as in Inner and Lower Carniola (Notranjska and Dolenjska). Those in Lower Carniola (Dolenjska) are the best researched, and it has been shown that the Urnfield inhabitants erected some unfortified settlements there, but life in them swiftly declined. In the 8th century BCE, new populations built larger settlements not only close to older ones, but also in completely new locations, usually reinforced by fortified stone walls. The Mirna valley was the most densely settled, while the main centers were Stična, Novo Mesto, Dolenjske Toplice, Veliki Vinji Vrh (Šmarjeta), Velike Malence, Libna, Podzemelj, Vače, and Magdalenska gora near Šmarje-Sap, in addition to many smaller settlements. The settlement at Molnik was situated in the periphery, already gravitating toward the Ljubljana Basin.[18] In Inner Carniola

---

18   Magdalenska gora: Sneža Tecco Hvala, *Magdalenska gora—družbena struktura in grobni rituali železnodobne skupnosti (Magdalenska gora—Social structure and burial rites of the Iron Age community)*, Ljubljana: Založba ZRC, 2012; Novo Mesto: Borut Križ et al., *Novo Mesto. 7, Kapiteljska njiva. Gomile I, XIV in XV/Barrows I, XIV and XV*, Novo Mesto: Dolenjski muzej, 2013; Molnik: Sneža Tecco Hvala (ed.), *Molnik pri Ljubljani v starejši železni dobi/The Iron Age site at Molnik near Ljubljana*, Ljubljana: Založba ZRC, 2017.

(Notranjska), new hillforts were established at Šmihel below Nanos and Trnovo, but for unknown reasons they disappeared in the 7th century BCE. This is surprising, as life in the Posočje and Lower Carniola (Dolenjska) communities was flourishing precisely at that time.

In the Posočje region, the first settlements were established mainly in the upper Soča Valley. The most significant were those in Kobarid, Bovec, Tolmin, and Most na Soči (formerly known as Sv. Lucija). The population spread out toward Upper Carniola (Gorenjska) during the 6th century BCE and settled in the Bohinj region, where rich iron ore deposits could be found. In Styria (Štajerska), some settlements had persisted since the Late Bronze Age, while the beginning of the Iron Age saw several new settlements established, such as Poštela on Pohorje and Ptujska Gora. However, in the middle of the 6th century BCE these settlements—like those in neighboring Hungary—disintegrated, possibly because of the Scythian incursion into the Pannonian basin and their frequent raids into neighboring territories. The consequences of the Scythian incursion were also reflected in Lower Carniola (Dolenjska), where local overlords adopted some Scythian weapons and equestrian equipment.

Not all communities buried their dead in the same way. In Posočje, the cemeteries were flat, extending along riverside terraces. At Most na Soči, more than 6,500 cremation graves have been excavated.[19] Flat burial grounds have also been discovered in Inner Carniola (Notranjska) where both skeletal and cremation graves were found, among them many urn burials.[20] In Styria (Štajerska), cremation burials were customary, either in simple earthen graves

---

19 The burnt bones were placed in simple pits, which were covered with stone slabs and sometimes encircled by a wreath of stones. Urn burials were rarer, and only the rich were interred in bronze buckets (situlas). The dead were cremated in their clothes, together with their jewellery and other goods, although goods acquired later were placed separately in graves. Weapons were usually not placed in these graves. See Biba Teržan, Fulvia Lo Schiavo, and Neva Trampuž-Orel, *Most na Soči (S. Lucia) II. Szombathyjeva izkopavanja/Die Ausgrabungen von J. Szombathy*, Ljubljana: Narodni muzej Slovenije, 1984 and 1985. The settlement: Drago Svoljšak and Janez Dular, *Železnodobno naselje Most na Soči/The Iron Age Settlement at Most na Soči*, Ljubljana: Založba ZRC, 2016; Janez Dular and Sneža Tecco Hvala (eds.), *Železnodobno naselje Most na Soči. Razprave/The Iron Age Settlement Most na Soči. Treatises*, Ljubljana: Založba ZRC, 2018.

20 Mitja Guštin, *Notranjska: k začetkom železne dobe na severnem Jadranu/Zu den Anfängen der Eisenzeit an der nördlichen Adria*, Ljubljana: Narodni muzej Slovenije 1979.

or in sepulchral barrows. The barrows had rectangular stone chambers in the center, which in some cases was reached by a paved corridor.[21]

The burial of corpses in barrows was characteristic of hillforts in Lower Carniola (Dolenjska) until the end of the 4th century BCE. Each settlement had several barrow cemeteries, each holding several dozen barrows. One of the large barrows near Stična, encircled by a wreath of stones, held 183 graves. This was an ancestral barrow that was used for several centuries, while barrows generally contained 20–30 graves. To accompany the deceased, who were buried in their garments, status objects (mainly arms, bronze vessels, and riding equipment) were also deposited in the grave.

Indeed, funeral attire and the manner of burials tell us most about the social structure. In the 8th and 7th centuries BCE, this structure was still highly traditional, since society was not yet very stratified. There were few warrior graves. Prosperity during the late Hallstatt period also introduced greater social differentiation (Fig. 5): there were at least six types of male attire (the most outstanding warriors, in addition to bearing two lances and battle axes, also wore a helmet and an armor), and ten types of female dresses. Women's costumes varied greatly because of the diversity of ornaments and the fact that Italian fashion was displacing domestic styles.[22]

Each settlement had a sufficiently large economic reserve for hunting, livestock, and cultivation on arable land; the plough was pulled by cattle. All types of agricultural implements have been found, especially in Late Iron Age graves and hoards. Although mainly cattle were being raised, the proportion of sheep, goats, and pigs had now increased. The inhabitants were already familiar with transhumance, the seasonal changing of grazing grounds, which played a large role in Roman times. They had relatively large numbers of dogs, but few horses. The smaller ones were work horses, while they rode valuable horses

---

21 Biba Teržan, *Starejša železna doba na Slovenskem Štajerskem/The Early Iron Age in Slovenian Styria*, Ljubljana: Narodni muzej Slovenije, 1990; Biba Teržan, "Štajersko-panonska halštatska skupina. Uvodnik in kratek oris/The Styrian-Pannonian Hallstatt Group. An Introduction and Brief Outline," *Arheološki vestnik* 70 (2019): 319–334.

22 One barrow at Stična, called the grave of the princess, contained remnants of her attire, with tiny bronze buttons sewn on, and a whole range of fibulae, bracelets, amber and gold necklaces, as well a golden diadem. See Stane Gabrovec, *Stična I. Naselbinska izkopavanja/Siedlungsaus-grabunge*, Ljubljana: Narodni muzej, 1994; Stane Gabrovec, *Stična II/1. Grabhügel aus der älteren Eisenzeit—Katalog*, Ljubljana: Narodni muzej Slovenije, 2006; Stane Gabrovec and Biba Teržan, *Stična. 2/2, Gomile starejše železne dobe. Razprave/Grabhügel aus der älteren Eisenzeit. Studien*, Ljubljana: Narodni muzej Slovenije, 2010.

imported from the east, whose remains have only been found in graves. The remains of wolves and foxes have also been discovered, suggesting that both were probably hunted for their skins.

The various crafts of the time were practiced on a limited scale, mostly at home. People worked with wood (carpentry and carving), wove baskets and fabrics, produced ceramic dishes, and made ropes. The blacksmith held an important function. Another craft, besides toreutics and metal casting, was the production of woolen and linen goods. This, too, is well documented, since ceramic and wooden spindles of from sticks, as well as pyramidal weights for threads on wooden looms, have been preserved. Decorative objects were fashioned from bone and horn, and even colored glass. Imported artifacts are evidence of links with foreign countries and the extent of trade at that time. The southeastern Alpine area was a transit region crossed by routes from the Balkans and Apennine peninsula. Most of the imported artifacts were found in graves. Some came from Etruria (for instance, the 8th-century BCE Etruscan equestrian equipment from Stična, a three-legged stool from Novo Mesto, and a bronze bowl from Črnolica), while others were Italo-Corinthian objects or possibly domestic imitations. Apulian ceramics and other vessels were imported from southern Italy, across the Adriatic, with the Histri and Liburni as intermediaries. Trading in horses, and graphite (for producing pottery) from present-day Czech and Moravian territories, is also documented. Most important, however, was the trade in amber and salt. Amber was brought here along the Amber Route, which connected the Baltic region with the northern Adriatic. Evidence of the salt trade is provided by a man's grave (6th century BCE) at the cemetery in Hallstatt (where salt was extracted). The remains were found dressed in a military outfit, characteristic of a warrior from Lower Carniola (Dolenjska), who had probably accompanied a trade expedition when he died in Hallstatt.

Imported objects also indicate that inhabitants of the southeastern Alpine regions cultivated relations with the Veneti, Etruscans, and other Italian peoples. At the end of the 7th century BCE, "Situla Art" objects were imported from the Etruscans via the Veneti and the Piceni, representing the height of artistic creation at that time. The finest products date from the 5th century BCE. "Situla Art" was a style that produced figural friezes with scenes from people's lives and plant and animal motifs on bronze vessels (mainly buckets known as situlae) and other bronze objects (mainly belt plates); the Vače situla being one of the most famous (Fig. 6).[23] Scenes may depict duels, including duels on horseback, or special

---

23 "Situla Art" has been documented from northern Italy to the Danube, although most of these decorated artifacts were found near Este and Bologna, the Southern Alps,

ceremonies related to rites and sacrifices. Associated processions of people in carriages, on horseback, and on foot are depicted, while their different social positions were reflected in their diverse attire. Banquets are also represented, as are flute players, erotic scenes, and other events. Perhaps they partially portray everyday life as well. However, the scenes must be interpreted mainly as depictions of cult performances and ceremonies accompanying religious festivals, the death of significant members of the community, or festivals in memory of a mythical ancestor honored as a hero, not unlike in the Greek world. Despite their links with Mediterranean cultures, the Hallstatt communities in the Slovenian regions did not rise above their prehistoric levels; they did not know cities, state order, or writing. Their development abruptly ended when, in about 300 BCE, the Celts arrived in this area, effectively ending a flourishing life in the fortified settlements. It was replaced by different patterns of settlement, religion, armament, and burial.

**Figure 5.** Golden female attire found in grave at Stična from the beginning of 6th century BCE.

in Nesactium, Istria, and Lower Carniola (Dolenjska). See Peter Turk, *Images of Life and Myth*, Ljubljana: Narodni muzej Slovenije, 2005.

**Figure 6.** The Vače situla.

## The Celts: The Great Conquerors

In the late 1st millennium BCE, much of Europe was under the influence of the Celts and their characteristic culture, which began to take shape north of the Alps in the mid-5th century. It encompassed large parts of France, northern Switzerland, southern Germany, and reached all the way to the Czech region. In that century, the Celts already exerted great influence on the Danubian territory as far as Slovakia and were trading their beautiful artifacts in parts of present-day Austria. They had developed a characteristic artistic style, based on close links with Greeks from Greek colonies such as Massalia (Marseilles), Etruscans, and other Italian peoples. They produced articles for everyday wear, ornaments, weapons and riding equipment, as well as bronze vessels and clay pottery. These items were found in the graves of Celtic nobility alongside products imported from the Mediterranean.

The Celtic migration began around 400 BCE and ended at the beginning of the 3rd century BCE. It completely transformed the face of Western and Central Europe, as well as the Balkans as far as central Asia Minor (Galatia), although it is not clear how large the groups of new settlers were. The historian Pompeius Trogus, of Celtic descent, mentions overcrowding, and particularly internal divisions and wars, among the causes for migration. Whether or not the Celts settled in Britain at that time is not certain. Attracted by the Mediterranean regions, they settled much of the Iberian Peninsula, almost all of France, and part of northern Italy, as well as the Eastern Alps, the Pannonian Plain, and the lower Danube, down to the Black Sea. At first they broke through to the Apennine Peninsula, forcing Etruscans and Umbrians to retreat from northern Italy. In 387 BCE, the Celts defeated the Romans at Allia, a stream in Etruria, and set fire to Rome, although they were unable to capture the capital. They extorted a costly ransom and withdrew from central Italy, although they managed to settle a large part of the Paduan Plain. In the first wave of migration toward the east, which took place along the Danube during the early 4th century BCE, the Celts reached lower Austria, Burgenland, southwestern Slovakia, and northern Transdanubia. By around 300 BCE, they settled the remainder of the Pannonian Plain and the Eastern Alps. The Celtic hordes invaded Macedonia, Greece and Thrace by incursion from the Balkans in the 3rd century BCE, forcing the Hellenistic world to coexist with its conquerors. Once the Celts had properly settled the conquered territories, they became a significant political factor and were incorporated into the world of Greek mythology, legends, and heroes. The Greek historian Appianus of Alexandria wrote that the one-eyed Cyclops Polyphemus and the beautiful sea-nymph Galatea had three sons, the mythic forebears of three important peoples of the time: the Illyrians, Celts, and Galatians.[24]

The Celts introduced not only a new religion and new cults, but also a new way of organizing life, evidence of which was found primarily in plains and riverside settlements. They also brought new technologies (the potter's wheel), a new type of warfare (battle chariots, the remains of which were found in two graves in Brežice), and swords, found at the Lower Carniola (Dolenjska) sites, which were already appreciated at the end of the Hallstatt period. Their apparel consisted of trousers, a light cape, the characteristic torque, and two-part iron belt chains. They set off for battle with bloodthirsty war chants and cries and the banging of swords upon shields. The enemy's weapons were offered in shrines to

---

24   *Illyriké*, 2. 3.

deities, while the heads of corpses were paraded as trophies. After they coming into contact with the Hellenic civilization and being influenced by the coinage of Philip II and Alexander, they too began to mint their own currency.

In approximately 300 BCE, the territory of eastern and central Slovenia began to be settled by the Taurisci, an East Celtic tribe, who colonized the Eastern Alps and the Pannonian basin. Western Slovenia was settled by the Carni; the boundary between them was located around Mt. Ocra (Nanos) in the region of Razdrto. In the 2nd century BCE, a Celtic kingdom known as the Norican Kingdom was formed in Austrian and Slovenian Carinthia and fostered friendly contacts with Rome. Although several princes minted their own money, the central role was nonetheless played by the king of the eponymous Norici tribe. Its religious center was perhaps somewhere in the region of old Virunum (Magdalensberg), which in all likelihood was a significant trade and crafts settlement of Roman and Italian traders at the beginning of the 1st century BCE. The Norican Kingdom, due to its excellent iron (comparable to steel), was an important political factor and a vital supplier of raw materials, arms, and other iron products to the Romans. As Caesar recorded, the powerful king of the Germanic Suebi, Ariovistus, had two wives, one of them the sister of the Norican king Voccio.[25] This indicates the political importance of the Norican Kingdom.

One of the main centers of the Taurisci was the seat of a prince who minted his own money in Celeia (Celje). Within the wide region of Celeia, the interests of the Taurisci and Norici met and may have led to clashes between them. They each pursued a different policy toward the Roman state. The Taurisci had settled on the transit territory next to the Amber Route, which led through the Slovenian region from Aquileia and Tergeste (Trieste), via Ocra (the Razdrto region), Nauportus (Vrhnika), Emona (Ljubljana), Celeia, and Poetovio (Ptuj), to the Danube and beyond. This was also the most dangerous route when it came to attacks on Italy. The Romans were therefore intent on securing it as rapidly as possible, which resulted in constant battles with the Taurisci and Iapodes, whose influence extended as far as Ocra.

Although the Celtic La Tène culture was more unified than the Hallstatt culture, the territory of Slovenia was not unified, because of earlier settlement by indigenous people. Four groups may be distinguished. The largest was the Mokronog group, named after the town of Mokronog in the Mirna valley. They may possibly be the same as the Taurisci, who came from the Pannonian basin, although one cannot always equate material cultures with a particular people

---

25   *De Bello Gallico*, 1. 53. 4.

or tribe.[26] Styria (Štajerska), which had previously been sparsely settled, came under intense Celtic influence, while in Lower Carniola (Dolenjska) the densely settled Hallstatt population continued to exist, although they had to leave their hillforts. The older, indigenous population still lived on in their settlements, which is indicated not so much by their burial customs or their attire—both had become distinctly Celtic—but rather by their ceramic vessels, which were hand crafted and decorated in the old style, for instance, at the Novo Mesto cemetery, on the "Kapiteljska njiva."[27]

One of the most significant sites of the Mokronog community is undoubtedly Mokronog itself, the site of some of the first La Tène graves in Slovenia to have been discovered. Also important were Mihovo below Gorjanci and Novo Mesto in the Krka Valley. The male graves of the Mokronog community contained large quantities of Celtic weapons, ranging from iron helmets with neck and cheek protectors which were worn only by persons of rank, to oval shields with iron bosses, and double-edged swords and lances (the Celtic warriors bore only one). Female attire was similar to the late Hallstatt dresses, although distinctly Celtic in form, similar to the Pannonian Celts. The fibulae were particularly characteristic, along with the iron and bronze belt chains, bronze bracelets with stylized animal heads at their tips, and glass and amber jewelry (Fig. 7).

---

26  The Mokronog group covered the territory of central and eastern Slovenia: the Ljubljana basin, Upper Carniola (Gorenjska), Carinthia (Koroška), Styria (Štajerska), Lower Carniola (Dolenjska), and northern White Carniola (Bela Krajina). The same culture was also documented in the neighboring regions of Croatia, Austria, and Hungary. See Dragan Božič, "Die Erforschung der Latènezeit in Slowenien seit dem Jahr 1964. (Raziskovanje latenske dobe po letu 1964)," *Arheološki vestnik* 50/1 (1999): 189–213; Dragan Božič, *Late La Tène-Roman cemetery in Novo Mesto. Ljubljanska cesta and Okrajno glavarstvo—Studies on fibulae and on the relative chronology of the Late La Tène period/Poznolatensko-rimsko grobišče v Novem mestu. Ljubljanska cesta in Okrajno glavarstvo—Študije o fibulah in o relativni kronologiji pozne latenske dobe*, Ljubljana: Narodni muzej Slovenije, 2008.

27  Due to the gradual Roman occupation of their territory at the end of the 2nd century BCE, the population moved back from the dangerous lowlands to the abandoned Hallstatt hillforts. Vinji vrh above Bela Cerkev is an important such site from this period.

**Figure 7.** Celtic necklace from Podzemelj made of glass beads.

The La Tène population, who lived in the Posočje region in western Slovenia, were characterized by the Idrija culture (named after Idrija pri Bači),[28] and the related communities in Inner Carniola (Notranjska) and the Karst were closely linked to Istria and Friuli. Their material culture bears strikingly few Central European Celtic characteristics. There was little Celtic jewelry (fibulae or glass beads), while the typical female attire included northern Adriatic fibulae and necklaces with three knots; they were not acquainted with glass ornaments. Roman sources

---

28   An interesting site was discovered in Kobarid: Miha Mlinar and Teja Gerbec, *Keltskih konj topòt. Najdišče Bizjakova hiša v Kobaridu/Hear the Horses of Celts. The Bizjakova hiša Site in Kobarid*, Tolmin: Tolminski muzej, 2011.

mention the Carni in this region. Several articles with Venetic inscriptions (e.g., two bronze vessels and a plaque from Idrija pri Bači) were found where the Idrija community lived, while a couple of graves at the same site contained helmets (a Celtic custom) and bronze drinking vessels that were frequently added to male graves. The most typical—and unusual—thing about these graves was the presence of considerable quantities of extremely diverse agricultural (and other) implements. At Gradišče above Knežak, a hoard of almost 400 Republican asses (the earliest from the mid-2nd century BCE) was discovered, and a similar find was made at Dutovlje (Karst). Both provide evidence of close trading and other links with Republican Italy. The area occupied by this community in the wide hinterland of Aquileia had probably come under Roman authority during the 1st century BCE. Roman Republican weapons discovered at several of the hillforts indicate that they had been occupied by force.

In central and southern White Carniola (Bela Krajina), in the Kolpa river valley, both Pliny the Elder (1st century CE) and Ptolemy (2nd century CE) mention the Colapiani. Mixed skeleton and cremation graves were excavated at the cemetery at Vinica by the Duchess of Mecklenburg, and were transferred after her death to a museum at Harvard University in Cambridge. Although the Vinica material culture produced several objects of its own, other finds reveal that people maintained close links not only with the indigenous inhabitants of Lower Carniola (Dolenjska), mixed with the Taurisci, but also with the Iapodes of Lika. Politically, links were probably closer with the latter, since Strabo, the Greek historian and geographer from the end of the 1st century BCE, refers to their territory as Iapodian.

# The Roman Empire: Conquest and Pax Romana

## The Founding of Aquileia

Our knowledge of the penetration of various Celtic tribes into the eastern Alpine area and their settlement of the Alpine and northern Adriatic regions contains enormous gaps. These tribes, as we have seen, were mainly known as Carni, Norici, and Taurisci. Their arrival and subsequent settlement during the late 4th and 3rd century BCE coincided with the Romans' consolidation of their position on the Apennine Peninsula and their first conquests across the Adriatic Sea.

The Romans' first military involvement with a kingdom on the opposite coast became known as the First Illyrian War against Agron and Teuta in 229 BCE. The Illyrian and Macedonian kingdoms were conquered soon afterwards in 168 BCE. The beginnings of the Norican kingdom—situated mainly in present-day Austria and parts of northern Slovenia—are unclear, but toward the end of the 2nd century BCE, Livy, one of the leading Roman historians of the early 1st century CE, referred to a powerful Celtic community on the other side of the Alps, governed by "elders." The kingdom was known for its mineral wealth, which was the main reason why Romans established friendly relations with it at some point in the 2nd century BCE. Livy reported that in 186 BCE certain "Gauls from across the Alps," 12,000 armed men, migrated peacefully to Venetia with no intention of plundering or waging war but with the aim of founding an oppidum in the region, twelve miles from the future Aquileia. Their town was destroyed by the Romans.[29]

A Roman ambassador, sent by the Senate across the Alps, was hospitably received by the "elders," and was told that the migrants had gone to Italy without the knowledge or permission of their tribal leaders. However, they did not intend to cause any harm, merely to settle in a desolate area because of overpopulation and the scarcity of farmland. Nonetheless, the Senate declared that they had been wrong to settle on foreign soil (even though the Romans were probably not legally justified to claim a right over the region). No Celtic king is mentioned in this instance, suggesting perhaps that his role within society had not yet been sufficiently distinguished. Both parties exchanged gifts and agreed that the Alps should be regarded as an impenetrable barrier.

---

29   *Ab urbe condita*, 39. 54 ff.

According to a widely accepted hypothesis, these "Transalpine Gauls" would have been the Taurisci, who had settled in present-day Slovenia and probably wanted to dominate trade across the Ocra pass on both sides of the Alps.[30] The Ocra pass is present-day Razdrto, below Mt. Nanos near Postojna (Fig. 8). It was part of the ancient Amber Route that connected the amber-rich Baltic regions to the northern Adriatic region, where amber was used in Venetic and Etruscan workshops. These Roman negotiations with the "elders" of a transalpine Celtic people may have been the first contacts between the Roman state and the Norican kingdom, although this is not certain. The Senate probably negotiated with the Taurisci, which was strategically understandable if the Taurisci dominated the easiest crossing between the Balkan and Apennine peninsulas, the Ocra Pass. The Taurisci certainly traded along the Sava and Ljubljanica rivers via Vrhnika (Nauportus) in the direction of Trieste (Tergeste), the village of the Carni.

Livy mentions that the Celts had descended into Italy along unknown roads to found their ill-fated town. This data presumably refers to the Taurisci and the road across Ad Pirum (Hrušica), which was a track at that time, later built under Augustus, since the road across the Ocra Pass could have hardly been regarded as unknown.[31] Under Emperor Augustus, the Alpine peoples who were conquered in 15 BCE by Augustus' stepsons Drusus and Tiberius, are mentioned in the inscription at La Turbie, above Monaco. Nevertheless, the Ambisontes are the only known hostile Norican tribe to have been listed among the rebellious Alpine tribes. Although the 2nd-century CE Greek geographer Ptolemy placed the Ambisontes in the south of the province, the suggestion that they inhabited the Aesontius (Soča) Valley must remain hypothetical.[32]

---

30 See Marjeta Šašel Kos, "The End of the Norican Kingdom and the Formation of the Provinces of Noricum and Pannonia," in: Bojan Djurić and Irena Lazar (eds.), *Akten des IV. intern. Kolloquiums über Probleme des provinzialrömischen Kunstschaffens/ Akti IV. mednarodnega kolokvija o problemih rimske provincialne umetnosti.* Celje 8.–12. Mai/maj 1995, Ljubljana: Narodni muzej Slovenije, 1997; relevant chapters of Verena Gassner, Sonja Jilek, and Sabine Ladstätter, in Herwig Wolfram (ed.), *Am Rande des Reiches. Die Römer in Österreich*, Vienna: Ueberreuter, 2002; cf. Karl Strobel, "Das frühe Stammesreich der keltischen Noriker in Kärnten—Ein Konstrukt der Wissenschaftsgeschichte," in: Renate Lafer and Karl Strobel (eds.), *Antike Lebenswelten. Althistorische und papyrologische Studien*, Berlin: De Gruyter, 2015, 28–152.
31 *Ab urbe condita*, 39. 45. 6; Riccardo Cecovini, *Galli Transalpini transgressi in Venetiam*: riepilogo degli studi precedenti e nuova ipotesi interpretativa, *Arheološki vestnik* 64/1 (2013): 177–196.
32 Jaroslav Šašel, *Opera selecta*, Ljubljana: Narodni muzej, 1992, 288–297.

The Founding of Aquileia    43

**Figure 8.** Ocra Pass below Ocra Mt., present day Mt. Nanos.

The identity of the enigmatic "Transalpine Gauls" (or Celts, as the Greek writers called them) is not certain, although the Taurisci seem to be the most plausible candidates. In any case, the Roman state decided to respond to their expansionist tendencies in 183 BCE by founding Aquileia, and the town was set up as a Latin colony in 181 BCE. From a flourishing emporium, the city soon developed into a prosperous metropolis and was, besides Rome, one of the most important cities in Italy. Aquileia's existence had far-reaching consequences for Romanization, the economy, and acculturation in Noricum and Pannonia regions. Its influence there remained significant until the very end of antiquity.[33]

The Celts, who lived in the northern Adriatic hinterland, were influenced by their proximity to the Veneti, with whom they must undoubtedly have had

---

33 'Romanization' is a useful concept that has not been adequately replaced yet. Giuseppe Cuscito (ed.), *Aquileia dalle origini alla costituzione del ducato longobardo. Storia—amministrazione—società*, Trieste: Editreg, 2003; Giuseppe Cuscito and Claudio Zaccaria (eds.), *Aquileia dalle origini alla costituzione del ducato longobardo. Territorio—economia—società*, I – II, Trieste: Editreg, 2007.

trade relations. Individual Roman merchants had also certainly been exploring the possibilities of Celtic markets even before the foundation of Aquileia. It is therefore understandable that the Celts, especially those tribes living in less fertile Alpine valleys, were attracted to the south. However, the Romans wanted to monopolize the northern Adriatic area. The Histri, who inhabited the fertile Istrian Peninsula south of Tergeste, were the first to feel threatened by Roman imperialism, and started a war. Some Celts sided with the Romans on that occasion: they were perhaps Carni, who may have wanted to take advantage of the defeat of the Histri. In 178 BCE, Catmelus (in the place of his tribe's king) commanded and led 3,000 Celtic soldiers alongside the Roman consul against the Histri, who were subdued a year later. At that time, Tergeste may have become known as a village of the Carni, while previously it may have belonged to the Histri.

## The Norican Kingdom

Unlike some Carni, both the Taurisci and the Ambisontes were hostile to the Norican kingdom and the Roman state. The relationship between the Norici and Taurisci is not entirely clear. Both minted their own coins, but even their coinage is insufficient evidence for defining both tribes in terms of politics, territory, and supposed supremacy (Fig. 9). In the region of the Taurisci, gold was discovered at some point in the 2nd century BCE, as reported by the Greek historian Polybius (and preserved by Strabo).[34] It was first extracted in collaboration with Roman entrepreneurs, and in such large quantities that its price dropped by a third throughout the whole of Italy. The Romans were consequently expelled by the Taurisci, who wanted to monopolize gold extraction and processing, as well as the gold trade. It is questionable whether they recognized the Norican king's authority at all in the 2nd century and early 1st century BCE, although the Norican kingdom had undoubtedly always endeavored to gain influence over the regions of the Taurisci.

The kingdom of Cincibilus and his brother are mentioned in the years 171 and 170 BCE, during the affair of the consul Gaius Cassius Longinus. Cincibilus, whom Livy called the King of the Celts, may be the same as the king of the Norican kingdom, or he may have been the king of the Taurisci. His brother, who is not named, intervened in the Senate in 170 BCE as the king's ambassador on behalf of his allies, the Alpine peoples, whom Longinus' consular army treated

---

34  Strabo, 4. 6. 12 C 208.

like enemies. These Alpine peoples were probably no other than the Taurisci; the Carni, Histri and Iapodes sent their own envoys to the Senate to complain about the same matter.[35] A year earlier, after the outbreak of the Macedonian War against the last Macedonian King Perseus, command in Macedonia had been assigned by lot—an usual procedure—to Publius Licinius Crassus. Longinus, against his expectations and wishes, had received Italy with Cisalpine Gaul, where he saw no possibilities for glory and enrichment. He arrived in northern Italy and decided to leave for Macedonia through the southeastern Alpine regions without the Senate's permission, wishing to proceed further across Illyricum.

Figure 9. Celtic silver coins from Celeia, 1st century BCE.

The name "Illyricum" was used for the Balkans during Livy's time, but Livy used it anachronistically when speaking of the early 2nd century BCE. During the Hellenistic period, Illyricum denoted only various Illyrian kingdoms in present-day southern Dalmatia, the Republic of Macedonia, and Albania. With the fall of the last Illyrian King, Genthius, the name gradually extended to embrace most of the Balkans. According to the historian Appian, who was active during the 2nd

---

35   *Ab urbe condita*, 43, 5 ff.

century CE, and was the only classical historian to devote a pamphlet to Illyrian history; the Illyricum extended over vast regions from present-day southern Germany to Bulgaria and included also the Balkans.[36] The story of Cassius Longinus, however, is most interesting for the history of the Celtic transalpine kingdoms, the Balkans, and their contacts with the Romans, since it illuminates the strategic and geo-political significance of the regions along the ancient Balkan trade route. Cassius Longinus evidently departed from Aquileia, which he left unprotected, and took thirty days' worth of provisions for his legions. He ordered that guides who knew the roads to Macedonia should be solicited from among the native Carni, Histri, and Iapodes. The Senate was informed of his departure by envoys from Aquileia, who diplomatically explained that their colony had remained insufficiently protected against possible attacks by the hostile Histrian and Illyrian peoples; they dared not accuse the consul directly. The senators referred them to Longinus and were incredulous to hear that he had left his province. Three senatorial envoys were dispatched that same day to pursue him and prevent him engaging in a war against any nation without the Senate's permission. Measures to protect Aquileia were postponed for fear of what might happen to the consul and the army.

The Senate's fears were more than justified, since Illyricum was then largely *terra incognita*. Not only was Longinus marching through rough and barely passable areas, but he had plunged himself and the army among unknown nations and tribes whose reactions could be unpredictable at best, and hostile at worst. Further fears were no less justified: by his action, Longinus might disclose the way to Italy to peoples settled along his route.[37] Some information about these regions was known thanks to trade routes which had operated for centuries across land and rivers, conveying goods and information to the eastern Alpine

---

36  That is, not only Dalmatia (coastal Croatia and its hinterland, and parts of Bosnia and Herzegovina), Pannonia (northern Croatia and Hungary with parts of Austria and Slovenia), and Moesia (Serbia), as in the age of Augustus, but also Noricum (Austria with parts of Slovenia) and Raetia (parts of Switzerland and southern Germany) during the Antonine emperors' period in the 2nd century CE. To illustrate why he called this large area Illyricum, Appian cited Illyrian customs collected in all of these provinces (Illyr. 6); see Marjeta Šašel Kos, *Appian and Illyricum*, Ljubljana: Narodni muzej Slovenije, 2005.
37  Marjeta Šašel Kos, "Cincibilus and the march of C. Cassius Longinus towards Macedonia/Cincibil in pohod Gaja Kasija Longina proti Makedoniji," *Arheološki vestnik* 65/1 (2014): 389–408.

areas. Yet, as far as geography was concerned, the distance between the Balkans and Italy had not been correctly estimated.

The Macedonian King, Philip V (Perseus' father), sought allies among the barbarian nations settled along the Danube River, hoping to persuade them to invade Italy. Livy adds that it would only have been possible to lead an army to Italy across the region of the Scordisci, a Celtic tribe at the confluence of the Sava and Danube around Singidunum, present-day Belgrade.[38] But Philip's utterly mistaken conception of the size of Illyricum is indicated by a story in which he climbed to the top of Mt. Balkan (Haemus), in the region of the Thracian Maedi, in order to see the Black Sea, the Adriatic, the Danube, and the Alps at the same time, which would help him greatly in planning the war against the Romans. Attacks across Illyricum were also planned by Hannibal and Antiochus of Syria toward the end of Hannibal's life, as well as by the King of Pontus, Mithridates the Great, in the 1st century BCE.

The southeastern Alpine region was strategically very important to Italy, since the route across the Ocra pass to the Apennine Peninsula was not only uncomplicated but also unprotected. However, the Romans did not control it at that time.[39] This is confirmed by the fact that Cassius Longinus' army had to be led by foreign guides. How far the consul actually got is questionable, but it was probably not as far as the Pannonian regions beyond Sisak in Croatia (Segesta/Siscia). Upon his return, he devastated various regions of the Iapodes, Histri, and Carni, as well as those of the Alpine peoples, and took large numbers of slaves— one of the main accusations made by Cincibilus' brother.

The Alpine peoples can probably be identified with various Tauriscan tribes, regardless of whether Cincibilus was a Norican or Tauriscan king. His brother could achieve no restitution of any kind for the damage done to his allies, while envoys of the other three nations with similar claims were unsuccessful too.

Cincibilus' kingdom was the only party to gain an advantage from the tragic affair. To silence them, the Senate dispatched two highly respected consular diplomats to Cincibilus and his brother. They bestowed rich, princely gifts upon them, including the right to export horses. Economic and political interests were shared at the highest level, and Rome wanted to maintain the best possible contacts with the kingdom across the Alps, since this mountain range was regarded as the most opportune frontier for protecting Italy. An agreement of mutual political hospitality (*hospitium publicum*), which is known to have

---

38   *Ab urbe condita*, 40. 57. 7.
39   Šašel, *Opera selecta*, 630–633 and *passim*.

existed between Rome and Noricum when the German Cimbri invaded in 113 BCE, may well have dated from the reign of Cincibilus.[40]

There may have been several other small kingdoms, but of lesser significance. One minor king was perhaps Balanus, who sent envoys to Rome in 169 BCE to offer military aid for the Macedonian war against Perseus. The Senate did not accept his offer, but rewarded him with costly gifts. The short, one-year difference between Cincibilus and Balanus would indicate the coexistence of another small kingdom rather than suggest that Balanus succeeded Cincibilus.

It may have been a grave mistake to assume that the conquest of these regions was mainly a story of rapid Romanization (a reciprocal process) with little force: much more fighting must have been involved in the conquest of these lands than existing sources reflect. There are several indications that northern Italy, called Cisalpine Gaul after its inhabitants, which was mainly comprised of Celts on this side of the Alps, gradually extended across the Ocra Pass. The Karst hinterland of Tergeste came within its sphere of influence, as did the Postojna Gate and Nauportus with the Emona (Ljubljana) basin. Cisalpine Gaul was a province from about 89 to 42 BCE, and then became part of the Roman Empire. However, the evidence of its expansion is scattered and scarce.

The conquest of the southeastern Alpine and Balkan regions was eventually completed under Emperor Augustus, and the Slovenian territory was divided among four Roman administrative units. The northern Adriatic and its hinterland as far as Nauportus and Emona belonged to the Augustan Tenth Region (later called Histria and Venetia) in Italy. Celeia (Celje) and its territory belonged to the province of Noricum, while Poetovio (Ptuj) and Neviodunum (Drnovo near Krško) were part of the province of Pannonia, which was still called Illyricum early in the 1st century CE. A small part of Inner Carniola (Notranjska) and the region along the Colapis (Kolpa) River, with no significant settlements, seems to have belonged to the province of Dalmatia (Map 1).

The conquest of these regions during the late Republican period (i.e., before Augustus' reign) involves a mysterious Cornelius, who is briefly mentioned by Appian alone. The former would have unsuccessfully fought against the Pannonians at some unknown time.[41] His utter defeat spread such fear of the Pannonians among Italians that in the future no consul dared start a campaign in Pannonia. Cornelius may have been either the consul of 159 BCE, Gnaeus Cornelius Dolabella, or the consul of 156 BCE, Lucius Cornelius Lentulus Lupus. They are the only two known Cornelii, who could have possibly fought

---

40  Šašel Kos, "The End of the Norican Kingdom."
41  *Illyriké*, 14. 41.

in Pannonia before 119 BCE, when Lucius Metellus (an unidentified member of the famous Roman Caecilii Metelli family) and Lucius Aurelius Cotta waged a campaign against the Segestani, which was also only mentioned by Appian. However, there is no mention in their careers of any such action.

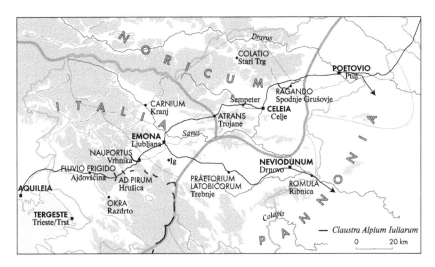

Map 1. Division of present-day Slovenia under Roman Empire.

Nonetheless, 156 BCE is usually noted as the first Roman encounter with the Pannonians. It supposedly occurred during the Roman war against the Celtic Scordisci, but, like many others, this is merely an ill-founded hypothesis.

The 119 BCE campaign, of which no details are known, therefore remains the first certain event in the history of the conquest of Pannonia, although Appian reports that the Romans had fought the Segestani twice before Octavian (the future Emperor Augustus). Aquileia, whose inhabitants must have been very intent on gaining influence and stability in their hinterland, was the starting point of any military action in the direction of Segesta/Siscia.

Ten years before Metellus' and Aurelius Cotta's military expedition against the Segestani, a campaign under Gaius Sempronius Tuditanus was launched against the Taurisci, Carni, Histri, and Iapodes (the latter of which settled in modern Lika, Croatia). When Sempronius Tuditanus was consul in 129 BCE, the Senate transferred to him the judicial powers of the agrarian commission established by Tiberius Sempronius Gracchus, which concerned the new division of land. However, realizing the difficulties of this work, he decided to leave Rome and "march against the Illyrians." Appian mentions that Sempronius Tuditanus and Tiberius Pandusa successfully fought against the Iapodes in the Alps.[42] Pandusa may have been governor of Cisalpine Gaul or Tuditanus' legate. In fact, Livy states that Tuditanus had first suffered defeat against the Iapodes, since they were only conquered by Decimus Iunius Brutus Callaicus, a famous Roman general who had successfully fought in Iberia against the Lusitani and Callaeci.[43] Pliny the Elder, who wrote a major natural encyclopedia under the Flavian emperors, also mentions Tuditanus' war against the Histri and the fact that he reached the Titius River (now the Krka in Dalmatia) in the region of the Liburni.[44] His triumphal inscription, of which two fragments were found in Aquileia, also mentions the Taurisci as a conquered people.[45] Apparently, Tuditanus, with the help of other generals, conquered several peoples during his campaign.

Aemilius Scaurus is known to have fought against the Carni in 115 BCE. During the century before the rise of Caesar, the economic development of the Norican regions and the strategic role of the Norici as protectors of Roman interests in the Alpine and transalpine regions were only briefly affected by the invasion of the Cimbri in 113 BCE. However, the Cimbri left instantly when they heard that the Norici had a special agreement with the Romans. In a battle near Noreia (not located), the Romans were defeated due to the treachery of the Roman consul, Gnaeus Papirius Carbo, who wanted to fight the Cimbri even though they had agreed to leave the country. The Norican kingdom's general prosperity and its economic importance for the Romans are best reflected in the growth of the Roman emporium at Magdalensberg (perhaps old Virunum). It became a significant trading post in the 1st century BCE, comparable to *mutatis mutandis*, the settlement of Italian merchants on Delos.

Parts of the region between Aquileia and Emona must already have been controlled, or even conquered, during the 2nd century BCE, like the important indigenous settlement at Grad near Šmihel below Mt. Nanos (Šmihel pod Nanosom) in the hinterland of the Amber Route and in the vast region of the Ocra Pass. It had probably been subdued in the early 2nd century BCE, as indicated by a hoard of weapons, which must have been used in battles between the Roman army and the indigenous population. Early Romanization of this area is also indicated by the finds of victoriatii (Roman Republican silver coins)

---

42  *Illyriké*, 10. 30.
43  *Periocha*, 59.
44  *Naturalis historia*, 3. 129.
45  Šašel Kos, "Appian and Illyricum," 321 ff.

from the early 2nd century BCE. The coins were discovered at several sites in the maritime and Karst regions, as well as in Inner Carniola (Notranjska), where tribes under the Carni rule must have settled. Republican victoriatii were also found around the Tauriscan settlements of Emona and Celeia.

The gradual expansion of Cisalpine Gaul to the northeast is archaeologically well documented. The earliest finds of 2nd-century BCE Greco-Roman amphorae are located at Sermin, a significant settlement in northern Istria not far from Tergeste.[46] Italic black glazed pottery and amphorae of the Lamboglia 2 type (the turn of the 2nd and 1st centuries BCE) were also found in the Razdrto area near the Ocra Pass.[47]

## Caesar and the Period of the Triumvirs

When Caesar was made proconsul in Cisalpine and Transalpine Gaul, as well as Illyricum, in 59 BCE, some strongholds had already been conquered in Cisalpine Gaul before he arrived. However, he had little time to dedicate to Cisalpine Gaul and Illyricum, since he was mostly absorbed in the Gallic War. Undoubtedly he also occupied certain strategic sites, because he planned a campaign against the Dacian state, ruled by Burebistas. The Geto-Dacian king had succeeded in extending his authority as far as Pannonia; by crushing the coalition between the Celtic Boii (who had settled in parts of Hungary, northern Austria around Vienna, and southern Slovakia) and the Taurisci, he threatened both the Norican kingdom and the Roman state. Nauportus, a settlement of the Taurisci, was a large Roman *vicus* by Caesar's time, and was governed by two freedmen, *magistri vici*. Two of these pairs are known by name; their masters probably belonged to enterprising trading families in Aquileia (the Annaei, Fabii, Petronii, and Fulginates).

A boundary stone between Aquileia and Emona from the early 1st century CE, discovered in the Ljubljanica River near Bevke (12 km southwest of Ljubljana), indicates that Nauportus belonged to the territory of Aquileia.[48] Other significant late Republican and Augustan finds have been located in the Ljubljanica since

---

46 Jana Horvat, *Sermin. Prazgodovinska in zgodnjerimska naselbina v severozahodni Istri/A Prehistoric and Early Roman Settlement in Northwestern Istria*, Ljubljana: ZRC SAZU, 1997.
47 Jana Horvat and Alma Bavdek, *Okra. Vrata med Sredozemljem in Srednjo Evropo/ Ocra. The Gateway between the Mediterranean and Central Europe*, Ljubljana: Založba ZRC, 2009.
48 Šašel Kos, "Appian and Illyricum," 481–482.

then, including Republican and Celtic coins. Italic pottery typical of the mid-1st century BCE was discovered together with native Tauriscan La Tène pottery at Nauportus and Emona. Italian settlers are also documented in Emona, perhaps as early as the period after Caesar's death (the Caesernii). The extent of these regions' dependence on the Norican kingdom remains unclear, though at some points its influence may have reached that far. In fact, much of Norican and Tauriscan history remains obscure, yet it seems clear that the Norican policy toward the Roman state was friendly, while that of the Taurisci was hostile.

Unlike Cisalpine Gaul, Illyricum at that time had not yet been organized territorially as a province, but was a *provincia* in the sense of a sphere of (military) action. It even seems that large parts of the territory between Illyricum and Cisalpine Gaul were not yet subject to Roman authority. Most of the Delmatae, Iapodes, and various Pannonian peoples and tribes, including the Segestani, had not yet been subdued by the Romans at the time of Caesar. Part of these regions first came under Roman influence during Octavian's Illyrian Wars in 35–33 BCE. After Caesar's death and several years of struggle for supremacy, the Triumvirate was established, but Lepidus soon lost power. With the pact of Brundisium in 40 BCE, Mark Antony and Octavian divided the Roman Empire along a border at Scodra (present-day Shkodër in Albania) in southern Illyricum.

Octavian's military campaigns in Illyricum in 35–33 BCE can be considered the next, and perhaps the most decisive, phase in the conquest of Illyricum, i.e., the future provinces of Pannonia and Dalmatia. It was decisive not so much because it involved the conquest of extensive new territories such as the central Balkans (specifically the regions of the Daesitiates, Maezaei, Ditiones, Breuci, and others), but rather because it continued Caesar's border policy in Cisalpine Gaul and Illyricum, strengthened the protection of northern Italy, and represented a systematic conquest of Illyricum. It was of immediate importance to Octavian to win glory as a successful general and exploit his military successes against Antony. In his Illyrian Wars he conquered many peoples and tribes of greater and lesser significance. Among them were the Carni and Taurisci, and, in an important campaign, the Iapodes and Segestani. Two memorable events were the fall of Metulum and Segesta/Siscia, and in 34–33 BCE he subdued the Delmatae and neighboring tribes.[49]

---

49  Ibid., 393–471.

## The Final Conquest under Augustus and the Founding of the Provinces of Noricum and Pannonia

Cassius Dio, a Greek historian who wrote a major history of Rome at the turn of the 2nd and 3rd centuries CE, reports that just prior to the Pannonian War, in 16 BCE, the Pannonians invaded Histria, together with the Norici, and were subdued by Silius and his legates.[50] It is generally accepted that Publius Silius Nerva had been proconsul of Illyricum in 17 and 16 BCE—or, if his province was the northern Italian Transpadana, it would have included Histria and Liburnia (northern Dalmatia). In the Liburnian city of Aenona, he was honored as a proconsul and patron. As a result of the invasion, the Norici became dependent on the Roman state, like the Pannonians who had been subdued earlier (under Octavian). Dio must be referring here to the Segestani and a few neighboring peoples, whose insurrection was probably why the Norican kingdom was annexed. The invading Norici must have been a people settled in close proximity to Histria. These may have been the Ambisontes, if their location along the Aesontius could be accepted (which is less likely), or rather the Norican Taurisci, mentioned by Strabo. Independent minting of Norican tetradrachmas ended in about 16–15 BCE. At the same time, certain northern and western Pannonian regions may also have been annexed. The latter had previously been more or less dependent on the Norican kingdom, and included the territory of Savaria (Szombathely) and Scarbantia (Sopron) in Hungary as far as Lake Pelso (Balaton), which were all Celtic regions.

The kingdom's dependence on the Roman state is additionally confirmed by Augustus' grant of immunity and Roman citizenship to one Gaius Iulius Vepo from Celeia, probably for special merits during the Augustan conquest of the southeastern Alpine area. These privileges would have been meaningless in an independent Norican kingdom. Vepo, who belonged to a well-to-do indigenous community in Celeia, had his tombstone erected during his lifetime in an entirely Romanized manner. It assumes that there must have been a stone-carving workshop and that Latin was well enough known. As early as the Augustan period, benefits resulting from acculturation and cooperation with the Romans were respected in Celeia, which meant that at least some members of the upper class in this Norican town had already been Romanized.[51]

---

50  Dio, 54. 20. 2.
51  Šašel, *Opera selecta*, 31–43.

The annexation of the kingdom is also proven by a small vexillation of the Pannonian legion, VIII Augusta, stationed at old Virunum (Magdalensberg) during the Augustan period. Together with a detachment of the *cohors Montanorum prima*, they constituted a small but sufficient garrison in the country, and the presence of Roman soldiers is also confirmed elsewhere in Noricum. Further evidence are two marble molds for producing gold bars, weighing 5.6 and 14.5 kg (17 and 44 librae) respectively, which were recently discovered at Magdalensberg and manufactured in the name of Emperor Gaius Caesar (Caligula, 37–41 CE).

Noricum became a province under Claudius (41–54 CE) at the very latest, when the presence of its first governor, the procurator, C. Baebius Atticus, is confirmed. However, this could also have happened under Tiberius (14–37 CE), as one could infer from the Roman historian, Velleius Paterculus, who wrote in Tiberius' time. Although he noted that the emperor added four new provinces to the empire (Raetia and Pannonia, the Norici and the Scordisci), this passage has been variously interpreted.

Illyricum came more firmly under Roman authority during the Pannonian War, which had begun in 14 BCE, after uprisings by the Pannonians. A year later command was given to the best Roman general, Marcus Vipsanius Agrippa. Tiberius assumed command after Agrippa's death in 12 BCE, and remained in power until 9 BCE, when most of the territories of the future provinces of Pannonia and Dalmatia came under Roman authority. Illyricum, which was the name of the undivided province, extended as far as the Danube. This was proudly proclaimed by Augustus in a long inscription in the sanctuary of the goddess Roma at Ancyra, in the province of Galatia (Turkey). The inscription lists the emperor's most significant achievements (*res gestae*), in the first person and from his point of view. Tiberius subdued the Pannonian Breuci and Amantini with the help of the Celtic Scordisci. In 11 BCE, he subdued the rebellious Delmatae, but not many details of this war are known.

The great Pannonian-Dalmatian rebellion in the years 6–9 CE once again completely shattered Roman authority in Illyricum. This most serious insurrection was also suppressed by Tiberius. Noricum does not seem to have been affected, apart perhaps for some parts later attached to Pannonia. The uprising, led by Bato, started among the Pannonian Daesitiates, who had settled in central Bosnia near present-day Sarajevo. They were joined by the Breuci under Pinnes and another Bato, and a number of other tribes. According to Velleius, their collective strength consisted of 200,000 infantrymen and 9,000 cavalry; moreover, some tribal leaders had served in the Roman army and were acquainted with Roman strategy and warfare. One of the main reasons for the uprising, according to Cassius Dio, was the ruthless exacting of tributes.[52] The rebels failed to capture Sirmium (Sremska Mitrovica in Serbia) after massacring Roman citizens. Their original plan was a triple offensive: detachments of their army would simultaneously invade Italy and Macedonia, while a third of their strength would remain in the center of the rebellious area. Daesitiatic Bato invaded Dalmatia instead and undermined their strategy. They should have captured Siscia before Tiberius arrived in time to rescue it. Augustus moved to Ariminum (Rimini) to be closer to the war zone should his advice be needed. The war itself was a most complicated affair, involving Roman armies from the East and from Germany, as well as detachments from Italy which were delivered to Tiberius by the historian Velleius Paterculus. Tiberius' nephew, Germanicus, was a subordinate commander in Dalmatia. The war ended after four years, and Suetonius called it "the most serious of all wars after the Punic."[53]

While life in Noricum developed in tranquility, Illyricum was devastated and had to recover under massive military supervision. However, both Noricum and the Pannonian part of Illyricum had been more or less densely settled before the Romans arrived. The age and importance of the existing settlements, and which ones deserved to be designated as towns, cannot be established with certainty, since many factors played a role in settlement and colonization. Economics played a major role in deciding where to build a village or town, but settlement patterns were also affected by other unknown factors. Some, such as natural catastrophes and epidemics, are only rarely taken into account. Epidemics may have affected life enormously in a given area, causing major depopulations, not unlike devastation due to pestilence in the Middle Ages. The "desolate kingdoms of shepherds" (*deserta regna pastorum*) mentioned by Virgil may be interpreted as large, deserted areas within some of the Norican eastern Alpine regions, in the hinterland of the Timavus, caused by an (animal) plague at the end of the 5th or in the 4th century BCE.[54] This is further confirmed by a complete lack of pre-Celtic names in certain areas, such as Upper Carniola (Gorenjska), several areas in the region of the Norici and the neighboring Taurisci (Poetovio), in the region of Savaria, Colapiani, the Latobici, and others.[55]

---

52  Marjeta Šašel Kos, *A Historical Outline of the Region between Aquileia, the Adriatic, and Sirmium in Cassius Dio and Herodian*, Ljubljana: Slovenska akademija zananosti in umetnosti, 1986, 152–191.
53  *Tib.* 16.
54  *Georgica*, 3. 470–481.
55  Šašel, *Opera selecta*, 514–521.

## Towns and Settlements

Economic factors to do with natural resources and trade were undoubtedly most important to the development of settlements, but so was the geopolitical nature of locations, such as settlements at important crossroads, confluences of rivers, or along an ancient trade route or navigable river. Roads and caravan tracks that had been important during an earlier period might have lost their significance as political situations changed, but certain vital lines of communication remained important throughout the age of antiquity. One was the route across Illyricum (i.e., the Balkans), which connected the Black Sea regions with Italy. It was largely a riparian route which began along the Danube, Sava, and Ljubljanica, and continued on land to Istria and Italy, which is mentioned by Strabo[56] and reflected in the legend of the Argonauts. Prehistoric trade of limited extent along these rivers, and poor geographical knowledge of these areas, may be regarded as the historical kernel of the legend. Settlements along these rivers, like Sirmium, Siscia, Andautonia (Ščitarjevo south of Zagreb), Emona, and Nauportus, were ancient prehistoric settlements, as their names indicate, even if archaeological finds do not always directly confirm their importance in prehistory.

There were four Roman autonomous towns in what is now Slovenia: Emona, Celeia, Neviodunum, and Poetovio. Emona had been a significant settlement since the Late Bronze Age (Urnfield Culture), as indicated by a large cemetery in the courtyard of the Slovenian Academy of Sciences and Arts. The corresponding settlement was discovered some years ago at Prule, on the right bank of the Ljubljanica (the river's ancient name was Nauportus), while traces of the subsequent one were excavated recently in the area of Gornji trg on the same river bank, opposite the site of the later Roman city. Its Late Iron Age settlers must have been the Taurisci, who are known to have inhabited nearby Nauportus, which was a more important settlement than Emona in the region (before Augustus). Its name is preserved in the accusative, *Pamporton* and *Nauponton*, in different manuscripts by Strabo, suggesting that the Romans had not accurately used the name "Nauportus" but adapted it to their linguistic sensibilities, which suggested "carrier of the ship." This led them to connect it with the story of the Argonauts, as reported by Pliny the Elder, who drew on ancient sources.[57] The ship Argo traveled down the Danube, Sava, and Ljubljanica as far as Nauportus, where it was transported to the Adriatic across the Alps. The settlement gained fame as a reloading station, where, according to Strabo, goods from Aquileia arriving by

---

56   Strabo, 4. 6. 10 C 207.
57   *Naturalis historia*, 3. 128.

wagon were reloaded onto boats to be transported to the Danubian regions.[58] Nauportus may have also been a Tauriscan customs outpost. The 1st-century CE Roman historian Tacitus called it "almost a small town."[59]

Toward the end of the 1st century CE, with the decline of Nauportus, some late Roman writers declared Emona to have been founded by Jason. Emona became a Roman *colonia*, at the earliest some years after Octavian's Illyrian Wars, or at the latest any time under Augustus (27 BCE–14 CE); it was named Iulia in his honor. Some veterans of the XVth and VIIIth legions settled there, along with many Roman traders and artisans from northern Italy, mainly from Aquileia. The presence of an indigenous population is poorly recorded in Emona, but extremely well documented in the nearby village of Ig, where most men must have worked as stone-cutters in local quarries with high quality limestone (intended primarily for Emona); this would explain why so many family tombstones were erected at Ig. The names of the Ig inhabitants are northern Adriatic (like Laepius, Plaetor, or Voltupar), or typically local (including Moiota, Buctor, and Buquorsa, Fig. 10), while only a few are Celtic (like Adnamatus).

**Figure 10.** Tombstone of Plaetor and Moiota, carved into natural rock at Staje near Ig.

---

58   Strabo, 4. 6. 10 C 297.
59   *Annales*, 1. 20. 1; Jana Horvat, *Nauportus (Vrhnika)*, Ljubljana: SAZU, 1990.

Emona's ground plan is rectangular, and the southern city walls are still preserved. A regular network of roads divided the town into so-called *insulae*, small plots of land on which houses were built. Houses were usually mostly restricted to the ground floor or one store, and were open to the inner court or atrium. Two dwellings have been excavated and are open to visitors in the southern part of the city.[60] The town's central area was occupied by the forum, the sanctuary and the main administrative building, the basilica. Various Roman and eastern gods were worshipped in Emona; the most numerous dedications were written to Jupiter, while the most significant indigenous goddess was Aecorna. Aecorna's sanctuary is attested at Nauportus; she was also worshipped in Savaria (Szombathely), by a community of Emonians. Aecorna must have been a polyvalent goddess, connected to the once extensive Ljubljana Marshes. Among the notable Emonian families were the Caesernii, probably active in metallurgy, the Barbii, who may have owned a building enterprise, as well as the Cantii, a merchant family. Glass and pottery were also manufactured in Emona.

In the late Roman period, an early Christian community settled in Emona, as confirmed by an early Christian center developed in and around the former baths. Although the church has not yet been unearthed, its existence is inferred from its baptismal chapel, whose preserved mosaic floor (including inscriptions of its donors) was excavated and presented as a monument *sub divo* (Fig. 11). In St. Jerome's time Christians had been well established in Emona, and in 376 and 377 CE he corresponded with Emonian virgins and a monk, Antonius. In 381 CE Bishop Maximus from Emona is known to have taken part in an ecclesiastical council at Aquileia.

The town's administrative reach was large, and Upper Carniola (Gorenjska) probably belonged to it. Villagers in small rural settlements tilled the land, worked in the woods, and bred cattle and small stock. Some places had ironworks. The main settlements were Nauportus, Ig, and Carnium (Kranj), which must have played a role as a military stronghold during the Augustan age. It was an important fortress in the late Roman period, and became more so during late antiquity and the Early Middle Ages. Another internationally significant prehistoric communication line was the old Amber Route with Nauportus and Emona as important posts. It led through Celeia, Poetovio, Savaria, and Scarbantia to Carnuntum (Petronell-Bad Deutsch Altenburg, not far from Bratislava on the opposite bank of the Danube), and across the Danube

---

60 Boris Vičič, "Colonia Iulia Emona: 30 Jahre spätter," in: Marjeta Šašel Kos and Peter Scherrer (eds.), *The Autonomous Towns of Noricum and Pannonia. Pannonia I*, Ljubljana: Narodni muzej Slovenije, 2003, 21–45.

to the north. These settlements must originally have belonged to the Norican kingdom. Celeia (almost certainly a pre-Celtic name, like Noreia, a major Norican settlement which has not yet been located) had been one of the centers of the Norican kingdom. It was probably situated at Miklavški hrib, where traces of an Iron Age hillfort were discovered, and a settlement developed at the site of modern Celje, when the Celts arrived. It was ruled by rich princes with their own mint, who competed for supremacy with other important centers of the kingdom. Roman rule seemed to have been imposed in a friendly way.

**Figure 11.** Partly preserved baptismal chapel at Emona. Archelaus and Honorata donated 1.8 m² of the mosaic floor.

The founding of towns and the creation of provinces went hand in hand. The former kingdom was most likely organized as a province under Tiberius (14–37 CE), or at the latest under Claudius (41–54 CE), when five large Celtic towns became Roman municipia,[61] i.e., Roman towns where the indigenous population predominated. These were Celeia, Virunum (on Zollfeld), Teurnia (St. Peter im Holz), Aguntum (Dölsach near Lienz), and Iuvavum (Salzburg). They all enrolled in the voting tribe

---

61   Pliny *Naturalis historia* 3. 146. See Marjeta Šašel Kos and Peter Scherrer (eds.), *The Autonomous Towns of Noricum and Pannonia. Noricum*, Ljubljana: Narodni muzej Slovenije, 2002.

Claudia. The native Claudii are attested even in the hitherto poorly Romanized areas, such as north of Zollfeld in the upper Mura Valley, and even north of the Alps. Solva (Wagna near Leibnitz) became a municipium under Vespasian (69–79 CE), the founder of the Flavian dynasty, and was thus named Flavia, while Ovilavis (Wels) and Cetium (St. Pölten) became municipia under Hadrian (117–138 CE) and were named Aelia after the emperor's family name. Urbanization was concluded under Caracalla (211–217 CE), who elevated a civil settlement outside the legionary fortress of Lauriacum (Lorch near Enns) to the rank of a municipium, while Ovilavis became a Roman colony. Villages and their native inhabitants were organized according to Roman ways throughout urban territories. The Norican mines were seemingly in imperial possession ever since Augustus.

Virunum was the provincial capital, but Celeia was undoubtedly the second most important town. At least at the very beginning of Roman rule, Celeia may have been the principal administrative center of the province,[62] and possibly the seat of the provincial governor, who was a procurator rather than a legate (of the army), since no legion was stationed in Noricum before the Marcomannic Wars.

Among the notable local families were the Vindonii, Varii, Bellicii, Spectacii, and Serandii. Some members of the municipal elite lived in a village at the site of present-day Šempeter in the Savinja Valley, where a cemetery with beautiful marble funerary monuments (some in the form of *aediculae*) was discovered. Four of the monuments with mythological reliefs (Europa, Ganymede, Iphigenia on Tauris) were reconstructed and are on display in the archaeological park. These tombs were preserved due to the catastrophic flooding of the Savinja River in the late 3rd century CE, when the late Roman Celeia was limited to the left bank of the Savinja's new course.

Water cults were, understandably, most important to Celeia's inhabitants. Adsalluta and Savus were worshipped in a nearby sanctuary at Podkraj near Hrastnik, as was Aquo, the divinity of the locally significant Voglajna River. Neptune, the supreme god of rivers and seas, was honored by all the town's inhabitants. Other gods who were worshipped included Jupiter of the High Peaks, known as Culminalis and Uxellimus, Jupiter Depulsor, the averter of evil, and the Celtic goddess Epona, as well as Noreia and Celeia, personifications of the province and town.[63] Forum and sanctuaries, built in a Celtic fashion, were

---

62   Irena Lazar, "Celeia," in: Marjeta Šašel Kos and Peter Scherrer (eds.), *The Autonomous Towns of Noricum and Pannonia. Noricum*, Ljubljana: Narodni muzej Slovenije, 2002, 71–101; Maja Bausovac, *Vivas felix. Celeia: arheološko najdišče Osrednja knjižnica Celje/ The Osrednja knjižnica Celje Archaeological Site*, Celje: Pokrajinski muzej Celje, 2014.

63   Marjeta Šašel Kos, *Pre-Roman Divinities of the Eastern Alps and Adriatic*, Ljubljana: Narodni muzej Slovenije, 1999.

discovered in the late 1990s and in the first years of the new millennium; before that, the remains of beautiful marble buildings, frescoes, and mosaics were mainly found during rescue excavations, since most of Celeia lies underneath the modern town. Two baths and sections of richly paved roads were also discovered. A large sanctuary, perhaps dedicated to the imperial cult, has been partly preserved at the foothills of Miklavški hrib.

Town walls with towers were probably built after the outbreak of the Marcomannic Wars in the late 2nd century CE. The city was an episcopal see in the late Roman period: an early Christian basilica was built, probably early in the 5th century, and a mosaic floor with inscriptions by donors and a baptismal chapel have been preserved. Ioannes, the Bishop of Celeia, was among the signatories to the protocol of the church synod at Grado, between 572 and 577. This meeting occurred after the Lombards came to northern Italy. Life in Celeia must have ceased after the mid-5th century, however, as indicated by the lack of archaeological discoveries.

The most significant minor settlement around Celeia was Atrans (Trojane), a road, post station and important pass (563 m) across the hills that divided the Emona and Celeia basins. It was located in the border area between Italy and the province of Noricum. Atrans was also an important customs post and a station of the *beneficiarii* (road and financial police). Its pre-Celtic name indicates that there must have been a prehistoric settlement nearby: Atrans itself was a Roman settlement, with no prehistoric finds. Many notable discoveries have emerged at Atrans, including fragments of one or two gilded statues of horses, which no doubt belonged to at least one equestrian statue of a Roman emperor. An important imperial building (a repaired *mansio*), documented on a fragmentary inscription, is dated to the reign of Marcus Aurelius and Lucius Verus.[64]

Urbanization in Pannonia began under Vespasian (69–79 CE), who founded the first municipia and colonies: Neviodunum, Andautonia, Siscia, Sirmium, and Scarbantia. They coincide, as expected, with settlements along two main communication lines, the old Amber Route and the river route across Illyricum, which is connected to the legend of the Argonauts. Neviodunum (now Drnovo near Krško) is the only one of the four Roman towns in the Slovenian regions that lost its importance after the fall of the Roman Empire, and its role was taken over by Novo Mesto in modern times. The province of Pannonia, based on civil self-government, was presumably not established before Vespasian. Until then, Pannonia seems to

---

64  For minor settlements of all four towns, see Jana Horvat, Irena Lazar, and Andrej Gaspari (eds.), *Manjša rimska naselja na slovenskem prostoru/Minor Roman Settlments in Slovenia*, Ljubljana: Založba ZRC, 2020.

have been a military district, known in official texts as Illyricum and placed under the command of a military legate. Three legions were stationed in Pannonia.

The territory of Neviodunum (meaning "the new town") was settled by the Latobici, probably one of the tribes formerly under Taurisci rule. After Octavian conquered the Taurisci during the Illyrian Wars (35–33 CE), the Latobici were probably organized as a civitas under the supervision of a Roman commander or a local tribal prince. In the early 1st century CE, two important settlements developed in the region: Neviodunum and Praetorium Latobicorum (Trebnje), the principal station along the main road which led across Illyricum to the east. Both towns were seats of the *beneficiarii*, and Praetorium Latobicorum was strategically of importance for its location near the border between Italy and Pannonia. On the plain along the Sava River, Neviodunum developed into a significant port city with everything needed for its functions—in particular, large military stores. Remains of rich buildings and baths were discovered, but the town had no walls. Most of its population remained Celtic, and one of the notable local families were the Eppii. A member of this family was a distinguished Roman knight and one of the co-mayors of Neviodunum. A teacher was also confirmed to have lived in the town; he was either an elementary or Greek teacher.[65] Pottery workshops and brick kilns were discovered on the outskirts of the town. One of the tombs, unearthed in a nearby cemetery, contained beautiful frescoes, representing a ritual family meal after the burial of a deceased member. Life in the town ended in the first half of the 5th century.

The last Roman municipal foundation in present-day Slovenia was Poetovio, where a legionary fortress had been built on the right bank of the Drava River early in Augustus' reign. Legion VIII Augusta was stationed there until it was replaced in about 45 CE by XIII Gemina from upper Germany. That legion was transferred to Vindobona (Vienna) under Emperor Trajan (98–117), and Poetovio became a Roman *colonia* called Ulpia Traiana. A Pannonian navy detachment was stationed there, and the town became one of the province's administrative centers.

Poetovio, too, bears a pre-Celtic name. It must have been a Tauriscan settlement in the Late Iron Age (perhaps that of the Serretes and Serapilli, who had settled along the Drava), and was probably situated at Panorama and the Ptuj Castle hill, along the Amber Route, at the Drava's crossing on its left bank. Indeed, the names of its known epichoric inhabitants are mainly Celtic. The

---

65  Milan Lovenjak, *Inscriptiones Latinae Sloveniae 1. Neviodunum*, Ljubljana: Narodni muzej Slovenije, 1998.

settlement was part of the Norican kingdom before the Roman conquest, but was soon transferred to Illyricum for geo-political and strategic reasons, since legions were only to be stationed in Illyricum and not Noricum.

Poetovio was the largest city in what is now Slovenia. When it became a colony under Trajan, many veterans from northern Italy and elsewhere came to settle there. There is evidence of soldiers, members of the municipal elite (Valerii, Aelii), high state officials, city magistrates, priests, freedmen, and slaves. Under Hadrian (117–138 CE), who repaired a stone bridge across the Drava, Poetovio became the seat of Illyrian customs. The first known Pannonian senator, Marcus Valerius Maximianus, came from Poetovio. He was admitted to the Senate (consisting of some six hundred members of the Roman ruling elite) under the joint reign of Marcus Aurelius and Commodus (177–180 CE) after greatly distinguishing himself during the Marcomannic Wars. He had temporarily occupied large parts of the Quadi territory (in Slovakia) and killed Valaon, the chieftain of the Naristi, with his own hands. During this time Poetovio was one of the main centers for military logistics. Under Emperor Gallienus (253–268 CE), several detachments of the two Dacian legions, V Macedonica and XIII Gemina, were stationed there.

Poetovio was also a significant trade center, where many artisans were active, and where stone-cutters' workshops and trade in Pohorje marble flourished. Glass, pottery, and bricks were produced there, while its land was most suitable for agriculture. Workshops were located on both banks of the Drava, as were various sanctuaries; a town quarter on the right bank was named after the goddess Fortuna. The forum and administrative quarters were built on the left bank, while cemeteries were situated along all main roads leading out of town, as was customary in Roman cities and villages. In the mid-3rd century CE, the Drava changed its course and probably damaged the location of the legionary camp. In the 4th century, the city became much smaller. Two small fortresses were built, one on Grajski grič, the other at Panorama, a previous location of the prehistoric settlement.[66]

Inhabitants worshipped gods from the Roman pantheon, as well as locally important gods Jupiter Culminalis and Depulsor. Among the indigenous gods were Marmogius, occasionally equated with Mars, and Nutrices, of Celtic origin,

---

66   Jana Horvat et al., "Poetovio: Development and Topography," in: Marjeta Šašel Kos and Peter Scherrer (eds.), *The Autonomous Towns in Noricum and Pannonia/Die autonomen Städte in Noricum und Pannonien—Pannonia II*, Ljubljana: Narodni muzej Slovenije, 2004, 153–189; Jana Horvat, Branko Mušič, Andreja Dolenc Vičič, and Anja Ragolič, *Arheološka najdišča Ptuja. Panorama/Archaeological Sites of Ptuj. Panorama*, Ljubljana: Založba ZRC, 2020.

documented exclusively in and around Poetovio; they are thought to have been birth goddesses, nourishers and guardians of children. The Mithras cult is well documented after the 2nd century CE, and Poetovio, where five Mithraic sanctuaries have been discovered so far, was one of the centers of its worship (Fig. 12). In the late 3rd century, an episcopal see was established in the city. Its first known bishop was Victorinus of Poetovio, a well-known Christian writer and martyr, who may have lived under Diocletian. An early Christian church has been confirmed at Panorama. The city population during late antiquity also consisted of Goths and other groups of barbarian tribes, who were trying to settle in the Roman Empire. In the second half of the 5th century, life in Poetovio gradually ceased.

**Figure 12.** Reconstructed third Mithraeum at Zgornji Breg in Ptuj, the Roman Poetovio.

## From Augustus to Antoninus Pius: The Period of Imperial Peace

The second half of Augustus' reign, after the Pannonian-Dalmatian rebellion had been put down in 9 CE, was more or less peaceful for the entire Roman Empire. It was very briefly interrupted when three legions stationed in the Pannonian part of Illyricum mutinied immediately after Augustus' death in 14 CE. The location of their camps is not known exactly, but Poetovio and Siscia almost certainly served as legionary fortresses, while Emona is less certain. These legions were in their summer camp, probably somewhere in the hinterland of Poetovio or Siscia, because Tacitus reported that military detachments sent to Nauportus to build roads, bridges, and other constructions had gone a long way from their camp.[67] Consequently, they could not have been stationed near Emona. The legionaries had various complaints, from the excessive length of army service to the land they received as retirement pay, which was unsuitable for cultivation. Since the governor, Quintus Iunius Blaesus, was powerless against the rebels, Emperor Tiberius dispatched his son, Drusus, to Illyricum. Drusus restored order in the recently conquered province as soon as he arrived, assisted by a lunar eclipse in the early hours of September 27.

Peace in the empire was disturbed for a brief period in the so-called Year of Four Emperors, 69 CE. After Nero's death, no other member of the Julio-Claudian house was destined to reign, and four emperors succeeded each other very briefly. They were Galba, Vitellius, Otho, and Vespasian, who eventually emerged victorious and founded the Flavian dynasty. It ended with the death of his second son, Domitian, in 96 CE.

Marcus Antonius Primus, the legate of legion VII Galbiana, decisively contributed to Vespasian's victory. His expedition across the Alps into northern Italy with some of the Roman troops occurred in the context of political and military unrest in several parts of the empire in 68–69 CE. In March 68, Gallia Lugdunensis, under Vindex, revolted against Nero's rule; in April, Hispania Tarraconensis, under Galba, and Lusitania, under Otho, followed suit. Soon afterwards the governor of Numidia revolted. In May, legions from Germany joined Galba. Nero committed suicide, and Galba was named emperor by the Senate. However, the army along the Rhine proclaimed Vitellius emperor, and in January 69 Galba was killed by the praetorians. Otho was declared emperor, but Vitellius broke into Italy with his army in April and defeated Otho at Bedriacum. Troops from Illyricum decided to support Vespasian who was also joined by the legions from Egypt, Syria, and Judaea.

Officers of the Illyrian army met at Poetovio in August that year, at the invitation of the Pannonian governor Lucius Tampius Flavianus. Antonius Primus convinced the Illyrian troop commanders, who favored Vespasian's accession, to let him lead the army from Poetovio to northern Italy even without the permission of Vespasian. In October he defeated Vitellius' army at Bedriacum and Cremona. These two decisive victories helped consolidate Vespasian's position in Italy, before Antonius Primus captured Rome.[68]

---

67   *Annales*, 1, 16 ff.
68   Šašel, *Opera selecta*, 332–344.

Peace was restored with the reign of the Flavian emperors, and was not disturbed by the Dacian wars of Domitian or Trajan. Dacia lay outside the Roman Empire; the two emperors conquered it, and Trajan eventually added it to the empire as a new province. Trajan also divided the province of Pannonia into upper and lower Pannonia; Neviodunum and Poetovio belonged to Upper Pannonia. The long-lasting peace ended when the Lombards and Obii invaded Pannonia under Marcus Aurelius in 166 CE. The invasion marked the beginning of the Marcomannic Wars, which lasted for almost fifteen years (167–180). They inflicted huge losses on the empire in general, and caused severe damage to Pannonia in particular.

# From the Marcomannic Wars to the Settlement of the Slavic Tribes

## The Marcomannic Wars

The short-lived incursion by 6,000 Lombards and Obii, who stormed across the Danube to invade Pannonia in 166 CE, brought an era of relative peace in the Roman Empire to an end. However, trouble was already on the horizon during the reign of Emperor Antoninus Pius (138–161 CE). Numerous emissaries from neighboring Germanic and Sarmatian tribes had been pleading with the emperor to incorporate their kingdoms into the Roman Empire, as protection from new barbarian tribes pushing down from the north. The Parthian War broke out after Antoninus Pius died, and Lucius Verus (161–169 CE), co-emperor with Marcus Aurelius (161–180 CE), maintained control, with the aid of troops sent to the eastern front from the Rhine and Danube frontiers. It was already being rumored that barbarian tribes were plotting a conspiracy along the Danube.

Aware of a potential incursion by neighboring Germanic tribes, which were under increasing pressure from other peoples' migration into their own regions, Marcus Aurelius assembled two new legions (II and III Italica) to be stationed for an offensive in Noricum and Raetia. However, the empire's security remained seriously neglected, leaving the Danube border particularly weak. Two years later, in 167 or 168 CE, the Quadi and Marcomanni were able to cross it, and invaded the empire from present-day Moravia and Slovakia and advanced as far as Verona in northern Italy. The invasion mostly passed through present-day Slovenia, but the tribes did not want to be delayed by besieging the towns, since their primary objective was reaching Italy. The most devastation therefore occurred in the countryside, while Celeia was probably the only major towns affected. In northern Italy, the Quadi and Marcomanni attacked Aquileia and razed Opitergium (Oderzo). To combat the unexpected barbarian incursions, the Romans established a defensive zone (*praetentura Italiae et Alpium*), which stretched roughly from the hinterland of Celeia to those of Tergeste (Trieste) and Tarsatica (Rijeka). Its center was a well-fortified base camp of the legion II Italica at Ločica, near Celeia, surrounded by a huge stone wall. The ground plans of numerous buildings in the fields, including the military headquarters and the field hospital, can still be traced from the air.

Emona was probably one of the centers of the defensive zone; it might have been precisely during the Marcomannic Wars that soldiers from the legion XIII

Gemina, who are mentioned in inscriptions, stayed there. Titus Varius Clemens, a Celeia-born Roman knight and high state administration official (who may later have been admitted among senators and who had the most intimate knowledge of the territory), probably participated in decisions about devising and organizing the defensive zone. He had held high office as imperial secretary during Lucius Verus' Parthian war, and ten honorary inscriptions, paying homage to him, have been discovered to date in Celeia.[69] The defensive zone, which later successfully protected the main route leading from the east and the Balkans through present-day Slovenia and the Postojna Gate to Italy, was maintained between 168 and 172 CE. TheLočica legionary camp was then abandoned and the legion assigned to Albing on the Norican *limes*. In 168 CE Marcus Aurelius and Lucius Verus personally oversaw the pre-war preparations. Their campaign, however, was delayed due to a massive plague outbreak that swept across the regions when Verus and his victorious army returned from the Parthian War to the Danube camps in 166 CE. The emperors thus sent for Galenus of Pergamon, the most famous physician of the time, to join them in Aquileia. They and others organized a health and sanitary service that extended to what is now Slovenia. A year later, Lucius Verus died of the plague while traveling to Rome.

Marcus Aurelius drew nearer to the Danube in 169 CE, first at Sirmium and then Carnuntum. He made his way to the Danube forts across Slovenian territory; an inscription at the Atrans road station refers to a large building, renovated by the two rulers and possibly used for lodging on their journeys. After a series of initial defeats (one of the most crushing ones in 170 CE allegedly resulted in as many as 20,000 fatalities), the Roman army finally prevailed. Marcus Aurelius negotiated with the Marcomanni and Quadi from Carnuntum. He frequently pitted various tribes against one another, but also allotted land to others in the frontier provinces of Dacia, Pannonia, Moesia, and Germania, and even in Italy. However, after some tribes, who had settled around Ravenna, rebelled and occupied it, Marcus Aurelius closed Italy's door on the barbarians and even exiled those who already settled there. He wanted to establish two new provinces on the opposite bank of the Danube (Marcomannia and Sarmatia), which would solve the frontier problem and ensure lasting peace.

Marcus Aurelius defeated the Marcomanni and Quadi in decisive battles, most probably in 172 CE. Their exact course remains largely unknown. The Roman

---

69 Šašel, *Opera selecta*, 206–219 (Varius Clemens), 388–396 (*praetentura*); relevant chapters in Péter Kovács, *A History of Pannonia during the Principate*, Bonn: Habelt, 2014.

victory was attributed to divine intervention, or miracles, which were mentioned in written sources (including the emperor's letter to the Senate) and represented in relief scenes decorating the column of Marcus Aurelius in Rome. The first sign of divine intervention supposedly came as an answer to Marcus' prayer, when a spectacular lightning storm caused a fire in the enemy camp, and the shower saved the besieged Roman army, which had been near death from thirst. According to the historian Cassius Dio, an Egyptian priest, Arnuphis, summoned the storm by calling upon Hermes, while Christian authors later attributed the victory to the prayers of Christian legionaries. Cassius Dio also reports that during peace negotiations in 175 CE, the Iazyges returned 100,000 prisoners to the Romans in order to demonstrate the undisputed power of the barbarian tribes. It has been hypothesized that the battles won by the Romans through divine intervention were commemorated with inscriptions (referring to June 11) in the temple of Jupiter on the Pfaffenberg hill at Carnuntum, but this is far from certain. Since warfare created a heavy financial burden for the Roman Empire, the emperor sold some of his property at auction. He overcame the shortage of soldiers by filling his units with slaves, gladiators, and even bandits from the Dalmatian and Dardanian hinterlands. Marcus Valerius Maximianus of Poetovio, one of the main military bases and supply centers of the time, gained prominence during these wars; he also led a navy division on the Danube overseeing the food supply for the Pannonian legions. Later, he became a senator and is still regarded as the first known member of the senatorial order from Pannonia to date. He also led a military detachment far inland into the territory of the Quadi, and personally killed one of the Germanic kings.[70]

Wars against various tribes continued until the death of Marcus Aurelius in 180 CE. His 19-year-old son and successor Commodus (180–192 CE) abruptly ended the warfare for reasons that are still unclear—or, according to Dio, because he hated work and loved the comforts of city life. Commodus concluded a truce with the Germanic tribes, renounced the conquest of new territories on the left bank of the Danube, and vacated all Roman forts in the land of the Marcomanni and Quadi beyond the imperial frontiers. In return, the Germanic tribes were required to contribute soldiers to the Roman army. These units, documented as *vexillationes peregrinae* in the army of Septimius Severus, were under the command of Lucius Valerius Valerianus of Poetovio. According to Cassius Dio, the death of Marcus Aurelius ended the golden age of the Roman Empire.

---

70   On the southeastern Alpine area during the Marcomannic Wars, see Šašel Kos, *A Historical Outline of the Region between Aquileia, the Adriatic, and Sirmium*, 218–255.

Italy and the Pannonian provinces suffered significant population losses due to the length of the Marcomannic Wars, and even more so due to a fatal plague epidemic. Contemporary authors described what had actually been a smallpox epidemic, which first struck the western half of the Roman Empire. Numerous hasty burials allow one to conclude that the pestilence also devastated the population in the Emona area. In Noricum, the outbreak of the disease is documented by inscriptions, e.g., on the tombstone of a certain Victorinus' family, which had lost several family members to the epidemic in 182 CE, including his 18-year-old wife and their baby. Another surviving document is a plaque listing members of a Mithraic cult association from the provincial capital Virunum (Zollfeld), who had convened an irregular meeting in 184 CE after five fellow worshippers died. The plague continued to spread across the empire during Commodus' reign.

In the postwar period, immigrants from the east mainly settled the two Pannonian provinces, which considerably changed the population's ethnic structure. This migration wave is documented by prosopography and onomastic studies based on tombstone inscriptions, and by an influx of fresh money from eastern mints which had not previously been present in this part of the empire. Very many farmers were ruined at this time, but the countryside had recovered by the 3rd century CE.

## The Reign of Pertinax and the Severi

After Commodus' violent death, the throne was filled by Pertinax (192–193 CE), who was killed the following year. His successor, Didius Julianus, proved utterly incapable of managing the perplexing political situation. In the year 191, Septimius Severus, a native of northern Africa, was appointed consular legate of the province of Pannonia Superior. He was stationed in Carnuntum, where the legion XIV Gemina recognized him as emperor as early as April 193 CE. Septimius Severus set out with his troops on an expedition to Rome, the imperial capital (hence the name *expeditio urbica*), through present-day Slovenia. Following the main route Carnuntum–Poetovio–Emona–Aquileia, Severus crossed the unguarded Eastern Alps, continued his march through Aquileia, and reached Rome seven and a half weeks after having been proclaimed emperor. After assassinating Didius Julianus, the army swore allegiance to the new emperor, who reigned until 211 CE. His "Master of the Horse," Lucius Valerius Valerianus, was probably one of the Poetovian Valerii, while the responsibility for the army's food supply rested with Marcus Rossius Vitulus, who was most likely a descendant of a Tergeste family.

The seeds of disintegration took root in the Roman Empire during the Severan dynasty's reign. Senatorial power faded, decision-making authority was increasingly transferred to the emperor and his associates, and increasing influence was gained by members of the equestrian order, from which Septimius Severus' family derived. The number of senators from eastern provinces grew, the bureaucracy became powerful, and yet nothing proved more fatal than the expanding role of the Roman army. Many financial and economic crises accompanied these developments, which were reflected in rising inflation and the decline of the old Roman religion. The worship of various eastern cults increased from the 2nd century CE onwards, while in the eastern part of Slovenia abundant evidence points to the Mithras cult spreading from the east. For instance, five mithraea are archaeologically documented in Poetovio alone.

After the successful conclusion of the Parthian Wars and the establishment of a new eastern province, Mesopotamia, Septimius Severus began his march over land through Asia and Thrace toward Italy in 195 CE. He stopped at Poetovio in the spring of 196 CE, according to an inscription by a Praetorian Guard tribune. That year the temple of Jupiter Dolichenus was also erected in Praetorium Latobicorum (Trebnje). This cult had spread from the town of Doliche (Duluk) in Asia Minor, and blossomed most widely across Pannonia under the Severan dynasty. 196–197 CE was marred by a civil war between Septimius Severus and the governor of Britannia, Clodius Albinus, whom the Gauls had proclaimed emperor in 195 CE. Buried hoards of coins found in present-day Slovenia and Austria suggest clashes between Severus' and Albinus' troops in the province of Noricum. In the spring of 197 CE, Albinus was defeated and killed in Gaul, while Severus headed back to the eastern frontier. The Parthians raided the newly established Mesopotamia, but were defeated decisively by Severus in 198 CE. Inscriptions from the period of the Severan dynasty, notably those from Praetorium Latobicorum, speak of *beneficiarii consularis* under the command of the governor of Pannonia Superior. These were police officials who oversaw border and road traffic, and cooperated in tax collection, tolls, and customs duties.

Elagabalus (218–222 CE) was proclaimed emperor by the army in the eastern part of the empire. Because of his extravagant behavior as the sun god, he was treated with contempt by the Roman aristocracy. A milestone near Celeia is evidence of Elagabalus and his army advancing from the east toward Rome, moving through Illyricum (Dio explicitly reports this too), along the road through Poetovio and Celeia. Elagabalus was succeeded by the young emperor, his cousin Alexander Severus (222–235 CE). The latter was able to bring relative peace to the region and restore the honor of the Roman Empire. He was murdered at Mogontiacum (Mainz) in 235 CE, and the imperial throne was assumed by Maximinus Thrax (235–238 CE).

## Maximinus Thrax and Other Roman Army Officers on the Throne

Maximinus Thrax, a native of Thrace, was the first so-called military emperor from Illyricum in the 3rd century. He fought various Germanic tribes, as well as the Sarmatians and Dacians, who threatened to overrun the empire. His military headquarters were located at Sirmium after 236 CE, where he spent most of his reign campaigning on the Rhine and Danube fronts. Numerous milestones from his era eloquently recount that he ordered the reconstruction of the road system across the entire Roman Empire, which had a significant impact on economic prosperity. He was also praised by inscriptions in Aquileia. Nonetheless, most of the Senate hated him, and while he was on campaign, news reached him in 238 CE that he had been usurped by the African proconsul, Gordian, and his son. After both Gordiani were killed, the Senate resolved to elect Pupienus and Balbinus as new co-emperors and pronounced Maximinus and his son, Maximus, public enemies. The Senate ordered all frontier towns in Italy to destroy their food supplies and to move the population out of harm's way; it also organized new troops and collected weapons in northern Italy. Maximinus was forced to take his army through the Balkans to Italy, in order to stamp out the revolt. He advanced from Sirmium through Mursa (Osijek), Poetovio, and Atrans, descending to the flatlands surrounding Emona. Coin hoards testify to his route and the panic that occurred among Pannonian and Italian frontier populations.[71]

Emona was the first significant Italian town on Maximinus' way to Rome. Its inhabitants deserted it, in line with the Senate's edict, after burning temple and house doors and destroying all the food supplies in the town and fields. The historian Herodian vividly reported these events. Later on, Maximinus and his army continued their unimpeded march to Italy, crossing the Alps and passing the strongholds of Ad Pirum (Hrušica; Fig. 13) and Castra (Ajdovščina). The first major town they reached in Italy was Aquileia. It was safeguarded by a magnificent wall, newly renovated by its inhabitants, which turned it into an invincible fort for Maximinus' advance guard. The main army crossed the

---

71 These and all other hoards are discussed in: Peter Kos, *The Monetary Circulation in the Southeastern Alpine Region ca. 300 BC–AD 1000*, Ljubljana: Narodni muzej Slovenije, 1986.

torrential Soča River, using a pontoon bridge made from wine caskets found in the fields, since the Aquileians had already destroyed the old stone bridge. Maximinus began to besiege the city, but later his own disgruntled soldiers murdered him and his son. The Aquileians did not open their gates until the newly elected emperor, Pupienus, had hastened there from Ravenna.

Something similar occurred in Moesia Superior in 253 CE. The army pronounced Marcus Aemilius Aemilianus emperor, following his decisive triumph over the Goths, who had been continuously raiding the empire along the Drava. In his attempt to surprise the legitimate ruler, Trebonianus Gallus (251–253 CE), and his co-ruler and son, Volusianus, Aemilianus launched a swift march, taking his army from the Balkans toward Italy through present-day Slovenia. His campaign took the emperors by surprise, as suggested by hoards of well-dated coins found in the area, and by the fact that Trebonianus Gallus was able to mount an unsuccessful counterattack only when Aemilianus' army had reached Umbria. Trebonianus Gallus and Volusianus managed to escape, but were later killed by their own army, while Aemilianus continued his march on Rome. There, he was killed by the troops of the subsequent ruler Valerianus (253–260 CE), who had hurriedly arrived with his men from Gaul and Germany.

**Figure 13.** Aerial view of the late-Roman Ad Pirum fort in the Claustra Alpium Iuliarum defensive system.

The latter half of the 3rd century was marked above all by a sense of insecurity caused by frequent barbarian tribe raids, as demonstrated by numerous coin hoards found in northeastern Slovenia. The coins had been buried during the Iazyges' and Roxolani's brutal invasions of Pannonia between the fall of 259 CE and the next spring. There is also considerable evidence to suggest that these invasions had a particularly devastating impact on the northeastern part of present-day Slovenia. Some of the population must have been slaughtered; otherwise they would have retrieved and used the hidden money. Others sought refuge in strongholds on hills. During Gallienus' military campaigns, troops of the legion XIII Gemina were stationed in Emona once again, as documented by local tombstones. Gallienus was the son of Valerianus and became the new emperor (260–268 CE), after Valerianus died in 260 CE in the east.

Numerous coin hoards also document the invasion of northern Italy around 270 CE by the Alemanni and Iuthungi. This invasion was launched when Aurelianus (270–275 CE) acceded to the throne, and brought destruction to a vast part of Noricum between the towns of Lauriacum and Aguntum. The evidence not only demonstrates how waves of the invasion struck southern Noricum, including present-day Slovenian territory, but also indicates one direction of the Germanic raid from Flavia Solva through Poetovio and Emona toward Italy. In the fall of 284 CE, the army in Pannonia pronounced the governor of Venetia, Marcus Aurelius Julianus, emperor. He set out with his troops toward northern Italy, through present-day Slovenia. Before his army was defeated at Verona in 285 CE by Carinus (283–285 CE), it had caused much anxiety and devastation during its advancement through Poetovio and the Vipava Valley. The evidence consists of the local inhabitants' buried savings.

The crisis of the Roman Empire came to a head in the mid-3rd century, when the threat of barbarian invasions from the east (notably from the Balkans) into the Apennine Peninsula became more serious. The empire responded by erecting defensive walls connecting watchtowers wherever access to the peninsula had been unobstructed, particularly along natural passages and passes in the Karst-Alpine area between Tarsatica, Tergeste, and Emona. This defensive system, initiated in the late 3rd century and completed during the Constantinian era, ran across mountainous areas between Tarsatica in Croatia and the Soča Valley in northwestern Slovenia. It was undoubtedly the most extensive Roman construction project undertaken in the territory of present-day Slovenia. Between Nauportus and Castra (Ajdovščina), with an important central fortress at Ad Pirum (Hrušica), the defensive system incorporated three barrier zones.[72] The walls were additionally strengthened by small garrisons in towers, and by small and large forts. The roads often crossed the defensive line through gate towers so that traffic could be completely supervised. Archaeological research of forts at Lanišče and on Martinj Hrib suggests that the second barrier zone was destroyed in a civil war that broke out in 388 CE between the legitimate emperor Theodosius (379–395 CE) and the usurper Magnus Maximus (383–388 CE), and was never restored to its former state. The defensive system existed until the end of the 5th century, when it became irrelevant. The 4th-century historian Ammianus Marcellinus named this late-Roman defensive system Claustra Alpium Iuliarum (the Julian Alps Barrier), based on the mountain range's name Alpes Iuliae (Julian Alps). They stretched from the Gail Valley in Austria to Učka in Croatia.

After the 284 CE civil war, the throne was occupied by Diocletian (284–305 CE), a high-ranking Dalmatian army officer, who radically reformed almost every significant administrative area of the empire. He introduced the Tetrarchy (i.e., a system of rule by four) to prevent violent usurpers, subdued the empire's external enemies, and initiated reforms to its administrative, military, taxation, and financial systems. He also launched a religious reform to settle accounts with Christians, which, like his monetary and tax reforms, eventually failed. The effectiveness of his governance was relatively short-lived. The same went for the Tetrarchy, which was abandoned soon after he retired to private life in 305 CE, and settled in his palace in Split.

Diocletian's religious reform triggered the persecution of Christians and ignited the great civil war that lasted from 306 to 312 CE. It ended with the victory of a Christian convert, Constantine the Great (306–337 CE), who proclaimed religious freedom in 313 CE.[73] Another civil war was fought in the hinterlands of Capris (Koper) between Maxentius (306–312 CE), the son of Diocletian's

---

72 The first barrier was formed by a wall from Reka to Rob at Velike Lašče and continued intermittently at Rakitna, Pokojišče, Verd, Zaplana, Vojsko, and Grahovo. The second barrier zone was represented by sections of walls between Lanišče and Martinj Hrib above Logatec, while the third consisted of the defensive system centered at Ad Pirum. The fortress at Castra was also built at that time. See Peter Petru and Jaroslav Šašel (eds.), *Claustra Alpium Iuliarum 1. Fontes*, Ljubljana: Narodni muzej, 1971; Peter Kos, *Ad Pirum (Hrušica) e i claustra Alpium Iuliarum*, Ljubljana: Zavod za varstvo kulturne dediščine Slovenije, 2015.
73 Rajko Bratož, *Il cristianesimo aquileiese prima di Costantino: fra Aquileia e Poetovio*, Udine and Gorizia: Istituto Pio Paschini, Istituto di Storia Sociale e Religiosa, 1999.

co-emperor Maximianus, and Licinius (308–324 CE). A stronghold manned by Maxentius' troops was overrun at the village of Čentur near Koper in 309/310 CE by Licinus' army, advancing from Tarsatica and Istria to Italy. The march was short-lived, since Maxentius re-conquered Istria and ordered the removal of inscriptions from monuments erected to Licinius. These events are accurately documented by a military chest containing a massive quantity of bronze coins buried in the Čentur fortress. The western part of the empire, including the western and central regions of Slovenia, came under Constantine's rule in 312 CE after he defeated Maxentius in a famous battle at the Milvian Bridge near Rome. The eastern part of Slovenia, still in Licinius' hands, was subjected to Constantine's rule by 316 CE, after a civil war provoked by the destruction of Constantine's statues in Emona. The ultimate rift between these two rulers occurred in 324 CE, when Licinius renounced his throne, in order to put an end to their long-standing hostilities. He was killed soon after his defeat.

## From Constantine to the Fall of the Western Roman Empire

Constantine frequently traveled across present-day Slovenia. His last visit to nearby Aquileia was in 326 CE, where he passed several important laws. During that time, he ordered the execution of his oldest son, Crispus (317–326 CE), in the Istrian town of Pola (Pula) for plotting against him, soon followed by an order to execute his second wife and Crispus' stepmother, Fausta, in Rome. After Constantine died in 337 CE, the Constantinian dynasty remained in power until 361 CE through the reigns of his sons Constantine II, Constantius II, and Constans, and his nephew, Julianus Apostata. In this period, however, the empire suffered from internal instability and religious hostilities within the Church. This escalated into several civil wars, notably in the territory between northeastern Italy and the middle-Danube basin, including present-day Slovenia and the immediate vicinity. Notably, Constantine's sons, Constans (337–350 CE) and Constantine II (337–340), fought a battle near Aquileia in 340 CE in which Constantine II was defeated and killed.

In January of 350 CE, Magnentius' army proclaimed him emperor at Augustodunum in Gaul (Autun in modern-day France). Magnentius assassinated the legitimate Roman emperor of the West, Constans, and immediately began an offensive against Constantius II (337–361), the Roman emperor of the East. At the end of the year, Gaul, Africa, Spain, and Italy were in Magnentius' hands. He continued his military campaign eastwards past the newly conquered towns of Aquileia and Emona and defeated Constantius II's rapidly approaching army at Atrans. Magnentius conquered parts of Pannonia, including the important city of Siscia, where he immediately minted his own money. He enhanced the defensive system (*Claustra Alpium Iuliarum*) when he crossed the Alps, as confirmed by written sources and discovered coins. Emona was finally conquered by Magnentius after several bloody battles, in which Constantius' military detachments were defeated. While Constantius' army suffered many serious losses, it was not until the next fall that Magnentius' army finally suffered utter defeat at the Battle of Mursa; he was forced to withdraw toward the west. Coin finds show that Magnentius lost Emona in August, 352 CE, after a series of heavy clashes. Soon afterwards, he was defeated at Ad Pirum on the Alpine defensive line. He lost Aquileia and committed suicide in 353 CE.

This tumultuous period, in which so much took place in present-day Slovenia, appears in detailed written sources. However, the transfer of Emona from Constantius II to Magnentius, and back to Constantius II two years later, is best illuminated by coin hoards, which often reflect calamitous events. The discovery of 22 gold multiples, one featuring the legitimate co-emperor Constans and 21 featuring Magnentius, which were minted in Aquileia and bore a strong propagandistic message (showing Magnentius as *liberator rei publicae*), is particularly famous. They were found in a house in Emona's forum and probably hidden by a high municipal dignitary just before Constantius II retook the city in 352 CE. Constantius II had his cousin, Constantius Gallus Caesar (351–354 CE), arrested at his residence in Poetovio in 354 CE. He was imprisoned for plotting a coup, and executed near Pola. Part of the army under Gallus' half-brother, Julian (360–363 CE), staged a revolt near Aquileia in 361 CE, switched allegiance to his enemy Constantius II, and started a siege of the city.

Regardless of the dramatic conditions during the reign of Valentinian I (364–375 CE), which saw barbarian tribes conduct waves of raids into Britain, Gaul, and the Danube basin, the Roman Empire was still a powerful state in the 4th century. It was temporarily consolidated under the rule of Emperor Theodosius (383–395 CE). The religious crisis was briefly alleviated by the triumph of orthodoxy over the Arian heresy in 381 CE, and any attempted usurpation was stifled relatively fast. Magnus Maximus (383–388 CE) was proclaimed the new emperor of the Western Roman Empire by a western garrison, to assume power over Britain, Gaul, Spain, and Africa. Only when he decided to conquer Italy in 387 CE did the legitimate emperor Theodosius I, whose seat was in the east, stand ready for an attack against the usurper. In 388 CE, the two armies engaged at Siscia and Poetovio. Maximus suffered defeat, but not decisively. Emona received the victorious emperor with enthusiasm, and Theodosius pursued Maximus through the chain of Alpine fortifications between Nauportus and Ad Pirum. Andragathius, Maximus' Master of the Horse, had fortified the defensive system, but made a tactical mistake due to misinformation, which led him to expect Theodosius to attack from the sea. As a result, he left the fortresses undermanned, which allowed Theodosius' army to pass through the defensive system practically unchallenged, burning down the least defended strongholds like Martinj Hrib and Lanišče. Magnus Maximus was captured and executed near Aquileia.

The next usurper was Eugenius (392–394 CE), who tried to seize the throne by pleading for the continuation of paganism. The ensuing war was therefore also a religious war. It reached its climax at the *Claustra Alpium Iuliarum* itself in a bloody battle on the banks of the Fluvius Frigidus (perhaps the Hubelj) in the Vipava Valley in 394 CE. Prior to the battle, Eugenius fortified the Alpine defensive system and erected statues to Jupiter on the surrounding hills. Theodosius initially suffered defeat, but overcame Eugenius on the second day and crushed the pagan aristocracy. The victory allegedly owes much to one part of Eugenius' army changing sides, and to divine intervention (fierce northeastern squalls turned projectiles launched by Eugenius' army onto his own soldiers). The entire state was subsequently reunited under one scepter.

Roman cities experienced their last boom in the 4th century. Some cities persevered until the end of the 4th century (Neviodunum), while others, judging by the remains of early Christian buildings, held on until the mid-5th century (Emona, Celeia). Poetovio, however, lasted well into the 6th century, as can be inferred (despite sparse archaeological finds) from a reference by the anonymous geographer of Ravenna. Other towns had gradually faded into obscurity since the mid-5th century, with only traces of temporary settlements recorded. According to archaeological finds in Emona and written sources, which report a migration to Istria, most of the population, including municipal administrators and clergy, moved. Furthermore, Roman country estates and rare forms of settlements were abandoned no later than in the first half of the 5th century. The population generally migrated from exposed lowlands to hilltop shelters, which had been set up in times of hunger during the 4th century. Some of them became permanent settlements in the next century. Consequently, by the end of the 5th century, the entire territory of present-day Slovenia—apart from the Prekmurje lowlands—was dominated by walled settlements on steep, not easily accessible hills.[74] Buildings were made of stone or wood, and the center of the settlement was most often occupied by one or more early Christian churches

---

74   Slavko Ciglenečki, *Höhenbefestigungen aus der Zeit vom 3. bis 6. Jh. im Ostalpenraum*, Ljubljana: Založba SAZU, 1987.

(Fig. 14). Some hilltop settlements had been established near roads, and were mainly occupied by military garrisons and their families, with the aim of protecting communications. Others served as ecclesiastical and administrative centers with churches, clerical buildings and facilities for the local population. Archaeological remnants indicate that the local population produced most of their own necessities, grew its own food, and engaged in some trading activities.

Figure 14. Aerial view of the late-Roman settlement at Ajdovski gradec near Vranje.

The Roman Empire faced another major crisis after the death of Theodosius I in 395 CE. Germanic tribes in the empire increased their influence within the army; the Church was undermined by various heretical movements and frequent disputes between the eastern and western halves of the state. The rapid decline of the western half meant that barbarian tribes could cross unimpeded. The Roman Empire was thus mostly invaded by Germanic tribes from "free Germania" and the Rhine-Danube border. This great wave of migration was set in motion by the Huns, a nomadic group of Asian horsemen, after they crushed the Ostrogothic state along the Black Sea in 375 CE. The Visigoths were also under pressure from the Huns, and in 401 CE, under Alaric's command, they easily crossed present-day Slovenia and the *Claustra Alpium Iuliarum*. They prevailed in the battle at the Timavus and secured access to Italy. After being forced to leave Italy years later, they set up provisional camps around Emona and Celeia, and peremptorily demanded a high payment from the emperor in Ravenna, in exchange for not invading Italy again. In 410 CE, the Visigoths plundered Rome. They settled in Aquitania after 418 CE, and later in Hispania.

During the late Roman era, the social structure was disrupted and differences among people faded, especially in the provinces. Free peasants became coloni, attached to the soil, and frontier soldiers became peasants. The number of slaves and rich civilians declined, and municipal administration decreased to be partly taken over by ecclesiastical authorities. By the 4th century, the Roman Empire had begun to recruit mercenaries from neighboring, mostly Germanic tribes to protect its border areas and communications. This cooperation was called *foedus*, and the non-Roman soldiers lived with their families in a completely Romanized environment. Small artifacts, discovered particularly in burial grounds, suggest that they were also living in present-day Slovenia. By the beginning of the 5th century, the garrison at the strategically important small fortress of the Ptuj castle, which protected the town and the Drava River crossing, included several Germanic soldiers. This was also true of the military unit stationed below the Predjama Castle at Postojna, which was controlling the section of road between Emona and Tergeste.

In about 400 CE, the Pannonian limes were destroyed, most Pannonian territory was lost, and the population fled to the south and southeast.[75] After the Visigoths had passed, the situation in Inner Noricum stabilized until the next crisis in approximately 430 CE, when the incursion of the Iuthungi raised dissatisfaction among the provincial population and could only be overcome by Flavius Aëtius. This uprising is only mentioned in the Roman literary sources and is not documented by coin hoards.

In the early 5th century, supreme power over the Western Roman Empire rested with the Roman commander Flavius Aëtius, who had been held hostage since 405 CE, by the Visigoth ruler Alaric, and later for several years by the Huns. He owed much of his future success to his Hunnish allies, whom he rewarded for their support with a portion of the Pannonian Plain along the Sava (433–434 CE). Vast ransom demands for Roman prisoners of war were a significant source of

---

75   Friedrich Lotter, Rajko Bratož, and Helmut Castritius, *Völkerverschiebungen im Ostalpen-Mitteldonau-Raum zwischen Antike und Mittelalter (375–600)*, Berlin, New York: De Gruyter, 2003, 156 ff.

income for the Huns. They attacked other towns in Illyricum and Pannonia from their base camps, as well as threatened the Norican-Pannonian area (including Poetovio), which bordered their territory.

The Poetovian area was then among the constant targets of Germanic and other tribes, particularly the Huns, who wanted to settle in Roman provincial territory or advance their invasion of Italy through Emona. In his report on the Eastern Roman mission to the court of Attila the Hun in 449 CE, the Byzantine 5th-century rhetor and historian Priscus of Panium in Thrace mentions a meeting with a Western Empire mission headed by *comes* Romulus, who was regarded (probably erroneously) as a Poetovian native. Priscus reports: "They embarked on their mission from the Norican town of Poetovio in order to mollify Attila."[76]

In 452 CE, the Huns launched an attack from Pannonia against the Western Roman Empire, which revealed itself in all its weaknesses. The Huns sacked Aquileia and stormed a number of other important Italian cities. This was also a period of great turmoil for the territory of present-day Slovenia, according to historical sources, while Hunnish raids left traces in oral folk tradition. On the other hand, there are hardly any archaeological finds conclusive of the presence of Huns.

After Attila's death in 453 CE, the territory left sacked by the Huns was rapidly settled by new peoples, particularly the Ostrogoths and Rugians, who worked as Roman *foederati*. According to the ecclesiastical writer Eugippius, there were still military garrisons in Noricum until 476 CE; they were disbanded when Odovacar deposed the last Roman emperor Romulus and became the first barbarian king of Italy. Thereafter, the population had to rely on its own defenses. All forms of civil or local administration, except the postal service, were replaced by ecclesiastical governance in the late 5th century. The postal service remained in operation, as is indicated by a recorded exchange of letters between Severinus in Noricum and Odovacar in Ravenna, and an upper-class Roman lady, Barbaria, who was living near Naples.[77]

The Western Roman Empire came to an end in 476 CE, when Odovacar, a Germanic warlord who had served as a mercenary in the Roman army, was proclaimed king by his fellow soldiers. He defeated the Roman general Orestes, who had governed the empire on behalf of his underage son, the emperor Romulus Augustulus (475–476 CE). Odovacar deposed the last Roman emperor of the West and deported him to Campania. Rome accepted

---

76　*Priscus*, Fr. 11.2 (ed. Blockley).
77　Rajko Bratož, *Severinus von Noricum und seine Zeit. Geschichtliche Anmerkungen*, Vienna: Österreichische Akademie der Wissenschaften, 1983, 14–15; 39 ff.

Odovacar as a new ruler, while Emperor Zeno (474–491 CE) of the East recognized his authority over Italy. Italy became the promised land for barbarian peoples. In 488 CE, Odovacar relocated part of the population of the Noricum Ripense to Italy. In 455 CE, the belligerent Ostrogoths had settled Pannonia and often clashed with neighboring Germanic peoples in the central trans-Danubian basin; Zeno directed the Ostrogoths toward Italy to rid himself of troublesome neighbors. In 489 CE the Ostrogoths crossed the Soča River with their families and all their belongings, defeated Odovacar, and fought to enter Italy. They established a kingdom centered on Ravenna in 493 CE, which survived until the mid-6th century. Other barbarian peoples were gradually settling down as well, while establishing their kingdoms in this period, such as the Jutes and Anglo-Saxons in England, the Visigoths in Spain, the Franks and Burgundians in France, the Gepids and Lombards in Pannonia,[78] and the Vandals in North Africa.

## Late Antiquity and the Coming of the Slavs

The Ostrogoths formed a vast state in Italy and the western Balkans as far as the Drina with a new administrative system, rule of law, road network, and uniform monetary system, which led to significant economic growth. Archaeological finds in present-day Slovenia indicate that they had garrisons at several most crucial fortresses—occupied by natives, e.g., at Zidani Gaber and Gradec near Velika Strmica (in the region of Novo Mesto), and in other areas in Slovenia. One of the most significant centers of native and Germanic peoples in the eastern Alpine area was the Tonovcov Fort (Tonovcov grad) at Kobarid, where the remarkably well-preserved remains of several buildings have been uncovered. Small churches in these fortresses next to major church buildings were probably built by the Ostrogoths, who belonged to the Arian branch of Christianity. Between 500 and 540 CE, Gothic troops were also stationed in the fort of Carnium (Kranj) to guard the road connecting with the upper Sava Valley (Fig. 15). Archaeological finds in Kranj and near Ljubljana suggest that Ostrogothic migration was soon followed by the Alemanni, who retreated from the Franks and migrated eastward through present-day Slovenia.

---

78   Slavko Ciglenečki, "Romani e Longobardi in Slovenia nel VI secolo," in: *Paolo Diacono e il Friuli altomedievale (secc. VI–X)*, vol. 1, Spoleto: Centro italiano di studi sull'alto medioevo, 2001, 179–199.

**Figure 15.** Ostrogothic and Lombard money, and an Ostrogothic fibula, 5th and 6th century CE.

The Franks had been strengthening their western position since the 6th century. By the end of the 8th century, they had expanded their kingdom into a powerful Christian state by incorporating numerous Germanic tribes. They thus headed a united Christian army, which warded off Muslim Arabs in the Battle of Poitiers after the latter had conquered Visigothic Spain after 711 CE. The center of the Roman state shifted to the east, where the Byzantine Empire was becoming more and more powerful. The reign of Justinian I (527–565 CE) in the East, ushered in an era of restoring the former Roman Empire. The eastern and northern Adriatic coast became the scene of long-lasting warfare between the Ostrogoths and the Byzantine Empire in 535–555 CE, which ultimately ended with the fall of the Ostrogothic kingdom. In 539 CE, the Byzantines conquered Istria and crushed Gothic rule in Pannonia. The final blow to Ostrogothic dominion came in 552 CE during the Byzantine march on Ravenna which was launched from Dalmatia and advanced along the Adriatic coast. The Franks, Byzantine allies at the time, temporarily invaded the eastern Alpine territory in approximately 540 CE. In 547/8 CE, Pannonia and part of Noricum were settled by Lombards, who had previously occupied the territories of present-day Hungary, part of lower Austria, and what is now the Czech Republic. Justinian had bestowed the lands on them to protect his western borders from the Franks, who were attempting to expand eastwards. In the territory of present-day Slovenia, the Lombards took over the administrative territories of Celeia and Poetovio and forts along the Pannonian frontier in present-day Lower Carniola (Dolenjska). They usually occupied the existing hilltop settlements and fortresses, which were mainly inhabited by native populations. New garrisons were rare, at least according to archaeological finds. By then, the Lombards had also reached the fort of Carnium: their presence has been documented by many artifacts found in male and female graves (Fig. 16). However, after the arrival of new peoples (especially the Slavs and bellicose Avars), the Lombards moved to Italy in 568 CE under King Alboin. They maintained individual garrisons in certain fortresses (e.g., Carnium) as advance guards for their new homeland in Italy until the end of the 6th century.

The territories abandoned by Germanic peoples in the 5th and 6th centuries were settled by Slavic tribes. The Slavs pushed toward the Oder, occupied vast territories north of the Danube, and flooded the Balkan Peninsula. Toward the end of the 6th century, Slavic tribes from the central Danubian basin gradually moved westwards, including into present-day western Slovenia.[79]

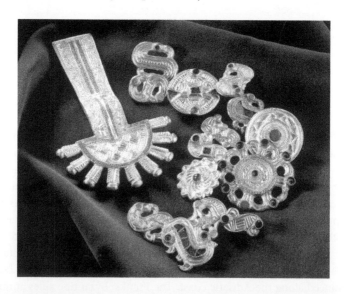

**Figure 16.** Lombard jewelry from middle and late 6th century found in the Lajh cemetery in Kranj.

---

79  Mitja Guštin (ed.), *Zgodnji Slovani. Zgodnjesrednjeveška lončenina na obrobju vzhodnih Alp/Die frühen Slawen. Frühmittelalterliche Keramik am Rand der Ostalpen*, Ljubljana: Narodni muzej Slovenije, 2002.

They reached the Italian borders in approximately 600 CE and began to invade Italy, undeterred by their defeat against the Byzantine army in 599 CE somewhere in northern Istria. Unlike the nomadic Avars, the Slavic tribes settled permanently, which destroyed the very last remnants of the ancient civilization.

\* \* \*

This brings us to the end of an age and at the same time to our introductory emphasis about the present-day Slovenian territory being on the periphery of the more developed countries of the time: the Mediterranean on the one hand and the Danube regions on the other. This again evokes the time when metallurgy, which enabled rapid economic development, was particularly significant in bringing about technological advancement and a socially stratified society. At the end of the Bronze Age and with the coming of the Iron Age, Slovenian territory witnessed a specific grouping of tribes, resulting in the ethnogenesis of various prehistoric peoples. Their names are not preserved in classical sources, but they must have left some trace in the genetic structure of the present-day inhabitants of Slovenia.

This territory first reached its peak in the Early Iron Age with the rich Lower Carniola (Dolenjska) cultural group, centered on Magdalenska gora, Stična, Vače, Novo Mesto, and several other, mainly hilltop settlements. One of its main characteristic features was "situla art," i.e., decorated small bronze buckets and belt buckles with images of major festivities. These were most likely connected with the cult of the dead and indirectly reflected scenes from everyday life.

The arrival of the Celts caused this population to decline, and the Slovenian territory became peripheral once again. The Norican Kingdom developed to the north of the Karawanks (Karavanke Mountains) while Celeia alone became part of the kingdom in the 1st century BCE. However, the Slovenian territory has always been a transit area of extreme strategic importance. Through the so-called Italo-Illyrian Gate at Postojna and across the famous pass below Mt. Ocra (Nanos) at Razdrto led the main route linking the Apennine and Balkan peninsulas, which was most convenient for traders and invaders alike. This was the Amber Route, connecting the Upper Adriatic with the amber-rich Baltic countries. It forked at Emona, and one branch led toward Siscia and further through the Balkans, following the direction of the Argonauts' river route, along which they had allegedly returned via the Danube, Sava and Ljubljanica, on the flight from Colchis to Greece.

Slovenian territory was never a center of any state-like formation; its position—although strategic—was always peripheral. In the Roman period, it was divided among four administrative units. The area around Emona belonged to Italy, the regions with Celeia as their center extending to Noricum, Poetovio and Neviodunum, belonged to Pannonia, while an insignificant small southern part was attached to Dalmatia.

After the bloody conquest of most of the Slovenian territory, with the exception of that which belonged to Noricum (which was peacefully annexed), the Romans imposed a higher civilization on these regions, which is termed self-Romanization in modern historiography. Indeed it meant acculturation of the country, through the blending of Roman and Celtic cultures, which brought about a typical provincial civilization. It sharply divided the provincial inhabitants from the barbarians living beyond the Roman frontier, i.e., beyond the Danube.

Peaceful life in the Roman Empire, between the reigns of Augustus and Marcus Aurelius, started to disintegrate during the outbreak of the Marcomannic Wars halfway through the 2nd century CE. Several military and economic crises alternated with relatively short periods of precarious stabilization and prosperity, until Diocletian introduced economic and administrative reforms. He hoped to prevent the disintegration of the empire, but the reforms proved to be short-lived, and by the end of the Tetrarchy, the fall of the Western Roman Empire was unavoidable. It occurred when the last western Roman emperor Romulus Augustulus was deposed in 476 CE. The eastern part of the empire was called the Byzantine Empire after its capital, the former Byzantium, which was then called Constantinople.

The emperors became increasingly incapable of exercising authority in the western part, which was governed by German princes: first by Odovacar and then by Theoderic the Great, the King of the Goths. When the Huns arrived in Pannonia in the early 5th century CE, the situation was already so dire that the provincial population could no longer remain in towns and villages. Some were killed, some emigrated, while some took refuge in hilltop settlements. Deculturation was in progress in all facets of life. The coming of the Avars and Slavs completely changed the population structure and settlement pattern, announcing the coming of the Middle Ages.

The Early Middle Ages

# The Early Slavs

The settlement of the Slavs—one cannot speak of Slovenes until well after the Early Middle Ages—in the Eastern Alps and the basins of the eastern Alpine rivers culminated in the final decades of the 6th century, though the process would continue until the beginning of the 9th century. The first Slavic groups probably came into the Eastern Alpine region from the north, from the area of the West Slavic language group. The first wave seems to have turned south around 550, leaving present-day Moravia and crossing the Danube between Traun to the west and Vienna to the east, first encompassing the territory of Upper and Lower Austria and then gradually spreading inland along the Alpine river valleys. A second wave of Slavic migration to the Eastern Alps from the southeast came somewhat later and was closely linked to the Avars. This nomadic people from the steppes took control of the Pannonian Plain after the Lombards had moved into Italy in 568, and attacked the Byzantine state across the Danube and Sava rivers. In 582 they captured Sirmium, the former capital of Illyricum, and started to move toward the northwest, accompanied by Slavs. The Slavic-Avar advance led to the collapse of ancient structures. It may be possible to trace the stages of the Slavic-Avar advance into the Eastern Alps through the synodal records of the Patriarchate of Aquileia, which reflect the fall of the ancient bishoprics in this area (Emona, Celeia, Poetovio, Aguntum, Teurnia, Virunum, and Scarabantia). By 588, the upper Sava Valley had fallen into Slavic-Avar hands, and by 59, they had taken the upper Drava Valley, where skirmishes with the northern neighbors, the Bavarians, soon began around present-day Lienz. In 592 the Bavarians were successful, but then in 595 they were heavily defeated in a battle decided by the Avar khagan with his cavalry. These battles, which flared up once more around 610, led to the development and consolidation of a border area that divided the eastern Alpine area for centuries: a Frankish-dominated western half separated from an Avar and Slavic east and southeast.

To the south, along the Soča River, and in Istria the border was established somewhat later. After conflicts with the Friulian Lombards at the beginning of the 8th century, the Slavs occupied the mountainous region to the west of the Soča, as far as the edge of the Friulian plain. This ethnic boundary has lasted, with minor changes, for over 1,200 years and into the present day. The advancing Slavs moved into Istria from the northeast via the Postojna Gate. First, until around 600, they settled the lands up to the peninsula's natural threshold to the south of the ancient Tergeste–Tarsatica (Trieste–Rijeka) road, where the Karst plain descends sharply down to the hinterland of Koper and Buzet. At the end of the 8th century, the local Frankish authority organized the colonization of Slavs into the unpopulated territories under the administration of the towns in the peninsula's interior.[80]

By the time the Slavs had settled the former provinces of Noricum and Pannonia, Roman organization had already disappeared from the area. It had survived for a considerable time, despite the major upheavals of the 5th century, as long as the area was still administered from Italy. This era came to an end in western Noricum with the Frankish occupation in 536/537. The eastern Norican and western Pannonian areas, ceded to the Lombards by the Eastern Roman Emperor Justinian in 548, were separated from the Roman ecumene by the Lombard migration to Italy in 568 and the establishment of the new Avar dominion in the central Danubian area. This does not mean that the new arrivals did not at least partly assume the heritage of antiquity. Older beliefs about the indigenous population abandoning the region completely or being forced out by the Slavs have long since been superseded. Numerous toponyms relating to the name Vlah, which the Slavs gave to the Romans and the Romanized population (e.g., Laško), as well as some hillforts that were continuously settled until the 7th century, indicate contact and cohabitation between the indigenous inhabitants and the Slavs. The Slavs also took up numerous ancient place and river names, as well as certain components of the indigenous inhabitants' everyday economic life, particularly Alpine dairy farming. Even though the area newly settled by the Slavs had been cultivated since antiquity, there were many changes in its structure at this time. The most evident was the collapse of ancient towns. A different social structure grew up alongside a new form of arable farming. The ecclesiastical organization collapsed completely, but the Christian cult did not: the indigenous Romanized population was able to preserve it at least in some areas, such as Upper Carinthia around Spittal an der Drau. There, the preservation of Deacon Nonnosus' gravestone from 532 in a monastic church from the end of

---

80  See Lothar Waldmüller, *Die ersten Begegnungen der Slawen mit dem Christentum und den christlichen Völker vom VI. bis VIII. Jahrhundert. Die Slawen zwischen Byzanz und Abendland*, Amsterdam: Adolf M. Hakkert, 1976, 180–187; Jaroslav Šašel, "Der Ostalpenbereich zwischen 550 und 650 n. Chr.," in Šašel, *Opera selecta*, 821–830; Friedrich Lotter, Rajko Bratož, and Helmut Castritius, *Völkerverschiebungen im Ostalpen. Mitteldonauraum zwischen Antike und Mittelalter (375–600)*, Berlin and New York: De Gruyter, 2003, 149–155; Peter Štih, *The Middle Ages between the Eastern Alps and the Northern Adriatic: Select Papers on Slovene Historiography and Medieval History*, Leiden, Boston: Brill, 2010, 87–197.

the 8th century in Molzbichl indicates the continuity of the Christian cult well beyond the initial period of Slavic settlement. The 8th-century Christianization of the Carantanians would later be connected with the spots of local Christian tradition.[81]

By the end of the 6th century, today's East Tyrol and Carinthia were already being described as the "Land of the Slavs" (*Sclaborum provincia*), while the presence of the Avar khagan indicates that this mountainous Alpine world was included in the Avar khaganate, the center of which lay between the Danube and the Tisza rivers in Pannonia. The Slavs were subordinated to their Avar masters, paying tribute and providing military service. Yet the attitudes and relations of the Avar warriors to the various Slavic groups varied according to time and geographical circumstances. Avar supremacy over the Slavs was undoubtedly more keenly felt at the center of the khaganate in Pannonia than on the periphery, in the hilly and heavily forested eastern Alpine and northwest Balkan areas, which weren't suitable for a nomadic lifestyle. Traces of the Avar presence in Carantania may be preserved in place names such as Faning and Fohnsdorf (Slov. Baniče and Banja Vas) deriving from the word *ban* (from the Avar name Baian). Avar rule lasted until the mid-620s, when two events—the start of Slavic resistance to Avar supremacy under Samo in 623 and the failed Avar siege of Constantinople in 626—ushered in major changes in the region.[82]

Before 626, barbarian peoples had already reached the walls of the great city on the Bosporus, but the Avars' was the first genuine attempt to conquer Constantinople, in alliance with the Persians and with their subjugated Slavic warriors. The failure of the siege proved a disaster for the Avar khaganate, almost precipitating its complete collapse. The catastrophe of 626 also improved the position of Slavs who joined the uprising under Samo. According to the Fredegar Chronicle, Samo joined the Slav revolt in the territory of today's Czech Republic and Slovakia in 623, which would mean that the breakaway from the Avar khaganate began before the great crisis of 626. Samo, a Frank by birth and perhaps a weapons trader, exploited the opportunity to take the fight to the Avars and become king of the first Slavic polity known to history. Its center was probably north of the Danube, but it seems to have included the area of the Eastern Alps later known as Carantania. In 631, the Frankish King Dagobert I unsuccessfully

---

81 Karl Heinz Frankl and Peter G. Tropper (eds.), *Heilige Nonosus von Molzbichl*, Klagenfurt: Verlag des Kärntner Landesarchivs, 2001.
82 Walter Pohl, *The Avars: A Steppe Empire in Central Europe, 567–822*, Ithaca, London: Cornell University Press, 2018, 117–162, 280–343.

attempted to destroy Samo's realm, uniting Frankish and Alemannic troops with Lombards, who had no option but to act against their Alpine Slav neighbors. In around 623/26, the Friulian Lombards had already wrested control of "the region of the Slavs, called Zellia" in the contact zone between Carinthia and Friuli, and the Slavs there were to pay tribute to the Lombard duke in Cividale del Friuli until around 740. Probably not without reason, the *Conversio Bagoariorum et Carantanorum* (The Conversion of the Bavarians and Carantanians), the most important source for the history of Eastern Alps and Pannonia in the 8th and 9th centuries, written in Salzburg in 870, links Samo to the very earliest Carantanian history.[83]

At that time, the Alpine Slavs, also known as Vinedi, who were probably connected with Samo's realm, had their own prince Vallucus, who ruled over "March of the Wends" (*Marca Vinedorum*). Vallucus and his Slavs were joined around 631/632 by a group of Bulgars led by Alzeco (Alciocus). They were a part of 9,000 Bulgars who fled from Pannonia to the Bavarians. After an initial welcome the Bavarians murdered several thousands of them on the order of the Frankish King Dagobert I. Only Alzeco's group escaped, and, fleeing once more, was received by Vallucus. Alzeco's Bulgars remained with the Alpine Slavs for around 30 years in the nascent Carantania before migrating after 662 to Benevento in Lombard Italy.

After Samo's death in 658, the Avars reestablished their supremacy over most of Slavic Europe, but not over the Carantanian Slavs, who—as the Alzeco episode indicates—were independent from all their neighbors: the Bavarians and the Franks, the Lombards and the Avars. The Avars also reinstated their supremacy south of the Karavanke Mountains and the khaganate once more extended to the borders of Italy. In around 664, at the behest of the Lombard King Grimoald, the Avars under direct command of the khagan attacked Friuli, defeating and killing the usurping Duke Lupus of Friuli. The battle probably took place in the Vipava Valley, where in 394 Christian Emperor Theodosius defeated his pagan rival Eugenius. Paul the Deacon, a Lombard from Cividale by birth, wrote of these events at the end of the 8th century and further reports that Lupus' son Arnefrit fled in fear of King Grimoald "to the Slavic people in Carnuntum, which

---

83 Herwig Wolfram, *Conversio Bagoariorum et Carantanorum. Das Weißbuch der Salzburger Kirche über die erfolgreiche Mission in Karantanien und Pannonien. Herausgegeben, übersetzt, kommentiert und um die Epistola Theotmari wie um Gesammelte Schriften zum Thema ergänzt*, Ljubljana: SAZU in Zveza Zgodovinskih društev Slovenije, 3rd ed., 2013.

is erroneously called Carantanum" (*ad Sclavorum gentem in Carnuntum, quod corrupte vocitant Carantanum*).

Although the term "Carantanum" does not belong to the time described by Paul, but rather to the end of the 8th century when he wrote his *History of the Lombards*, this is the oldest undisputed reference to the name. Paul's report clearly shows that the Carantanians' ethnic name—probably first mentioned by the Anonymous cosmographer of Ravenna as Carontani—was derived from the regional name for the area in which they lived. The name was originally associated with the area around Zollfeld (*Carentana*) and Ulrichsberg (*Mons Carentanus*) north of Klagenfurt, where the *civitas Carantana* (Karnburg) and *ecclesia sanctae Mariae ad Carantanam* (Maria Saal) stood. The name came to refer to the entire area ruled by the prince of Karnburg and is not only pre-Slavic but also pre-Roman in origin. Etymologically the name Carantanians probably means "people from Caranta." The root Car- is typical of the wider Alpine-Adriatic area, and is also found in names such as Carnia and its derivative Carniola, as well as the name Karst.[84]

## Carantania

The fundamental unit of political, social, and legal life as conceived in the Early Middle Ages was the tribe or the people (gens, *rod*, ethnos). Extensive research on the ethnogenesis of Germanic, Slavic, and steppe nomadic peoples in recent decades clearly indicates that the peoples of the Early Middle Ages were not communities of shared origin but polyethnic communities, identified by their belief in a common origin, as well as customs and legal practices that these heterogeneous groups participated in and recognized as their own.

The Carantanians had polyethnic roots too, but the Slavs were the dominant group, which is why the neighbors referred to the entire population as Slavic. In the above-mentioned *Conversio*, the most important source for Carantanian history, they were described as "Slavs called Carantanians" (*Sclavi qui dicuntur Quarantani*). Beside Slavs, indigenous Romans and Croats were also involved

---

84  Harald Krahwinkler, "Ausgewählte Slawen-Ethnonyme und ihre historische Deutung," in Rajko Bratož (ed.), *Slovenija in sosednje dežele med antiko in karolinško dobo. Začetki slovenske etnogeneze/Slowenien und die Nachbarländer zwischen Antike und karolingischer Epoche. Anfänge der slowenischen Ethnogenese 1*, Ljubljana: Narodni muzej Slovenije, SAZU, 2000, 413–418.

in the ethnogenesis of Carantanians, with the additional possibility of small numbers of Avars, Bulgars, Dulebians and Ostrogoths among them.[85]

The principality of the Carantanians was the oldest early medieval polity formed in the eastern Alpine region. It started to emerge at least in 620s, when the Slavs in present-day Carinthia already had their own prince—*dux Vallucus*—and therefore their own political organization, described as *marca Vinedorum*. According to Paul the Deacon, there already lived a specific Slavic people (*gens Sclavorum*) in Carantania in 664, later named Carantanians (Lat. *Carantani*; Slav. *Horutane*). In any case, the Carantanian ethnogenesis came to an end before 743, when in a turmoil of dramatic events that had a decisive impact on the future the Carantanians clearly passed into history.

At that time their prince was Boruth and the Carantanians were being seriously threatened by the Avars. Finding himself in a difficult position, Boruth turned to the Bavarians and their Duke Odilo for aid. Together, they defeated the Avars, but it came with a price for Carantanians—they were forced to submit to the lordship of Frankish kings. Carantanian loyalty was guaranteed by hostages, including Boruth's son Cacatius (Slov. Gorazd) and his nephew Hotimir, who were taken to Bavaria and raised as Christians. These fateful events occurred before 743, as by then the Carantanians were already marching in the Bavarian army against the Franks.

In 749, after the death of Boruth, the Bavarians acquiesced to Carantanian requests for Cacatius to be sent home and made their prince. But three years later, Cacatius died and was succeeded by his cousin Hotimir. Both became princes with the permission of the Frankish king. Hotimir was accompanied home by the first Salzburg priest to come to the Carantania. Pope Zachary already included Carantania in the Salzburg diocese before 752. The Bishop of Salzburg at that time was the erudite Irishman Virgilius (746/47 or 749–784). Hotimir asked him to personally lead the Christianization in Carantania, but instead he sent the regional bishop Modestus as his envoy (*episcopus missus*). A large number of churches were consecrated in Carantania during the Salzburg mission, which lasted until roughly the end of the 8th century, although only three churches consecrated by Modestus can be specifically named. Modestus

---

85  For the history of Carantania, see Wolfram, *Conversio*, 109–165; Hans-Dietrich Kahl, *Der Staat der Karantanen. Fakten, Thesen und Fragen zu einer frühen Slawischen Machtbildung im Ostalpenraum (7. –9. Jh.)*, Ljubljana: Narodni muzej Slovenije, SAZU, 2002; Štih, *Middle Ages*, 108–122; Paul Gleirscher, *Karantanien—Slawisches Fürstentum und bairische Grafschaft*, Klagenfurt/Celovec: Hermagoras Verlag/ Mohorjeva založba, 2018.

remained in Carantania until his death in 763 which unleashed the first rebellion of Carantanian opposition against the Christian faith and the prince so closely associated with it. In 765 there was another revolt, but again Hotimir quickly quashed it. His death in 769, which may have been connected to the extinction of the princely dynasty, led to the third and most violent revolt. In the three years that followed, there were no longer any priests in Carantania. It was only by the direct military intervention of Duke Tassilo III of Bavaria that the Carantanian rebels were crushed in 772 and the previous order restored. Bavarian contemporaries compared Tassilo with the Emperor Constantine, and his victory to Charlemagne's destruction of the Saxon sanctuary Irminsul.

However, Tassilo had initially intended to resolve matters peacefully. To this end, in 769 he established a monastery in Innichen at the source of the river Drava, on the border with Carantania. The monastery had an explicitly missionary purpose, "that the faithless Slavs be brought to the path of truth." A similar task was entrusted to the oldest Carinthian monastery in Molzbichl near Spittal an der Drau, recently discovered by archeologists and most probably founded after Tassilo's victory over the Carantanians in 772, which ended the political and religious crisis in Carantania. Carantania's new prince was Valtunc, now even more closely linked to Bavaria. The links with Salzburg were reestablished, and by the time of Virgilius' death in 784, six groups of missionaries had come to Carantania. His successor Arn continued the work of Virgilius and after his elevation to archbishop in 798, he and the Bavarian prefect Gerold, Charlemagne's brother-in-law, introduced in 799 a regional bishop Theoderic in Sclavinia, which covered Carantania and the newly conquered Pannonia north of the Drava. This restored the institution of regional bishop, to which Virgilius had appointed Modestus and which was maintained until the mid-10th century, with an interruption in the third quarter of the 9th century. The institution also served as a model to Archbishop Gebhard of Salzburg in the founding of the first Carinthian bishopric in Gurk in 1072.

Salzburg's role in the Christianization of the Carantanians was decisive, but not unique. The *Conversio Bagoariorum et Carantanorum*, which praised and monopolized Salzburg's achievements in the Carantanian mission, suppressed the role of other ecclesiastical centers such as Aquileia and Freising in this undertaking. The most convincing arguments suggest that Freising's properties in Upper Carinthia were the place where three short but invaluable religious texts were written down around the turn of the millennium. The texts, known as the *Freising Manuscripts* (Slov. *Brižinski spomeniki*), are the most important pieces of evidence of the earliest Christianization of the Slavs and their religious practices. Freising Manuscripts (referred to below as FM I, FM II, FM III) are written in Carolingian minuscule and preserved in a Latin codex which is kept in Munich. They are the oldest Slavic texts in the Latin alphabet and were a part of a "vademecum" used by Bishop Abraham, who acquired estates for the Church of Freising in two lands populated by Slavs, Carinthia and Carniola. Modern linguists sometimes refer to the language of the manuscripts as Old or Early Slovene, while contemporary writers referred only to the Slavic language (*lingua Sclavanisca*). In terms of content, FM I and FM III are general confessional formulas, while FM II is a rhetorically complex sermon on sin and a call to repentance and confession. Texts, which were copied around the turn of the millennium, date back to at least the middle of the 9th century, though hypotheses on their origin and sources differ significantly. However, regardless of the many theories, the indisputable fact remains that the preserved texts were used among the Carantanian Slavs.[86]

By the beginning of the 8th century at the latest, Carantanian society was organized under the lordship of a prince. In terms of position, a prince of this kind, generally referred to as *dux gentis* in Frankish sources, was king (*rex gentis*). This is supported by the Slavic word *knjaz* (Slov. *knez*), which refers to a Slavic ruler of the Early Middle Ages and derives from the Germanic *kuningaz*, meaning a king who ruled over a "small area." Eight Carantanian princes from the second half of the 8th century and the first third of the 9th century are known to us by name. At that time, Carantania was already subject to the Bavarians and Franks as a tributary or client principality, one of many on the eastern and southeastern border of the Frankish empire. Internally, the principality retained its gentile (tribal) constitution, one of which was the installation of the Carantanian princes. It was the Carantanians themselves, albeit with the permission of their Frankish king, who made Cacatius and then Hotimir their princes. The first three Carantanian princes known by name were related, and princely authority was therefore hereditary within a ruling family. The Carantanian prince was installed as such with a ceremony that involved him being placed on the Prince's Stone which once stood at Karnburg, the political center of the principality. The Prince's Stone—actually the base of a Roman column turned

---

[86] See France Bernik (ed.), *Brižinski spomeniki. Znanstvenokritična izdaja*, Ljubljana: SAZU, 1993; Jože Pogačnik (ed.), *Freisinger Denkmäler, Brižinski spomeniki, Monumeta Frisingensia*/Literatur, Geschichte, Sprache, Stilart, Texte, Bibliographie, München: Trofenik, 1968; Janko Kos, Franc Jakopin, and Jože Faganel (eds.), *Zbornik Brižinski spomeniki*, Ljubljana: SAZU, 1996.

upside down—is today displayed in Klagenfurt and it is believed to be the oldest preserved symbol of power in the eastern Alpine region. Thus, in the mid-8th century, the Carantanians were among the first to adopt a constitutional model that combined a tribal constitution with the lordship of the Frankish king; a model that would become widespread in the 9th century among Slavic and other peoples along the eastern and southeastern Frankish border.[87]

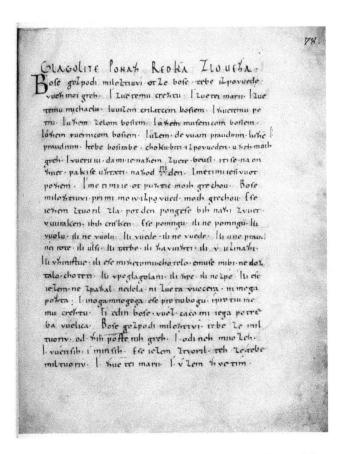

**Figure 17.** Glagolite ponaz redka zloueza ("Say after us [these] few words"). Beginning of first Freising Manuscript.

---

87   Štih, *Middle Ages*, 169–189.

**Figure 18.** The oldest known illustration of the Installation of a Carantanian Prince in Leopold Stainreuter's Österreichische Chronik von den 95 Herrschaften. The original is kept in City Library, Bern, Switzerland, Cod. A. 45, from c. 1480.

Beside the prince and his family, there was another high-ranking social class: the nobility. The 9th and 10th century Bavarian sources indicate the existence of Slavic nobility in Carantania already before the end of the 8th century and provide evidence of its survival into the later period. For example, in 830, a certain Carantanian Slav (*de genere Carontania Sclavaniorum*) Baaz bequeathed properties he had held in Bavaria to the Church of Freising. There are many reasons to support the idea that Baaz was a descendant of one of the "noble" hostages who had accompanied the son and nephew of the Carantanian prince Boruth to Bavaria in 743. This and other similar examples indicate that groups existed within the Slavic community whom the neighboring nobility considered of sufficient standing to be accepted into their ranks and even their families.

A specific social group of the Carantanians and Alpine Slavs were *kosezi*. They were a specific Alpine Slav feature, as the places related to them lie between the upper Enns River to the north and the Kolpa River to the south. They appear relatively late in written sources, with the oldest mention (*Koséntzes*) dating from the mid-10th century. According to Croatian tradition, recorded in *De Administrando Imperio* of the Byzantine Emperor Constantinus Porphyrogenetus, *Koséntzes* was the name of one of the brothers who led the Croats to the hinterland of the Dalmatian cities. The Bavarians called them *Edlinger* (noble people), while in Aquileia and in accordance with the Lombard tradition in Friuli, kosezi were called *arimanni*, who were a special military class of the Lombards (*exercitales*). Both terms indicate that kosezi were a privileged social group. However, the evidence available in sources from the High Middle Ages onwards suggests that the group's social status continually declined. In the 12th and 13th century documents, the kosezi are still presented as equal to the lower nobility (ministerials), but by the Late Middle Ages they were merely peasants with some privileges. The peasant who installed the duke of Carinthia in the Late Middle Ages was a *kosez*. At the same time, there is no doubt that the kosezi class goes back to the Early Middle Ages. The origin of the name kosezi, which is preserved only in the Slovenian language, is not Slavic, and there are many theories about its etymology. The Slovenian name precludes a Germanic origin of the word, because *Edlinger* couldn't be translated as kosezi but rather as *plemeniti* (the Slovene word for "nobles"). The kosezi had therefore already existed when the Bavarians took them to be nobles during their first contacts with the Alpine Slavs. Furthermore, the ceremony in which a peasant-kosez on the Prince's Throne symbolically handed over power in the Duchy of Carinthia to a new duke dressed in peasant clothing—therefore symbolically equal to him— could only have originated in pre-feudal times. The direct contact between the kosezi and the Carantanian prince and/or the later Carinthian duke, and their connection with military service makes the theory that they were originally a military retinue (Slav. *družina*, Ger. *Gefolgschaft*) of the Carantanian prince quite acceptable. Of course, there are many other possible explanations.[88]

## Carniola

The polyethnic but predominantly Slavic principality of the Carantanians was not the only polity to form in the Early Middle Ages within the Slavic-settled area in the Eastern Alps. Almost all the Slavic-settled area of modern Slovenia remained outside of Carantania. This was particularly the case for the Sava Valley, which was separated from Carantania by the Karavanke Mountains. The upper Sava Valley was known in the Early Middle Ages as Carniola. The name, derived from the territorial name Carnia, means nothing more than Little Carnia. The ancient homeland of the Carnians, *patria Carnium*, lay on the other side of the continental watershed, in the mountainous world north of the Friulian plain. It is therefore no coincidence that a Friulian Lombard Paul the Deacon was the first to use the term Carniola, derived from Friulian geographical terminology, to describe the Slavic land (Carniola, patria Sclavorum) to the east of Friuli. The ethnic name of the Carniolans (Carniolenses) derives from the territorial name and is only recorded in 820 in the *Royal Frankish Annals*. All Early Medieval sources that mention Carniola and the Carniolans (though few in number) clearly distinguish between two Slavic communities north and south of the Karavanke Mountains. Alongside Carantania, Carniola was the second Slavic entity in the Eastern Alps. Its center was Carnium (Kranj), a fortified settlement on the confluence of the rivers Sava and Kokra, where the biggest cemeteries from the migration and Slavic period in present-day Slovenia were discovered.

The mention of the Carniolans in the Royal Frankish Annals of 820 reflected great social and ethnic changes after the collapse of the Avar khaganate. Individual Slavic peoples were starting to form during this period among most of the Slavs in the former Avaria, attested by the appearance of new Slavic ethnic names such as Czechs, Moravians, Guduscans, Timocians, Abodrites, and Croats. Similarly, judging by the name Carniolenses, a separate Slavic people began to form in the upper Sava Valley by the end of the 8th century at the latest, taking its name, like the Carantanians, from the area in which they lived. As the direct

---

88   Sergij Vilfan, *Rechtsgeschichte der Slowenen bis zum Jahre 1941*, Graz: Leykam, 1968, 58–62; Štih, *Middle Ages*, 156–165.

eastern neighbor of Friuli, Carniola came under Frankish rule in the course of Charlemagne's Avar war between 791 and 796. It was included in the expanded Duchy (March) of Friuli and, as it seems, retained its tribal constitution until 828 when it was transformed into a Frankish county.[89]

---

[89] Štih, *Middle Ages*, 123–135.

# The Carolingian Period of the 9th Century

After the Franks subjugated the Lombard kingdom in 774, they gained a direct border with the Avars for the first time. The border ran through expansive forests of the mountainous and hilly region along the watershed between the Sava and the Soča rivers in the west of present-day Slovenia. Twelve years later, in 788, when the last Bavarian Duke Tassilo III was deposed by Charlemagne, subjecting both the Bavarians and the Carantanians to his direct rule while also bringing Byzantine Istria under Frankish authority, the two most important Central European powers faced each other along a line extending from the Danube to the Adriatic.

That same year, the Franks achieved first successes against the Avars along the Austrian Danube in the north and in Friuli in the south. These two areas were the springboards for a large offensive in 791 which officially began a war. The main Frankish army, commanded by Charlemagne himself, moved along the Danube to the Rába River, while the southern battlefront, where the Franks pushed into the upper Sava Valley, was a secondary theater. This was in contrast to later events, when in 795 and 796 Friuli became the starting point for a wide offensive toward the very heart of the Avar khaganate. The Franks' decisive military success came in 795, when Duke Eric of Friuli sent his army, led by an enigmatic Vojnomir the Slav, into Pannonia between the Danube and the Tisza River. There they pillaged the very core of the Avar khaganate, called the "Ring." The Avars' fate was sealed in 796 by another Frankish army led by Charlemagne's son Pippin, the king of Italy. Frankish control was extended far toward the east, to central Danube. However, the victory did not lead to immediate pacification of the newly conquered lands. In 799, Gerold and Eric, the two highest-ranking commanders on the Franks' Avar flank, were killed. Gerold, responsible for Bavaria and the northern part of the border, met his death somewhere in Pannonia, while Eric, responsible for Friuli and the southern section of the border, died in an ambush near the town of Tarsatica (Trsat) in Liburnia, not far from what is now Rijeka in Croatia. In 803 and 811, the Frankish army had to intervene again in Pannonia.[90]

The vast new Frankish territory had to be organized administratively and ecclesiastically. The ecclesiastical endeavors were primarily focused on rapid and successful Christianization. As early as the summer of 796, a synod was held in

---

90  Pohl, *Avars*, 376–389.

a Frankish military camp on the Danube in Pannonia, where Patriarch Paulinus of Aquileia, Bishop Arn of Salzburg, and other bishops discussed questions relating to the mission. On the same expedition, King Pippin defined the Drava River as the border between the Salzburgian and Aquileian missionary spheres in Pannonia, which was confirmed by his father Charlemagne in 803. In 811, Charlemagne also defined the Drava as the ecclesiastical border between Salzburg and Aquileia in Carantania. This division formed the basis for the ecclesiastical organization of Slovenian territory for almost a thousand years, until the church reforms in the second half of the 18th century.

The administrative framework of the newly conquered territories had been suggested to follow the two main routes of the Frankish military offensives against the Avars, which set out from Bavaria and Friuli. In the east of Bavaria, the Bavarian Eastern March (Lat. *Marchia orientalis*, Ger. *Ostarrîchi*) was organized under its own prefect, while in the south the power of the duke of Friuli was extended far eastward. Bavarian Eastern March encompassed the Traun region (Traungau to the west of the Enns River in the then Bavaria), Pannonia north of the Drava River, and Carantania. To the south, the extended Friuli March or Duchy covered a huge area from present-day Slovenia, Istria and the Dalmatian hinterland to the Drava River in Pannonia. During the uprising of Liudewit, Duke (Prince) of Lower Pannonia (Croat. Ljudevit Posavski, 819–823), Duke Baldric of Friuli, who was charged with putting down the uprising, extended his command over Carantania, parts of which had joined the rebellion.

In addition to territories which came directly under the administration of the two prefects, there were many client principalities throughout this large region which retained a relative amount of internal independence under Frankish lordship, while also providing the first line of defense for the Carolingian empire. In the Bavarian Eastern March, these were the principalities of the Carantanians and, from 805, of the Avars, who submitted to the Franks and were settled around the Lake Neusidl. In the March of Friuli, there were Carniolans living in the upper Sava Valley, Guduscans in the hinterland of Dalmatia, and the Slavs between the Drava and Sava rivers (present-day Slavonija), ruled by Duke Liudewit of Lower Pannonia.[91]

In 818, Liudewit sent emissaries to Emperor Louis the Pious to bring charges of "cruelty and intolerance" against Duke Cadaloh of Friuli, the successor of Eric, who was killed in 799. The next year, Liudewit launched an open rebellion (*rebellio*

---

91  Herwig Wolfram, *Grenzen und Räume. Geschichte Österreichs vor seiner Entstehung. Österreichische Geschichte 378–907*, Vienna: Uebereuter, 1995, 212–225.

*Liudewiti*) that soon became a war (*bellum Liudewiticum*), in which he initially had the upper hand. Liudewit's rebellion, backed by the military successes, had an integrating impact on neighboring peoples, bringing together Carniolans, part of the Carantanians, and the Timocians, who abandoned the Bulgars and at first wanted to join the Franks. Even the rather unfortunate Patriarch Fortunatus of Grado, whose ecclesiastical province in Istria and Venetia was divided between the Frankish and the Byzantine Empire after the Treaty of Aachen in 812, sent— perhaps at the behest of the Byzantines—craftsmen and masons to Liudewit to build fortifications. Liudewit inflicted significant damage against the Frankish ally Borna, the Prince of Guduscans (later Duke of Liburnia and Dalmatia), but also suffered major defeats himself due to concentric Frankish attacks, which destroyed the economic foundations of his power. He was finally forced to flee to Dalmatia, where he was killed in 823.[92]

Liudewit's Carantanian and Carniolan allies actively engaged in battles against the Franks. In 819 and 820, the Carantanians frequently faced the Frankish army along the Drava, but without success. In 820, the Slavs again demolished Maximilian's monastic cell in Pongau (Bischofshofen), south of Salzburg. In the same year, Duke Baldric of Friuli, who had succeeded Cadolah on his death in 819, again brought the Carantanians and Carniolans "who live along the Sava River" under his authority. But when the Bulgars attacked Pannonia in 827 and "banished Slavic princes, replacing them with Bulgar governors," the blame for this blow was laid at Baldric's door. He was removed at a diet in Aachen in 828 and the immense area under his authority was divided between four counts.

The consequence was that the Frankish administrative structure along the southeastern border changed and Frankish counts replaced indigenous princes. Two counties were established on the territory of the Avars' tributary khaganate around Lake Neusidl between the Rába and the Danube. In Carantania, its last prince Etgar (Rastislav?) was replaced by count Helmwin, who came from Bavaria, along with count Salacho, who is documented in Carniola in the 830s. Even the Slavs in Pannonia did not escape this fate, and in 827 they fell under the rule of the Bulgars, who replaced the Slavic princes with their own governors. When the Franks regained this territory, they made Pribina, a Slavic prince in Nitra who had defected to the Franks, the Count of Pannonia (north of the Drava) in 848, continuing the process started by the Bulgars.

---

92 Harald Krahwinkler, *Friaul im Frühmittelalter. Geschichte einer Region vom Ende des fünften bis zum Ende des zehnten Jahrhunderts*, Vienna, Cologne, Weimar: Böhlau, 1992, 186–192.

The 828 reform abolished the March of Friuli, which after the Avar war extended all the way to the Drava River. Its territory was significantly reduced to the benefit of the Bavarian Eastern March, which was now entrusted with defending the entire southeastern border. It also incorporated Carantania, as well as Carniola and Pannonia on both banks of the Drava River. The Italian border was once more moved to the Karst passes, as before the fall of the Avar khaganate. The establishment of new counties did not affect the Slavs under Frankish lordship in Dalmatia and Slavonia, as their princes are recorded well beyond this period.[93]

The reform of 828 also did not affect Istria, which remained part of Italy. It had been a Byzantine province until 788. It had come under the rule of the eastern emperor (where it remained for more than two centuries) between 535 and 544, during the *renovatio imperii* of Justinian I. Administratively, it was governed by a *magister militum*, subject to the Exarch of Italy. The migration of the Lombards from Pannonia to Italy in 568 split the tenth Italian region *Venetia et Histria* into two parts: the Byzantines retained control of Istria and the Venetian lagoon, while the Lombards held continental Venetia. The Three-Chapter Schism, which Pope Gregory the Great directly referred to at the end of the 6th century as the Istrian schism, further divided the northern Italian area, since in 607 two patriarchates—in Byzantine Grado and Lombard Aquileia—were established. Around 788 Charlemagne occupied Istria and in 791 its unnamed duke already participated in a military expedition against the Avars as a Frankish vassal.[94]

Perhaps he was the same duke, Duke John, against whose regime the representatives of Istrian towns and smaller fortified urban settlements protested to emissaries (*missi*) of Charlemagne and his son Pippin in 804. The record of this judicial diet, which took place near the Rižana River in the hinterland of Koper, is the most important document for the history of Istria in the Early Middle Ages. The Istrians' main complaints against Duke John were that he was taking

---

[93] Peter Štih, "Priwina: slawischer Fürst oder fränkischer Graf?," in: *Ethnogenese und Überlieferung*, Vienna, Munich: Oldenbourg, 1994, 213–215; Peter Štih, "Integration on the Fringes of the Frankish Empire: The Case of the Carantanians and their Neighbours," in: Danijel Dzino, Ante Milošević and Trpimir Vedriš (eds.), *Migration, Integration and Connectivity on the Southeastern Frontier of the Carolingian Empire*, Leiden, Boston: Brill, 2018, 103–122; Herwig Wolfram, *Salzburg, Bayern, Österreich. Die Conversio Bagoariorum et Carantanorum und die Quellen ihrer Zeit*, Vienna, Munich: Oldenbourg 1995, 306–310.

[94] Jadran Ferluga, "L'Istria tra Giustiniano e Carolo Magno," *Arheološki vestnik*, 43/1 (1992): 175–190.

control of the taxes the towns were paying to Charlemagne, that he was forcing free citizens into socage and giving them extra duties, that he had brought Slavs into territories under the administration of the towns, who were now grazing their animals and cultivating their fields, and that he had introduced new administrative forms and appointed his own military commanders (centarchs). At the Diet of Rižana, John had to repeal his innovations and ensure that the Slavs remained where they would not harm the towns.[95]

The changes in Istria under the new Frankish regime and the consequent dissatisfaction of the local elites were largely the consequence of war between the Franks and the Avars. Lasting, with interruptions, for over ten years (791–803), the war dictated the tone of events between Italy and the mid-Danube region: it exhausted the people and regions, demanding all available manpower, the adaptation of the economy and the centralization of power. Istria could not avoid these changes, and the measures that were so unpopular with its inhabitants were largely the result of adapting the local institutions and economy to the demands of war. But as Frankish influence in Venice increased and Charlemagne came into conflict with the Byzantines, the growing dissatisfaction in Istria threatened to destabilize the entire region, thwart Frankish plans to seize Venice, and perhaps even harm their authority in Istria itself. Therefore, resolving the problems in Istria became Charlemagne's priority for the northern Adriatic. Reverting most of Duke John's measures passed at the Diet of Rižana in 804 calmed relations in the peninsula. One consequence of the return to the institutions and customs which were in use under the Byzantine rule was that it took a very long time for the new Frankish order to establish itself.[96]

In the first phase of migrations, around the end of the 6th and beginning of the 7th century, Slavs only settled in Istria as far as the large Karst ridge just south of the Trieste–Rijeka road. Slavic, Avar, and Lombard plundering of the peninsula during that period forced the population to retreat behind the walls of fortified towns and smaller castles. Some areas were abandoned economically. The curtain had been drawn on the wealth of the 6th century described by Cassiodorus in his letters—praising a peninsula rich in wine, oil and grain, and claiming that Istria was rightly known as *Ravennae Campania* in the sense that it was as important

---

95   Harald Krahwinkler, "'In territorio caprense loco qui dicitur Riziano,' il 'Placito' di Risano nell'anno 804," *Quaderni Giuliani di Storia* 27/2 (2006): 255–330.

96   Peter Štih, "Imperial Politics and Its Regional Consequences: Istria between Byzantium and the Franks 788–812," in: Mladen Ančić, Jonathan Shepard, and Trpimir Vedriš (eds.), *Imperial Spheres and the Adriatic: Byzantium, the Carolingians and the Treaty of Aachen (812)*, London, New York: Routledge, 2018, 57–72.

to the Ostrogoth Ravenna as Campania had been to imperial Rome. When the Franks assumed control of Istria at the end of the 8th century, Slavic immigration was strongly encouraged. For economic and probably military reasons, the Slavs were now settled in areas that had only been extensively used since the beginning of the 7th century, to make better use of them as arable land and to increase the income that would partially be destined for royal coffers. These areas belonged to towns (*civitates*), which were also episcopal sees, as well as to smaller fortified settlements (*castella*), which with their territories formed the peninsula's basic administrative network.

Since 607, Istrian bishops were suffragans of the patriarch in Byzantine Grado. Frankish occupation of the peninsula, which was confirmed in 812 by the Treaty of Aachen, led to a split of the metropolitan province of Grado between Byzantine and Frankish territory. The Mantua Synod of 827 granted the ecclesiastical authority over Istria to the patriarch of Aquileia in Frankish Friuli. But supported by the doges of Venice, the Patriarch of Grado, who would later transfer his see to the Rialto and become the Patriarch of Venice, continued the dispute with Aquileia, which ran on for centuries before reaching resolution in 1180. Only then did the Patriarch of Grado give up the claim of ecclesiastical authority over Istria.[97]

In 840, the Frankish Emperor Lothair and the Doge of Venice agreed the first known treaty between Venice and its Frankish neighbors—with the Istrians and Friulians being the first to be mentioned. The Venetians committed themselves to helping the emperor "against the inimical Slavic tribes," which allowed them to trade with towns in the Frankish kingdom of Italy. Conflicts between the Venetian fleet and Croatian pirates who looted the towns of western Istria during the reign of Doge Orso I (864–875) keep with the tradition of the *Pactum Lotharii* from 840. The pact (*promissio*) of 932, with which the people of Koper committed themselves to an annual supply of wine in exchange for permission to trade with Venice, provides a clear picture of the gradual extension of Venetian influence across Istria. The next year, the Venetians agreed a treaty with the Istrian margrave, and the representatives of Pula, Poreč, Novigrad, Piran, Koper, Muggia and Trieste, which allowed Venice to run its Istrian possessions without hindrance. The Venetians' effective weapon in achieving this agreement was a trade embargo, which indicates how important economic ties with Venice were

---

97   Rajko Bratož, "La chiesa istriana nel VII e nell'VIII secolo (dalla morte di Gregorio Magno al placito di Risano)," *Acta Histriae* 2/2 (1994): 65–77; Giorgio Fedalto, *Aquileia, una chiesa due patriarcati*, Città Nuova: FSCBA, 1999, 171–271.

for the Istrian towns already in the first half of the 10th century. Thus began the developments that led to all the western Istrian towns (save for Trieste) coming under the rule of the Republic of St. Mark in the Late Middle Ages and to a political division of the peninsula that would last until the end of Venetian state in 1797.

Friuli, like Istria, had always had ties with its neighboring regions to the east as well. Aquileia, founded in 181 BCE, was the starting point for the important routes connecting Italy with the central Danube area. The region maintained its ties with the east both during and after the Early Middle Ages. In 568, the Lombards migrated from Pannonia into Italy and founded their first and initially most important duchy in Friuli. The Lombard Friulian dukes often carried out a very independent policy, opposing the king in Pavia. Despite the political border separating the Friulian Lombards from the Slavs and Avars, their interest in the east did not fade. It is this interest that generated a great deal of information of inestimable value to Slovenian history, recorded by Paul the Deacon in his *History of the Lombards*. In 611, the Avars sacked Cividale del Friuli, taking many women and children with them. One of these was an ancestor of Paul the Deacon who later managed to escape from Avar captivity and, arriving exhausted into Slavic territory on his way home, was helped by an old woman. In 664, Arnefrit—son of the rebellious Duke Lupus of Friuli, killed in the Vipava Valley—found political refuge among the Slavs of Carantania. With their assistance, he even hoped to return to power in Friuli. In around 737, the deposed Friulian Duke Pemmo wanted to flee to the Slavs in Carniola. The Lombards who participated in the quelled Friulian uprising of 776 against Charlemagne's new Frankish authority also found political asylum in the neighboring Avarian khaganate. One of the most prominent of these fugitives was Aio, who was later present at the Diet of Rižana in 804 as Charlemagne's emissary (*missus*). Charlemagne's son Pippin "found" Aio during his campaign in Avaria in 796, and he became a loyal servant of the Frankish king and emperor. The political unification of the two areas under Frankish rule, the extension of the Friulian March far to the east, and the start of the Aquileian mission were all to strengthen the connections between Friuli and its Slavic neighbors.[98]

However, while soldiers and missionaries were traveling east, pilgrims were heading to Friuli. The aim of their pilgrimage was to reach San Canziano d'Isonzo, east of Aquileia, where in Frankish times a gospel book was kept that was thought to contain an autograph of St. Mark the Evangelist. In the margins

---

98   Štih, *Middle Ages*, 190–211.

of this codex, known today as the Gospel book of Cividale after the place where it is now kept, the Slavic, German, Roman and Christian names are written of numerous pilgrims from the late 9th and the early 10th century, "who came to this monastery" from all over the Alpine-Adriatic-Danube region. Among the many distinguished names of pilgrims that clearly illustrate Friuli's connecting role in the Early Middle Ages, one finds, for example, Witigowo, a Carantanian count from around 860, the Slavic prince and count Pribina from Lower Pannonia, as well as Pabo, Richeri and Engilschalk, who held significant positions in the Bavarian Eastern March at the same time. Names from Bulgaria were also recorded in the codex, including Michael—the Bulgarian Khan Boris—who assumed the name of his godfather, the Byzantine Emperor Michael III, when baptized in Constantinople in 864. One also finds the names of the emperors Louis II (850–875) and Charles III (the Fat). In 884, after having reached a peace treaty with the Moravian prince Svatopluk at Tulln (on the Danube in modern-day Austria), Charles the Fat traveled via Carantania and Friuli to Pavia, which may well have been when he entered his name in the Gospel book. The same holds for Braslav, a Slavic prince between the Drava and Sava at the end of the 9th century; his name was also written in the codex, and he too was at Tulln in 884. From the Dalmatian area, the name of Trpimir, the first attested Prince of the Croats from the middle of the 9th century, was also written down in the gospel book, which thus became a sort of *liber vitae*.[99]

After the reform of 828, the Franks' permanent conquests from their wars with the Avars became part of the Bavarian Eastern Prefecture, and hence also part of the Bavarian Kingdom (*regnum*) of Louis the German, grandson of Charlemagne and son of Louis the Pious. This kingdom was based on the *Ordinatio Imperii*, a constitution from 817 that divided the Frankish realm between the emperor's three sons. Bavaria and the predominantly Slav-populated territories stretching to its east and southeast were given to Louis the German. Along with Bavaria itself, it was Carantania that provided the power base within the Eastern March that enabled first Louis' son Carloman (in 876) and later his grandson Arnulf (in 887) to claim the title of Eastern Frankish King.

It was into this Bavarian Eastern March that Pribina, a Slavic prince from Nitra in present-day Slovakia, fled in 833 from the north of the Danube with

---

99 Krahwinkler, *Friaul*; Uwe Ludwig, *Transalpine Beziehungen der Karolingerzeit im Spiegel der Memorialüberlieferung. Prosopographische und sozialgeschichtliche Studien unter besonderer Berücksichtigung des Liber vitae von San Salvatore in Brescia und des Evangeliars von Cividale*, Hanover: Hahnsche Buchhandlung, 1999, 175–236.

his son Kocel and a military entourage. Pribina came to its prefect Ratbod, who presented him to Louis the German in Regensburg. He was baptized on his orders. Pribina's excellent relations with the Bavarian aristocracy and his contacts with the Salzburg Church stemmed from his time as the (pagan) ruler of Nitra. There are many indications that Pribina's wife came from the Bavarian noble family of Wilhelminers. However, Pribina soon had a falling-out with the powerful Ratbod, and he was forced to flee again. Together with his son and retinue, he now took refuge with the Bulgars—probably in Syrmia. From there he went to the Slavic prince Ratimir, who was an institutional successor of the rebellious Liudewit of Lower Pannonia. This meant that he returned to the Frankish territory under the jurisdiction of prefect Ratbod. In 833, Ratbod took up arms against Ratimir, who in turn withdrew, while Pribina's group moved northwest, crossed the Sava River and came into the territory ruled by Count Salacho—i.e., into Carniola, which had been part of Bavarian Eastern March since 828. Pribina's lengthy odyssey came to an end with a reconciliation with Ratbod, arranged by Salacho. He finally established his homeland in Pannonia in 840, when Louis the German granted him a large territory, west of Lake Balaton along the Zala River, as a fief.

After the victorious conclusion of the Avar war, Pannonia formed the "wild east frontier" of the Frankish realm—a land offering great opportunities for personal advancement. Pribina was one of those who exploited those opportunities. He built his capital where the Zala River flows into Lake Balaton. The marshy environment gave the name to the fortified settlement, which contemporary sources report in Slavic, German, and Latin forms: Blatenski Kostel, Moosburg, and Urbs paludarum. Blatenski Kostel became the center of Pribina's seigneury, where he began "to gather tribes from all around and multiply them on that land." In addition to the groups of Avars, Slavs and even Gepids already settled there, who had survived the collapse of the Avar khaganate, numerous new colonists began to arrive in Pannonia from the Slavic world in the Eastern Alps and from both sides of the Danube as well as from Bavaria.

Only after Pribina's position had been consolidated and the structures of power established did Pannonia truly open its doors to Salzburg, which had officially held ecclesiastical power over it since 796. At least 17 churches were consecrated in Pribina's "principality" during Archbishop Liupram's reign alone (836–859), and another 14 under his successor Archbishop Adalwin (859–873). The locations of most of these churches cannot be identified, but those that can indicate that Pribina's authority stretched from the Rába River in the northwest, to Pécs in the southeast, and to Ptuj in the west. The reward for Pribina's successful consolidation of Frankish Pannonia, and for his "zeal for the work of

God and king," came in 848, when Louis the German granted Pribina lordship over all the lands he had previously held as a fief and appointing him count, which made him the king's representative in Pannonia north of the Drava River. Only the Pannonian properties of the Salzburg Church were exempt from his rule, owing to the immunity the Church enjoyed. This unique dual position—similar to the status of tribal princes in Brittany, who were also the counts of the Breton March—would characterize Pribina's power in Pannonia from then on. In addition to the office of count, he continued to be the prince of his people. In a document by Louis the German from February 860, where Pribina was last referred to as living, he is described as a prince (*dux*) and his territory as a principality (*ducatus*). Sources from this time also use dual titles to refer to his son, Kocel. He is referred to as the Count of the Slavs (*comes de Sclauis*) and a Pannonian prince (*knaz' panon'sky*).

**Figure 19.** Ptuj, panorama with castle.

Pribina was killed around 860/861 by the very Moravians he had fled from many years before. His death was related to the turmoil which was sweeping across the Bavarian Eastern March at that time. In 854, the powerful prefect Ratbod was deposed due to disloyalty. Two years later, Louis the German replaced him with his son Carloman, who began to pursue a policy of much greater independence. In 858, he formed an alliance with the Moravian Prince Rastislav. Having protected his rear, Carloman began to openly resist his father. In order to create an independent realm (*regnum*), he drove all the counts still loyal to Louis the German out of the Eastern March between 857 and 861, occupying the "Pannonian and Carantanian border" with his own supporters. The first victim of this policy was Pabo, a Carantanian count, who was forced to flee to Salzburg. His fellow counts did not fare any better. Among fugitives were Count Witigowo, also from Carantania, Richeri, count of the Szombathely region, and probably also Kocel, who was in Regensburg in spring 861. The worst fate was reserved for Louis the German's ever-loyal Pribina, who was killed.[100]

As part of Louis the German's attempts to regain the lost power, he granted extensive holdings within his rebellious son's territory to the Bavarian Church and to nobles. The most important of these gifts was received by the Salzburg Archbishopric in November 860. With this Magna Carta, Salzburg gained numerous manors (*curtes*)—i.e., estates organized for economic use—which extended from Melk on the Danube through what is now southeast Lower Austria, Burgenland and Hungarian lands west of Lake Balaton, to Carantania. This document also bears witness to the growing economic and political importance of Carantania within the Eastern March. Carloman's efforts to achieve political independence were accompanied by an (unsuccessful) attempt by Oswald, a regional bishop of Carantania, to carve out an ecclesiastical province independent of Salzburg. This much at least may be inferred from letters he addressed directly to Pope Nicholas I (858–867) in order to avoid his superior, the Archbishop of Salzburg. Despite the failure of Carloman's efforts—in 865 the rebellious son was finally brought back into the fold by his father—there remained a strong emphasis on the special status of Carantania under the Carolingian lordship of Carloman, and later his son Arnulf.

Archbishop Adalwin of Salzburg and Kocel celebrated the Christmas of 865 together in Kocel's capital, which was a clear sign that the political situation in the east had calmed, but it was not to last. Since 863, two eminent Byzantine missionaries, the brothers Constantine and Methodius from Thessaloniki, had been working north of the Danube in Moravia. Both were highly educated, had already proven themselves as missionaries by the Khazars in the Crimea, and were familiar with the Slavic language of their Macedonian homeland. This led Constantine to create a Slavic alphabet, the Glagolitic, which they used to translate liturgical texts. In Moravia they instituted a Slavic liturgy, but the

---

100  Štih, "Priwina," 214–222; Wolfram, *Conversio*, 166–213.

resistance from the Franks' Latine priesthood led Constantine and Methodius to withdraw from Moravia via the Danube to Kocel, "who took a great liking to Slavic books," by the beginning of 867 at the latest. From there they left for Venice in the same year, before going to Rome at the invitation of Pope Nicholas I, where they were received by a new pope, Adrian II (867–872). The interests of the two brothers aligned with the new eastern policy of the Roman Curia. Rome demanded unrestricted ecclesiastical jurisdiction over former Illyricum, which included Pannonia. In February 869, Constantine, who had entered a monastery and took the name Cyril, died in Rome. The same year the pope sent Methodius to the princes Rastislav, Svatopluk and Kocel as his legate "to the Slavs" and as a bishop. The old enemies united over the question of an independent Slavic Church. Methodius visited only Kocel, who sent him back to Rome, and the pope made him Archbishop of the Pannonian and Moravian Slavs.

The success of the work of Methodius and his followers in Kocel's Pannonia was so notable, and in such a short time (869/870), that the Salzburg Church had to pull back despite a presence which had lasted three quarters of a century, as Methodius "supplanted the Latin language and Roman teaching and well-known Latin letters." This could not have happened without Kocel's overt political support. His decision to support Methodius was a break with his father's pro-Frankish policy, and it was exclusively down to Kocel that the Slavic mission remained alive despite serious threats from 870 to 873. In 870, Methodius' Bavarian opponents took him prisoner and, at a synod held before Louis the German at Regensburg, found him guilty of intruding into a foreign diocese. This was probably what prompted the composition of the work on the *Conversio Bagoariorum et Carantanorum* (*Conversion of the Bavarians and Carantanians*), a reminder of what the Salzburg Church had achieved in the region and at the same time a dossier making the case against Methodius. Since Salzburg's position in Carantania was steadfast, given that three popes in the second half of 8th century had confirmed its ecclesiastical jurisdiction in the area, the document was an attempt to present the Pannonian mission as the continuation of the Carantanian one, in order to give legitimacy to Salzburg's Pannonian aspirations. The *Conversio*, though it also provides priceless information on the earliest history of Carantania, was therefore a means for Salzburg to fulfill its objectives in Pannonia. Methodius' capture coincided with changes in the political situation, as Svatopluk moved over to Carloman's side and betrayed his uncle Rastislav to gain lordship over the Moravians. It was only by the vigorous intervention of Pope John VIII that Methodius' release was secured in 873. He had spent a short time with Kocel who was therefore subject to threats from Bavarian bishops. After 874 at the latest, when the previously hostile Franks and

Moravians reached a modus vivendi with the Peace of Forchheim, Methodius moved to Moravia, where he worked until his death in 885. Kocel could not maintain his position within Frankish Pannonia. He is last mentioned in 874, when Archbishop Theotmar of Salzburg consecrated his church in Ptuj. By 876 he had disappeared from the historical record and Carloman's son Arnulf had assumed control of Pannonia.[101]

In 871, Louis the German placed the administration of the Danube counties on the Moravian border in the hands of the Margrave Aribo. This changed the power structure that had been in place since 828, when government of the entire Eastern March had been united in the hands of a prefect or a royal prince. By 876 at the latest, after Louis the German had died, Arnulf had assumed the lordship of his father Carloman in the east. It included Carantania, Carniola and Pannonia north of the Drava, as well as areas south of the river, where the Slavic prince and Frankish vassal Braslav—successor of Liudewit and Ratimir—had his *regnum*. The counties along the Danube were excluded from this complex. By 884, according to the *Annals of Fulda*, these lands were known as "Arnulf's kingdom." Carantania represented the center of Arnulf's power. From here, Arnulf was able to acquire not only lordship over Bavaria (after 880), but also in 887—with military aid from "Bavarians and Slavs"—lordship over the Eastern Frankish Kingdom. Even after that date Arnulf remained closely connected to Carantania. He celebrated Christmas in 888 in Karnburg, the former seat of the Carantanian princes. After 887, when Arnulf became King of the Eastern Frankish Kingdom, members of the Bavarian high nobility assumed power over Carantania as counts. Among them was also Luitpold, a relative of Arnulf's on the maternal side, and the founder of the Bavarian ducal dynasty, the Luitpoldings. He is first mentioned as a margrave in Carantania in reference to a gift of 895, with which Arnulf transformed the land of Waltuni—an ancestor of St. Hemma—in modern-day Austrian Carinthia and Styria, and possibly Carniola, from a feudal into an allodial possession.[102]

In the final quarter of the 9th century, Pannonia underwent a difficult period. A bloody war broke out between Arnulf and Svatopluk of Moravia that lasted for three years (882-884). During this war, Pannonia and places along the Danube

---

101  Franz Grivec, *Konstantin und Method. Lehrer der Slawen*, Wiesbaden: Otto Harrassowitz, 1960; Maddalena Betti, *The Making of the Christian Moravia (858-882)*, Leiden, Boston: Brill, 2014.

102  Heinz Dopsch, "Arnolf und der Südosten—Karantanien, Mähren, Ungarn," in: Franc Fuchs and Peter Schmid (eds.), *Kaiser Arnolf. Das ostfränkische Reicham Ende des 9. Jahrhunderts* (Munich: C. H. Beck, 2002), 143-185.

suffered the most. There, Svatopluk "slaughtered murderously and fiercely like a wolf, destroyed much with fire and sword." The peace that Arnulf reached with Svatopluk (885) helped him assume power over the Eastern Frankish Kingdom (887). Five years later, Arnulf decided to attack Svatopluk, and in the summer of 892 he pillaged Moravia with Frankish, Bavarian, and Alemannic troops. He was also supported by the nomadic Hungarians, who were viewed in the west as the new Avars.

Frankish sources first recorded the Hungarians in 862, when they were probably involved in the tumultuous events in the Danube River basin related to Carloman's uprising and Rastislav's aspirations for independence. They had definitely entered the region by 881, when they battled the Bavarians at Vienna. In 894, the year when the Moravian Prince Svatopluk died, they managed to cross the Danube and "devastated all Pannonia unto destruction." This changed the status of Hungarians from Arnulf's allies into enemies, threatening the very existence of Frankish Pannonia. The situation became critical soon thereafter, when the Hungarians occupied the Pannonian basin east of the Danube. In 896, Arnulf strengthened the defense of the southeastern Frankish border by handing Pannonia north of the Drava to Braslav, a Slavic prince and Frankish vassal from the south of the river. A huge territory, ranging from Sisak in the south to the Danube in the north, came under the command of Braslav, who had already participated in preparations for the war with Moravia in 892. The present-day Slovakian capital city Bratislava is probably first mentioned in 907 as Brezalauspurc (Braslav's castle). Even so, Braslav's activity and Arnulf's defensive measures did not stop the Hungarian horsemen. Their main objectives were Bavaria and the wealthy northern Italy, which they first reached in 899. The following year, they also pillaged Bavarian territories west of the Enns River, and soon after that, Carinthia. By this time, they had probably already occupied Frankish Pannonia around Blatenski Kostel, while the Bavarian-Frankish administration stood firm in the Danube area west of Mautern near Krems. This was the furthest extent of the customs regime inaugurated between 904 and 906 in Raffelstetten near Sankt Florian at the request of Louis the Child, Arnulf's son and the last Eastern Frankish Carolingian ruler.

However, the devastating Bavarian defeat at Bratislava at the beginning of July 907 led to the fall of Carolingian order in the southeast. The Pannonian-Danube area up to the Enns came under Hungarian control, while the Slovenian territory along the old Italo-Pannonian road became a transitional area for Hungarian raids into Italy and descended into turmoil. Hungarians crossed the Slovenian territory more than 25 times before suffering the decisive defeat at Augsburg in 955 that signaled the end of their pillaging and the start of their adaptation to the western ways of life. The settlement of Vogrsko, near modern-day Nova Gorica on the Slovenian-Italian border, is one such reminder of the Hungarians (*Ogri* in old Slovene). Similar toponyms have been retained in Friuli, which lay on the Hungarian incursion routes and was also devastated in their attacks. In the spring of 1001, Emperor Otto III made a grant to the Aquileian Church of "half of the castle called Solkan and half of the village known in the Slavic tongue as Gorica," specifically mentioning the damage caused by the Hungarians. The other half was granted to Count Werihen of Friuli in the same year. In the late 10th century and even more so in the 11th century, a period of great renewal in Friuli under the leadership of the patriarchs of Aquileia saw numerous Slavic colonists arriving in Friuli from Carniola, and probably also from Carinthia. The first evidence of the new immigrants is from 1031, when the settlement Mereto di Capitolo near present-day Palmanova was referred to as the "village of Slavs." The predominantly Roman majority had assimilated these colonists by the end of the Middle Ages, but traces of them remain in Friuli in place names such as Sclavons or Belgrado.[103]

---

103  Wolfram, *Grenzen und Räume*, 248–273; Jochen Giesler, *Historische Interpretation*, vol. 2 of *Der Ostalpenraum vom 8. bis 11. Jahrhundert. Studien zur archäologischen und Schriftlichen Zeugnissen*, Rahden/Westf.: Verlag Marie Liedorf, 1997, 55–76.

# Feudalism

# Reorganization of the Marches and a Shift of Ethnic and Language Borders

Shortly after conquering the Magyars, the restored Holy Roman Empire continued to expand toward the southeast. In socio-economic terms, the restoration could not simply resume where the Carolingians left off, because the Carolingian order broke down during the Magyar incursions south of the Karavanke Mountains, while the remaining population hung onto old Slavic customs. The entire region of the Eastern Alps was divided anew into one duchy and several marches. Along the Danube, the Eastern March was restored to bolster defenses against the Magyars. The most important unit was the Duchy of Carantania, part of the Duchy of Bavaria. The latter was in 952 united with the marches of Friuli and Verona. The Duchy of Bavaria thus covered the entire region inhabited by the predecessors of the modern-day Slovenes. King Otto I inaugurated his brother, Henry I, as the first ruler of this "super-duchy." He was succeeded by his underage son Henry II, but the united region was actually led by his mother Judith, the daughter of the former Bavarian Duke Arnulf. In 976, Henry II rose against Emperor Otto II and was deposed, while Carantania was separated from Bavaria to become an independent duchy. The detached duchy kept its marches, from the middle course of the Mura River to Verona. This was very important for the political situation in northern Italy, where the Italic kingdom was already disintegrating into a series of smaller and mutually antagonistic political entities. A large Carantania ensured not only the defense of the southeastern borders of the empire, but also the maintenance of links with the still vitally important Italian Peninsula. Consequently, up until the 15th century, the northern Italian regions were much more closely linked with the empire than with the southern part of the peninsula. This connection exposed the eastern Alpine region to the severe effects of the political and social agitation, such as the Investiture Controversy and the struggle between the papacy and the empire until the 13th century.

Understandably, the large and heterogeneous Duchy of Carantania could not achieve political unity, which caused its quick dissolution. In 1002, the king separated the duchy's marches between the Mura and Sava rivers, which were established after 955, during the counter-Magyar offensives. The march covering the territory between Bruck an der Mur and Radgona was first mentioned in 970 (later known as the Carantanian March); the Drava March, between the Pohorje range and Ptuj, and the Savinja March, between the Savinja and Krka river basins, were first mentioned in 980. Present-day Upper Carniola and Inner Carniola belonged to the Carniola March with its center in Kranj, first mentioned in 973. The eastern part of the Friulian March incorporated the upper Soča basin and the Vipava Valley; the Istrian March encompassed the Istrian Peninsula and the Karst region extending up to the Javornik range south of Postojna. Later, the marches were transformed into the provinces of Carinthia, Styria, Carniola, Gorizia, and Istria, but these were of different sizes and invested with different rights.

The western feudal order could only be restored in these provinces by pursuing planned colonization and by developing the network of feudal estates. In the Eastern Alps, feudal estates began to develop in the 9th century, but to the south of the Karavanke Mountains their number did not increase until the end of the century, when Arnulf started granting crown lands as fiefs or allods. The approach to restoring the feudal order was rather simple: since land in the newly liberated regions was plentiful during the 10th and 11th centuries, the king granted fiefs to his most loyal subjects. Therefore, it is not surprising that during the Investiture Controversy some decades later, these landlords, and even the Patriarch of Aquileia, were as a rule reliable allies of the empire; it was not until 1100 and the death of Emperor Henry IV that the Papal party predominated in the Eastern Alps. The Carinthian duke, in contrast, could not count on the loyalty of secular lords or margraves, because they were invariably the vassals of the king. As a matter of fact, as many as three Carinthian dukes appointed by the emperor in the mid-11th century could not enter Carinthia because they were opposed by the high nobility.

The first to receive ample royal gifts were the dioceses. Not by coincidence: in the Ottonian and Salian administrative system, the (arch)dioceses were the ruler's most reliable supporters. In addition to pastoral tasks, they also performed many important political, judicial, and economic functions. They enjoyed commensurate royal support, so church estates could comprise several hundred square kilometers, including lands in the new marches—quite a substantial amount of land, given the varied topography of the region. One of the oldest large estates in Carantania was held by the Archdiocese of Salzburg (from 860). In the mid-11th century, it obtained the crossroads and its surroundings in Ptuj and a large contiguous tract of land in the lower Sava basin. The Diocese of Freising received from Emperor Otto II in 973 an integral territory in western Upper Carniola, which later evolved into the Škofja Loka estate; by the 11th century it had been granted other large estates in Carinthia and Istria. From 1004, the Diocese of Brixen in Tyrol established its own estate with Bled at its center. Emperor Otto III granted the Patriarch of Aquileia half the lands between the Soča River, the Vipava Valley, and the Trnovo Plateau in 1001, with a further addition of Inner Carniola in 1040. The ecclesiastical lords had public (judicial) power on their estates and were immune from secular courts; they were the mainstays of the emperor whenever he needed to restrain the self-willed dukes and margraves.[104] The Patriarch of Aquileia took the most advantage of this situation: in 1077, Emperor Henry IV granted him the marches of Istria, Friuli, and Carniola. The patriarch hence became the emperor's nominal deputy and the secular prince in control of these entities, as well as head of the entire metropolitan see.

The sovereign's generosity made the southeast of the country attractive for the secular nobility. The high nobility from the central parts of the empire did not make permanent residences on their new estates, but they kept a watchful eye on the colonization, increased allodial holdings, performed public functions, and acquired judicial sovereignty. From Saxony came the Weimar-Orlamünde family, who for decades before 1077 had exerted control over Istria and Carniola. The Eppensteins and the Spanheims, both from the Duchy of Franconia, became hereditary dukes in Carinthia: the former obtained properties along the upper reaches of the Mura River in what is today Austrian Styria, and the latter in the territory in the wider surroundings of Maribor, central Carinthia, along the lower reaches of the Savinja River, and in the Ljubljana basin. From the 12th century onwards, they added the territories around Kostanjevica in the southeastern part of the empire, which they obtained through wars and colonization at the expense of the Kingdom of Croatia.

Still, the majority of noblemen came from Bavaria. The lords of Auersperg may have settled in Inner Carniola as early as the 11th century. The Counts of Bogen and the Andechs settled in the region early in the 12th century. After most of the Weimar-Orlamünde estate was divided, the Counts of Bogen received the regions in western Carniola around Vipava and in Lower Carniola, while the Andechs family obtained lands in central Upper Carniola.

In the 12th century, the Andechs family adopted the title *dux Meraniae*, the name of the small region in the Kvarner Gulf. Other high-ranking noblemen who arrived during the early 12th century were also of Bavarian origin: the lords of Žovnek (Ger. Sannegg), who settled in the upper Savinja region, the Counts of Ortenburg in Carinthia and the lords of Ptuj (Pettau), who as the ministerials

---

104 Peter Štih, "Ursprung und Anfänge der bischöflichen Besitzungen im Gebiet des heutigen Sloweniens," in Matjaž Bizjak (ed.), *Blaznikov zbornik/Festschrift für Pavle Blaznik; Loški razgledi*, Ljubljana, Škofja Loka: Založba ZRC, 2005, 37–54.

of the Archdiocese of Salzburg managed its estates in the Drava basin while expanding toward the east across the border of their own accord. The Traungaus (Otokars), Margraves of the Carantanian March, also came from Bavaria, as did the Counts of Gorizia, who in the early 12th century adopted that name from the seat of their estates in eastern Friuli, along the middle reaches of the Soča River.

The most important noble family in the Eastern Alps until the end of the 12th century was descended from Countess Hemma of Friesach in Carinthia. At the beginning of the 11th century, her family held the vast united territories for several decades; the territories came from Hemma's side (the Luitpolding dynasty and the Slavic stock of Svetopolk and Preslav) and from the side of her husband, William II, Margrave of the Savinja March (heirs of Count Pribina since the 9th century). Their united territory extended from Friuli to the Sotla River (on the border with Croatia), and from the Danube to the Krka River in Lower Carniola. The central, allodial part of that estate in the Savinja Valley closely corresponded to the boundaries of the Savinja March (Ger. Mark an der Sann, Mark in der Sanntal). With William killed in 1036 and both their sons deceased, Hemma became one of the wealthiest women of her time. In 1043, she founded a convent at Gurk in Carinthia and showered it with land grants. The Archbishop of Salzburg dissolved the convent in 1072 and established on the same site his first suffragan diocese in the Eastern Alps, which soon became one of the biggest landholders: it held estates in Lower Styria (the Sava basin, the Savinja Valley, and the Sotla basin), and in Lower Carniola (on the Krka River). The rest of Hemma's estates were divided after her death among the Ascuin branch of her family (later the lords of Attems, Plains, Prisis, Counts of Weichselburg, etc.) who had largely died out by the beginning of the 13th century.[105]

Therefore, by the end of the 11th century, the basic structure of common land was divided throughout the Slovenian regions, while crown lands almost disappeared. The feudal lords constituted an efficient lower ruling structure that provided an economic framework. On the other hand, the network of estates, and thus province borders, continually changed according to new endowments by the king. The pre-feudal Slovenian noblemen had exerted control over their properties from (wooden) fortified manors called *dvori* (*curtis*). Once landed property developed in the 10th century, these manors became part of feudal lords'

---

105   Heinz Dopsch, "Die Stifterfamilie des Klosters Gurk und ihre Verwandtschaft," *Carinthia I*, 161 (1971): 95–123; Ljudmil Hauptmann, "Grofovi Višnjegorski," *Rad JAZU 250, Razreda historičko-filologičkoga i filozofičko-juridičkoga*, 112 (1935): 215–239.

estates, but they could not fulfill the new demands of colonization and military campaigns. There was a need for new, well-fortified, and above all stone-built military-feudal centers, or castles, which would also have a socially exclusive character. One of the first such castles was Richenburch (Slov. Rajhenburg) in the lower Sava basin, the construction of which was probably concluded sometime after the mid-10th century, although it was first mentioned in 895. Its very name points to its significance for the border of the empire (*Richen-*), its outstanding size, and bulwark features (*-burg*). Such castles, as central fortifications of large feudal estates, were mentioned in early written records dating from the end of the 10th century onward: Bosisen at the Škofja Loka estate held by the Freising bishops (973), Solkan, the fortification of the Patriarch of Aquileia and Count Werihen (1001) near Gorizia, and Bled, the center of the estate owned by Brixen bishops (1011). Some later sources indicate that the castles in Ljubljana, Kranj, Celje, and Ptuj also belonged to this group of the earliest central castles. These castles and estates were not the permanent residences of their owners. The work was performed by domestic and foreign vassals and unfree ministerials.[106]

The 10th century witnessed epochal changes that affected the entire population. Until then, the lands in the Duchy of Carantania had been cultivated primarily by native Slavs, who used more or less extensive methods. Some peasants were free men who lived on the countryside manors, and others were the population of interrelated villages (*župa*) who sustained themselves by working the land which had already been individually assigned. Under the feudal system, common land—the most widespread form until the 9th century—was considered land without a known owner, so it became part of the crown lands. Village inhabitants were therefore mutually connected through deeper interests, although they could belong to different lords. The majority of farms, particularly those more remote from estate centers, were under the jurisdiction of landlords' offices, some of which covered the territories corresponding to the former organizational unit called *župa*. This Old Slavic term denoted an autonomous group of people interconnected through kinship or economic links and living in neighboring villages. Its leader was called *župan* (first mentioned as *jopan* in 777 in the Slavic area by the Enns River in Upper-Austria), whose function endured under the system of the feudal estates; as a result, during the High Middle Ages, the original Slavic *župa* became a basic organizational unit of landed lords, while the *župan*

---

106 Peter Štih and Vasko Simoniti, *Slovenska zgodovina do razsvetljenstva*, Ljubljana: Mohorjeva založba, 1995, 132–134.

thereafter became an intermediary between the villagers and the lord.[107] Initially, *župan* worked a larger farm and was exempted from paying taxes. The *župan*'s role diminished with the development of feudal estate administration, and by the 15th century, his status was reduced to almost the same as that of his fellow villagers. As leaders of village communities and administrative officials of the landed nobility, they retained their posts as members of patrimonial and district courts until the 18th century.

Until the 13th or 14th century, feudal lords, particularly ecclesiastical lords, held a large portion of property in demesne, worked by farmhands and local peasants obliged to carry out labor service. However, this form of agrarian production did not prevail because of the late feudalization of the Slovenian territories. The most intense colonization of the Slovenian territories took place between the 10th and 12th centuries, when all lowlands and hill regions were colonized and many new villages established. The structure and size of these cluster settlements (typical of the hill regions with individually delimited farms within the village fields) and roadside villages (in the lowlands, with all of an individual's fields set out in strips behind the house) remained essentially unchanged until the beginning of the 20th century. The most widespread type of land cultivation was the open-field system, which had already been extensively tested in the western parts of the empire. The colonizers were usually given larger, semi-free farms not burdened by labor service. In addition to possessing individual property (mostly farms, orchards, and meadows), the peasants also used common parish land (waters, pastures), while woodlands, reserved for hunting, were mainly in the possession and under the administration of landlords.

During the pre-feudal period, peasants were freemen, but under the feudal system most of them lost both their land and freedom. By the 13th century, various types of colonization had produced a whole series of (un)free peasant statuses graded according to various degrees of their hereditary properties. The lowest was the status of unfree workers on estates held in demesne. Semi-free statuses varied greatly among themselves, because they were based on the types of dues and obligations of the holder and the right to buy land or move from the farm, rather than on inheritability. Toward the end of this period, the open-field system converted personal dependence into land dependence: the legal status of a peasant was dependent exclusively on the legal status of a farm. Accordingly,

---

107 For more information about the attempt to reconstruct the transposition of the early Slavic *župa* into the feudal system using the example of the surroundings of Bled between the 7th and 15th centuries, see Pleterski, *Župa Bled*, 112–146.

only those *kosezi* who lived on "kosez farms" (so-called *koseščina*) retained their free status throughout the Middle Ages, while those who lived on ordinary farms were transformed into serfs in the 11th and 12th centuries.[108]

Local colonization and immigration from German lands transformed the cultural landscape of the Eastern Alps. By the 14th century, two ethnic (language) blocs had been formed: to the south of the Hohe Tauern range, within the mountainous areas of Carinthia, in Carniola, and to the south of Graz in Styria, the language was predominantly Slovene. North of this demarcation line, German prevailed. This linguistic border moved further south in the 15th century and it remained unchanged until the 19th century.[109] A similar bipolarity could be found in the west, where the border between the Italian and Slovene language areas ran roughly along the Soča River and across the hinterlands of the coastal towns in Istria. The border between the Slovene- and Croatian-speaking population in the south and east was the border between the empire and the Hungarian-Croatian Kingdom, running along the Kolpa River, across the Gorjanci range, and along the Sotla and Mura rivers. Here and there these areas contained smaller islands of foreign-speaking settlers, who were assimilated into the majority population by the end of the Middle Ages, or in some cases by the 18th century. The most important German speaking enclaves within the territory of present-day Slovenia were in the Škofja Loka estate held by the Diocese of Freising in the 12th and 13th centuries (Bavarians and Tyrolians), and from the 14th century onwards in the consolidated estates of the Counts of Ortenburg in the Kočevje region in Lower Carniola (Thüringians, Franks, and Carinthians). The remoteness of the Kočevje region and the concentration of the German-speaking population kept this German cultural community alive as late as World War II.[110]

---

108  Sergij Vilfan, "Koseščina v Logu in vprašanje kosezov v vzhodni okolici Ljubljane," in: *Hauptmannov zbornik*, Ljubljana: SAZU, 1966, 179–210.
109  In Carinthia, the border extended across the Gailtal Alps to Lake Ossiach, north of Zollfeld toward the Sau Alps and further across Slovenske gorice to Radgona. See Bogo Grafenauer, *Oblikovanje severne slovenske narodnostne meje*, Ljubljana: Zveza zgodovinskih društev Slovenije, 1994.
110  Sergij Vilfan, "Die deutsche Kolonisation nordöstlich der oberen Adria und ihre sozialgeschichten Grundlagen," in: Walter Schlesinger (ed.), *Die deutsche Ostsiedlung des Mittelalters als Problem der europäischen Geschichte*, Sigmaringen: Jan Thorbecke, 1975, 567–604.

# From Autonomy to the Unification of the Alpine and Danube Basin Regions

In the historical territories, today inhabited by the Slovenes, the development of "provinces" (Lat. *terra*, Slov. *dežela*, Ger. *Land*)—the typical medieval and early modern administrative, territorial, and socio-political formations—took place between the 12th and 16th centuries. In the 15th century, when territorial fragmentation was at its peak, the territory of present-day Slovenia was divided into several provinces and counties belonging to three countries (the Holy Roman Empire, the Venetian Republic, and the Hungarian Kingdom). The provinces that fully or in part covered the territory of present-day Slovenia, i.e., Styria, Carinthia, Carniola (plus its two associated smaller provinces), Istria, and Gorizia, were only finally shaped under the Habsburgs in the 15th century. They lasted until 1918 as political units inside the Habsburg hereditary lands, but in the popular imagination they still figure as traditional provinces today.

At the beginning of the 12th century, the Duchy of Carantania/Carinthia and its marches were fragmented into judicially and administratively independent individual feudal estates held by secular and ecclesiastical lords. As early as the 11th century, intermarriages among important dynastic families caused an increase in the size of family estates, the expansion of their public judicial rights, the subjugation of smaller landlords, and the like. In general, dynastic families endeavored to convert their consolidated estates into political units over which they would have administrative and legal control. Consequently, the influence of the state or empire within the marches declined in inverse proportion to the rise in power of the "lord of the province." Usually, the transformation of unified estates into a province was effected by a dynast who held the hereditary title of a margrave or a duke.

Other paths were possible, too: dynasties without public office but possessing large private territory and jurisdiction sometimes tried to segregate the territory from the border march and shape their own provinces. The former political units were no longer connected either through common administration or through common legal order: since dynasts were of diverse origins, they adhered to their respective tribal laws, and so did their ministerials. While gaining independence from their lords during the 12th century, the ministerials, pursuing their own objectives, backed up the development of universal provincial law under the authority of the provincial lord and with noblemen's courts, under which all noblemen would be equal regardless of their hereditary status or their affiliation to a particular lord. Therefore, the lower nobility was the main driving force behind the formation of provinces: the success of a dynastic ruler and the establishment of provincial governing bodies depended on the lower nobility's loyalty alone. In addition, dynasts could also count on support from urban settlements whose trade across the borders of urban estates was severely hampered by territorial fragmentation.[111]

After it was renamed Carinthia in the 11th century, the name of Carantania disappeared from its inhabitants' memory within just a few decades. It took several centuries for a new name, one that would cover the linguistically and culturally related inhabitants of various political units, to gain currency. In the 19th century, this was quite an old supra-provincial term, which had not been used in practice for a long time—the Slovenes. From 16th century on, their awareness of individual, social, linguistic, and cultural affiliation stemmed from their affiliation to a political entity, which was a melting pot that attempted to shape regional awareness as contrasted to national consciousness. Regional awareness thus meant that the province of Carniola, for example, was equally home to its Slovene-speaking inhabitants as well as the German or Italian speaking settlers and their "Carniolanized" children. Even with such a mindset, both the common people and the elites already entertained the idea of linguistic and political borders coinciding. This could explain why the 10th- and 11th-century Ottonian Empire placed all the Alpine Slavs under one common administration and why Istria was attached to the Duchy of Carinthia rather than Friuli, despite Friuli being much closer to it geographically (but not culturally or linguistically).[112]

## Styria

The first and for a long time the only true province in the Eastern Alps was Styria. It developed from the Carantanian March, extended to the south and the east, and easily absorbed the neighboring counties. The Carantanian March occupying the middle course of the Mura River separated from the Duchy of Carinthia in 1035. The rulers of the former became the Counts of Traungau (also Otokars), who later adopted the name "Margraves of Styria" after their

---

111 Andrej Komac, *Od mejne grofije do dežele. Ulrik III. Spanheim in Kranjska v 13. stoletju*, Ljubljana: Zgodovinski institut Milka Kosa, ZRC SAZU, 2006, 27–37.

112 Štih and Simoniti, *Slovenska zgodovina*, 79–80; Kočevar, Vanja. "Ali je slovenska etnična identiteta obstajala v prednacionalni dobi? Kolektivne identitete in amplitude pomena etničnosti v zgodnjem novem veku" (parts 1, 2), *Zgodovinski časopis* 73/1/2/3/4 (2019): 88–116 and 366–411.

seat at the castle of Steyr, while the name "March of Carantania" disappeared. Through association with neighboring counties, through inheritance (especially of the property of the Eppenstein family, who died out in 1122), and thanks to the emperor's benevolence, by 1160 the Margraves of Traungau had acquired the status of provincial lords within their territories and laid the groundwork for the provincial law that became influential beyond the borders of the province. Their first acquisitions in the territory occupied by Slovenia today were the Spanheim estates in the Drava March, inherited by Margrave Otokar III in 1147. By 1180, when the border march was elevated to a duchy, the ministerials of the Margraves of Traungau were already well emancipated.[113] In 1186, they forced Duke Otokar IV to endorse the rights that made it possible to form a unified provincial nobility. These provisions were part of the "Georgenberg Pact," by which Otokar made the Austrian Babenberg dukes his heirs and stipulated the unification of the two provinces of Styria and Austria if he died without descendants.[114] This happened in 1192, and the Austrian Duke Leopold V took over Styria.

Under the Babenbergs, Margraves of the Eastern March since the late 10th century, both provinces flourished economically and culturally. For the Babenberg dynasty, Styria opened the door to Carniola, where after 1228 they took over the estates of the quasi-provincial lords the Counts of Andechs. In 1246, the Babenberg dynasty met a similar end to its predecessors: the last male descendant, Frederick II (1230–1246), was killed in a battle against the Hungarians.[115] There ensued a long war for the Austrian-Styrian succession, ending in 1261 when the Bohemian King Otokar II Přemysl emerged victorious over his competitor, the Hungarian King Bela IV. To legitimize his rights, Otokar married Frederick's sister, but the decisive factor in his victory was the loyalty of the provincial nobility. Through such maneuvering, and without seeking approval from the emperor, Otokar managed to bring all the provinces between

---

113 For more on the process of ministerials' emancipation, see Friedrich Hausmann, "Die steirischen Otakare, Kärnten und Friaul. Besitz, Dienstmannschaft, Ämter," in: Gerhard Pferschy (ed.), *Das Werden der Steiermark*, Graz, Vienna, Cologne: Veröffentlichungen des Steiermärkischen Landesarchives 10, 1980, 225–275; Ljudmil Hauptmann, "Mariborske studije," *Rad JAZU 260, Razreda historičko-filologičkoga i filozofičko-juridičkoga*, 117, Zagreb (1938): 57–118.
114 See the text of the Pact in: Karl Spreitzhofer, *Georgenberger Handfeste*, Graz, Vienna, Cologne: Styria, 1986, 12–19.
115 Karl Lechner, *Die Babenberger. Markgrafen und Herzoge von Österreich 976–1246*, 3rd ed., Vienna, Cologne, Graz: Böhlau, 1985, 192–217.

Bohemia and the Adriatic Sea under his rule: as early as 1251/1252, he ruled over Austria; in 1253 he became King of Bohemia, and in 1269 he inherited the Spanheim estates in Carinthia and Carniola. He was expelled in 1276, and the politically fully united Styria was given as a hereditary fief to the House of Habsburg, whose reign was to last for more than six centuries.

## Carinthia

Between 976 and 1077, Carinthia was governed by a series of feeble dukes from the central parts of the empire.[116] They lacked any vested interest in the Alpine duchy, and their potential appetite for profit was curbed by the special supervisors of crown lands, and even more so by secular dynastic families (the Aribo family) and ecclesiastical princes (from Salzburg and Bamberg). Accordingly, Carinthia's transformation into a province was much slower than Styria's, although Carinthia, as the main successor to the Carolingian-era Carantania, had nominally been a duchy since 976. The situation improved between 1077 and 1122, when the hereditary dukes were the Eppensteins. They were the first larger property holders in the region, and occasionally performed the duties of the Friulian and Istrian margraves. After their line died out in 1122, the title of duke was inherited by the Counts of Spanheim, who had moved in several decades earlier from the Duchy of Franconia. However, since they were not the heirs to the Eppenstein property in Carinthia, the northern counties of Carinthia were alienated and came under the control of the Counts of Traungau (the heirs to the Eppensteins). The Spanheims' power as dukes was therefore even more restricted than the Eppensteins' in the central regions between Klagenfurt, St. Veit (the seat of dukes), and Völkermarkt.

The greatest obstacle preventing the Spanheims from establishing the province were the large estates of the ecclesiastical lords, who were directly subject to the crown. These were part of the duchy, but the Spanheims had no authority over them whatsoever. Most annoyingly, these estates controlled the important transport routes between Bavaria, Friuli, the Danube basin, and Carniola. A similar obstacle was posed by the estates of the Counts of Gorizia in present-day east Tyrol (later the Front County of Gorizia), which remained outside the provincial authority of the Carinthian dukes (the Habsburgs) as late as the end of the 15th century, when the lines of the Counts of Gorizia died out.

---

116  For the most detailed presentation of Carinthian politics and the building of the province, see Claudia Fräss-Ehrfeld, *Geschichte Kärntens 1. Das Mittelalter*, Klagenfurt: J. Heyn, 1984.

Map 2. Map of "Great Carantania" and its Marches, about 1000.

Only Duke Bernhard von Spanheim (1202–1256) managed to expand ducal authority in Carinthia, but huge obstacles and military defeats led him to shift his attention to Carniola. After the death of his son Duke Ulrich III in 1269, the Spanheims lost the Duchy of Carinthia to the Bohemian King Otokar II Přemysl, who obtained it through an inheritance agreement and retained it until 1275. After the brief reign of the last Spanheim Philipp, Patriarch of Aquileia, Carinthia came into the hands of the Habsburg dynasty in 1279. In 1286, King Rudolf gave Carinthia as a hereditary fief to Meinhard IV of the Tyrol line of the Counts of Gorizia, who was installed in the traditional ceremony of the Carantanian princes. Yet the unfinished story of the formation of the province of Carinthia repeated itself. Meinhard was a tenacious duke, but he had support only within his rich province of Tyrol, which had long since been internally consolidated. Although he had practically no property in Carinthia, he nevertheless attempted to subdue the local nobility. As was to be expected, within just a few years he came into conflict and even open war with them.

Owing to persistent particularism and a family split into two almost non-collaborating and non-related branches, neither Meinhard nor his sons managed to unite Carinthia into a province, although the duchy already had provincial law. After the death of Henry II, the last member of the Tyrol-Carinthian branch of the Counts of Gorizia, the duchy was taken over by the Habsburgs in 1335. Yet throughout the 15th century, even the Habsburgs were not able to eliminate dynastic antagonisms, despite strong support from the lower nobility and towns. Moreover, Carinthia came close to being obliterated as a territorial unit after the Counts of Gorizia and Counts of Celje extended their positions. The Counts of Gorizia consolidated their holdings there in 1415, forming a separate county directly subject to the crown. The ambitious Counts of Celje first acquired a large property in Carinthia formerly owned by the Counts of Heunburg (whose line had become extinct in 1322) and then, in 1418, the lands of the extinguished line of the Counts of Ortenburg. The Habsburgs could unite the province only after the Counts of Gorizia lost the battle for the heritage of the Counts of Celje in 1460 (when they acknowledged the Habsburgs' rule over the "Front County of Gorizia") and their line died out in 1500. The province was finally shaped in 1535, when the Habsburgs subdued the last independent estate owned by the Diocese of Bamberg. Although Carinthia lacked political unity, provincial awareness among Carinthian nobility was quite developed by that time, stemming from the peculiarities of Carinthia—its relation to historical Carantania and the installation ceremony.

## Carniola

The central Slovenian province, the March of Carniola, was first mentioned as a march in 973, and retained the traditional name of the tribal principality of Carniola (Slov. Kranjska, also *krajina*, "march").[117] The March of Carniola needed a long time to develop into a unified province. At the beginning of the 11th century, it encompassed only the central part of what later became Upper Carniola, the Ljubljana basin, and the eastern part of Inner Carniola. Until around 1000, it was subject to the Duke of Carinthia, and then directly to the king. At that time, the margraves of Carniola came from the Bavarian family of Sempt-Ebersberg, which had been connected to the Sava basin since the end of the 9th century. After 1036, they expanded their authority to the east and southeast, because the Savinja March had come under the jurisdiction of the Margrave of Carniola after the assassination of Count Wilhelm II. However,

---

117  In a deed by Emperor Otto II granting the estate in Carniola to the Diocese of Freising: "in comitatu Poponis comitis quod Carniola vocatur et quod vulgo Creina marcha appellatur," in: Theodor Sickel (ed.), *Die Urkunden des Otto II. MGH DD.*, Hanover, 1999 [1888], 56.

Carniolan margraves did not own property in these lands, so could not assert themselves against the local lords.

In the mid-11th century, the united Carniola and Savinja marches[118] and the March of Istria fell under the control of the Weimar-Orlamünde family, who also wrested some parts of the Kvarner Gulf from the Croatian Kingdom. In 1077, Emperor Henry IV granted the marches of Carniola, Istria, and Friuli to his loyal ally, the Patriarch of Aquileia, who after 1093 retained Carniola and Friuli. The patriarchs did not own any sizable property in Carniola (they held some properties in the upper Savinja region and in Inner Carniola), unlike in other marches, so the patriarch transferred the authority as a fief to his deputy.

In the mid-12th century, Carniola was already too fragmented to be transformed into a province by anyone relying on the powers invested in margraves, apart from the patriarch. It was only toward the end of the century that the Bavarian counts from the Andechs family amassed enough power to be able to assert themselves as provincial lords on their estates, independent of the nominal margrave. The Andechs family, who at that time became related to the French and Hungarian royal families by marriage and who were reliable allies of the emperor, could then unite margravial powers in the Savinja region and Carniola proper. By so doing, they created an entirely new formation within the Carniola and Savinja marches that territorially did not have much in common with the 11th-century March of Carniola.

The bedrock of Andechs' power was the inherited territory between Tržič and Motnik in Upper Carniola with its center in Kamnik. Count Henry IV expanded it when he married the sole heiress to the Counts of Weichselburg (Slov. Višnja Gora) in 1209, by adding estates in the Zasavje region and Lower Carniola. He also acquired the land to the southeast of Carniola, which the lords of Pris and the Counts of Weichselburg had obtained from the Croatian Kingdom by pushing the border from the Krka River to the Kolpa River. In 1208, at the peak of his power, Henry IV was ostracized under suspicion that he had participated in the assassination of King Philip II. He lost the titles of the Margrave of Istria and deputy of Carniola, which were retaken by the patriarch. However, since Henry's brother Berthold became patriarch in 1218, Henry was able to maintain

---

118    Nevertheless, the common name "Carniola and the Windische March" ("Kranjska in Slovenska marka") persisted for a long time, even though after the 13th century Carniola included only those parts of the Savinja March to the south of the Sava River and the lands between Zagorje and Motnik in the eastern part of Upper Carniola.

his rule (but not his title) over his estates in Carniola, where he began to exercise the authority of a provincial lord. This brought him into conflict with his main competitor, Bernhard, Duke of Carinthia.

When Count Henry died in 1228 without issue, a bitter struggle for his legacy ensued among his relatives—the Dukes of Merania, the Babenbergs, the Spanheims, and the Counts of Gorizia. Duke Frederick II of Austria-Styria had the most legitimate claim, since he had married Agnes of Andechs, so in 1232 he proclaimed himself the "Lord of Carniola." Frederick's death in 1246 opened the way for Spanheims' to the provincial lordship of Carniola, although Carniola (like Styria and Austria) was again returned to the crown and governed by a crown deputy. Ulrich III Spanheim was able to legitimize the inheritance by marrying Frederick's widow. Ultimately, it was his successful subjection of the Andechs family's ministerials that enabled him to unite the Spanheim estates, the lands held by the Andechs family and the Counts of Weichselburg, and the Church fiefs, and turn this consolidated territory into a new political unit and a fully shaped province of Carniola.[119]

As in Carinthia, the vast Spanheim estate passed after Ulrich's death in 1269 to the Bohemian King Otokar II, who in Carniola also usurped large feudal estates owned by the Patriarch of Aquileia. Otokar found support for his reign among the lower nobility and townspeople, while the higher nobility perceived him as a threat, because various state functions were evidently taken away from them in favor of royal deputies from Bohemia. In 1274 the Reichstag deprived Otokar of the Babenberg and Spanheim inheritance, so by next year his governance over these provinces was in shreds. When Otokar died in 1278 at the decisive Battle of Dürnkrut, the victor, King Rudolf of Habsburg, could start forming his dynastic empire in the Eastern Alps. Rudolf knew that his dynasty would not be able to establish firm rule over the provinces unless he secured the support of the nobility and the Church. To this end, he first won their favor and then distributed the available counties and the border march among his sons. To Count Meinhard IV of Gorizia-Tyrol, who in 1279 received Carniola as a pledge, he gave Carinthia, so that the territory once ruled by Otokar was divided between the Habsburgs and the Counts of Gorizia-Tyrol. The alliance between the Habsburgs and Meinhard's successors soon dissolved. In 1306, Meinhard's son Henry II even went to war with the Habsburgs over the Bohemian Kingdom. Henry was defeated in 1311 and lost the eastern parts of Carniola to Habsburg Styria. After Henry's death in 1335, the Habsburgs obtained Carinthia and

---

119  Komac, *Od mejne grofije do dežele*, 47–240.

Carniola almost without obstacle. In 1338, the Habsburgs secured important internal support for the internal unification of the two provinces by granting privileges to the nobility of Carniola and Carinthia.[120]

## Habsburg Rule

The Habsburg emperor after 1358, Duke Rudolf IV (1358-1365), endeavored to win as much independence from the empire as possible for his family properties and to obtain the title of electoral prince. During his reign, Carniola was de facto turned into a duchy and Carinthia into an archduchy without an imperial charter. Rudolf won the County of Tyrol for his dynasty, and also had ambitions to penetrate Friuli and northern Italy and move further toward the Adriatic coast. The Habsburgs had long been apprehensive about the expansion of the Venetian Republic in light of the Patriarchate of Aquileia's crumbling temporal power. But before the major confrontation with Venice, Rudolf had to cleanse western Carniola of independent feudal estates. He first subordinated the lords of Auersperg and the estates of the patriarchate in Inner Carniola. Between 1360 and 1362, he occupied the patriarchate's estates in Cerknica, Postojna, Lož, and Slovenj Gradec, and in 1366 his successors took supreme control of the estates of the Duino lords in Karst, along the Gulf of Trieste, and the port of Rijeka in the northern part of the Gulf of Kvarner. With these new acquisitions, Carniola obtained two outlets to the Adriatic Sea. In 1374, after the Istrian line of the Counts of Gorizia died out, the Habsburgs took over the County of Pazin (Pisino) in inland Istria and attached it to Carniola as a separate unit. These gains created the frontline between the Habsburg lands and Venice that remained restive for several centuries.

The first larger conflict involved Trieste, the only major town on the northern Adriatic coast which was still free, and which had long been vacillating between the two powers. Ultimately, the nearby Venetian Koper appeared to be too strong a competitor, and with Venice's hegemony an undesirable option, Trieste chose the lesser evil and acknowledged Habsburg rule in 1382. However, the rule was not absolute. The city continued to be an independent political subject with powerful self-government and direct subjection to the lord of the province; in 1463 it even made peace with Venice on its own. Nevertheless, Venetian obstruction prevented Trieste from increasing its economic significance until

---

120   Alois Niederstätter, *Die Herrschaft Österreich, Fürst und Land im Spätmittelalter*, Vienna: Ueberreuter, 2001, 67-144.

the early 18th century, when it finally became a "free port" and outstripped the dwindling Venice within just a few decades.[121]

In 1374, the Habsburgs did not yet have control over the entire territory of the former March of Carniola. In Inner and Upper Carniola, large consolidated estates were in the hands of the Counts of Ortenburg, who in 1395 became direct subjects of the crown. The Istrian branch of the Counts of Gorizia held another separate province that was granted (in 1365) special provincial privileges: the County in March and Metlika, in Lower and White Carniola. Even when the Habsburgs inherited this small province in 1374, it was only added to Carniola, but not merged with it. Only later, in 1441, did Duke Frederick V (as Emperor Frederick III) amalgamate Carniola, the County in March and Metlika, and the County of Pazin, dividing the entire territory into four administrative districts.[122]

With the death of Duke Rudolf IV in 1365, the Habsburg dynasty not only inherited the difficulties with Carinthia and Venice, but was also plunged into bitter and century-long family disputes. However, this struggle did not affect the affiliation of the nobility and hereditary lands, nor were they threatened by the division of hereditary lands among several branches of the Habsburg dynasty. The first such division occurred in 1379: Albert III received the counties of Lower and Upper Austria, while Leopold III received everything else, including all the lands in the Eastern Alps. Following Leopold's death in a battle against the Swiss in 1386, Albert became his orphans' guardian. Later his son Albert IV and Leopold's son William agreed on a joint administration of all the Habsburg provinces. William's younger brothers also had their own vested interests.

Since in 1355, Duke Albert II had obliged the nobility to protect the unity of the Habsburg provinces. In case the members of the dynasty came into conflict, the provincial assembly intervened in the dispute among the brothers in the early 15th century and later even acted sporadically as the guardian of underage princes. After another division of the provinces in 1414, one branch took over

---

121 Ferdo Gestrin, *Trgovina slovenskega zaledja s primorskimi mesti od konca 13. do konca 16. stoletja*, Ljubljana: SAZU, 1965, 73–88; Miha Kosi, *Spopad za prehode proti Jadranu in nastanek "dežele Kras". Vojaška in politična zgodovina Krasa od 12. do 16. stoletja*, Ljubljana: Založba ZRC, 2018.

122 Even so, as late as the 16th century, the nobility from the former estates of the Gorizia family in Lower Carniola participated in the meetings of the Carniolan provincial assembly in Ljubljana as a separate group, whose provincial privileges were separately approved by the Habsburgs. See Peter Štih, "Dežela Grofija v Marki in Metliki," in: Vincenc Rajšp and Ernst Bruckmüller (eds.), *Vilfanov zbornik. Pravo, zgodovina, narod*, Ljubljana: Založba ZRC, 1999, 123–143.

Styria, Carinthia, Carniola, Istria, and Trieste; this conglomerate later came to be known as "Inner Austria." At the end of the 16th century, during yet another division, Inner Austria for several decades constituted a separate Habsburg dominion with its own prince and its seat in Graz in Styria. The conflicts within the Habsburg dynasty only came to a temporary halt in 1463, when Emperor Frederick III's nephews died and he again united all provinces (save Tyrol) under his rule. The first Duke of Inner Austria, Ernest, probably motivated by the imperial throne having been taken by the Habsburg competitor Sigismund of Luxembourg, was extremely keen to point out the peculiar historical features of his province. Like all the dukes of Carinthia before him, he was installed in the ancient ceremony at Zollfeld. His installation in 1414 was the last to use this ritual.[123]

## The County of Celje

In the 14th century, when the reshaping of the eastern Alpine regions under Habsburg rule seemed to be over (except in Carinthia and Gorizia), a new dominium, the County of Celje (Ger. Cilli), began to spread across all of these provinces and in the first half of the 15th century jeopardized Habsburg rule in the Eastern Alps. The new dominion reached such proportions that it developed into a separate province. It was composed of legally disparate and territorially unconnected estates owned by the Counts of Celje in Styria, Carinthia, and Carniola. Their significance increased through the ingenuity of the counts, who cunningly manipulated the eternal struggle between the House of Habsburg and the House of Luxembourg for the title of emperor. The House of Luxembourg made some formal Habsburg vassals direct subjects of the crown (the Counts of Ortenburg in 1395, and the Counts of Gorizia's estates in 1415), reducing their dependence on the Habsburgs.

---

123  Niederstätter, *Die Herrschaft*, 172–250.

**Figure 20.** "Mantled Virgin Mary—the Protector," relief from c. 1410 in the Church of the Mantled Virgin Mary—the Protector on Ptujska Gora (near Ptuj): under Virgin Mary's mantle, the artist portrayed the church's founder Bernard of Ptuj with his wife, the Patriarch of Aquileia, Hungarian King Sigismund with his wife Barbara of Celje, Bosnian King Tvrdko, Count Herman II of Celje, and other dignitaries from the broader region.

Thanks to shrewd and homogeneous family politics, well-suited marriages, and inheritance, as well as smart financial transactions, the Counts of Celje rose in just over a century from feudal lords with their seat in Žovnek castle in the upper Savinja Valley to the owners of a vast estate of Celje in 1333. In 1341, the Habsburgs' rival Emperor Louis IV bestowed on them the prestigious title of the Counts of Celje, which Emperor Charles IV confirmed in 1372. Once officially elevated, they linked with European elites through dynastic marriages: first to the Bosnian and Polish royal dynasty, and eventually to Sigmund (1387–1437), King of Hungary and Bohemia and Holy Roman emperor, whose life was saved in 1396 in the Battle of Nikopol (Nicopolis) by Count Herman II (1385–1435). This opened the counts' way into Hungarian, Bohemian, Bosnian, and Croatian politics and estates.

In 1423, the Habsburgs relinquished their lordship over the County of Celje, and in 1436 Sigmund granted the counts the title of princes of the empire (without seeking the Habsburgs' consent) and transformed the counties of Celje and Ortenburg-Sternberg in Carinthia into imperial fiefs. From then on, the Counts of Celje could start to transform their estates in the Eastern Alps into a separate province with a complete set of regal and judicial attributes. For Carinthia, Styria, and Carniola, this meant a risk of disintegration, which increased when certain neighboring feudal lords (e.g., the Walsees in Karst) imitated the Counts of Celje. These developments led to a long war with the Habsburgs. It ended in 1440 with the Habsburgs' acknowledgment of the counts' countship and with the signing of a mutual inheritance agreement in 1443. The agreement turned out to be in favor of the Habsburgs in 1456, when conspirators among the Hungarian nobility killed the last count, Ulrich II, in Belgrade. The counts' entire property passed into the hands of the Habsburg family, which ultimately completed their governance as the provincial lords in the Eastern Alps. Since the topic of the Counts of Celje features prominently in Slovenian history, one of the following subchapters gives a more detailed account.[124]

## The County of Gorizia

Even older than the Celje province, although less intimidating in the eyes of the Habsburg dynasty due to its peripheral location and isolation, was the County of Gorizia. The Habsburgs had to wait until 1500 to add it to their hereditary lands. Its rulers, the Counts of Gorizia, had come to Karst and the central basin of the Soča River at the beginning of the 12th century. They also had estates in Tyrol and Istria, where they were in perpetual conflict with the Patriarchate of Aquileia. Their county in the Soča basin developed from the estates and courts which were between 1001 and the 14th century, with the exception of Gorizia, officially only fiefs held by the patriarchate's advocatus (its representative in secular matters) in the Friulian March.

In 1253, Count Meinhard III added the Tyrol region to their estates through inheritance. In 1271, his property was divided between his two sons, the respective founders of the Tyrol and Gorizia lines. In 1286, the Tyrol line received the Duchy of Carinthia in fief, but lost it shortly afterwards in 1335, and the Habsburgs took possession of Tyrol in 1363. The Gorizia line retained the old family estates in the Soča region, Friuli, Istria, Carinthia, east Tyrol, and southeastern Carniola. The two lines supported each other until the end of the 13th century, but then went their separate political ways. Ever since their arrival in the region, the Counts of

---

124 See "The Stars of Celje," this chapter.

Gorizia had worked toward shaping a province by taking good advantage of the title of hereditary advocatus of the Aquileian patriarchate and of the governor of Friuli, as well as their royal rights, until 1365 when the Gorizia line was elevated to the title of prince of the empire—much earlier than the Counts of Celje. With this, Gorizia seceded from the Friulian March, but it retained a slightly modified Friulian provincial law.

The problem preventing the Gorizia line from establishing a province was their estates' geographical and administrative dispersion and the family's division into two branches, insufficiently linked for relatives of the surviving branch to take over the other's estates when it died out in 1374. Consequently, the two branches formed four separate units. The County of Pazin in (inner) Istria and the County in March and Metlika were ruled by the "Istrian line" until 1374. The counties were then inherited by the Habsburg family and not the relatives of the Gorizia family. Only in 1456 did the counts attempt to reestablish the connections between the Front County of Gorizia in east Tyrol and Carinthia, with its center in Lienz, and the Back County of Gorizia situated along the middle course of the Soča River and in Karst, under the control of the main family line. However, after defeat in the struggle for the inheritance of the Counts of Celje, they were forced to cede all of their property in Carinthia to the Habsburgs in 1460.

After the extinction of the main and the last surviving line in 1500, the remaining property became part of the provinces of Tyrol and Carinthia, while the principal territory in the Soča region and Karst was preserved by the Habsburgs and remained a separate unit. The Habsburgs even expanded it somewhat in the 16th century by adding the lands that Emperor Maximilian I had wrenched from the Venetian Republic: the united county of Gorizia and Gradisca survived until 1918.[125] The territories of today's Slovenia, with the exception of the lands east of the Mura River belonging to the Hungarian Kingdom and the coastal Istria under the authority of the Venetians, were thus again united under one ruler, the Habsburg dynasty, for the first time since King Otokar's era.

## The Istrian Peninsula

The development of the Istrian Peninsula was quite different from that of continental regions. Socially and culturally, it was divided into the coastal

---

125  Peter Štih, *Studien zur Geschichte der Grafen von Görz. Die Ministerialen und Milites der Grafen von Görz in Istrien und Krain*, Vienna, Munich: Oldenbourg, 1996; Vojko Pavlin, *Goriška—od zadnjih goriških grofov do habsburške dežele*, Nova Gorica: Pokrajinski arhiv, 2017.

Roman-Venetian part (organized into communes) and the inner, Slavic part (organized feudally) belonging to the empire. In the 10th century, Istrian towns' aspirations to autonomy became too troublesome for Venetian hegemony, whereupon the Venetian Republic more or less forced the ill-disposed towns into loyalty agreements (the first was made with Koper in 932). In this way, Venice was able to reinforce its trade monopoly in the Adriatic Sea and ensure the undisturbed management of its estates in Istria. Officially, ever since the Ottonian era, the entire peninsula had belonged to the March of Istria, which from the beginning of the 12th century was under the authority of the Aquileian patriarch, and from then until 1209 under the authority of the Spanheim and Andechs families. In 1209, the patriarch restored his control over Istria, although by that time governance was already largely divided among Istrian dioceses and towns.

Aquileia's endeavors to curb the autonomy of the towns, restrict the powers of the Counts of Celje and Venice, unite the march, and transform it into a "province" met with partial success until the mid-13th century thanks to the support of Emperor Frederick II, who wished to see Italy become part of his empire. Then the Counts of Gorizia, in alliance with some town communes, forcefully removed all the patriarchate's strongholds in the inland peninsula and established their own County of Pazin formed around the estate of the Diocese of Poreč, having been its advocati since the 12th century. Venice, for its part, also took advantage of the quick collapse of the patriarch's authority. As a growing power, it already represented a threat to Aquileia in Friuli, and by the mid-12th century had also extended its influence to the hinterlands of Istria. Since the autonomous Istrian towns did not perceive Venice as much of a danger in the mid-13th century compared to the feudal lords (the Aquileian patriarchs and the empire), between 1267 and 1284 the Venetian Republic was able to subdue all coastal towns except Trieste, Pula, and Muggia, despite the resistance of the largest towns such as Koper. Pula eventually came under Venetian rule in 1331, and Muggia in 1421.

**Figure 21.** "Dance of Death," fresco by Janez of Kastav from 1490 in the Church of the Holy Trinity at Hrastovlje (Istria).

Venice allowed the Istrian towns to follow their own communal law. The republic guarded its interests by installing its political-military delegate (*podestato*). Subject to some restrictions, it also allowed all Istrian towns (especially Koper as its political center, and Piran as a trade center) to trade with ports in the Adriatic Sea and the eastern Mediterranean. Nevertheless, Istrian towns did not reconcile themselves to Venetian rule until the late 14th century; but their resistance was in vain. The largest rebellion was mounted in and around Koper in 1348, but was soon put down by the Venetian government. In 1420, the Venetian Republic occupied Friuli and removed the secular rule of the Patriarchate of Aquileia. This brought the Venetian Republic to the borders of the lands held by the Counts of Gorizia in the Soča region and the Habsburgs' lands in Istria. Between the 15th and 17th centuries, the conflicting interests of these big players in the northern Adriatic region created an atmosphere of latent hostility in the area, occasionally escalating into full-scale wars.[126]

---

126 Bernardo Benussi, *L'Istria nei suoi due millenni di storia*, Trieste: G. Caprin, 1924; Darko Darovec, *A Brief History of Istra*, Yanchep: ALA Publications, 1998.

# A Land under the Hungarian Kingdom (Prekmurje)

The present-day Slovenian territories east of the Mura River shared their destiny with the Hungarian Kingdom between the 10th century and 1918. The territory was politically divided between the districts of Vas and Zala, and a similar division existed between the dioceses of Győr and Zagreb. The modern Slovene name Prekmurje, which first appeared in the sources of the Diocese of Zagreb, later encompassed all Slovenian territories east of the Mura, which in 1918 became part of the Kingdom of Yugoslavia.

# "Tres Ordines Slovenorum": Society, Economy, and Culture

## The Church

After the incursion of the pagan Magyars, the traditional diocesan and parish organization was preserved only in the biggest Istrian coastal towns (the dioceses of Koper, Poreč, and Pula). In Pannonia, where the Bavarian bishops had struggled against Archbishop Methodius during the late 9th century, Salzburg-style ecclesiastical organization collapsed, never to be restored. In less exposed Carantania, where it relied on the provincial bishop, it survived until the mid-10th century. The situation was different in the areas south of the Drava under the jurisdiction of Aquileia, as parishes could only be founded after the end of the Magyar incursions. The early parishes were large and were established in former missionary centers, for example in Gornji Grad or St. Hermagoras, or next to the main settlements such as Ljubljana, Kranj, Mengeš, Radovljica, Škofja Loka, Cerknica, and Šempeter in the Savinja Valley. Another route to establishing parish networks, which was most frequent in Carantania during the 9th century, was the foundation of proprietary churches. These were entirely in the care of their founders, i.e., landlords.

The early parishes fragmented into smaller units during the 11th and 12th centuries, and by the 13th century all proprietary churches had become part of parish networks. From then on, there was no longer any need to maintain the lower (missionary) Slavic tithe, so it was replaced with the canonical tithe. Sloppy spiritual service in unintelligible Latin and the frequent absence of priests, who left uneducated assistants behind as their deputies, were responsible for the lingering pagan customs (e.g., the veneration of trees and wells), which survived in the remote Alpine regions as late as the 14th century. In Istria, language-related predicaments were resolved by allowing the use of Slovene in church services and the Glagolitic alphabet in written texts. In 1467, there was even a theological seminary teaching Glagolitic established in Koper for Slavic priests. Consequently, in some places in Istria and Dalmatia, Glagolitic was used for ecclesiastical purposes as late as the 19th century.

The suffragan diocese in Gurk, Carinthia, was established in 1072 by the Salzburg Metropolite. Archbishop Eberhard II established two smaller dioceses, one with its seat in Seckau near Graz in 1218, and the other in St. Andrä in the Lavant Valley in Carinthia in 1228. His ostensible reason was to satisfy the spiritual needs, but a more likely motive was his fear that his ownership of Gurk might be disputed. The bishops of these dioceses were appointed and consecrated by the Archbishop of Salzburg rather than the Pope, a quite peculiar custom within the Catholic Church. However, since these suffragan dioceses were small, the archdiocese maintained two Carinthian and two Styrian arch-deaconates. Two further arch-deaconates were responsible for the parishes in Prekmurje: one was under the auspices of the Diocese of Zagreb, the other under the Diocese of Györ.

The patriarchs of Aquileia were not in favor of suffragan dioceses. They already had enough trouble with subordinate bishops in Istria: its northern part belonged to the small Diocese of Koper and the Diocese of Trieste, while the western part of Inner Carniola and Karst belonged only to the large Diocese of Trieste. In 1237, Patriarch Bertold made a lone, unsuccessful effort to establish a diocese that would cover the territory south of the Drava, with its seat in Gornji Grad, the site of a Benedictine monastery since 1140. The first diocese established within the eastern part of the patriarchate was the Diocese of Ljubljana, founded in 1461. It was not under the authority of the patriarch, but of its founder, the Pope, and the Habsburgs. A network of Aquileian archdeaconates instead of dioceses had been established east of the Soča River by the 13th century. These archdeaconates had some episcopal powers, but not the most important one, the judicial power. This was reserved for the patriarch's office in Udine, the patriarchal seat since 1236.[127]

---

127  In 568 the seat of the patriarchate was moved from Aquileia to Grado. From 737 to 1027 the seat was in Cividale; then it was returned to Aquileia and in 1238 it was moved to Udine, where it remained until 1751.

The Church 149

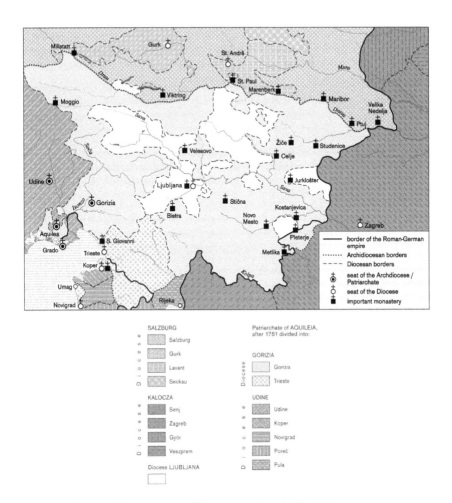

Map 3. Ecclesiastical organization of Slovenian regions before 1777.

Archdeacons convened ecclesiastical assemblies, paid visits to parishes, announced patriarchal rulings, interdicts, and excommunications, supervised, presented, and led the investiture of priests, conducted notary duties, and the like. Until the 12th century, the region that is today central Slovenia was under the jurisdiction of the Archdeaconate of Carniola, which then split into the archdeaconates of Carniola and Savinja Valley (covering the region of Styria south of the Drava). In the 14th century, the Archdeaconate of Carniola further split into the archdeaconates of Upper Carniola and Lower Carniola, and after 1420 into several smaller archdeaconates encompassing the parishes of some large monasteries (Ribnica). The westernmost Slovenian territories were under the authority of the archdeaconates of Tolmin and Lower Friuli, and the region between the Karavanke Mountains and the Drava River was under the archdeaconate of Carinthia. Episcopal duties in these territories were taken over in the 14th century by the general vicars of the patriarch (usually bishops from Istria and northern Italy), since the patriarch had spent a few years in exile on the estate of the Counts of Celje and as a subject of the hated Venetian Republic was not allowed to enter the Habsburg-ruled territories after 1420.[128]

In the territories with a fully developed ecclesiastical network to the west and south of the Soča River, the monasteries could contribute productively to religious activity in the Early Middle Ages. In Friuli, several monasteries were erected: the monastery of St. John (San Giovanni) of Duino above Trieste was established as early as the beginning of the 7th century. By the 9th century, Carolingian Friuli had twelve monasteries. A convent in Koper was established in 908 (the oldest known convent in modern Slovenia). Carantanians first came in contact with the monasteries in the 8th century when they were Christianized. The monasteries in Innichen and Molzbichel were established precisely for the needs of Christianization. A further eight Benedictine monasteries emerged across the wider Carantania region during the 11th century.[129]

The fact that the first monasteries in the central part of present-day Slovenia were only founded in the 12th century reflects the level of religious life and organization, but also the smaller density of settlements there. The first was the

---

128 Dušan Kos, *Zgodovina morale 2. Ljubezenske strasti, prevare, nasilje in njihovo kaznovanje na Slovenskem med srednjim vekom in meščansko dobo*, Ljubljana: Založba ZRC, 2016, 21–45.
129 The most important Benedictine monasteries in eastern Friuli were situated in Rosazzo and Moggio; in Carinthia and Styria the monasteries were located in Gurk, St. Georgen on Lake Längsee, Ossiach, Millstatt, Arnoldstein, St. Paul, Admont, and St. Lambrecht.

Cistercian monastery in Stična, founded by Peregrine I, Patriarch of Aquileia, in 1136. The monastery became the economic, cultural, and religious center of wider Carniola and southern Styria. It incorporated more than ten parishes and its estates were among the largest in Carniola. The Cistercian monks encouraged colonization and agriculture even more than the Benedictine monks. They brought new cultivated plant species from the west and are said to have introduced the iron plough to Slovenian lands. Additional Cistercian monasteries were established by the Spanheims in Viktring near Klagenfurt in 1142, and in Kostanjevica in Lower Carniola in 1234. In 1140, Peregrine established the only Benedictine monastery in Gornji Grad.

Soon afterwards, in 1164, the Margrave of Styria, Otokar III, established in Žiče the first Carthusian monastery in the German part of the empire. The first monks came from Chartreuse, the center of the Carthusian order. When the Carthusian establishments in French and Spanish provinces acknowledged the authority of the Pope in Avignon during the 14th and 15th centuries, with Carthusians in Italian and German provinces staying loyal to the Pope in Rome, the monastery in Žiče was the seat (1391–1410) of the general prior of the Carthusian order of Roman obedience; in the period between 1398 and 1410, this was Stephen Maconi, formerly the secretary of Catherine of Siena. The Carthusian monastery in Jurklošter in the lower Savinja Valley was established in 1173 by Bishop Henry of Gurk. The third Carthusian monastery in Slovenia was established in Bistra near Ljubljana before 1260 by the Carinthian duke and the lord of Carniola, Ulrich III of Spanheim. The last Carthusian monastery in the Slovenian regions was established by Count Herman II of Celje, in Pleterje in Lower Carniola. The flourishing cities and the colonization of the countryside in the 13th century also encouraged the arrival of preaching and knightly orders, of which there were eight in these regions.[130] Their number further increased in the 14th century, especially when new convents were built, but between the 15th and 17th centuries many of them closed down.

---

130 The Dominicans arrived in Ptuj in 1230; the three Dominican female convents (in Studenice, Velesovo, and Marenberk) and the three strongholds of the Order of the Teutonic Knights (Deutsch-Ritterorden) in Ljubljana, Velika Nedelja, and Metlika were established during the same period. The six Minorite monasteries were located in the most important towns: Gorizia (established in 1225), Ptuj (1239), Ljubljana (1242), Celje (before 1250), Maribor (around 1250), and Koper (around 1260).

**Figure 22.** Remains of Carthusian Monastery in Žiče.

During the 14th century, religious life and Church organization in Slovenian lands, much like elsewhere, became beleaguered by decadence. In remote regions which had never been incorporated fully in parish networks and where early conversion to Christianity had only been perfunctory, superstition, heresies, and paganism recurred time and again, and the priests were never able to fully eradicate them. Interestingly though, the first witch hunt in this region did not take place until the 17th century. The main problem leading to decadence lay with the ecclesiastical staff: the majority of priests who served under the Patriarchate of Aquileia came from Friulian or local noble families. As a rule, they did not have close contact with their flock, preferring to live in patriarchal centers, mostly in Cividale and Udine. Most of them were not sufficiently educated, and in addition, their pay was too low to spark their interest in providing good religious service. The situation further deteriorated in 1420, when the Venetian Republic deprived the patriarchate of its secular powers. Patriarch Louis II Teck sought asylum with the Counts of Celje. From his sanctuary, he managed to win over the majority of the local clergy and became the ardent leader of the anti-papal party at the Basel Council, since Pope Eugenius IV was Venetian by origin. The situation calmed down only after the patriarch's death, when Emperor Frederick III made an agreement with Pope Eugenius in 1446 in return for broad rights to exert control over the Church in his lands. This enabled the Habsburgs to legally interfere with ecclesiastical administration and to gradually develop a state-controlled church system that reached its peak toward the end of the 18th century under Emperor Joseph II.

Attempts at reform in the 15th century proved futile, and the resolutions of the Provincial Synod in Ljubljana in 1448 did not change religious life and clerical discipline.[131] Even the monasteries were enveloped in decadence. Illiterate monks were not a rarity, and education was also deficient in convents inhabited mainly by noble women. The scandalous situation in the Gornji Grad monastery forced the authorities to close it down in 1473. Religiosity thus became a personal matter, with people relying increasingly on fraternities and making pilgrimages. This was also fueled by continual frictions with the clergy. Noblemen, too, were in conflict with the clergy over the church tithe, taxes, judicature, and patronage of the churches. Therefore, toward the end of the 16th century, the situation in the Slovenian regions was much like the situation elsewhere, conducive to the rise of the Reformation, which engulfed all social classes.

## Peasants and the Countryside

Within what is now Slovenia, peasants were by far the most numerous segment of society until the early 20th century. Estimates for the 15th century suggest 80 percent—of a total of less than half a million people—were peasants. Although the main wave of colonization was concluded in the 13th century, hamlets and isolated farms continued to crop up across forested areas, hilly regions, and eastern swampy lowlands as late as the 16th century. This secondary, mainly local colonization was supported by landlords through reduced labor service and

---

131   For more on the state of religion in the Slovenian lands and on the establishment of the diocese in Ljubljana in 1461, see Josip Gruden, *Cerkvene razmere med Slovenci v XV. stoletju in ustanovitev ljubljanske škofije*, Ljubljana: Leonova družba, 1908.

lower taxes. The total number of settlements, many of which were located above the present-day highest settlements, was even greater than it is today. During that period, the population as a whole increased, but growth was uneven across different regions as it was affected by frequent epidemics, wars, and weather calamities. During the early 14th century, laborious cultivation caused by the balance upset in the proportions of forests, pastures, meadows, and settlements led to the extensive and protracted devastation of the cultural landscape in the lower basins of the Sava and Drava rivers. In addition, a number of farms in southern Carniola that lay on the path of Turkish marauders were abandoned, as revealed by 15th-century estate tax registers. The landlords populated their estates partly with refugees from the Balkans and partly with other estates' serfs, attracted by competitive terms. On the other hand, in the regions unaffected by Turkish raids, the peasant population grew to such proportions that local villages ran out of land to accommodate new farms. Since small Slovenian towns could no longer sustain the surplus population, the 15th-century lords set about resolving the problem by granting the peasants the last available plots of land on the estates they held in demesne; holdings in demesne were hence preserved only by the monasteries. When even this did not help, the existing farms were partitioned into smaller legal and cultivation units. One part of the impoverished population went to live in huts (Slov. *kajža*), which gave their name to this village social group.

Farms, as the basic economic units, were given a variety of legal statuses depending on the manner of their establishment. Various statuses of personal bondage were equalized during the 14th century, so only a peasant's economic power determined the legal status of the farm. This gave rise to a homogeneous serf class whose personal non-freedom, confinement to the farm, was only formally its main characteristic; in reality, its main attribute was the peasants' commitment to the lord's patrimonial court and their payment of land and judicial taxes that circumscribed it. Toward the end of the Middle Ages, farms in Slovenia were divided mainly into "burgher," "rental," and "free" farms. "Burgher" farms were considered more favorable for peasants because his descendants could inherit them and a lord's obstinacy in collecting inheritance taxes for passing a farm to an heir was thus limited. A peasant could sell such a farm with the lord's consent, and the lord was entitled to part of the profits from the sale. This status entailed somewhat higher regular taxes. The most numerous were "rental" farms. Initially, they were not inherited automatically, but this changed during the 14th and 15th centuries because of the lack of peasants. Rental farms were more burdened by unpopular labor service, although labor service was generally not as onerous for Slovenian peasants as it was, for example, in the eastern parts of Europe: from the 12th century onwards, labor service was mainly replaced with dues paid in money or kind. The two legal farm statuses were almost equated in practice in the 17th century. "Free" farms implied freedom only on the part of a feudal lord who could dismiss a peasant at his discretion. Only a few peasants were in the possession of truly free farms that were actually bought from a landlord; these farms were also subject to separate land law. Free peasants survived only in the hilly regions of the western border parts of Slovenia (Gorizia, Istria). Some developed comprehensive village self-government with their own courts. In Istria and Gorizia, a special status developed, particularly among winemaking peasants, called *colonatio*: a peasant had individual freedom, while the land was rented for a limited term without ownership rights attached. The taxes associated with this type of farm were very high.

Economic and consequently social position of peasants primarily depended on the amount of taxes. A tax paid to a landlord was usually calculated based on the size and the quality of the farm and amounted to approximately 20 percent of produce. Added to this were the church tithe, occasional judicial-legal taxes (e.g., when the farm changed hands), and extraordinary provincial taxes (e.g., for defense against the Turks), which became quite common in the course of the 15th century. In the 13th century, when the ministerials moved up the social ladder and established their own estates, taxes increased everywhere, since land rents were the primary source of finance for noblemen's and the everyday needs and military campaigns ecclesiastical lords. With the strengthening of towns and the monetary economy, payment in money came to be preferred over payments in kind. The process was not general, though, since landlords preferred nearby farms to pay in farm produce, e.g., crops and wine, which they then sold. This practice continued well into the modern period.

The commutation of levies initially corresponded to the actual price of produce, but in the 15th century the price favored peasants. In an effort to resolve the perpetual financial crisis, the landlords sometimes chose to increase the levies and the amount of labor service arbitrarily, or to convert rental farms into more expensive buying farms, although the provincial lords were explicitly against the excessive exploitation of peasants. In the meantime, social differentiation among the serf class had already occurred: a class of prosperous peasants emerged who opposed the obstinacy of the lords and the restrictions imposed by the towns on non-agrarian activities. Peasants' self-confidence was further bolstered by the awareness that their defense against the Turks was entirely in their hands, that they alone were funding the defense of the province as a whole, and that they were skillful merchants no longer oppressed by the lack of individual freedom. They were joined by the growing peasant proletariat without land that was left with no other choice but to move to towns, work as haulers, engage in village crafts, or work as hired farmhands. All this led to the great peasant revolts in the 15th and 16th centuries: the first broke out in 1478 in southern Carinthia.[132]

**Figure 23.** "Holy Sunday," fresco from c. 1460 (workshop of Janez of Ljubljana) on façade of the Church of the Annunciation at Crngrob near Škofja Loka; scenes of various occupations that may not be performed on Sundays.

---

132 Pavle Blaznik et al. (eds.), *Družbena razmerja in gibanja*, vol. 2 of *Gospodarska in družbena zgodovina Slovencev. Zgodovina agrarnih panog*, Ljubljana: DZS, 1980, 481–502.

From the 13th century onwards, peasants were under the jurisdiction of their lords' "patrimonial courts" for civil and small criminal offenses; serious crimes were prosecuted by the district courts, presided over by the owners of larger estates. In some places, e.g., Istria, small communal matters were decided by a "village court" (Slov. *veča*) chaired by a landlord. Peasants had a limited right of appeal to the provincial governor's courts. Vineyards in Styria and Lower Carniola had a special legal status: the special "highland law" (or "vineyard law") was codified in the 16th century and pertained to higher elevation vineyards and workers. It came with separate courts that handled all types of matters (except criminal offences) involving vineyard workers. These courts were composed of reputable vintners, while the chairman of the court was appointed by the owner of the vineyard.

## The Nobility

One important development with serious implications for provincial lords was the emancipation of the lower nobility (i.e., ministerials), who until then had been unfree. Ministerials first settled in the Slovenian territory during the 11th century. They came from abroad with landlords, for whom they performed military, administrative, and court services. By the 12th century, this class had also absorbed the remnants of the once-free Slavic nobility and even a certain number of *kosezi*, although kosezi generally preserved their privileged peasant status.[133] Ministerials were treated as objects: lords could give them away as gifts or exchange them. The most fateful implications of such an absence of freedom were the restrictions on the possession of private property and marriage, and the dividing of the children of ministerials who belonged to different landlords between their respective lords. Studies show that ministerials in the service of the high nobility came from a rather small number of families, but all these families were large, had many branches, and maintained close kinship and business ties. This played an important role when it came to the inheritance of extinct lines and resulted in the accumulation of property in the hands of a handful of individuals.[134]

The ministerials' indispensability in consolidating provinces and feudal estates, their acquisition of hereditary fiefs, and the extinction of various lines of the high nobility (which at any rate were few in number), enabled provincial

---

133  Pleterski, *Župa Bled*, 91–112.
134  Dušan Kos, *In Burg und Stadt. Spätmittelalterlicher Adel in Krain und Untersteiermark*, Vienna, Munich: Oldenbourg, 2006, 48–52.

ministerials first to obtain privileges and then to fill the places of extinct families. This first happened in Styria, where in 1186 Margrave Otokar IV allowed his ministerials to marry without restrictions, to inherit a fief if the owner died intestate (women could also be heirs *ab intestato*), and to handle their possessions without restrictions.[135] Emperor Frederick II confirmed these rights in 1237, so the ministerial noblemen were legally equated with free noblemen. The personal bonds that tied ministerials to their lords were replaced with bonds making them subject to the provincial lord, i.e., to "the province." Some other members of the high nobility followed suit. For example, the Counts of Heunburg granted similar rights to the ministerials on their estate in Lož in Inner Carniola. In Carniola and Carinthia, the emancipation of ministerials ended later than in Styria, but it was concluded by 1276 at the latest, i.e., by the time of the temporary peace imposed by Rudolf of Habsburg.[136] Uniform law did not mean that provincial noblemen all merged into a homogeneous class. What was essential for the nobility was that the homogenization of statuses, let alone of property, never occurred. Moreover, until the mid-14th century, certain members of the high nobility (e.g., the archbishops of Salzburg) continued to keep their vassals in bondage by placing restrictions on their marriages. Later on, the supremacy of lords over ministerials was transformed into their supremacy over the client lower nobility. Intra-class marriages were still observed; what underlay them now was no longer the absence of freedom, but the promise on the part of the noblemen that nuptial isolation would be compensated for through profitable jobs.

The nobility underwent important changes during the Habsburg dynasty rule. In 1338, Duke Albert II granted the nobility of Carinthia and Carniola privileges similar to those of Styrian noblemen, and in 1365 Count Albert V of Gorizia granted them to the nobility of the County in March and in Metlika. All members of the nobility within the province were therefore equated and placed under the jurisdiction of the central provincial court that replaced the courts of feudal lords. Furthermore, the lord of the province surrendered certain privileges (e.g., taxation)[137] in favor of the nobility, radically reduced the military obligation on the nobility, and allowed them to participate in the governance of the province. The nobility skillfully exploited the continual financial difficulties and conflicts

---

135 Spreitzhofer, *Georgenberger Handfeste*, 14–17.
136 Sergij Vilfan, *Pravna zgodovina Slovencev od naselitve do zloma stare Jugoslavije*, Ljubljana: Slovenska matica, 1961, 200–203.
137 The earnings of the lord of a province came from rents and from special rights and taxes paid by the towns belonging to the lord. The nobility and the Church were exempted from tax payments.

within the Habsburg dynasty and succeeded in obtaining the right to form a kind of standing council. It came to be known as the "provincial assembly" (Ger. *Landesstände*; the participating nobility were called *Landherren*), with regular meetings starting around 1410. The provincial assembly, composed of noblemen, townspeople, and clergy, appointed special bodies that, together with the governor (a deputy of the provincial lord), managed the province until the 18th century. During the centralization and modernization of the state administration, provincial assemblies lost much of their power.[138]

The material and social advancement of the lower nobility was reflected in the cultural landscape. Particularly conspicuous was the flourishing architecture of castles. In the territory of modern Slovenia, castles began to be built between the early 12th and late 13th centuries. Castle building was related to the territorial consolidation of feudal estates, the development of provinces, and the peak of colonization. As early as the 12th century, every estate had at least a fortified ministerial court, if not a stone-built tower. A castle became almost a synonym for lordship, and the point of identification for the nobility, who began to turn the names of their castles into hereditary "surnames." Until the 13th century, the construction of a castle depended on the power of a lord, so castle owners were almost exclusively influential secular or ecclesiastical lords. Given that the golden age of castle building in Slovenia coincided with the shaping of provinces and with the liberation of the unfree nobility, the lords of provinces endeavored to curb the castle-building mania among the unruly nobility, because it limited their strength both militarily and symbolically. By the end of the 13th century, their efforts yielded results. The province was unified, so they were able to contain the arbitrary dispute resolution among the nobility, which until then had been the main motive for building castles. Consequently, only a few new castles were built in the 14th and 15th centuries, and even those had to have explicit (written) approval beforehand. By this time, the lord of the province had at his disposal effective mechanisms for sanctioning any unlawful new construction, as is shown by records reporting the demolition of castles.

---

138 Andrej Nared, *Dežela—knez—stanovi. Oblikovanje kranjskih deželnih stanov in zborov do leta 1518*, Ljubljana: Založba ZRC, 2009, 55–184.

**Figure 24.** Remains of Šalek castle near Velenje, c. 12th century.

The majority of the approximately 300 strongholds built during the High Middle Ages in what is today Slovenia were located on elevated and naturally protected sites. Castles were also erected above or inside urban settlements where town lord deputies and their retinues had their seats. Their layout was determined by the configuration of the land. As a rule, these were small castles (at least compared to the average size of a castle in Western Europe) with living quarters across two floors, an enclosed courtyard, and a protruding defensive tower. Another type was a fortified tower to which other necessary facilities were only appended later. Obviously, only rare noble families or ecclesiastical lords could afford the construction of bigger castles, while the ministerials had modest strongholds passed down from generation to generation until the line died out or the owner experienced economic collapse. Nevertheless, even these modest fortifications distinguished their owners (inhabitants) from the lower nobility who were not castle owners. Until the middle of the 15th century, the existing castles were continually expanded by adding new premises, but then the strategy changed. Bigger Renaissance mansions and urban palaces thereafter began to appear in lowlands, and their number further increased during the 16th century. By that time, the threat of Turkish raids had disappeared, while the non-institutional armed settlement of disputes among the nobility was no longer an option because judicial powers were concentrated in the hands of the provincial lord and a regular army had been established under the lord's authority. This eliminated the main reason for maintaining those most cramped castles perched on high points, which began to deteriorate rapidly.[139]

## Towns and Townspeople

Besides mining and minting, the establishment of towns was possibly the most reliable indicator of the increased flow of money and the emancipation of crafts and trade from agriculture. In historical provinces, excluding Istria, towns emerged during the 12th and 13th centuries when the European economic boom reached the Eastern Alps, albeit with some delay. The centers of crafts and trade were towns and market towns. When using the term "town," a measure of caution seems to be in order: during the Middle Ages (and later), the inhabitants of towns never abandoned their agrarian activities, and not a single really big town was established in these regions. Even at the end of the Middle Ages, the majority of towns only had around 1,000 inhabitants or less, and only in the biggest towns (e.g., Ljubljana, Trieste, Koper, and Piran) did the population exceed 4,000. The relatively favorable geo-strategic position between the middle course of the Danube (Vienna) and the northern Adriatic (Venice) was not in itself a sufficient stimulus for the towns to flourish or for the economic significance of this region to increase. One reason was the structure of distant trade, particularly trade in raw materials. Another important factor was an underdeveloped consumer market without the necessary capital to buy luxury products with added value or to manufacture of interest to the European market. Local merchants primarily exported agricultural produce (wine, honey, cattle, wax), raw materials, and simple products (iron, fur, leather); all luxurious goods, exotic fruits, and sweet wines were imported.[140]

Until the 11th century, even the Istrian coastal towns that boasted ancient tradition (Trieste, Koper, Poreč, Pula), as well as those with no such continuity (Izola, Umag, Rovinj), were only local centers for specific agrarian activities (salt harvesting, wine making, olive oil production). By that time, the few larger towns

---

139 Kos, *In Burg und Stadt*, 22–164; Michael Mitterauer, "Burg und Adel in den österreichischen Ländern," in: Hans Patze (ed.), *Die Burgen im deutschen Sprachraum. Ihre rechts- und verfassungsgeschichtliche Bedeutung II*, Sigmaringen: Jan Thorbecke, 1976, 360–380.
140 Gestrin, *Trgovina*, 36–59.

(Trieste and Koper) had also developed maritime trade extending to the eastern Mediterranean. A common trait shared by all Istrian towns was the spread of their governance far into the rural hinterlands, which effectively meant that the Venetian part of Istria was, until the collapse of the republic in 1797, partitioned among many towns. Under Byzantine rule, these towns enjoyed a certain degree of autonomy. During the Carolingian era, the powers of Friulian margraves in Istrian towns were concentrated in town deputies (judges). Before the year 1000, Istrian towns (municipalities) had introduced shared decision-making invested in assemblies of free citizens (*arenga*). When these towns came under the authority of Venice in the 13th century, the municipal governing bodies were organized according to the Venetian model. The organization included many municipal offices unknown to inland towns. The supreme political and military representative of Venice was called *podestà* (captain). The main municipal body was the assembly of noblemen (patricians) called the Great Council, which elected other bodies and clerks. During the 13th century, the Great Council membership became hereditary and reserved exclusively for patrician families (whose number was not very large). These proved successful in fending off immigrants as well as their less prosperous and non-noble fellow citizens from the levers of decision-making.[141]

The inland towns and market towns (apart from Ptuj) did not have ancient traditions, even though some stood on the sites of ancient cities (Ljubljana, Kranj, Celje). From the 10th century onwards, the more important semi-agrarian settlements situated at strategic locations and around important castles gradually evolved into true urban communities.[142] These towns were evenly distributed across the region. In terms of their legally defined urban areas, they never grew beyond urban centers encircled by limited suburbs. The toll protected them against the competition created by foreigners, countryside artisans, and merchants. These towns developed the custom of regular weekly fairs, in contrast to Istrian towns where fairs had never become a formalized practice. The network of 27 Slovenian inland "towns" was supplemented by around 70 "market towns" that included certain functional elements of cities, but most were not walled and their self-governance was largely dependent on the town lord.

---

141 Sergij Vilfan, "Stadt und Adel—Ein Vergleich zwischen Küsten- und Binnenstädten zwischen der oberen Adria und Pannonien," in: Wilhelm Rausch (ed.), *Die Stadt am Ausgang des Mittelalters*, Linz: J. Wimer, 1974, 63–74.

142 Fran Zwitter, "K predzgodovini mest in meščanstva na starokarantanskih tleh," *Zgodovinski časopis*, 6–7 (1952/1953): 218–245; Miha Kosi, *Zgodnja zgodovina srednjeveških mest na Slovenskem. Primerjalna študija o neagrarnih naselbinskih središčih od zgodnjega srednjega veka do 13. stoletja*, Ljubljana: Založba ZRC, 2009.

Since the establishment of the towns was a royal right, until the 11th century, urban settlements were only established upon royal approval. Thereafter, the actual importance of royal consent, although still officially required, diminished until 1232. During the 12th century, the establishment of craft and fair centers was planned by powerful lords who used them to consolidate their territories in the struggle to become lords of the province. Between the early 12th and 15th centuries, several mints operated between Carinthia and Croatia.[143] Only a small number of towns in Carinthia were established, more for the needs of trade than for political goals. Carinthia was on the route of the most important road connecting Vienna and the Venetian Republic and it also had several silver and lead mines. The first town to experience economic growth was Friesach in Carinthia. Established by the Archbishop of Salzburg, it was a market town between 1090 and 1106, and was granted municipal rights by 1130. St. Veit, Villach, Klagenfurt, and Völkermarkt were several decades younger.[144] Ljubljana, Kamnik, and Kranj in Carniola were formed as urban settlements in the 11th and 12th centuries and acquired municipal privileges during the next century. Together with Kostanjevica in Lower Carniola, these cities were the most important strongholds of the provincial lords in Carniola during and after the 14th century. In 1365, Duke Rudolf IV established Novo Mesto in Lower Carniola to boost trade in this region. In contrast, the development of Škofja Loka, the feudal and municipal center of the bishops of Freising, took a completely different course, as did the development of Metlika and Črnomelj in White Carniola, whose lords were the Counts of Gorizia. The market towns of Krško, Kočevje, Višnja Gora, and Lož were only granted municipal rights under the authority of the provincial lord during the 1470s, because of the Turkish threat. However, in an economic sense they never developed beyond trading peasant centers. In Lower Styria, the most important and the most influential town was Ptuj, an ancient traffic center between the Pannonian Plain and the Adriatic. It retained that role throughout the Early Middle Ages. Later, Ptuj became an important transit site in the cattle trade between Hungary and Italy. Maribor and Slovenj Gradec, and the slightly younger Brežice, Slovenska Bistrica, and

---

143 The first attempt at the central minting of coins was made by the Habsburgs in the 14th century, but they could not establish a monopoly until the collapse of the County of Celje a century later. Within the Slovenian territories, Venetian coin was a serious competitor to coins minted in Graz and Vienna until the 17th century.

144 Alfred Ogris, *Die Bürgerschaft in den mittelalterlichen Städten Kärntens bis zum Jahre 1335*, Klagenfurt: Verlag des Kärntner, 1974.

Ormož all developed during the 12th century. Celje only obtained municipal status in the first half of the 15th century.[145] In the western part, the only town was Gorizia, first mentioned in 1001, which acquired municipal rights in the 14th century.

Continental towns obtained urban attributes (fortified walls, the right to hold fairs and judicial rights) and self-government bodies only on the approval of the lord of the town, i.e., the owner of the land, the formal founder, and the conferrer of citizenship rights. Unlike in Istria, inland towns achieved autonomy and legal status at a much slower pace, by successively adding new economic and administrative freedoms. Consequently, these towns did not have coded laws, while their administration was in many respects dependent on common law. The only exception was Ptuj, which in 1376, in accordance with the will of the lord of the city (the Archbishop of Salzburg), obtained a general municipal statute, although in terms of accuracy and completeness it was far behind the statutes of Istrian towns dating from the 13th century or later. Since the rights a lord bestowed on the towns under his supremacy were quite similar to one another, the towns owned by a specific lord constituted a network of similarly organized places with like privileges: Ptuj, for example, had laws akin to those in force in the other two towns of the Diocese of Salzburg, Ormož and Brežice. Legal order in Škofja Loka much resembled that of the Bavarian cities under the jurisdiction of the Diocese of Freising. The same internal similarity existed between the laws of the Lower Carniolan towns Metlika, Črnomelj, and Novo Mesto, which in the 14th century obtained municipal rights modeled on those granted to Kostanjevica. Kostanjevica itself was granted privileges around 1300 by the lord of the province, the Carinthian duke from the family of the Counts of Gorizia-Tyrol.

The representative of a town lord in continental towns was a "town judge" who was, however, obliged to collaborate with the assembly of all citizens (commune). In the 13th century, the assembly's decision-making was transferred to a group of twelve representatives, who as a rule were prosperous citizens. Their functions were not hereditary, as was the case in Istrian towns. By the 14th century, this 12-member body had evolved into collegiate bodies: the most important was the "external council." In bigger towns, the external council appointed an executive body, the "internal council," from among its members, while retaining a supervisory function. In the Carniolan towns, under the authority of the Counts

---

145 Norbert Weiss, *Das Städtewesen der ehemaligen Untersteiermark im Mittelalter. Vergleichende Analyse von Quellen zur Rechts-, Wirtschafts- und Sozialgeschichte*, Graz: Historische Landkommission für Steiermark, 2002.

of Gorizia-Tyrol from 1279 to 1335, the position of the town judge (and with it all the concomitant earnings) could even be leased to as many as three judges. From 1370, the position of town judge was within the citizens' domain and no longer subject to the will of the town lord. This change was first instituted in Ljubljana, with other towns soon following suit. Thus, relatively early on, the town judge became the actual instrument of urban autonomy in continental towns, although the town lord retained the formal right to give consent to his election. This form of self-government survived in the continental towns until the end of the 18th century, when all the Habsburgs' hereditary lands underwent radical administrative reform. In the process, towns were deprived of a large portion of political autonomy and were subordinated to the supervision of state administration.

In terms of language and ethnicity, urban inhabitants formed a heterogeneous group, since the present-day Slovenian region occupied the meeting point of diverse European cultures—Roman, German, Magyar, and Slavic. In the towns of Carniola, the Slavic population that moved in from the agricultural countryside was definitely in the majority. In Ljubljana, the largest and (from the 14th century) the capital town of the province, Slovene-speaking inhabitants accounted for at least 70 percent of the total population of 6,000 at the time of the Reformation. On the other hand, in Istrian and continental towns, the official languages used in government, education, and high culture were Italian and German, respectively. Numerous examples indicate that knowledge of the official languages was never taken for granted or deemed necessary for candidates competing for official functions, including those of the highest rank. Foreigners accounted for only a fraction of the urban population, but their role in the municipal and provincial economies was incomparably greater: the Florentines and the Jews (the latter since around 1320) were the bankers in many Istrian and continental towns, while wholesale merchants were mainly of Italian and German origin.[146]

The relatively rapid development of towns and market towns slowed down in the mid-14th century. From that time on, the region's border location and vulnerability to Turkish raids, as well as perpetual conflicts among the Habsburgs, the Venetian Republic and the Hungarian Kingdom, hampered progress. In 1360, the Habsburgs closed the roads to the south of the Karavanke

---

146 Janez Peršič, *Židje in kreditno poslovanje v srednjeveškem Piranu* (Ljubljana: Oddelek za zgodovino Filozofske fakultete, 1999); Josip Žontar, "Banke in bankirji v mestih srednjeveške Slovenije," *Glasnik Muzejskega društva za Slovenijo*, 13 (1932): 21–35.

Mountains for Venetian merchandise such as copper, redirecting trade routes to Vienna and the Alps. During the 14th and 15th centuries, all inland towns were involved in disputes over the restrictions on trade in wine and iron. In 1461, Trieste won the right for all goods traveling from Carniola to Venice to pass through Trieste. This had serious implications for grain merchants in Istria and Carniola. General circumstances (e.g., plague epidemics) and the overall economic situation also had critical consequences that aggravated the supply of agrarian products to towns: peasants circumvented the municipal rules on the upper price limits by selling their produce elsewhere, mainly abroad. The price of food was further increased by numerous tolls. The number of urban artisans and merchants stagnated, not only because of restrictions imposed by the guilds, but also because peasants could satisfy part of their demand for craft products by producing those themselves. In rural areas, the growth of crafts and trade intensified significantly in the 14th century as a result of pressure from landlords to replace tax payment in goods with payments in cash (peasants could only earn cash by selling their products in towns), and because part of the peasant population that became destitute turned to supplemental activities as an additional means of survival. The landlords supported such extra activities, thus exacerbating the enduring conflict with the towns. The agreements between the towns and the nobility (in Carniola such an agreement was concluded in 1492) that restricted artisanship to the towns' immediate vicinity and only allowed trade activities in rural areas during local religious holidays were not of much help; these limitations did not affect the wholesale trade in iron, wine, salt, cattle, and grain with Italy that formed the basis of the robust peasant trade and haulage that for centuries outstripped urban trade.[147]

The citizens of Slovenian towns, faced with local peasant and foreign competition, did not have enough capital to engage in profitable activities, such as the production of expensive goods or distant trade. One reliable indicator that points to the existence of such difficulties is the fact that inland Slovenian towns never saw the formation of a real patrician class. Those rare urban dwellers who achieved noble status during the Middle Ages did not become noblemen because they engaged in urban activities, but because they had purchased estates that turned them into landlords. The majority of local merchants traded with Istrian towns; as regards major undertakings, from the mid-15th century

---

147 Vlado Valenčič (ed.), *Ljubljanska obrt od srednjega veka do začetka 18. stoletja* (Ljubljana: Mestni arhiv, 1972), 6–19; Gestrin, *Trgovina*, 41–99; Kosi, *Potujoči srednji vek*, 30–58.

onwards, local merchants began to be squeezed out by trading associations from southern German cities (Augsburg, Nuremberg). Despite the anger of the local population, the lord of the province did nothing to restrict these associations. On the contrary, in 1495 and 1515 he even bowed to their pressure and expelled the Jews from Carinthia, Styria, and Carniola, where they had meanwhile taken over the financial transactions. Even the entrepreneurs who invested in new mines (e.g., a mercury mine in Idrija in 1493) and ironworks were usually foreigners and well-off noblemen organized in financial associations, who no longer limited their business to individual provinces or took notice of the provisions on the division of work among social classes. They organized production through investment contracts between skilled workers and financial associations. At any rate, these early capitalist approaches further reduced the competitive ability of local townspeople, so, even more than in the past, they resorted to outdated guild restrictions in order to survive. As a consequence, the medieval towns in the Slovenian regions never evolved into centers that could offer essential novelties in production to elevate the culture to a higher level.[148]

## Culture

The border regions of the Holy Roman Empire and the Eastern Alps were inevitably affected by European artistic trends. Even if their luster was somewhat dimmed, Romanesque and Gothic features were the standard elements of visual images of one's closeness to God in these regions as much as they were elsewhere. Art arrived in the southeastern part of the empire with delays, and the selection of artists and the splendor of artworks were adapted to the financial standing and the needs of the people who commissioned works of art. On the other hand, such a situation created more opportunities for regional painters and sculptors, some of whom possessed extraordinary talents and were sensitive to northern Italian and south German art trends.

High written culture was for a long time the domain of the Church; initially, the language was exclusively Latin, and not only for texts related to the church. At the end of the 10th century, the church in Maria-Wörth am Wörthersee in Carinthia boasted a library that included 42 codices. Even more exclusive collections could be found in certain monasteries: toward the end of the 12th

---

148 For details, see Ferdo Gestrin, *Slovenske dežele in zgodnji kapitalizem*, Ljubljana: Slovenska matica, 1991; Janez Mlinar et al. (eds.), *Mestne elite v srednjem in zgodnjem novem veku med Alpami, Jadranom in Panonsko nižino*, Ljubljana: Zveza zgodovinskih društev Slovenije, 2011, passim.

century, the scriptorium of the Cistercian monks in Stična produced the biggest collection of authentic illuminated codices in the region following northern French trends. At the same time, the clergy also catered to "lower" folk culture. Thanks to the pastoral role of the Stična monastery, during the early 15th century the scriptorium also produced shorter religious texts in Slovene (e.g., the *Stična Manuscript*). Probably the biggest 15th-century library in the region was that of the Carthusian monastery in Žiče, containing more than 2,000 codices and incunabula. The local scribes mainly copied the most interesting classical works (encyclopedias, religious, philosophical, and scholastic texts).[149] Local authors were few, and even they were educated and worked abroad. For example, the scholar and astronomer Herman de Carinthia, who lived in the first half of the 12th century, worked in France and Spain. He was one of the first promulgators of Islamic culture: he translated from the Arabic Euclid's *Elements* and the Quran. Several foreigners worked in local monasteries: at the Carthusian monastery in Jurklošter the most renowned were Siegfried, a monk from Swabia (the 13th-century author of a rhymed poem about Duke Leopold VI), Michael of Prague (14th century), and Nicolas Kempf of Strasbourg (15th century). In the early 14th century, the monk Philip, who came to Žiče from northern Germany, reworked in German a Latin epic about the life of St. Mary, which was a real hit in Germany for two centuries.[150] John, abbot of the Cistercian monastery in Viktring in Carinthia, came from Lotharingia (Lorraine). He wrote the Latin chronicle *Liber certarum historiarum*, considered one of the most important works of historiography of the 14th century. A Minorite from Celje wrote the *Chronicle of the Counts of Celje* in German soon after 1456. The Carinthian priest Jakob Unrest wrote three chronicles, of Austria, Carinthia, and Hungary, at some time before 1500.

Koper had an episcopal school at the end of the 12th century, and it is not surprising that this most important Istrian town was the first to experience the flourishing of Humanism and Renaissance in the 15th century. It had a school for nobles and an academy (of knightly games). It is also believed that in the mid-15th century the first attempts at printing in Koper were made. The most important continental town, Ljubljana, obtained its first official school only after the establishment of the diocese in 1461. Private schools had previously existed in Ljubljana (since 1291), in Kamnik (since around 1300), in Klagenfurt (1325), in Maribor (1452), etc. The most serious competitors of these schools

---

149  For a list of the majority of preserved medieval manuscripts of diverse origin kept in Slovenian libraries, see Milko Kos, *Srednjeveški rokopisi v Sloveniji*, Ljubljana: Umetnostno zgodovinsko društvo, 1931.
150  Jože Mlinarič, *Kartuziji Žiče in Jurklošter*, Maribor: Pokrajinski arhiv, 1991, 466–497.

were monasteries, particularly convents in which the nuns were mainly noble women (Velesovo, Mekinje, Studenice, Marenberk, Škofja Loka). The first nearby university was in Graz, founded in the 16th century. The first university in the territory of modern Slovenia was the university in Ljubljana, established in 1919.

Little is known about medieval secular literary culture, and our knowledge of courtly culture among the nobility is only marginally better. Around the mid-13th century, there were three local Minnesingers whose songs in German were preserved in the 14th-century *Codex Manesse* (*Große Heidelberger Liederhandschrift*).[151] Noblemen, particularly those related to the Habsburg court, were also keen proponents of courtly culture during the 15th century. Around 1500, at the court of King Maximilian, the knight Caspar Lamberger and several other knights from Carniola, Styria, and Carinthia were among the most ardent fans of tournaments and restored courtly culture in Central Europe.[152] Although in 1430 the Count of Celje, Ulrich II, set out to Santiago de Compostela in the spirit of medieval knights and initiations accompanied by a numerous retinue (in Spain he had audiences with Alfonso V of Aragon and Juan II, King of Castile), soon after this journey the Counts of Celje had the first Renaissance court built and gathered important humanists around them in the manner of great patrons of the arts.[153]

The continual immigration of foreign noblemen (mainly German and Italian-speaking) led to a mingling of locals and foreigners, whereby the latter, once they settled in the region, lost their sense of foreignness. Since after the introduction of the feudal order the language of the nobility was German and the clergy's was Latin, Slovenian literature could not develop. German was the main language used in all official matters after 1300, while in Istria and Friuli from the 15th century, the language was Italian. Rare written documents in Slovene nevertheless indicate that the high nobility spoke and understood Slovene, and it even appeared in

---

151 Anton Janko and Nikolaus Henkel, *Nemški viteški liriki s slovenskih tal. Žovneški, Gornjegrajski, Ostrovrški/Deutscher Minnesang in Slowenien. Der von Suonegge, Der von Obernburg, Der von Scharpfenberg*, Ljubljana: Znanstveni inštitut Filozofske fakultete, 1997. The fragments of the novel by Parzival of Wolfram von Eschenbach, the copying of which was commissioned in the 13th century by an unknown nobleman of Carniola, have also been preserved: Janez Stanonik, *Ostanki srednjeveškega nemškega slovstva na Kranjskem*, Ljubljana: Filozofska fakulteta, 1957.
152 Dušan Kos, *The Tournament Book of Gašper Lamberger/Das Turnierbuch des Caspar von Lamberg*, Ljubljana: Viharnik, 1997, 108–136.
153 Ignacij Voje, "Romanje Ulrika II. Celjskega v Kompostelo k Sv. Jakobu," *Zgodovinski časopis*, 38 (1984): 225–230; Primož Simoniti, *Humanizem na Slovenskem in slovenski humanisti do srede 16. stoletja*, Ljubljana: Slovenska matica, 1979, 15–38.

some works of high culture.[154] Especially the lower nobility, who lived and worked in regions where Slovene was the language of the majority, had to know and use it. As a consequence, they were also familiar with Slovene oral culture, especially with folk songs, which even into modern times still strongly reflected the spirit of the organization and the customs of pagan, Old Slavic society.

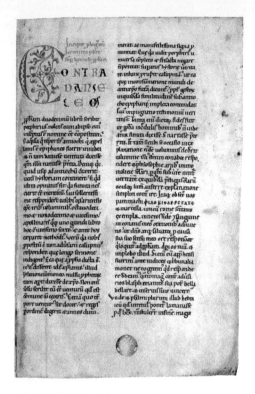

**Figure 25.** Cod. 685 in Österreichische Nationalbibliothek, Wien (S. Hieronimus, Commentarius in prophetas), produced in the Stična Monastery at the end of 12th century.

---

154 This is mentioned in the novel *Frauendienst* by a Styrian adventurer Ulrich von Liechtenstein, who says that upon his arrival in Carinthia in 1227, clothed as goddess Venus, Duke Bernhard von Spanheim greeted him in Slovene: "Bog vas sprejmi, kraljica Venera!" ("God welcome you, Queen Venus!"): Ulrich von Liechtenstein, *Frauendienst*, transl. Viktor Spechtler, Klagenfurt: Weiser Verlag, 1999, stanza 592. The Tyrolian poet and adventurer Oswald von Wolkenstein (1367–1445) used several Slovene phrases in two love poems that were written in several languages.

# The Stars of Celje

The rise of the Counts of Celje began in 1396, a crucial year in European history. The fate of the crusade, the crushing defeat of the Western European knights at the Battle of Nikopol under the command of Emperor Sigismund of Luxembourg, King of Hungary, transformed the Turkish Empire from a regional power center into a continental superpower. The sultan's intentions and ambitions thereafter concerned not only the Balkans but the wider Danubian region. Meanwhile, the last tangible opportunity to save Byzantium faded from the horizon of global politics.

The military engagement on the wetlands along the lower Danube River was not as spectacular as the epic Battle of Kosovo seven years earlier, and even its precise date—somewhere between September 25 and 28, 1396—was eventually forgotten. The catastrophe did not endanger Sigismund's kingdom, although it all but dethroned him, a political manipulator who never quite accomplished his rigorously elaborated plans. That he escaped the clutches of the Turkish army was nothing short of a miracle.

The major role in rescuing Sigismund from his unenviable position, and subsequently from the thick of the battle, was played by Count Herman II of Celje (Ger. Cilli).[155] Afterwards, the enterprising Herman also helped Sigismund to restore his severely undermined authority across the lands of the Crown of St. Stephen. The emperor, whose legitimacy came under question following the death of his wife Mary (1395),[156] extended his gratitude to the supportive count by marrying his youngest daughter, Barbara (1408), and bestowing upon him numerous feudal estates in Croatia and Slavonia. Herman also became a member of the elite "Order of the Dragon" and, before long, prided himself on the title Ban of Slavonia, Croatia, and Dalmatia (1406). This ushered in the rapid

---

155 Milko Kos, *Srednjeveška kulturna, družbena in politična zgodovina Slovencev*, Ljubljana: Slovenska matica, 1985, 262.
156 Sigismund of Luxembourg ascended to the Hungarian throne by marrying its legitimate heiress Mary of Anjou—the daughter of the late King Louis the Great and Elizabeth Kotromanić—in the fall of 1385. Seething with discontent, the magnates from the lands of the Crown of St. Stephen captured and imprisoned him in 1401. Sigismund was rescued from his predicament by his allies in Hungary and neighboring lands who were associated with Herman II of Celje. See Jörg K. Hoensch, *Kaiser Sigismund. Herrscher an der Schwelle zur Neuzeit 1368–1437*, Munich: Beck, 1996, 103–106.

rise of the Celeian dynasty from a family of mere local significance to the heights of Europe's most prominent patricians. For the first time since the disintegration of Carantania, which had been enlarged with marches (at the onset of the 11th century), a center of power was thus established in the Slovenian provinces with which even the broader Alpine-Danubian region had to reckon.

**Figure 26.** Celje.

Count Herman II was the most outstanding representative of the ancient line of the free lords of Savinja (Sovne) or lords of Žovnek (Sonekke), who were first mentioned in historical sources in the 12th century. The landed property held by this aristocratic dynasty was long limited to a small area surrounding their home castle in the southeastern foothills of the Alps. Yet, through prudent acquisition of advocate's rights on ecclesiastic estates and through a series of favorable marriages, the lords of Žovnek had become noteworthy actors in the Savinja Valley and across the entire Sotla River basin by the beginning of the 14th century. The expansion of their landed property, as well as their power, accelerated the downfall of several noble families with whom they had been mingling on the "marriage market." However, the Žovneks had not established direct familial ties with the most distinguished high-medieval dynasties in the eastern Alpine region (i.e., the Andechs-Meran, Babenberg, and Spanheim dynasties), and without inheritance rights they could not actually draw any substantial benefit from those dynasties' rapid decline. At the time, the branches of their family tree would mainly reach as far as Middle Styria and the Drava basin. This significantly increases the possibility that the lords of Žovnek were indeed an aristocratic family, with the Savinja Valley as their autochthonous domicile. The only exception that would potentially allow a tentative link with the side branch of the St. Hemma dynasty was that the names of their representatives appeared in the early period.[157]

In 1306, a bitter contest for the Bohemian throne began between the Counts of Gorizia-Tyrol and the Habsburgs; it subsequently became an outright war, spilling over to Slovenian soil. Driven by their self-interest, the free lords of Žovnek sided with the Habsburgs. Only two generations earlier, the counts of Celje and Gorizia had split the provinces between the Danube and the Adriatic among themselves after the collapse of King Otokar II Přemysl's rule (1276–1278), but now their strengths and ambitions were no longer even remotely comparable. The Celeian dynasty had already started to turn its eye toward the throne of the medieval Roman Empire under the rule of the unscrupulous Rudolph I, and was daringly entering continental politics. The Counts of Gorizia, so disunited that one branch stood against the other in the struggle over the throne of St. Wenceslaus between 1306 and 1308, had to content themselves with regional importance.

Ulrich II of Žovnek trusted in the plans of the Habsburgs, who had soon made their ambitions in the eastern Alpine regions clear after consolidating their power in Austria and Styria (1282). Like the Babenbergs and Otokar II Přemysl before them, their main goal was to establish a strong dynastic territory extending from the middle Danubian basin in the north to the Adriatic coast in the south. During the battles for the Bohemian crown, Ulrich II voluntarily transformed his alliance with the Habsburgs into vassalage (1308). He handed them the Žovneks' entire landed property—only to be immediately re-granted it as a fief. He thus committed himself to enforcing Habsburg policies "in the field," while having justified confidence in his lords' appreciation for the limited scope of his actions. Ulrich's plans proved to be well conceived in 1311, when the Savinja basin was annexed to Styria, since the new rulers of the province

---

157 Peter Štih and Vasko Simoniti, *Slovenska zgodovina do razsvetljenstva*, Graz, Ljubljana: Mohorjeva družba; Korotan Ljubljana d.o.o., 1995, 108. Such an explanation is considerably challenged by the fact that later the lords of Celje were not associated with the dynastic policy of the secondary branch of the family of St. Hemma, i.e., the Counts of Višnja Gora.

where he held his estates considered him a man of his word. Therefore, it is not surprising that his son Frederick had already been appointed provincial governor of Carniola when the Habsburgs were still preparing to take actual possession of it (nominally, the province had been in their hands since December 1282, but they had later rented it to the Counts of Gorizia-Tyrol).

In the early 14th century, the lords of Žovnek won not only reputations but also new estates in the Savinja Valley. From their late relatives, the Counts of Vovbre, they inherited the city of Celje (1322–1333), where they established their new center. The rapid growth of their power and significance was soon also manifested in the formal advancement of the family's status, when Emperor Louis IV of Bavaria granted them the title of Counts of Celje.[158] Thenceforth, the rise of this vital and enterprising aristocratic family from the Savinja Valley would outgrow its dependence on one single ruling dynasty. Their support was now sought by a growing number of crowned heads that nourished their interests in the eastern Alpine region.

By the middle of the 14th century, the Celeians had not taken any overt stand against the Habsburgs. Firstly, they must have been aware that they still lacked the strength to enter a matching rivalry with their lords. Secondly, at that time the two sides' political orientations were not in competition. And thirdly, both were pursuing expansion through similar means. In the late 14th century, the Counts of Celje expressed intense interest in Eastern Europe (for which the Habsburgs had thus far showed no appetite) through a sequence of intermarriages. Around 1360, Frederick's younger son Herman I married Catherine, the daughter of Stephen II Kotromanić, Ban of Bosnia. But as she was also the sister-in-law of Louis of Anjou, King of Poland and Hungary, the counts immediately found themselves at the very center of international politics. While initially onlookers rather than protagonists, they did not miss an opportunity to further enhance their standing. Therefore, by the end of 1380, William, the nephew of Count Herman I, had married Jadwiga, the daughter of "the Last Piast," Casimir III the Great.

This particular arrangement served the interests of the cunning Louis of Anjou: his carefully calculated matchmaking policy most elegantly fortified the position of his daughter and heiress Jadwiga in Poland. Jadwiga, the oldest daughter and descendant of the former Polish royal dynasty (which had died out on the male side), thus joined the family of Louis' distant married-in relative with an immense dowry of 20,000 golden florins, permanently removing her

---

158 The Žovneks and Bavarian Wittelsbachs were also blood relations (through the Pfannbergs). Their family contacts were reestablished in the 15th century.

from Louis' kingdom.[159] The maneuver nonetheless elevated the first lady of Celje to monarchic heights in the next generation, when William's daughter Anna, named after her mother, became the second wife of the Polish-Lithuanian ruler Władysław II Jagiełło in 1401 or 1402. Although not particularly pleasing to the eye according to contemporary reports, she excelled as a queen, and took over the Polish throne while her husband decisively defeated the army of the Teutonic Knights in the Battle of Grünwald.

The counts' attention eastwards meant that the Habsburgs, with their focus on Central Europe, did not perceive them as noteworthy rivals for a long time. The vassals from the Savinja Valley constituted no imminent threat to the Habsburg expansion toward the Adriatic Sea, which was accomplished in September 1382 when the city of Trieste voluntarily placed itself under the patronage of Leopold III the Just, Duke of Austria, Styria, Carniola, and Carinthia. Significantly, the lords did not even attempt to forestall the rise of the Celeians. Following the end of the reign of Louis IV of Bavaria, the Habsburgs had even agreed that Emperor Charles IV of Luxembourg should re-affirm their vassals as counts (1372). Yet only one generation later, in the wake of the Battle of Nikopol, both families had become bitter enemies. From then on, the Celeians would no longer be of service to the Habsburgs as provincial governors in their duchies.[160]

Following the deaths of his father Herman I (1385) and second cousin William (1392), Count Herman II became sole master of the House of Celje. He thoroughly redefined its political priorities and overhauled its resources to help achieve the dynasty's vision within a reasonable period of time. The emphasis shifted from mercenary warfare, a key source of the Celeians' wealth throughout the 14th century and a resource for establishing familial ties, toward the counts' own economic and political initiatives. It is no coincidence that Herman II despised the Jews, who had built up a vast trading and banking network throughout the eastern Alpine region during the High Middle Ages.[161] But after strengthening his

---

159   Janko Orožen, *Od začetka do leta 1848*, vol. 1 of *Zgodovina Celja in okolice*, Celje: Council of Culture and Science of the Celje City Municipality Assembly (Celjski zbornik, special edition), 1971, 139.

160   In 1390, Count Herman II of Celje was last recorded as provincial governor of Carniola. See Bogo Grafenauer, *Doba zrele fevdalne družbe od uveljavljanja fevdalnega reda do začetka kmečkih uporov*, vol. 2 of *Zgodovina slovenskega naroda*, Ljubljana: DZS, 1965, 397.

161   Franjo Baš, "Celjski grofi in njihova doba," in: Franjo Baš, *Prispevki k zgodovini severovzhodne Slovenije. Izbrani zgodovinski spisi*, Maribor: Založba Obzorja, 1989, 329.

close friendship with King Sigismund, he was undoubtedly capable of managing dynastic policy under his own direction. Consequently, this had a significant impact on the formation of a quadrilateral of forces across the entire Central European region. The Polish and Hungarian rulers (i.e., Władysław II Jagiełło and Sigismund) became brothers-in-law precisely through their Celeian wives. Moreover, the typical feudal "thirst for land" practiced by numerous generations of Herman's ancestors in southern Styria and Carniola was frequently reflected in petty and avaricious estate profiteering. However, it finally assumed considerable significance at the turn of the 14th and 15th centuries, as it provided a solid departure point for the count's ambitious territorial designs in the middle Danubian basin. What had effectively changed was merely the function of the Celeians' dynastic estate. Instead of a strong bulwark for their Habsburg lords in the south, it became the southwestern anchor of an independent political enterprise aimed at an accession to the throne.

Herman II's designs grounded him firmly in the eastern Alpine region delimited by the dynastic marriages of his house. By far the most crucial constant in his political calculations was Sigismund of Luxembourg, whose court was somewhat reminiscent of Archimedes' point of the then Central Europe. Should Sigismund, the (second) son of Charles IV, ultimately mount the throne of the medieval Roman Empire as well, as seemed more and more likely, this would also solve the ongoing problem of the Celeians' vassalage to the Habsburgs, which could only be "painlessly" abolished through the intervention of the crown. This intense interest occupied Herman's mind until his old age. There is every indication that Herman did not perceive the estates he had inherited from his ancestors as ordinary feudal possessions. Even though he had continuously been consolidating his power in Croatia and Slavonia since 1397, and had advanced to Count of Zagorje and even Ban (i.e., Viceroy) of Croatia and Slavonia, he remained firmly moored in the Žovnek-Celje core of his estates. This core was obviously the primary source and most unfailing guarantee of his power. He calculated that a considerable length of time would have to elapse before his—albeit immense—territorial acquisitions within the extremely unstable Hungarian Kingdom could be acknowledged as equal in worth and significance (although as early as c. 1405, there was no striking difference between the value of the Celeians' estates in Styria and in the Hungarian crown lands). Herman was also most certainly aware that it was exactly in Europe's southeastern region, the triangle between the Eastern Alps, the Pannonian Basin, and the Balkans, that the greatest short-term prospects for wealth, influence, and perhaps coronation lay. It was thus no coincidence that his oldest son Frederick married the well-to-do Princess Elizabeth of the House of Frangepáns, an extremely influential dynasty in Croatia (around 1405), or that his grandson Ulrich married the daughter of Đurađ Branković Smederevac, despot of Serbia (1433). Even the edict issued by Tvrtko II Kotromanić, King of Bosnia, stipulating that should he die childless, his throne was to be passed to his Celeian relations, clearly implies Herman's exceptional political strength and reputation.[162] Indeed, no other man could match the monarchs' power in the early 15th-century borderland between Central Europe and the Balkans. Therefore, Herman's name was self-evidently at the top of the Order of the Dragon's list of members, while the city of Celje appeared to become not only the birthplace of the wives of rulers but also the cradle of future kings.

As time passed, Herman completely freed himself from Habsburg vassalage. Sigismund, who by 1410 also held the crown of Rome, vested in him the right to blood feud in the County of Celje in April 1415. Herman therefore began to obtain the attributes of a feudal lord closely connected to the highest authority. In 1418, the Celeians inherited from the extinct Counts of Ortenburg a vast estate in Carinthia and Carniola. The inheritance gave them control over as much as three quarters of the southern one of the two Habsburg duchies. The most essential of the latest acquisitions was the state seigneury Ortenburg-Sternberg in Carinthia, which enabled the new owners to establish a direct legal relationship with the crown. Moreover, Herman could now exert pressure on the Habsburgs from every direction, after his second son's marriage restored his family contacts with the Wittelbachs of Bavaria. As a corollary, Duke Ernest the Iron, head of the primary branch of the Habsburg dynasty, eventually relinquished his feudal superiority over the Celeians in 1423.

Herman and his family were then able to concentrate their efforts on a speedy ascension to princes of the Holy Roman Empire. This would also give them equal status to the Habsburgs, whom they had then rivaled almost equally in Carniola, Carinthia, and Styria. In 1430, the Celeians fell short of their aim, as the draft charter to grant them the title of princes stayed in the archives: perhaps their former lords still remained too reluctant. The perfect moment for their elevation to princedom came in 1436, when Frederick V, Duke of Styria, set off on a pilgrimage to the Holy Land and Sigismund—now also crowned King of Bohemia (1419) and Emperor of Rome (1433)—issued a charter endowing his relatives with the longed-for title. Royal prerogatives accompanied the

---

162  Grafenauer, *Doba zrele fevdalne družbe*, 398. Later, however, embittered opposition in both Hungary and Bosnia prevented the Celeians from actually taking over Tvrtko's throne.

charter, granting them the freedom to mint their own currency, manage their own mining industry, and set up their own so-called noble court of justice. This initiated the establishment of a new (although not yet completely territorially coherent) Province of Celje, which seceded from Styria and Carinthia. Shortly before his family was actually raised to princedom, Count Herman II, the man most responsible for these developments, passed away in Bratislava on October 13, 1435.[163] He was buried in the Carthusian monastery in Pleterje, which he had built in 1403 with a view to making it his last resting place. As Herman II became a prudent feudal lord, he attended to his soul and memory as well: under his patronage, the *Chronicle of the Counts of Celje* was initiated, which contained colorful accounts of his heroic deeds.

Count Herman made countless enemies during his life. Less successful rivals envied his meteoric rise, while those who had lost out as a result of his diplomacy conspired against him. He was also deeply concerned about his descendants, since they included no true successor, only heirs. His sons, Herman III and Louis, connecting him through their wives with the Wittelbachs and Ortenburgs, had died. His firstborn, Frederick II, aged around 50, became a widower in 1422 or 1423—most likely by his own hand—and soon afterwards married Veronika of Desenice in conspicuous haste.[164] The old count made sure that the unwanted daughter-in-law, who was reportedly even endowed with the gift of witchcraft, disappeared from history as quickly as possible, and the tumultuous affair was eventually suppressed with the assistance of Sigismund's court. The licentious count Frederick had initially tried to oppose his father politically, seeking support from Venice, but had met with utter defeat and was even imprisoned for a while by Herman. As an ally of Louis II of Teck, Patriarch of Aquileia—whose temporal

---

163  There is a slight possibility that the Celeians were (first) raised to princedom as early as September 27, 1435, i.e., shortly before the death of Count Herman. The record "indicating" this was contained in a document that today is no longer available and is most probably a falsification. What remains certain is that the members of the Celeian dynasty obtained their princely title in Prague on November 30, 1436. See Peter Štih, "Celjski grofje, vprašanje njihove deželnoknežje oblasti in dežele Celjske," in: Vincenc Rajšp, Ferdo Gestrin et al. (eds.), *Grafenauerjev zbornik*, Ljubljana: Znanstvenoraziskovalni center SAZU, Filozofska fakulteta, Ljubljana and Pedagoška fakulteta, Maribor, 1996, 242–244.
164  According to all narrative sources, both favorable and unfavorable to the Celeians, it was generally believed that Count Frederick murdered his first wife, Elizabeth of Frangepán (from whom he had lived apart for several years).

power in Friuli was crushed by the Venetians in 1419 and 1420—Herman could not kneel to La Serenissima.[165]

Even later, Frederick II would fail to demonstrate any noteworthy political wisdom, but paid all the more attention to conviviality. Even the German humanist Hartmann Schedel's renowned *Liber chronicarum* (*Book of Chronicles* or *Nuremberg Chronicle*) depicts him as an incorrigible Epicurean. When asked whether his pilgrimage to Rome in the Holy Year of 1450 had been of any avail to him, the sybaritic count, already on the wrong side of 80, reportedly gave the sarcastic answer: "My cobbler still continues making boots upon my return." Not long afterwards, Aeneas Sylvius Piccolomini, who held the Celeians in utter abhorrence as both the representative of the Habsburg court and Bishop of Trieste, attributed to the following, downright pagan epitaph:

> Hereon I descend to Hell. What fate awaits me there, I cannot tell; I know only what I am bequeathing. I wallowed in luxury and am taking nothing with me—save the spirits I have drunk and food I have eaten and save the countless pleasures in which I have relished.[166]

Although these lines were an overstatement, Frederick II certainly did not care as dearly about the Catholic Church and religious sentiment as his father, who had legitimized his bastard Herman and paved his way to become Bishop of Freising (1412) and Trent (1421). Religious skepticism, which had already taken over the late medieval scholarly and aristocratic elite, was unmistakably also echoed in Celje.

Herman II's relationship with his youngest daughter Barbara, the wife of Sigismund, was only slightly less problematic. After the birth of her daughter Elizabeth (c. 1408), their paths and interests hardly ever crossed. For many, the vivacious Queen and Empress Barbara was the personification of depravity, a second Messalina, yet others saw her as a genuine embodiment of Venus. Subtler witnesses even observed her interest in alchemy and her reconciliatory stance toward various Christian sects in Bohemia (considered a nest of European heresies during the 14th century). Since she did not express any particular affection toward her "Teutonic" son-in-law Albrecht II of Habsburg, several

---

165 After the collapse of his temporal reign in Friuli, Patriarch Louis even took refuge with Count Herman; between 1420 and 1430, he ruled over his ecclesiastical province from Celje. King Sigismund did not intend to abandon him; however, the overall instability in Hungary and the Roman Empire had prevented his troops from engaging in a military confrontation with the Venetian Republic.
166 See Orožen, *Od začetka do leta 1848*, 238; Baš, "Celjski grofi in njihova doba," 331.

authors even accused her of harboring animosity toward the Germans.[167] Yet, having inherited her father's strong personality, Barbara was bound to have her own judgment of affairs and people.[168] Herman and Sigismund, whose mutual relations were characterized by complete unanimity over four decades (1396–1435), had to disregard Barbara's "contrariness" on several occasions. The two men were unmistakably more related through interest than blood.

After Herman II's death, the policy-making of the dynasty was passed to Ulrich II, who would soon demonstrate his competence when Frederick V, Duke of Styria, Carniola, and Carinthia, declared a war to re-annex the Principality of Celje. Prince Ulrich had been safely installed in the court of King Albrecht II of Habsburg since 1438, and had succeeded to the Roman, Bohemian, and Hungarian thrones of his late father-in-law Sigismund. At first, Ulrich managed to somewhat appease the situation in the eastern Alpine region; the new ruler would by no means side with his relative Duke Frederick V, but would continue to pursue the Luxembourgian political agenda "on a lesser scale." The resulting conditions were thus slightly more favorable for the Celeians. This advantage was partly also owed to the fact that Prince Ulrich had found a competent general for his army, Jan Vitovec, in Bohemia, where he also served as regent. The latter mastered the terrifying Hussite tactics, for which the army of Sigismund, emperor and king, was no match, and the mercenary could therefore match the appreciably stronger Habsburg forces. Therefore, a truce soon brought an end to the skirmishes that broke out in limited areas (in the Savinja basin and Carniola) and occasionally intensified.

Prince Ulrich was able to take advantage of temporary breaks in fighting with Frederick V. The suspension of hostilities lasted long enough for the Celeian to join the contest for the Hungarian crown upon the sudden death of Albrecht II in 1439. He offered fervent support to his cousin, Queen Elizabeth, a widowed mother-to-be. The Court party, in which Ulrich had the final say,

---

167 Barbara had considerable reservations regarding Sigismund's wish that his Bohemian throne be taken over by Albrecht II of Habsburg, who was not fluent in any other language than German. After her husband's death on December 9, 1437, the widowed empress even sought refuge in Poland (1438–1441). This lay the ground for speculations that she intended to place Władysław III, King of Poland, or his brother Casimir, on the Hungarian and Bohemian thrones.
168 Baš, "Celjski grofi in njihova doba," 333, 334. There are indications that during Sigismund's last days, Barbara made a series of attempts to intervene in his political affairs. Disagreeing with her active involvement, Sigismund had her imprisoned, which also kept her from attending his funeral.

soon took hold of the Holy Hungarian crown: on the night of February 20–21, 1440, this distinguished royal symbol, kept under lock and key at Visegrád, was brazenly stolen by Elizabeth's lady-in-waiting, Helen Kottanner, who then delivered it to her queen. The next day finally saw the birth of Albrecht's son, who entered history by the name of Ladislaus V the Posthumous. On May 15, 1440, Prince Ulrich arranged a completely legitimate coronation of the infant in Szekésfehérvár. It was he who held the crown of St. Stephen over the head of the not-yet-three-month-old ruler's head during the ceremony.[169]

It hardly needs to be said that numerous Hungarian magnates vehemently opposed to being governed by an infant king. That effectively meant the reign of the Court party and its backbone of noble families, which Sigismund of Luxembourg had placed under his patronage at the turn of the 14th and 15th centuries to consolidate his throne. Since the opposition comprised several foreign dynasties, its members could act against Queen Elizabeth and her son as representatives of their autochthonous homeland interests. Furthermore, what the Hungarian patricians expected from their ruler was swift action in the war against the Turks—which a king in swaddling clothes could not deliver. Therefore, the opposition predominantly supported the idea of crowning Władysław (Vladislaus) I/III. By 1442, they had won a complete victory,[170] yet soon afterwards their chosen leader embarked on a reckless crusade against the Turks only to lose both his army and his life at the Battle of Varna on November 10, 1444. Thereupon, Ladislaus the Posthumous was generally recognized as King of Hungary, while Ulrich of Celje, as his relative and benefactor, hoped to assert his universal power not only in the Hungarian crown lands but also in Bohemia and the Austrian archduchy. Almost on the eve of his death, his

---

169   Helen Kottanner, who held the infant King Ladislaus during the coronation ceremony on May 15, 1440, also "eternalized" her bold actions in her memoirs, which she dictated to an unknown scribe. See Igor Grdina and Peter Štih (eds.), *Spomini Helene Kottanner*, Ljubljana: Nova revija, 1999.

170   At the end of 1442, Elizabeth and Władysław I/III signed a peace treaty conceding the throne to Władysław. At that time, mutual confidence was established between the Jagiellons and the Albertinian line of the Habsburgs (connected with the Luxembourgs and the Celeians), which was also reflected in their family ties. Casimir IV, King of Poland, who had succeeded his brother Władysław I/III, married Elizabeth's daughter (and namesake). The Jagiellons thus asserted themselves as the most prominent late medieval dynasty in Eastern Europe. After the death of Matthias Corvinus in 1490, Vladislaus, son of Casimir and Elizabeth, ascended the Hungarian throne and entered history as the incompetent King Dobzse (King Alright), who agreed to whatever was suggested to him.

ambitious plans had come true. Nevertheless, he still had to end the war with Frederick V, who was crowned King of the Romans (as Frederick IV) in February 1440 and Emperor of the Holy Roman Empire (as Frederick III) in March 1452. Conditions in the field were appreciably more favorable to the Habsburgs now than they had been upon the conclusion of the truce.

Both Ulrich and Frederick (who had taken over the administration of the Principality of Celje after Herman II's death) had thoroughly prepared their armies to resume the war. Despite their significantly higher potential, the troops of the Holy Roman Empire largely remained on the defensive, although interestingly recorded much greater successes than at the very beginning of the war. Armed resistance to the throne had gradually seen the Celeians lose their momentum, and they opted for a lasting settlement. In 1443, both sides agreed to a compromise peace: the crown was willing to reaffirm Frederick and Ulrich as princes, provided that the cores of their estates would no longer possess any essential attributes of autonomous provinces. (But at least the Celeians would continue minting their own coin—even though the Habsburg ruler had most probably denied them such a royal prerogative.) Furthermore, a mutual inheritance contract was concluded in case one of the dynasties should die out.[171] Thus the Habsburg monarch finally restored his authority, while Prince Ulrich was left undisturbed to focus on his ambitious plans with Hungary. Moreover, as a feudal lord in Styria, Ulrich enjoyed both provincial and state protection from possible intrusions by his enemies from the comitati of the crown of St. Stephen into his estates within the boundaries of the Holy Roman Empire. Yet another reason for the reconciliation between the Celeians and King Frederick IV was the fact that the Habsburg monarch had become the guardian of Ladislaus the Posthumous in November 1440. Thus, if Ulrich was to place his Eastern European politics into the hands of the royal infant—who would most probably never have been crowned without his endeavors—he first had to reach some kind of agreement with his keeper.

Having reconciled with King Frederick, Prince Ulrich redoubled his political efforts in the Hungarian crown lands, where the Transylvanian magnate and commander-in-chief János Hunyadi was obtaining more and more power under Władysław I/III. Even though Hunyadi, who had won glory through his merits in numerous battles with the Turks, was of Vlach descent, his supporters regarded

---

171 Štih, "Celjski grofje, vprašanje njihove deželnoknežje oblasti in dežele Celjske," 247–253. It should not be overlooked that by 1443, King Frederick was still childless, while the future of the Celeian dynasty appeared secured.

him as the only true defender of Hungarian interests. As Frederick IV refused to send King Ladislaus the Posthumous to Hungary, Hunyadi was made governor in June 1446. This regency could thus not boast of undisputed legitimacy.

At least initially, this seemed to guarantee an indirect advantage to Prince Ulrich, who held many trumps simply because of his family's ties with the Ottoman Empire (his wife Catherine was the sister of Sultan Murad's most influential wife). The Hunyadis' and Celeians' armies, which had already engaged in a brief but embittered confrontation in 1446, were too equally matched for either side to prevail. By the middle of the 15th century, both sides (briefly) acknowledged this fact. The two rapidly rising dynasties, which spared no resources in their ascent to the pinnacle of power, now desired to settle their mutual relations as elegantly as possible. The enormous efforts of Ulrich's father-in-law, Đurađ Branković Smederevac, led to the signing of two agreements that would finally end the hostilities between the Court party and Homeland party in Hungary. Đurađ, despot of Serbia, desperately sought steadfast support from the northern Christendom under increasing Ottoman pressure from the south. Both agreements were supported by marriage proposals: in the first agreement, Ulrich's daughter Elizabeth was to marry the Hungarian governor's older son (Ladislaus), and in the second she was to be given in marriage to his younger son (Matthias).

However, Hunyadi suddenly changed his mind and the agreements never came into effect. The Hungarian warlord obviously believed that he still had the upper hand in the grand contest for full sovereignty over the Hungarian crown lands, and overtly tried to double-cross his adversary by agreeing with Frederick IV, King of the Romans, that Ladislaus the Posthumous was to remain with his keeper until he came of age. (The thick-skinned Habsburg also concluded a similar agreement with George of Poděbrad, Regent of Bohemia.) This would considerably prolong the gubernatorial government in Hungary, with the Court party acting solely on the nominal ruler's behalf. Since the Court party was unable to execute the king's will without internal obstruction, its authority was considerably compromised.

Prince Ulrich responded to Hunyadi with long-term moves that would guarantee his primacy in the middle Danubian region within a reasonable period of time. Firstly, he helped to incite the revolt of the Austrian provincial assemblies against Frederick IV. Ulrich aimed to free Ladislaus from his guardian and designate himself as the sole patron of the infant king's interests. Frederick, who had exerted every effort to be crowned Holy Roman emperor, initially underestimated discontent in the province along the Danube River. By August 1452, however, he found himself encircled and under siege by the enemy's army in Wiener Neustadt. The emperor also had to face opposition from his perpetually displeased younger brother, Albrecht VI. On September 4, Frederick was forced to hand Ladislaus over to his second cousin, Prince Ulrich. The 12-year-old king, whom Austrian provincial legislation had already recognized to be of age, could now succeed to the throne of his father, Albrecht II. Even though his regents in Bohemia and Hungary remained in office, his nominal kingship finally came to an end. Ladislaus and Ulrich, who had become a "rerum suarum director," saw themselves as having mutually compatible interests. Their harmonious relationship was fairly reminiscent of that between Emperor Sigismund and Count Herman. Due to his scheming enemies in Ladislaus' court, the Celeian fell from grace in fall 1453, although he was restored by February 1455.[172] Without his second cousin, the infant king would certainly have been stripped of his authority.

Ulrich, a shrewd diplomat and skillful warrior, had gradually accumulated even more power than his grandfather Herman. He not only publicly humiliated the King of the Romans but self-confidently continued minting coinage engraved with his own coat of arms and name. He also obtained the title of Ban of Slavonia, Croatia, and Dalmatia. By then, even János Hunyadi must have understood that his power was in irreversible decline. In summer 1455, he eventually chose to revive the unrealized marriage agreement between the two disputing dynasties and sent his son, Matthias, to the royal court in Buda. However, the reconciliation between the Court party and the Homeland party was this time rendered impossible by the intervention of nature: Ulrich's daughter Elizabeth had died a sudden, unexpected death.

The year 1456 witnessed the climax of tensions between the Hunyadis and Prince Ulrich as a result of a severe foreign policy crisis. The Turkish sultan Mehmed II the Conqueror had initially followed the advice of his father, Murad II, and on ascending to the throne expressed his sympathies to the Celeians and even offered them support. But by now, and in the wake of the fall of Constantinople, he wanted to lay his hands on the key to the Hungarian crown lands—the fortress in Belgrade at the confluence of the Sava and Danube rivers. That same summer, the Ottomans started a siege of the city with an immense land force and a strong Danubian fleet. Surprisingly enough, the Christian warriors, led by János Hunyadi and Johannes Capistranus (an ascetic preacher and impassioned propagator of the great campaign of the Cross against the Crescent), arrived in

---

172 Peter Štih, "Ulrik II. Celjski in Ladislav Posmrtni ali Celjski grofje v ringu velike politike," in: Igor Grdina and Peter Štih (eds.), *Spomini Helene Kottanner*, 31–40.

Belgrade first. Mehmed's onslaught on the city was repelled, yet resulted in a Pyrrhic victory for Hunyadi and Capistranus, who both died from the plague. The Hungarians and crusaders who had joined the fight were in a desperate need of a new leader. Ulrich of Celje could not have asked for a better moment to assert his rule in the lands of the Crown of St. Stephen. At the beginning of September, he concluded an agreement with George of Poděbrad to protect himself against any surprises that might arise from the Bohemian Kingdom. At the head of new crusading troops and accompanied by King Ladislaus he then hurried south to be named new regent and commander-in-chief at the Diet of Futog by the ruler and magnates (among whom the Court party gained a temporary majority on Hunyadi's death). His purpose had been achieved: he not only guided the steps of the young monarch, but also controlled the most important land of the crown.

These events infuriated Hunyadi's son, Ladislaus, who had become the new leader of the Homeland party. Prince Ulrich was instantly caught up in a tempest of defamations and pronounced the most formidable internal enemy of Christendom.[173] Many rumors have also echoed throughout the historiography of the Corvinian Renaissance. The Celeian was even accused of attempting to suppress the Hungarian language and nation[174] and plotting with Đurađ Branković Smederevac, who was already weakened by the fight against the Turkish incursions.

A full-blooded *homo politicus*, Ulrich had anticipated a harsh reaction from Hungarian "patriotic" magnates and tried to appease Ladislaus Hunyadi. He assumed the role of a statesman—and adopted his rival. But Ladislaus never took the adoption seriously, knowing that Ulrich had outlived all his children and was left without a successor: he could therefore do away with his princely rival with a simple murder. From Futog he hastened to Belgrade, where he concocted the plot in alliance with his uncle Mihály Szilágyi, who commanded the fortress's fleet. On October 8, 1456, King Ladislaus and Prince Ulrich entered the city castle

---

173 Kos, *Srednjeveška kulturna, družbena in politična zgodovina Slovencev*, 269. On the events unfolding in Belgrade, the papal legate cardinal wrote the following to Alphonse, King of Naples: "There is no doubt that the Count [Ulrich of Celje] was killed by Ladislaus, son of János Hunyadi, the one who had slain so many Turks. The son was believed to be defending Christendom no less by murdering the count than his father had by warding off Mehmed; because both Mehmed and the count were enemies of Faith, Count Ulrich being the enemy within and Mehmed the enemy without."

174 See Elisabeth Galántai and Julius Kristó (eds.), *Johannes de Thurocz Chronica Hungarorum. I Textus*, Budapest: Akadémiai Kiadó, 1985, 274.

accompanied by their closest retinue, when the gates behind them suddenly closed. The crowd of crusaders—the author of the *Chronicle of the Counts of Celje* wrote of as many as 40,000 warriors—remained outside the castle walls. The final confrontation came the very next day, when during the morning mass Ladislaus Hunyadi called the new regent to consult with him "on a matter most urgent." Ulrich followed him, only to be first reproached for his pathological greed and possessiveness and then brought to the point of Hunyadi's sword. Ladislaus and his conspirators attacked the Celeian, and killed him after a lengthy fight in which Ladislaus himself was also injured. Ulrich's head was severed from his body.

Bereft of his counselor and regent, the infant king first feigned ignorance; he silently accepted the conspirators' explanation that Prince Ulrich had provoked the attack. Yet he soon took his revenge on the insidious "patriotic" magnates who were determined to use him as their political stooge. In March 1457, Ladislaus Hunyadi was brought to court as the author and prime mover of the great Belgrade conspiracy. According to the *Chronicle of the Counts of Celje*, several executioners first inflicted many wounds on him, so that he would die a similar death as Prince Ulrich. Ladislaus' younger brother, Matthias, was imprisoned.[175] The only event that somewhat dampened the Court party's ultimate triumph was the death of King Ladislaus eight months later. The crusade that would press home the advantage of János Hunyadi's final victory the previous August was forgotten. Amid its internal turmoil, Hungary failed to employ the masses of European soldiers installed across the Danubian plain. The venturesome policy of Sigismund of Luxembourg and Władysław I/III to oust the Turks from Europe was abandoned. Despite his indirect links with the Ottoman court, Prince Ulrich could have become its last pursuer, by not having sent the crusaders home once Mehmed's attack on Belgrade had been repelled.

The dramatic death of the last of the Celeians was a major turning point in the history of the Slovenian territory. Thereafter, the Habsburgs' quest for Vienna on the one side and the Adriatic Sea on the other could proceed more or less unopposed. They succeeded in forming a "territorial concentration," which had been the chief political aim of the Babenbergs and Otokar II Přemysl. By gaining control over Carinthia (1335), as well as over the Celeians' estates (1456–1457), and later uniting their own dynastic territories under one family branch (1490) and one ruler (1493), the Habsburgs at least roughly recovered the complex of their hereditary lands in the eastern Alpine region. The once-prominent Counts

---

175 Franz Krones, *Kronika grofov Celjskih* [The Chronicle of the Counts of Celje], transl. Ludovik Modest Golia, Maribor: Založba Obzorja, 1972 [1883], 45–47, 52.

of Gorizia had become a shadow of their former selves by the 15th century, when for a period of time their destiny was being decided behind the castle walls of Celje.[176] Their once remarkable independent power gradually dwindled until they became nothing more than pawns on the Central European chessboard. For several centuries to come, the north-south political-geographical axis in the Slovenian provinces fully prevailed over the east–west axis. The heartrending cry "The Counts of Celje today and never again!" resounding at Prince Ulrich's funeral heralded the transformation of the region between the Pannonian Plain, the northern Adriatic, and the Alps into a *political* province. Never again would it be the center that would decisively determine the fate of Central and Southeastern Europe.

---

176 Elizabeth, the oldest daughter of Herman II, married the weak and sybaritic Count Henry IV of Gorizia. Their children were raised in Celje. On Prince Ulrich's death, the Counts of Celje engaged in a dispute with the Habsburgs concerning his inheritance, only to lose their cause. In 1460, they were forced to sign a very harsh peace agreement at Pussarnitz, which significantly fortified the position of Emperor Frederick and his family in Carinthia. See Hermann Wiesflecker, *Maximilian I. Die Fundamente des habsburgischen Weltreiches*, Vienna, Munich: Oldenbourg, 1991, 25.

# The Bloody Fall of the Middle Ages

In the days of Count Herman II and Prince Ulrich, the Celeian House had many opportunities to influence European politics. Their diplomatic tentacles reached as far as Castile, while their blood was mixed with several royal dynasties, particularly the Jagiełłonian and Luxembourgian ones. Their audacity at intervening in Hungarian and Austrian affairs was very reminiscent of the gambits of Italian Renaissance rulers. After the deaths of his children, Prince Ulrich came to the realization that he was the last remaining representative of the distinguished Celeian lineage. With his time running out, he allowed himself to embark on extreme ventures. Although his mid-15th century policy-making was downright risky and frequently impromptu, it was by no means futile. The last of the Celeians found his own ways to further Sigismund of Luxembourg's endeavors to create a center of power in the middle Danubian basin and neighboring regions that would dominate a broader European territory. This political conception was first briefly materialized by Matthias Corvinus at the turn of the 16th century, and much later by the Habsburgs.

The Celeians also left a lasting and indelible legacy in the territory of their origin. After the collapse of the Aquileian patriarch's temporal reign, their steadfast support to their relations by marriage, the Counts of Gorizia, prevented the Venetian Republic from spreading its rule over the entire Soča River. They bestowed generous gifts on numerous churches and monasteries, adding to the impressive image of several buildings (e.g., the Carthusian monastery in Pleterje) as well as statuary art and sculpture (e.g., in Celje and at Ptujska Gora). Also remarkable was their relatively flexible stance on religious matters. On the one hand, Count Herman II shared his opinion with his royal son-in-law Sigismund and spoke fervently in favor of a unified Catholic Church at the council in Constance. On the other hand, Ulrich's wife Catherine, who descended from the Serbian Branković dynasty, was permitted to retain her Orthodox faith upon settling in her new homeland. (Matthias Corvinus' ill-fated bride, Elizabeth, also received an Orthodox upbringing.) It would similarly be wrong to ascribe the Celeians' great benevolence bestowed upon the monastery in Žiče—which in 1391–1410 was the seat of Stefano Macone, the prior general of the Carthusian Order of the Roman Obedience—to sheer dogmatism. It was more likely part of their pragmatic approach: to take advantage of every opportunity. In the Middle Ages, such a center of spiritual power undeniably bore not only major religious but also political significance. The influence that could be acquired through supporting it far outweighed the value of the gifts and services awarded to the monastic community.

Although Prince Ulrich gained enough power after his victory over Emperor Frederick at Wiener Neustadt in 1452 to immediately transform his numerous estates across the Holy Roman Empire and the Hungarian crown lands into a unified territory, his political strategies predominantly revolved around establishing broader connections and designing long-term aims. The territorial unity built up by Count Herman's diplomacy during the reign of King Sigismund of Luxembourg soon all but vanished from history. At first Ulrich's widow, Catherine, strived to retain the Celeian estates. However, their heirs—the Habsburgs in particular—proved too strong opponents. Those castellans in her service who would not be lured by Frederick III's promises and offers were ultimately faced with the force of Habsburg arms. A brief war of succession broke out, initially marked by a series of dramatic turnabouts. The crafty commander-in-chief Jan Vitovec, having first sided with the Habsburgs, attempted to capture his emperor in 1457, when Frederick was visiting Celje. However, the emperor had left the city castle before he could reach him. The fighting spilled over not only into the Savinja Valley but also into Carniola and Carinthia (where the Counts of Gorizia tried to lay their hands on the estates of the extinct dynasty). But since Frederick's most immediate adversary, King Ladislaus the Posthumous, had been felled by a plague epidemic before the end of the same year, the question of succession to the Celeian legacy was practically resolved. The Holy Roman emperor, whose glory did not exactly rest on his shrewd political and military leadership, eliminated his most dangerous rivals by simply outliving them.

The Celeians' policy on cities and serfs was not as much in line with contemporary tendencies in Central Europe as was their ambition to conquer new lands and states. They apparently sought to retain ultimate control over economic flows in their hands. Count Herman II certainly did not expel the Jews from his estates because of his profound devoutness, as is claimed in the *Celje Chronicle*.[177] The real reasons were his economic interests. (The same could be said for the provincial nobility in Styria, Carinthia, and Carniola, which had negotiated a similar measure from the Habsburgs in 1496–1515 for the territories

---

177   Ignacij Orožen, *Celska kronika*, Celje: J. Jeretin, 1854, 16. As long as Herman could profit from Jewish bankers, he conducted all kinds of financial arrangements with them without prejudice. His zeal in the matters of faith had therefore little to do with religious fanaticism. For instance, in 1425 he negotiated from the Aquileian patriarch, Louis II of Teck, permission to indulge in meat, milk, and eggs at his table during Lent in the company of ten people. See Orožen, *Od začetka do leta 1848*, 253.

of those crown lands.) Herman's lack of initiative in establishing new cities and market towns also speaks volumes about his aspirations to hold a monopoly over regional economic flows. The control over trade, which had experienced notable growth before the Ottoman incursion into the Pannonian Plain (in some years up to 20,000 oxen would travel from the Hungarian crown lands to Italian seigneuries through the region encircling the southern foothills of the Karavanke Mountains), provided a continuous and reliable source of revenue.[178] It is therefore little wonder that Celje acquired its city rights only during the rule of the frivolous Frederick, in the middle of the 15th century. In a similar vein, the Celeians showed a reluctance to establish market towns; the only one with its origin (in the 14th century) linked to their dynasty was Šoštanj.

The Counts and Princes of Celje had even less compassion for their serfs, so it is hardly surprising that they were memorized in Slovenian popular narratives solely as evil tyrants and ruthless, depraved men. As early as 1558, the Protestant reformer Primož Trubar wrote:

> And still today one can see and hear from far and near how the nobles, grand and small, ill-treat their peasants and serfs; they are enthralled by force or fraud, overwhelmed by dues and toils until dragged to the bitter end; their land does not last to the third heir, their folk and blood will soon cease to be; as for the Princes of Celje, no matter how many cloisters and curacies they erected, how many journeys they made to Rome, before they all met their end or were done to death, they forced themselves upon their peasants' daughters, dishonored them and committed other injustices by force or fraud.[179]

In the Slovenian collective conscience, the memory of the Celeians might have been overcast by dark shadows even prior to their final demise, since both Frederick's and Ulrich's ascensions to princedom were strongly connected to a long series of bloodsheds that quite often seriously hindered the everyday life of common people. But the starkest times were yet to come. The intensified restoration or erection of city walls, as well as the adoption of defensive measures for their respective lands in the 15th century, were the signs of increasing popular anxiety turning even monasteries into fortresses.[180] People of all standings were

---

178 Štih and Simoniti, *Slovenska zgodovina do razsvetljenstva*, 147.
179 Mirko Rupel, *Slovenski protestantski pisci*, 2nd ed., Ljubljana: DZS, 1966, 97. The excerpt is taken from Trubar's postil printed in 1558.
180 Paolo Santonino, who traveled the Slovenian provinces during the late 15th century, described the Carthusian monastery in Žiče as follows: "This monastery is surrounded or lies embraced by hills on a rather flat stretch of land, so that it cannot be seen from the wayside until one reaches its walls. It is all in all well [built] and encircled by a high wall and moat as a castle—and for a good purpose, too, since

gripped by the unremitting "Turkish scare," which "East of the West" was not the result of collective hysteria but had very realistic causes. The onslaught of Islamic warriors was universally perceived as God's punishment, and it was precisely this sentiment that greatly helped the cause of Protestant reformers to gain momentum in the 16th century.

The Turks first entered Europe in 1356, conquering the eastern Balkans within just a few decades: in 1371 and 1389 they defeated the Serbian army, in 1393 they overpowered Bulgaria, and in 1396 at Nikopol they destroyed the united crusader army led by the Hungarian King Sigismund of Luxembourg. They then spilled over Hungary and one Turkish raiding party penetrated as far as Ptuj. This first Turkish incursion into the Slovenian territories was followed by regular assaults for years to come.

After a lengthy lull following their earliest advancements in Carniola and Styria between 1408 and 1415, the Turks only renewed their incursions under the aggressive Mehmed II the Conqueror. Furthermore, the fall of the Bosnian Kingdom in the spring of 1463 gave them a decisive starting point for their campaigns: the sultan's northernmost fortresses were now less than 100 km from White Carniola's capital, Metlika. The Turks, who had mainly been exerting pressure on the "Christian oecumene" with light cavalry, returned to the Slovenian provinces in 1469. This time they did not merely loot at random—as they had done in the aftermath of the Battle of Nikopol—but carried out a series of meticulously organized campaigns to systematically undermine the economic power of individual Hungarian comitati, Venetian provinces, and Habsburg provinces. The sultan's soldiers refrained from attacking fortified cities and castles, but inflicted devastating damage on market towns and villages that in the "maelstrom of war" were left to their own devices. The attackers frequently pitched provisional camps in the middle of enemy territory, remaining sometimes for weeks and accumulating their loot and captives to eventually be taken away to the southeast. Such were the Turkish tactics to break the will to resist in individual areas and prepare favorable ground for their ultimate conquest. The most severe damage and suffering was sustained by peasants, since the attackers, being typical plunderers rather than conquerors, did not waste time besieging fortified towns or castles.

---

it otherwise could not be defended against pillaging and burning during so many Turkish raids." Paolo Santonino, *Popotni dnevnik 1485–1487*, Klagenfurt, Vienna, Ljubljana: Mohorjeva založba, 1991, 86.

The warriors, who mostly rode under the banner of the Bosnian beglerbeg,[181] numbering only a few true Ottoman soldiers, drove thousands of people out of the Slovenian provinces. (According to contemporary reports, there were approximately 15,000 exiles recorded in 1471 alone, when the beglerbeg's army was rampaging through Carniola for no less than three months.) They destroyed several hundred villages, dozens of churches and market towns, and did not even spare monasteries. Only in 1483, after around 30 major advancements into Carniola, Styria, and Carinthia, did the Turkish pressure ease somewhat.[182] Nevertheless, as the defensive efforts had not had any substantial results, a great number of provinces had been severely stricken by then. Most effective were the intelligence and warning services.[183] Woodpiles had already been stacked on individual hilltops in peacetime, and once the sultan's cavalry was spotted on the horizon they were set alight to warn the local population of the coming danger. The very danger of Ottoman conquest had a decisive influence on the centralization of defensive efforts which were increasingly occurring at a supra-provincial level. The Habsburgs increasingly started to treat the territories under their direct control as a specific and complete unit.

According to the calculations made by the Carniolan, Styrian, and Carinthian provincial assembly, the Turks exiled no less than around 200,000 people during the first hundred years of their incursions. This caused severe population and economic crises, which could be only partially mitigated by the settlement of refugees from pashaliks of the Ottoman Empire. In individual areas, particularly the Sava basin and the Lower Styrian Drava area, as well as the Slovenian Littoral, 30–50 percent of farms were laid to waste after having been revived during the High Middle Ages. The drastic drop in population was also affected by natural disasters—particularly epidemics, earthquakes, floods, and the rampaging swarms of locusts devastating crops.[184]

---

181 The governor of a province in the Ottoman Empire (transl. note).
182 The most devastating Turkish incursions into Carniola, Styria, and Carinthia took place in 1473, 1476, 1478, 1480, and 1483. For a more detailed account of Turkish raids and their effects on the Slovenian lands, see Vasko Simoniti, *Turki so v deželi že. Turški vpadi na slovensko ozemlje v 15. in 16. stoletju*, Celje: Mohorjeva družba, 1990, 8–81; Ignacij Voje, *Slovenci pod pritiskom turškega nasilja*, Ljubljana: Znanstveni inštitut Filozofske fakultete, 1996, 81–190.
183 At first, the Venetian intelligence service was more effective than its Habsburg counterpart. By the end of the 15th century, the Turks had also raided Venetian Friuli via Carniola, Istria, and Gorizia.
184 Bogo Grafenauer, *Doba prve krize fevdalne družbe na Slovenskem od začetka kmečkih uporov do viška protestantskega gibanja*, vol. 3 of *Zgodovina slovenskega naroda*,

A trifle less belligerent than the Turks were the Christian rulers and potentates, who were continuously embroiled in disputes. Emperor Frederick, who began to fancy that his dynasty was destined to perform a worldly mission, spent much time at odds with the new Hungarian King Matthias Corvinus. The Habsburg obviously believed himself to be the rightful successor of Ladislaus the Posthumous, although he lacked both the ability and the power to fulfill his aspirations. At the beginning of 1459, the Archbishop of Salzburg, Sigismund I of Volkersdorf, indeed placed the crown of St. Stephen on Frederick's head—a provocative symbolic gesture to spark off longstanding tensions between the two leading Central European powers.

The emperor was also engaged in a bitter rivalry with his ambitious brother, Albrecht VI, and the Venetians. The war waged against Venice in 1463 was mostly fought on Slovenian soil. Trieste ultimately remained in Habsburg hands, but after cessation of hostilities fell under even greater pressure from Venetian-controlled southeastern Istria. Before long, the thick-skinned emperor also had to tackle a violent outburst of general discontent in his empire: in 1467, the Styrian nobility closed ranks and established an alliance with the aim of toppling the archduke of Austria and emperor of the Holy Roman Empire. Later on, the agitated nobility even deployed its own mercenary troops under the command of the battle-hardened professional warrior Andrej Baumkircher. Encouraged by the keen support of Matthias Corvinus, the nobles seized control of several cities and market towns, including Maribor, Slovenska Bistrica and Slovenske Konjice. Frederick III could not suppress this dangerous rebellious movement until 1471.

Individual "troublemakers" were rising up in other places as well. In Carniola, for example, Erasmus the Knight, another protégé of Corvinus, had shut himself away in his picturesque Predjama castle near Postojna and renounced his obedience to the emperor. After a long siege, the army of Frederick III finally managed to break Erasmus' rebellion in fall 1484,[185] though this did not help stabilize the wider situation. The war between the Hungarians and the emperor, which had started in 1477 with a view to succeeding the deceased Bohemian King George of Poděbrad and continued without interruption due to the politically

---

Ljubljana: Kmečka knjiga, 1956, 35, 47. In the Brežice area, highly exposed to the Sava and Krka flood basins, the population crisis had started even before the great plague epidemics. In this particular area no less than 70 percent of farms were already depopulated by the beginning of the 14th century.

185 According to a traditional belief passed down through generations by historiographic and literary sources, there are indications that Erasmus was killed by a precise catapult shot through his toilet window.

motivated conflict to seize the Esztergom or Salzburg archbishopric, showed no signs of abating, even though one side would occasionally prevail over the other. In the end, however, the scales started to tip in favor of King Matthias. Soon, the cities between Vienna (1485) in the north and Slovenj Gradec (1489) in the south acknowledged his rule and authority. Although Emperor Frederick's reputation was almost completely eclipsed, he still cherished the vision of his family's exalted mission, so he too saw to his successors' prosperity. He arranged a marriage to Mary of Burgundy for his son Maximilian, named after the late-Antique Celeian martyr,[186] thus laying magnificent Western European prospects before the Habsburgs. While other royalties were delivered their kingdoms by Mars, the Austrian dynasts were bestowed with theirs by Venus. The warriors, indeed, would all too often forget to ensure the prosperity of their able heirs. The same was true for Ulrich of Celje and Matthias Hunyadi, King of Hungary.

Almost until the very end of the Corvinian epoch (1490), Lower Styria, Carinthia, and Carniola (as far as Ljubljana) were still the ultimate destination of Hungarian expansion. While Frederick's mercenaries would often claim late payments from the vassals of their royal debtor, Corvinus' Hungarian soldiers, in contrast, greatly improved order in the newly conquered regions, and Corvinus even brought the Turks to conclude a peace treaty with the Hungarian king in 1483. He was impressed in Slovenian consciousness as the legendary "dobri kralj Matjaž" (good King Matthias). During such apocalyptic times, it only took a glimmer of improvement to raise the hopes of the underprivileged masses that the eternal defender of their rights would finally come to their rescue. According to popular belief, King Matthias never died, but sits asleep under Mt. Peca in Carinthia (the Germans' popular imagination has treated Emperor Frederick I Barbarossa in a similar manner); once his beard will have wound around the stone table for the seventh time, he will rise awakened in the darkest hour to bring redemption to the poor.

## The Peasant Uprisings

The systematic destruction of the rural economy during the Turkish invasions and Frederick III's wars led to severe economic regression in the Slovenian provinces. Since the collapse of an independent political center with the demise

---

186 Hermann Wiesflecker, *Kaiser Maximilian I. Das Reich, Österreich und Europa an der Wende zur Neuzeit. Band I. Jugend, burgundisches Erbe und Römisches Königtum bis zur Alleinherrschaft 1459–1493*, 66. St. Maximilian supposedly protected Emperor Frederick III from the worst when he was nearly captured by Jan Vitovec in 1457.

of the Celje and Gorizia dynasties, the priorities of the entire territory extending from the Alps and the Adriatic Sea were defined by the House of Habsburgs, Matthias Corvinus, and the Venetians, whose interests also targeted other regions. Gloomy conditions were further worsened by spiraling tax burdens imposed to finance the ceaseless warfare, and general uncertainty threw trade into decline. Walled cities could do little but vegetate, while the countryside was left to its own devices. In time, the subjects developed their own defensive system: from about 1460, walls were erected around churches and/or other buildings providing shelter to villages and their surrounding areas. Before the end of the Middle Ages, no less than 350 such strongholds, called *tabori* by the local inhabitants, were set up across the Slovenian ethnic territory.[187]

Peasants, with their newfound combat experience in warding off the Turks, soon grew in self-confidence. Since they were occasionally also called to arms by the provincial government (as the so-called "black army") during major onslaughts, they also seemed to have become an increasingly important factor in public life. Enterprising campaigns of individual subjects also grew: since the constant threats of war had made the growth of cities and market towns uncertain, an ever-larger share of trade was transferred into the hands of individuals.

Nonetheless, the situation for peasants was worsening, especially in comparison to their feudal lords. The progressive consolidation of the provinces in the Late Middle Ages brought about a stable structure of relations among nobles, and each crown land (Styria, Carniola, and Carinthia in 1414) obtained its own collection of privileges for the nobility.[188] This enhanced the legal status and political role of the overlords, and above all the importance of their corporations, which sought to determine formulation and functioning of the increasingly complex judicial and administrative systems of the respective provinces. Feudal loyalty to one's lord and monarch thus gradually took on a new dimension: it was no longer to be regarded as a personal commitment to an entity holding authority over a certain territory, but also contained an awareness of belonging to a community defined in terms of territory and traditions. In practice, the concept of homeland took on a very tangible political substance. The aristocracy, the driving force in the diets,[189] invested increasing efforts into extending their rights, while their

---

187  Štih and Simoniti, *Slovenska zgodovina do razsvetljenstva*, 162.
188  Sergij Vilfan, *Pravna zgodovina Slovencev od naselitve do zloma stare Jugoslavije*, Ljubljana: Slovenska matica, 1961, 201, 202.
189  Diets were established in the Late Middle Ages. In Slovenian diets, the aristocracy had the last word, simply because there were almost no notable ecclesiastical centers

significance with respect to defensive power was declining as classical feudal and knightly armies were superseded by mercenary troops.

The increasingly distressed peasants, who ever more frequently perceived the politicizing and demilitarizing nobility as a parasitic social stratum or even an anomaly of the Lord's creation, began contemplating their own alliance. The latter would not only unite them in their struggle against the Turks, but also protect the interests of their everyday life, which was extremely arduous, even though the non-agrarian economy sector was reaching its full swing (rural commerce and trade). Between 70 and 80 percent of peasants were living on the verge of the existential minimum.[190]

Severely affected by rising obligations to their overlords and by an escalating state tax burden, the subjects made persistent claims for the "old justice," a return to the traditional feudal system and taxes. But since neither the aristocracy (which was already undergoing structural changes) nor the archduke and the emperor would comply with the peasants' demands, the Slovenian provinces also entered an era of violent peasant uprisings in the latter half of the 15th century. These uprisings would later determine the history of the provinces between the Alps and the Adriatic Sea until the Enlightenment. In their own way, they would even leave their imprint on the March Revolution of 1848 (e.g., the attack on Ig Castle on March 21–22, 1848, echoed across the entire Habsburg Monarchy). These peasant uprisings—up to 180 of which took place between the Alps and the Adriatic Sea from the mid-15th to the late 16th century[191]— cannot be regarded as a specifically Slovenian feature. Nonetheless, it would be almost impossible to find a region in Europe where the outcries of discontent among serfs were equally crucial; every now and then their revolts even spilled over the Slovenian borders. Moreover, the very first printed words in the Slovene language referred to these voices: a poem written by German mercenary soldiers who were engaged in suppressing the Great Peasant Uprising in 1515, published in a special leaflet, noted the two peasant slogans "Old Justice" and "Unite, Unite, You Poor Slaves."[192]

---

of power. Cities were also relatively weak. For a certain period, peasants held the right to have their representatives in the Gorizia Diet.

190 Štih and Simoniti, *Slovenska zgodovina do razsvetljenstva*, 164.
191 Zdenko Čepič and Dušan Nećak, *Zgodovina Slovencev*, Ljubljana: Cankarjeva založba, 1979, 284, 316, 317. In the 14th century, Slovenia recorded only one revolt, in the area surrounding the town of Stična.
192 Jože Koruza, *Slovstvene študije*, Ljubljana: Filozofska fakulteta, 1991, 84, 85. The "journalistic" poem depicting the battle at Celje during the summer of 1515 was printed in Vienna shortly afterwards.

Five major uprisings occurred in Slovenia in this period. The first began in Carinthia (1478); the second spilled over to Lower Styria and Carniola (1515); the third one took place in the Sotla basin, part of the Drava and Sava basins, as well as the Croatian Zagorje region and the Kolpa basin (1573); the fourth one spanned across the Savinja Valley and the neighboring areas in the Kozjansko region, the Drava Field and Upper Carniola (1635); and the fifth one broke out in the Tolmin and Gorizia areas (1713). They involved between 3,000 (1478) and 80,000 people (1515), and lasted up to several months. They shared a resistance to the changes progressively introduced into the feudal system. Some of them even aimed at undermining the very structure of relations generated by the estate-based "division of mankind." For this reason, it is not surprising that in numerous areas the rebellious peasants also received support from miners, ironworkers, and rural tradesmen, while market towns and cities sympathized with their claims.

**Figure 27.** Peasants' uprising.

The rebels, especially at the beginning, had a mystical view of the emperor. They firmly believed that their ruler understood the pleas of the "oppressed and offended," and that he would not only appreciate their demands but also meet them. In the 15th and 16th centuries, the subjects recognized their opponents solely in individual feudal lords and, at most, provincial assemblies embodying a predominantly aristocratic corporation. As time passed, however, their illusions crumbled. In 1635 the peasants organized fierce opposition to the clergy, who fervidly supported the feudal system. In the borough of St. Jurij beneath Rifnik, they even slaughtered a parish priest because he had preached to render unto Caesar what is Caesar's, and unto God what is God's.[193] In some instances, the serfs' outrage went hand in hand with religious disillusionment. In 1515, for example, a farmer known as Klander waged an impassioned campaign across Upper Carniola, claiming that the Holy Spirit was speaking through him. He complemented the claims of his mystical mission by blessing crucifixes and the rebels' war banners.[194]

The preparations for the peasant insurrection in Carinthia, which broke out after Easter 1478, took several years. The peasants had already made an alliance immediately after the Turkish incursion over the Karavanke Mountains and the Drava River in 1473. In the end, the archduke and emperor, respectively, prohibited any such self-organization outside the established institutional framework (this prohibition also extended to the nobility, owing to the emperor's wretched experience with Baumkircher). The alliance, which initially recruited its members voluntarily and later on employed coercion, refused to abide by the order of Frederick III, who was held in lesser esteem with each passing day. Moreover, the alliance would even open its door to priests, burghers, and individual noblemen. The introduction of an equitable social system in Carinthia was placed under the jurisdiction of newly instituted courts with the final say belonging to the peasants. At the time, the rebellion did not yet call for a complete abolition of taxes to feudal lords, and refrained from manifesting its full power with fierce attacks on castles.

The anxiety of the ruler and the provincial assembly regarding quelling the revolt was finally dispelled by the Turkish incursion into Carniola and Gorizia. When the rebels learned of their approach, they resolved to confront the enemy at the entrance to the Gail Valley. Two thousand five hundred rebels gathered in the gorge at Coccau, yet the majority deserted the battlefield before the fighting commenced. This left only 600 men who persevered under the leadership of a peasant named Matjaž and tried to prevent the Turks from advancing into the heart of Carinthia at the end of June 1478, perishing almost to a man. Their heroic resistance was to no avail. In the next three weeks, the sultan's army

---

193  Bogo Grafenauer, *Kmečki upori na Slovenskem*, Ljubljana: DZS, 1962, 299.
194  Grafenauer, *Doba prve krize fevdalne družbe*, 68.

mercilessly plundered the areas where the rebel alliance was strongest. The few surviving rebels were hunted down and captured by their overlords and handed over to the provincial courts.[195]

However, the ferocious peasant uprising of 1515 was much more dangerous and far-reaching. General living conditions remained dire, although the situation had slightly improved under Frederick's successor Maximilian I. From 1491 onwards, the Turks renewed their incursions in Carniola and Styria (albeit not as fiercely as under Sultan Mehmed II the Conqueror). At the same time, the Slovenian Littoral became the scene of a bitter war between the Habsburgs and the Venetian Republic. Although the operations were most intense at the very outset, the armed engagements inflicted new wounds on the Gorizia, Tolmin, and western Carniola areas and the city of Trieste: understandably, wreaking havoc across enemy territory was a fundamental strategic objective in medieval warfare. The war stemmed from a combination of material political interests and prestige. The new King of the Romans, Maximilian, who had successfully restored Habsburg rule over the lands and provinces along the Hungarian border, also received the title of emperor in 1508. Venice, anxious about the rising strength of the Habsburg monarch at its eastern and northern borders, prevented his visit to Rome to attend his own investiture. The Venetian decision was partially due to indignation that Maximilian had taken possession of the entire legacy of the deceased Counts of Gorizia.

The course of the war initially seemed uncertain for the emperor: as soon as hostilities broke out, the Venetians seized the Soča Valley, Trieste, Vipava, and Rijeka. By 1509, however, Maximilian's army had pushed them to the edge of the Friulian plain, where the border was subsequently settled after the cessation of military operations in 1516 and the peace of 1522.[196] The Habsburgs then obtained Aquileia and the Tolmin area, which was annexed to their County of Gorizia. Their subjects could not have been particularly enraptured with their rulers' successes, which had only been made possible through their own contributions in the form of state taxes. The feudal lords also aspired to draw maximum profit from the "state of emergency" within a broader rear of the war zone so as to improve their financial position, while the emperor could only implore his patricians not to stand in the way of his war with the external enemy. The peasants, on the other hand felt the increased burden and started to form their own alliance.

---

195   Ibid., 56–59.
196   Ibid., 65, 66.

The call for the "old justice" voiced at the beginning of 1515 resounded from southern Carinthia and around Graz all the way to the Kolpa and Soča basins. Even though the uprising, with its core in Carniola and Lower Styria, also reached several German-speaking areas in the north, it was universally recognized as genuinely Slovenian from its very outset. In 1515, the rebellious subjects, united in a vast but disunited peasant alliance, attacked and seized control of numerous castles. The rebels, calling themselves "the poor," would occasionally not even hesitate to lay their hands on the emperor's property. The movement embraced great masses of followers from the start: in Styria alone, its leadership numbered no less than about 300 members. Moreover, the peasants were well organized in several areas: in Carinthia, for instance, they not only elected their leaders and military commanders, but also designated "orators" and "counselors" (who obviously saw to the morale and effective planning of future actions). The most hated overlords and representatives of the privileged elite, who blighted their subjects' lives by increasing taxes and servitude, found themselves in mortal danger, and a considerable number of them paid with their lives for their greed and authoritativeness. At the height of the peasant uprising, the patricians, whose insufficiently fortified castles could not protect them from a persistent siege, were forced to seek shelter in fortified cities—in Carniola's Ljubljana and Kamnik, in Styria's Maribor, and in Carinthia's Villach.[197]

The peasants did not need long to make their demands clear. Once their alliance had been established early in 1515, they embarked on negotiations with the authorities in the respective provinces and even sent a special delegation to the emperor in Augsburg. Yet they took up arms again almost at once: the first victims had fallen in March of that same year. The growth and radicalization of the peasants' outrage was essentially a matter of chance: when three suns rose in the sky of iridescent colors on February 10, 1515, unlearned peasants interpreted the extraordinary phenomenon as a herald of stark times ahead, prompting them to react swiftly. Their agitation was soon transformed into the first actions against their overlords.[198] Both the rebels and the nobles were inflamed by the fall of the imposing Mehovo Castle in Lower Carniola on May 17. This victory also dramatically demonstrated the symbolic side of this enduring struggle for the "old justice": the rebels beheaded both masters of the castle and threw their bodies over the castle walls, while capturing the lady of the house, dressing her in peasant rags, and sending her to work in the fields.

197 Štih and Simoniti, *Slovenska zgodovina do razsvetljenstva*, 177.
198 Mirko Rupel, *Primož Trubar: Življenje in delo*, Ljubljana: Mladinska knjiga, 1962, 12.

While the first Carniolan seigneuries were falling to the rebels, the delegation of Slovenian subjects had reached the emperor. In contrast to the provincial nobility and its sinister reminders of the bloody end of the peasant revolts in Hungary and Württemberg the previous year, the ruler was not arrogant toward the representatives of "the poor," and a peaceful solution seemed to be within arm's reach. Yet the imperial court had long since lost its clear grasp of the developments. On the one hand, the king had prohibited the establishment of the peasant alliance (even prior to the arrival of their representatives in Augsburg), but on the other hand, he did not voice any unequivocal support to the nobility. Emperor Maximilian, who aimed to centralize the administration of the hereditary lands of the Habsburgs, was not particularly pleased about the excessive concentration of power in the hands of the provincial assemblies. Therefore, he sought to reaffirm himself through his commissioners not only as an arbiter but also as the sole effective authority in the respective crown lands. However, the intervention by the ruler's representatives proved to be an utter failure with both the nobility and the peasants. In the end Maximilian clearly expressed his support for the aristocracy, which by the end of spring had already fully raised its own army.[199]

Despite protracted and carefully designed preliminaries, the first blow aimed at the rebels missed its target. In the middle of June, the nobility and their troops (including armed Croats coming to their aid from the Hungarian Kingdom) found themselves trapped at the town of Brežice. They were encircled by approximately 9,000 peasants, who then took control of the town in a single attack. The unscrupulous feudal lords—who had wanted to sell 500 wives and children of the rebels to Dalmatia as slaves—were killed. Only then did the advantage shift to the nobility. In Carinthia, both provincial and imperial troops which had gathered in Villach started to crush the revolt. An increasing number of towns were coerced to abandon the peasant alliance, and the unity of the rebels was already faltering. The vortex of the battle was then shifted toward the east. Jurij Herberstein, who had rallied his army of Styrian, Carinthian, and Carniolan provincial assemblies around Graz, headed south, swiftly advancing into the rebel territory. On July 5, his camp in Celje was raided by the peasants. The onslaught turned into a vicious battle that lasted until July 10, when the rebels ultimately suffered a calamitous loss. Herberstein's communication to the emperor indicated that the battle had taken the lives of no less than 2,000

---

199  Grafenauer, *Kmečki upori na Slovenskem*, 100–125.

insubordinate subjects.[200] Two weeks later, the army reached Carniola, receiving a jubilant welcome from the nobility while making it clear to the peasants that they would never be able to match its force.

The defeated rebels were brutally punished, while the anti-Turkish forts (*tabori*), which had also evidently been used in their struggle for the "old justice," were razed to the ground. The provinces through which Herberstein's army passed were not much different from those that had suffered the fiercest Turkish onslaughts several decades earlier. Many rebels were hanged or impaled, while the luckier ones were merely beaten up and robbed by "peacemakers." The nobility's vengeful justice also provided for the subsequent executions of the peasants in Kranj and Graz (161 rebels were condemned to death in the Styrian capital alone). The all-Slovenian peasant uprising of 1515, which inspired the great peasant revolt across the German provinces of the Holy Roman Empire in 1524, remained in Slovenian memory long thereafter through special taxes, such as the so-called "rebellion pfennig."[201]

---

200 Bogo Grafenauer, *Boj za staro pravdo v 15. in 16. stoletju na Slovenskem. Slovenski kmečki upor 1515 in hrvaško-slovenski kmečki upor 1572/73 s posebnim ozirom na razvoj programa slovenskih puntarjev med 1473 in 1573*, Ljubljana: DZS, 1974, 101, 102.
201 Ibid., 109.

# The Early Modern Period

# From Humanism to Reformation

The Late Middle Ages may have been the most exhausting period for the Slovenian territory since the migration of peoples, but it nevertheless planted the first seeds of future progress. Emperor Frederick III could not calm his anxiety over retaining the estates on the southeastern flanks of the Holy Roman Empire even by assuming possession of the Celeian heritage. In 1461, he founded the Ljubljana Diocese, mainly to restrict the ecclesiastical impact of the Patriarch of Aquileia, who depended on Venice and had been residing in Cividale del Friuli and Udine ever since the outbreak of the first war between La Serenissima and the Habsburgs.[202] The existence of the new diocese was affirmed by the emperor's former secretary, Aeneas Silvius Piccolomini, in 1462. Initially, the diocese was absurdly small: while it was undoubtedly allotted no more than three parishes, thirteen others were designated merely as sources of income for sustaining its dignitaries. This led to constant wrangling with the Aquileian authorities. Nonetheless, the bishops of Ljubljana were soon elevated to princes in 1533, and before the beginning of the 17th century they were also able to impose their authority over the entire territory, which one way or another was continuously mentioned in relation to the establishment of their diocese. Moreover, striving to settle the controversies in the late medieval Church (the decline of the "Council" party and the ascendancy of the "Papal" party), Emperor Frederick managed to attain the right to appoint bishops of Trieste, further entrenching his position along the northern Adriatic coast.

The Aquileian patriarchs succeeding Louis II of Teck had almost entirely lost their influence over developments in the eastern part of their ecclesiastical province, which strengthened the significance of smaller religious centers in Slovenia. Still, the disputes between the emperor and the Venetians not only fortified the border between Friuli and the Habsburg hereditary lands but also erected a high spiritual barrier. Similarly, with the demise of the Counts of Gorizia (1500), the two rival powers were no longer separated by a neutral zone. In the 16th and 17th centuries, the border areas between the Habsburg lands and the Venetian territories were the scene of two bitter wars between 1508–1522 and 1615–1617. The Venetian Slovenes, who inhabited the mountainous area surrounding the city of Cividale, were crucial to the republic as defenders

---

202    Štih and Simoniti, *Slovenska zgodovina do razsvetljenstva*, 169.

of the border (which was at last settled along the mountain peaks west of the Soča River). They therefore enjoyed a considerable degree of autonomy. Their administrative system consisted of two mayoralties with 36 villages. The adjacent localities were governed by elected representatives (deans). The two mayors were even granted the right to exercise blood feud.[203]

These developments considerably hindered the sort of cultural interrelations that were fostered among Italian seigneuries, where the Renaissance civilization flourished during the 15th century. On his visit to Ljubljana in 1444, Aeneas Silvius Piccolomini, who in many respects embodied the ideal humanist scholar, felt as if he had found himself in the middle of "barbaric and uncivilized" provinces.[204] The sudden ruin of the Celeian House and the downfall of the Counts of Gorizia further reduced the chances of forming any noteworthy cultural centers that would align the Slovenes' intellectual endeavors with contemporary shifts in Europe's "republic of spirits." Even the formation of the Ljubljana Diocese failed to bring forth any substantial initiatives, as its overpastors too often meddled in purely political matters: for instance, Christophorus Raubar, who presided over the Ljubljana Diocese between 1488 and 1536, was not only a grand diplomat and successful warlord but also Lord Chamberlain. The Habsburg dynastic policy, which during the reign of "the Last Knight" Maximilian I (1486/93–1519) started to transform into a (proto-)absolutist state through establishing common bodies and institutions, simply had no regard or concern for the specific needs of the Alpine-Adriatic region. The centralization and concentration of administrative authority strictly took place according to the interests and priorities dictated by the imperial court. Thus in the Late Middle Ages and at the beginning of the 16th century, only a few Slovenian personalities could win esteem and recognition for their achievements. An appreciable number of them even became established in various political and cultural centers across Europe.

Besides Raubar, another man of distinction was Thomas Prelokar from Celje, who was awarded his doctoral degree in Padua in 1466 and later became Chancellor of the University of Vienna. As a diplomat of Frederick III and tutor of the Emperor Maximilian, he was highly cherished for his merits by the ruling Habsburg dynasty. No less prosperous was his episcopacy in Constance. The careers of his close countrymen from Slovenian Styria, Brikcij, Preprost and Bernard Perger were nearly as brilliant, and both were appointed deans of the

---

203  Čepič and Nećak, *Zgodovina Slovencev*, 326, 327.
204  Alfonz Gspan et al., *Zgodovina slovenskega slovstva I*, Ljubljana: Slovenska matica, 1956, 168.

University of Vienna. The former won glory as a scholiast on Cicero's writings, while the latter embarked on reforming the teaching of Latin in the humanist spirit. At the beginning of the 16th century, the imperial court of Maximilian I recognized the orator, diplomat, and Chancellor of Vienna University, Paulus Oberstain from Radovljica; and Jurij Slatkonja from Ljubljana, who subsequently became the first true Bishop of Vienna. During the preparations for the emperor's meeting with both Jagiełłonian kings, Vladislaus II (King of Hungary and Bohemia) and Sigismund Senior (King of Poland and Lithuania), in July 1515, the imperial court in Vienna even expressed interest in the Slovene language. Oberstain, a brilliant polyglot, composed a Latin ode in praise of Maximilian two years before the meeting of "the great three," in which he stressed that the emperor could learn the Slavic language from him.[205]

The conditions that brought about the successful Habsburg-Jagiełłonian governmental "congress" in Vienna[206] apparently influenced the first notable affirmation of the Slovene language since the ritual of installing Carinthian dukes ended in 1414. The Habsburgs, having secured their position in the lands of the Spanish crown once Philip the Handsome had married Joan the Mad, now also aspired to exert a decisive influence over the Slavic East. It was not by chance that Emperor Maximilian surrounded himself with people from the Slovenian territory,[207] and they were also the winning bid in his dealings with Moscow, which he saw as an ally in any possible disputes with the Jagiełłonian rulers in Bohemia, Hungary, Poland, and Lithuania. In 1516, Maximilian sent a Vipava-born patrician, Žiga Herberstein, to Vasili III, Grand Duke of Moscow (his first choice for an envoy, Christophorus Raubar, Bishop of Ljubljana, had refused to set out on the long journey). Herberstein completed the mission successfully, undoubtedly also because of his fluency in the Slovene language.[208] The imperial

---

205 Primož Simoniti, *Humanizem na Slovenskem in slovenski humanisti do srede 16. stoletja*, Ljubljana: Slovenska matica, 1979, 193. In his ode, Oberstain speaks of how Maximilian ordered him to compile a dictionary of the Slavic language; it was believed that the Slavic language was only one.
206 The Habsburgs and Hungarian-Bohemian Jagiełłons then made a double family bond and concluded an inheritance agreement. In the wake of the Battle of Mohács on August 29, 1526, Emperor Maximilian's successors greatly profited from this gesture, since they could take hold of the Bohemian Kingdom as well as several comitati within the Hungarian crown lands.
207 By the late 15th and early 16th centuries, the Slovenes were the only Slavic nation under Habsburg rule. Only after the ignominious death of Louis II in 1526 did the dynasty also exert dominion over Croats, Bohemians, and Slovaks.
208 Žiga Herberstein, *Moskovski zapiski*, Ljubljana: DZS, 1951, 5.

court's new requirements overcame the general prejudice against living languages, which started to find a place alongside humanist Latin throughout Central Europe.[209] It was no wonder that when Petrus Bonomo, Bishop of Trieste and a longstanding close companion of Maximilian, returned to his episcopal see in 1523, he began interpreting classical texts in Slovene (rather than exclusively in German or Italian). He did not share the negative attitude of many rural nobles toward the language. (Žiga Herberstein also reported in his autobiography that several peers in the late 15th century had scorned him for learning Slovene.[210]) Bonomo's vast humanist scholarship and devotion to the Slovene language also inspired the Protestant reformer, Primož Trubar, whose character was shaped within the Bishop of Trieste's circle.

Maximilian I's centralizing policy paved the way for the rise of individual Slovenian courtiers and also helped reduce the significance of provincial borders. Therefore, the onset of the Reformation in the Alpine-Adriatic region enabled the introduction of an integrationist language approach that to a large extent exceeded the legacy of politically motivated territorial division. Maximilian, still fascinated by long-gone chivalric ideals and striving to place the Habsburg dynasty atop a pyramidal worldly monarchy through a deliberate policy of arranged marriages, actually accelerated administrative modernization in Slovenia and neighboring areas. His visions may have been anachronistic, but his statesmanship laid the foundations for the later Austrian Monarchy. To illustrate, he resolved to transform the estate-dominated Habsburg hereditary lands into as coherent a political and territorial unit as possible. Central offices (a financial chamber and a regiment) were established for the Upper and Lower Austrian crown lands, with the Slovenian provinces belonging to Lower Austria. The Innsbruck libels were issued in 1518 after a joint session of the representatives of the Habsburg crown provincial diets, and enhanced the military links between them in particular.[211]

Cooperation between the provinces in the Alpine-Adriatic territory was also expanded. The major threat posed by Turkish expansionism further encouraged the joint defensive efforts of Carniola, Styria, and Carinthia (and also Gorizia

---

209 Upon receiving nobility, Jurij Slatkonja obtained a coat of arms with an emblem of a golden horse. This means that Maximilian's imperial court had semantically derived his family name from the Slovene language (according to the principle of folk etymology).
210 Ludovik Modest Golia, "Herbersteinovo življenje," in: Herberstein, *Moskovski zapiski*, 188, 189.
211 Wiesflecker, *Kaiser Maximilian I*, 300.

after 1500). Inner Austria, comprising the above listed hereditary crown lands, became a genuinely "indivisible body" in the 16th century, i.e., during the reigns of Emperor Ferdinand I and Archduke Charles.[212] Following the death of Louis II in 1526, the last Jagiełłonian King of Bohemia and Hungary, Inner Austria also encompassed western Hungary and *reliquiae reliquiarum*, or "the remnants of the remnants" of Croatia. These lands found it increasingly difficult to withstand the Turkish onslaught. The cities and provinces in the Pannonian Basin, Slavonia, and Dalmatia fell one after another to Suleiman the Magnificent, who had succeeded to the Constantinople throne in 1520, and over the next decade (1522–1532) the Inner Austrian provinces feared the return of disasters from the times of Mehmed II the Conqueror. This period also marked the only incursion of the main Ottoman army onto Slovenian soil, in 1532.[213]

The Habsburgs' major military operations against the Turkish forces did not yield the desired results until the very end of the 16th century. The Inner Austrian provinces were particularly badly affected by military disasters across Slavonia in 1537. The Habsburg armed forces, led by several aristocrats, including Ivan Kacijanar, Governor of Carniola (the nephew of Žiga Herberstein and the brother of the Bishop of Ljubljana), were wiped out almost to the last man. Even though Kacijanar was a battle-hardened warrior who had steadfastly defended Vienna from Suleiman's invading army, his attempt to repel the invaders from the city of Osijek (Hun. Eszék) was crushed. After his flight from the battlefield, court circles accused him of high treason and he was eventually murdered as a potential instigator of the anti-Habsburg opposition from Carniola and neighboring areas.[214] The poorly coordinated, overly impromptu, and, above all, futile defensive efforts shattered the morale across the Inner Austrian provinces, to the benefit of the advancing Ottomans, and a myth of the sultan's invincibility was built up. Moreover, until the late 17th century, much of Central Europe lived in the belief that the grass never grew back where Turkish hooves had trod.

Even before the Habsburgs had assumed the Hungarian and Bohemian thrones, the Inner Austrian provincial assembly concluded that it would be better to defy the Ottoman invaders outside their own territory. They had been

---

212  Štih and Simoniti, *Slovenska zgodovina do razsvetljenstva*, 184.
213  The sultan's retreating army, which aspired to conquer Vienna in 1532, passed through Slovenian Styria after its defeat at Kőszeg (Slov. *Kisek*). The little town in western Hungary was saved by the Governor of Carniola, Nikolaj Jurišič, who was born in the Croatian city of Senj.
214  Ivan Kacijanar, born at the Upper Carniolan castle of Kamen (1491/1492–1539), was appointed joint commander-in-chief of the Inner Austrian armed forces in 1531.

taking over individual fortresses in the Hungarian crown lands since 1522; by 1578, they had seized 88 of them. Since then, almost 5,000 soldiers had been permanently stationed in them as garrisons and mobile units. Between 1537 and 1578, defense costs rose almost sixfold in Carniola and more than fivefold in Styria, though only 250 percent in Carinthia. By 1613, the Inner Austrian provinces had invested nearly 25 million florins into the maintenance of fortresses and defense organization in Croatia, while the Imperial Treasury contributed roughly eight times less.[215] Since there were no significant sources of income across the entire Alpine-Adriatic region, except the mercury mine in Idrija (It. Idria), which started operating at the end of the 15th century, tax burdens grew dramatically. Ecclesiastical institutions were willing to sacrifice a portion of their wealth to the cause as well. After 1579, the lion's share of the funds was invested into the construction of the imposing fortress in Karlovac, which occupied the key position at the confluence of the Korana and Kolpa rivers (the westernmost point of the Turkish Empire) and guarded the passage from Bosnia to Carniola. The new urban settlement, encircled by state-of-the-art fortifications, proved to have played its defensive role well: after 1579, the sultan's cavalry stopped raiding central Slovenian territory.

However, improving the defensive capacity of the Inner Austrian provinces and the adjacent provinces in Croatia required a reversal of depopulation as well as the enhancement of fortifications in the relatively wide border area. The war-torn land could no longer sustain the military garrisons and mobile units, so the Habsburg rulers and their commanders began to welcome the (predominantly) Orthodox Christian population from the inland Balkans which had fled westwards from the Islamic conquerors. The fugitives were mainly stockbreeders struggling to escape from the long arm of the sultan's authority. In the Croatian and Slovenian provinces, these settlers were known as Uskoks. In 1535, Ferdinand I granted them freedom and land; in return they were expected to perform compulsory military service. The region of Žumberk in easternmost Carniola was so overcrowded by the new settlers that its ethnic composition completely changed, although in the neighboring White Carniola and elsewhere the immigrants assimilated into the general population.[216]

A specific territorial unit outside the administrative structure of the Habsburg kingdoms and crown lands was therefore gradually formed in the 16th century along the western border with the Turkish Empire. By the end of Ferdinand

---

215   Štih and Simoniti, *Slovenska zgodovina do razsvetljenstva*, 183.
216   Ibid., 180–184.

I's reign, this so-called Military Frontier, which in Croatia extended from the Adriatic Sea to the Drava River, stood as an effective bulwark against the Turks for the southern flanks of the Habsburg hereditary lands. Due to its primarily defensive function, the frontier was under direct control of the imperial court in Vienna and the archducal court in Graz. Carniola therefore gradually lost contact with Žumberk, which, upon the final abolition of the Military Frontier in 1881, was annexed to Croatia within Hungary. The Uskoks, who were exempt from feudal taxes, swore unflagging allegiance to the emperor and thus also played a crucial role in internal affairs. Also remarkable was their loyalty to military commanders on the Military Frontier, who came mainly from the Inner Austrian nobility, and rebellious peasants consistently failed to win them over to their cause even in the late 16th century. On the contrary, during the great Slovenian–Croatian uprising in winter 1537, the Uskoks were decisive in swiftly suppressing the revolt.[217]

The major external threat, like at the end of the Middle Ages, left a unique mark on developments across Carniola, Styria, Carinthia, and Gorizia in the 16th century. It not only influenced their mutual integration, necessary to enhance their defense, but also affected the adoption and fate of Reformation initiatives in an extremely geo-strategically volatile zone. The proximity of the Turkish border forced the devoutly Catholic duke of Inner Austria, who had held the right to determine the faith in his crown land since the Religious Peace of Augsburg in 1555, to concede to Lutheran ecclesiastical organization. Had he opted for less compromising religious and political measures, he would most probably have ignited a major upheaval in the already tenuous balance across Inner Austria and fatally undermined its defensive power.

The dire situation of the Church was matched elsewhere across the Holy Roman Empire in the 15th century. The collapse of the Aquileian patriarch's temporal power was immediately followed by the first major and protracted crisis in which the Church could hardly control its priests, discipline was lax, and clerical celibacy and asceticism were no longer synonymous. Profane practices, which spread among individual servants of the Holy Church before the mid-15th century, indicated the profound crisis that gripped the European West, and the general conditions had worsened since the "marathon" Council in Basel (1431–1449) failed to settle the religious disputes within the Catholic "oecumene." A number of ordained men, including in the Slovenian provinces, took concubines, opened taverns, played dice, and hunted. The Holy Sacraments

---

217  Čepič and Nećak, *Zgodovina Slovencev*, 284.

were administered only on exceptional occasions and even then strictly for payment.[218] Christian rigidity was rapidly fading from the spiritual map of Central Europe. In certain places even monasticism greatly distanced itself from the ideals of the early- and high-medieval *vita contemplativa*. Intellectual decline had already become widespread among monastic communities by the end of the 15th century, and the Alsatian Nicholas Kempf (1397–1497), who resided in Pleterje and Jurklošter, was the last notable medieval monastic writer in the Slovenian provinces. Within the centuries-old walls of venerable monasteries, shameless and overtly profane acts were often performed. For instance, in his itinerary covering 1485–1487, Paolo Santonino depicted the reception given to Pietro Carli, Bishop of Caorli and visitant to the Aquileian patriarch, by the Dominican nuns of the Studenice monastery in Lower Styria:

> The Mother Superior and her sisters gave us a kind-hearted reception [that surpassed] any description. His Eminence Bishop [...] all dripping with sweat from the journey, was immediately presented with a long silk-and-gold collared robe reaching to his ankles. He clothed himself with eagerness and thus attired much resembled a German primate [...]. Then the Bishop was led to a rather secluded corner of the monastery. There, in the presence of the entire chapter of nuns, the youngest and fairest washed his hair, while another covered and wrapped his head in a warmed garment. And the master of the monastery himself poured water over the honorable guest's body. Our Bishop patiently endured all these services bestowed on him: and yet, who in heaven could refuse such graciousness from virgins of beauty untold? Nonetheless, this never led to any form of indecorum, at least not in deed; while the spirit and desire were most probably brought to the very brink of the unbearable.[219]

However, conditions were downright anarchic in the Velesovo monastery in Carniola, which also belonged to the Dominican order. According to Santonino:

> The Most Reverend Bishop was delivering sermons to the Mother Superior and other nuns on many teachings. And when it came to his knowledge that one of them had broken her obedience to that same Mother Superior, she was ordered to confess her transgression and guilt [...]. Thence, [the nuns] were questioned on their observance of the commandments of the order, management of spiritual and temporal goods, disposal of revenues, performance of God's service and many other suchlike matters. The nuns then admitted that they would every now and again interrupt their reclusion by leaving the monastery to visit their relatives, that nearly all owned personal property and that, furthermore, they would neither dine nor sup together in the refectory. The Reverend Bishop admonished them with comforting and kind words to restrain from committing

---

218 Milko Kos, *Zgodovina Slovencev od naselitve do petnajstega stoletja*, Ljubljana: Slovenska matica, 1955, 353.
219 Santonino, *Popotni dnevnik*, 80, 81.

such breaches in the future and leading men into the monastery. For carrying out the Divine Service in their church, the reverend nuns have four chaplains, young, may I add, who dwell outside the monastery, yet near enough.[220]

The situation did not improve at the beginning of the 16th century; indeed, the ordained servants of the Church started to accumulate more and more services and functions. Priests, requested to celebrate more Divine Services, developed a habit of sending ill-educated vicars to give mass in their stead. Ecclesiastical princes had their hands full with worldly matters: Christophorus Raubar, Bishop of Ljubljana (having been elected overpastor at the age of 12), also served as provincial governor of Carniola (1529–1536) and vice-regent in Lower Austria (1532–1536). Petrus Bonomo, who became chancellor of the Imperial Council of the Lower Austrian provinces in 1521 after the death of Emperor Maximilian I, had an equally brilliant temporal career. Bonomo, who had been named Bishop of Trieste with the Pope's blessings in early April 1502, was only able to devote his undivided efforts to the diocese two decades later.

The great profanation of the Church, together with frequent Turkish raids, which ordinary people perceived as God's punishment for their sins, profoundly shook their faith in the worldly order, and Martin Luther's cry to return to the original Christianity—i.e., "the Holy Gospel without human additions"— echoed among Slovenes. Luther, who resolutely opposed the sordid trade in "indulgences" in 1517, won widespread public acclaim among the nobility, the priesthood, burghers, and peasants, although for very different reasons. What became evident to all was that the Renaissance Church, increasingly reminiscent of a vast mercantile establishment trading in man's faith, hopes, and fears, could suit the requirements of only a few.

In an era of everyday distress, worries, and insecurities, as well as eschatological angst, Luther's adherents understandably held a monopoly on criticizing the current state of affairs. Ordinary people, too, were expressing their discontent with religious and everyday life, through a profound desire for pilgrimages and the building of new churches, although this was met with fierce opposition from Luther's supporters. Moreover, even those priests who had not joined in Luther's stark criticism of the papacy would sometimes find themselves in conflict with lay zealots devoted to building Houses of God (the so-called *štiftarji*). Another key role appears to have been played by the rising number of Anabaptists, whose center was in Münster, Westphalia (until the city fell into Catholic hands in June 1535): Styrian prisons were already full of Anabaptists in 1530, except the

---

220   Ibid., 42.

prisons of Celje County, which had still not been completely incorporated into the unified administrative structure of the province. The Anabaptists, spreading southwards, reached Carniola somewhat later when they settled in Ljubljana and Kamnik, while Styria recorded their appearance in Klagenfurt, Villach, and several other cities in the entirely German-speaking part of the crown land.[221]

Unlike the vocal church-builders—who at least stirred up unease among the Catholic hierarchy—and the "heretical" Anabaptists, the Lutherans at first refrained from openly revealing their doctrine. Their first successes in the southern part of the Inner Austrian provinces, in the towns of Villach, Slovenj Gradec, and Radgona, were recorded even before 1530. In several places, their protest against the situation in the Church began as a culmination of humanist considerations about religion. The Bishop of Trieste, Petrus Bonomo, who firmly supported the papacy against Luther, gradually began to change his views, and late in his life discovered fresh ideas in the writings of Erasmus and in Calvin's book *Christianae Religionis institutio* (1536). Jurij Slatkonja from Ljubljana, installed in the Viennese episcopal see, came to the same conclusion, and shortly before his death in spring 1522 allowed the followers of the new religious concepts to deliver sermons at St. Stephen's Church. The Roman Catholic Church, reeling under Lutheran criticism, was losing ground in a rapidly growing number of provinces roused by the Protestant thought.

One of these provinces was Carniola, where a group of Lutherans began to form in scholarly theological circles. In the absence of Bishop Raubar (who was said to have received the Eucharist in the Protestant spirit on his deathbed under both bread and wine), the three canons of the Ljubljana Diocese (vicar general Lenart Mertlic, Pavel Wiener, and Jurij Dragolič) came very close to the Reformation ideal of a "pure, simple and faithful" proclamation of the Holy Gospel.[222] In a similar vein, several laymen—particularly the influential patrician Žiga of Višnja Gora and the state scribe Matija Klombner—had realized by 1530 that Protestantism provided the only answer to the challenges of the era. Clearly, the Lutheran circle in Ljubljana was theologically distinguished: for instance, the Kranj-born Pavel Wiener was installed as rector of the parish in Sibiu (Ger. Hermannstadt), and in February 1553 he was even appointed the first superintendent of the Protestant Church in Transylvania.

---

221  Štih and Simoniti, *Slovenska zgodovina do razsvetljenstva*, 195.
222  Rupel, *Primož Trubar*, 54–58.

The members of the Inner Austrian provincial assembly, who held more or less regular local diets (from 1412 in Styria and from 1431 in Carniola),[223] generally supported the Protestant principles. In Slovenia, Lutheranism largely won support of the aristocracy (although it was also close to the townspeople). To many, a modest Church that would not trade in salvation seemed to embody the word of God far more convincingly than the Catholicism of the extravagant Renaissance popes. On the other hand, the Habsburg ruler Ferdinand I, Maximilian's second imperial grandson, refused to renounce his loyalty to Rome and strove to prevent the spread of Protestant ideas. Even so, political realism restrained him from the most severe actions against the Protestants: he could not risk an overt attack on them because the Turks, who tolerated various Christian confessions in their own as well as in vassal territories (e.g., Transylvania), posed too serious a threat to his crown lands. Nor had Maximilian's centralizing reforms broken the will and power of the aristocracy in individual crown lands to (co-)decide on political matters. Ferdinand's imperial court was thus left with no other alternative than to prepare for an agreement.

Under this agreement, the Lutherans would recognize the primacy of the Pope, while Rome would allow the Eucharist to be administered under the appearances of both bread and wine (as was already done in the case of Czech Utraquists). Two bishops of Ljubljana, Franc Kacijaner (1536–1543) and Petrus Seebach (1558–1568), who refused to fan the flames in a volatile period, actually followed the emperor's policy with considerable loyalty. This significantly facilitated the mission of the most important Slovenian reformer, Primož Trubar, who had been appointed vicar and preacher in the Carniolan capital under Bishop Raubar, presumably in 1533. In Trieste, where he received the "education of the heart" in the humanist spirit of Bishop Bonomo and where he was also ordained, Trubar became convinced that every Christian should, first and foremost, follow "the Holy Gospel without human additions." For him, the tradition of the Church could no longer provide unfaltering guidance for the pursuit of religious life.

## Primož Trubar and the Reformation

Primož Trubar was born in the village of Raščica near Velike Lašče in 1508. As a talented young man, he went to school in the city of Rijeka, which then still belonged to the crown land of Carniola. He later traveled to Salzburg, where

---

223   Kos, *Zgodovina Slovencev*, 345. After 1453, the land estates also started to hold sessions in the "General Provincial Diet" for Styria, Carniola, and Carinthia.

he became acquainted with Luther's teacher, Johann von Staupitz. Trubar was already Bishop Bonomo's protégé in May 1528, when he enrolled at the University of Vienna; he broke off his studies the next year due to the Turkish advance toward the region, and retreated to his native country. Bonomo later ordained him a priest and secured him a decent income. Trubar soon became renowned for his stark opposition to the widespread building of churches and worship of saints, which was enough to make him a Protestant in the eyes of many. In the wake of the denunciation presented to the Provincial Governor Nikolaj Jurišič in 1540, Trubar returned that same year to the episcopal court in Trieste, which became a gathering place for Protestants from both Italian seigneuries and the Habsburg lands.[224]

During this time, Trubar had developed his own views on the fundamental issues of religious doctrines and practices, and advocated them passionately until he was ultimately exiled from his native country in the middle of 1565. A master of Latin, German, and Italian, he could draw valuable knowledge from reading and discussion with humanists and reformers, and built a vast overview of current developments across Christendom. His work attracted increasing attention, and he became known as one of the most vocal supporters of Lutheranism along the northern Adriatic coast. Petrus Paulus Vergerius the Younger, the Bishop of Koper, who had once met Luther as the Papal Nuncio in 1535, even brought charges of heresy against Trubar—but only to divert attention from his own secret attempt to convert to Protestantism. Nonetheless, Trubar was appointed Canon of Ljubljana in 1542 and chosen as Bishop Franc Kacijaner's personal confessor. However, the situation in Carniola became unbearable for Trubar when Kacijaner was succeeded by Urbanus Textor (1543–1558), who resolved to insist on strict adherence to the first decisions of the Council of Trent. With a warrant issued for his arrest, Trubar finally decided to seek refuge in the predominantly Protestant North of the Holy Roman Empire in 1548.[225]

However, Trubar did not forget his native country despite living in Germany. Once appointed a preacher in Rothenburg (upon the recommendation of Martin Luther's and Philipp Melanchthon's mutual friend Veit Dietrich), he started to compile the first two Slovene books, *Catechismus* and *Abecedarium*, printed in Tübingen at the end of 1550. Later he also undertook to complete the Slovene translation of the New Testament, published in several parts (1555–1577; republished in 1582), and of the Old Testament Psalms (1566). He also arranged for the printing of several songbooks, a calendar, and dogmatic writings, of

---

224 Gspan, *Zgodovina slovenskega slovstva I*, 208.
225 Rupel, *Primož Trubar*, 58–60.

which the most extensive was the translation of Luther's *House Postil* (published in 1595), which he finished on his deathbed. By 1561, at the invitation of the Carniolan provincial assembly, Trubar triumphantly returned to Ljubljana as the first superintendent of the Protestant Church in his native country. The much broader objective of his mission, bringing Lutheranism to all his fellow countrymen,[226] took him on a grand tour of the western Slovenian provinces, which echoed deep into Friuli (in Gorizia he also delivered his sermons in Italian). When visiting Križ, where "the entire [...] Vipava Valley had rallied, with many priests among the crowd," he made an especially strong impression on the common people by entering the town riding on a little donkey.[227] After the Slovene publication of the *Cerkovna ordninga* (Church ordinance) in 1564, Trubar was permanently banished from the country by Archduke Charles, who had succeeded his late father, Ferdinand I, in the Inner Austrian provinces, since the right to issue such a document could only be exercised by a prince.

Trubar moved to Württemberg, but never relinquished his "fateful book," and even in 1575 was still guiding the Carniolan and Carinthian Protestants in understanding and adhering to its provisions.[228] For the rest of his life, he placed himself in charge of Slovene printing and the education of young fellow countrymen who came to study at the University of Tübingen. His greatest protégé was Jurij Dalmatin (c. 1547–1589), who translated the entire Bible into Slovene and started publishing it, part by part, in 1575. With the financial support of the Carniolan, Styrian, and Carinthian provincial assemblies, the complete translation was printed in Wittenberg (1583–1584) in an edition of 1,500 copies—an astonishing number for the Slovenian nation of half a million people. Dalmatin's *opus perfectum* remained in use by the Ljubljana Diocese and its counterparts across Carniola, Styria, and Carinthia for the next 200 years (until the publication of the first Catholic translation of the Holy Bible between 1784 and 1802), and played a central role in furthering literary continuity between the Reformation and the Counter-Reformation and Baroque.[229] The major achievement of Primož Trubar and his followers, who had anchored their native

---

226 Regardless of the political borders, Trubar saw the Slovenian territory as a unit. In 1555, he already mentioned the "Slovenian land" where he had preached until 1547. This means that the characterization pertained to the whole domain inhabited by Slovenes, since Trubar worked in Carniola, as well as in Styria and Trieste.
227 Jože Rajhman, *Pisma Primoža Trubarja*, Ljubljana: Slovenian Academy of Sciences and Arts, 1986, 173.
228 Primož Trubar, *Zbrana dela Primoža Trubarja*, vol. 2, Ljubljana: Rokus, 2003, 326, 327.
229 Gspan, *Zgodovina slovenskega slovstva I*, 240–143.

language within the cultural map of Europe in less than 50 years, also earned recognition in other parts of the continent. It was certainly no coincidence that Elias Hutter's polyglot Bible included a Slovene version, based on Jurij Dalmatin's translation.

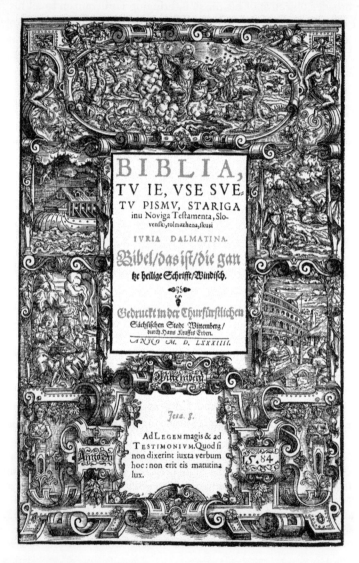

**Figure 28.** Cover of Dalmatin's *Biblia, tu je, vse Svetu pismu, Stariga inu Noviga testamenta* (facsimile).

Regarding cultural issues, Trubar was a stern realist. Based on his intimate knowledge of a wide range of Slovene vernacular dialects, he created a standard literary language that even the Catholics started to use during the next generation. He rejected the idea of Vergerius (who became the Duke of Württemberg's diplomat after his conversion to Protestantism) to write for the entire South Slavic region, as that would deviate from the Reformation ideal of the intelligibility of God's word.[230] A remarkably talented autodidact, Trubar had acquired vast theological knowledge and a significant humanist education (by 1557, he had already stressed that his mother tongue, like Ancient Greek, used a grammatical dual form), and he was fully aware of the immensity of his actions. Slovene was only occasionally written down before Trubar's time, but his efforts developed it into a written language. By bringing the biblical message closer to his contemporaries, Trubar helped make the Slovene language the chosen tool for salvation. The reformer's opus not only disproved the general belief that Slovene, like Hungarian, was unsuitable as a written language,[231] but also nourished hopes that the translations of his writings into Croatian and Serbian would play a crucial role in converting the Turks to Christianity.

Trubar was esteemed among the Protestants in the Holy Roman Empire as the highest authority in the European southeast,[232] and his reputation was not unfounded: in 1567, he paid an illegal visit to Bosnian captives held in the Ljubljana castle dungeons to gather the most reliable information possible on the Holy Quran and Islam.[233] He had already founded a bibliographical institution in the German town of Urach (under his management between 1561 and 1562), which published Protestant literature in Slovene, Croatian (in both the Glagolitic and Latin alphabets), Serbian (in the Cyrillic alphabet), and Italian, amounting to approximately 25,000 volumes altogether.[234] Trubar's adversaries were fully aware of the significance of his work—his opus was placed in the Vatican's Index of Forbidden Works in 1596.

---

230 Igor Grdina, *Od Brižinskih spomenikov do razsvetljenstva*, Maribor: Založba Obzorja, 1999, 113–122.
231 In 1582, Trubar wrote: "For it is widely known that 34 years ago there was neither a letter nor a register, even less a book, in our Slovene language, as they held the Slovene and Hungarian languages too coarse and barbaric to be either written or read." See Rupel, *Slovenski protestantski pisci*, 274. The first book in Hungarian was published in 1541 and the translation of the entire Bible in 1590. As is the case with Slovene, the first books in Hungarian were printed by Lutherans.
232 Grdina, *Od Brižinskih spomenikov do razsvetljenstva*, 108.
233 Rupel, *Slovenski protestantski pisci*, 194.
234 Ibid., 183.

Trubar's views were consistently characterized by a tendency to seek the golden mean in Reformation theology, and he unsurprisingly did everything in his power to prevent the extremist version of Lutheranism, advocated by the Istrian Matthias Flacius Illyricus, from sweeping over the Slovenian provinces. The latter was a renowned theologian of exceptional learning who had laid the foundations of Protestant historiography (even though he could not pride himself on a realistic conception of *hic et nunc*), and also originated from Carniola. Trubar thus faced an even more arduous task, but his labors to bring the Inner Austrian Protestants to accept the Formula Concordiae (in Carniola and Styria in 1580 and in Carinthia in 1582) were crowned with overwhelming success. All Flacianist writings, which Sebastian Krelj (1538–1567) strove to assert, were removed from the official repertoire of the Slovenian Church under the elegantly contrived pretext that their orthography was inappropriate. Theological schisms among the Protestants were to be concealed, otherwise the disputes would be turned to the advantage of the Catholics, who still largely prevailed among the Slovenes.[235]

Another characteristic of Trubar was his emphasis on a balance between the submissiveness of subjects and the benevolence of the nobility. He found Luther's instigation of feudal lords against rebellious peasants alien to his own views, even though his own works were in fact financially supported by aristocrats. His ideal was universal justice, as demonstrated by a monumental thought from 1577: "In summa: for every single tyrant and Mameluke, to the very last one, who ever hated the true children of God, persecuting, tormenting and murdering them from the very beginning of this world, a bitter demise was their portion."[236] All Trubar's published writings showed a deep compassion and commitment to common Slovenes, whose background he shared. He treated the frustrating dogmatic issue of trans-substantiation with utmost providence, although he was evidently more inclined toward the Zwinglist views than Lutheran doctrine. It was thus of no surprise that he kept up a regular correspondence with the Swiss reformer, Heinrich Bullinger. Similarly characteristic seems to have been his drafting of a relatively independent reproduction of the Augsburg Confession (*Articuli*, 1562) for the Slovenian Church. Even so, Trubar could not support pure Lutheranism without constraint until he had been permanently banished from his native country by Archduke Charles. The reformer was fully aware that Protestantism in Inner Austria could persevere only with steadfast support

---

235 Grdina, *Od Brižinskih spomenikov do razsvetljenstva*, 193–196.
236 Rupel, *Slovenski protestantski pisci*, 264.

from the northern regions of the Holy Roman Empire. Through his influence on the ducal court in Stuttgart (Duke Christopher and his son, Louis the Pious) and on Tübingen theologians (e.g., Jacobus Andreae), his main achievement was to ensure that Slovene literary endeavors would still be given sympathetic consideration in the Principality of Württemberg even after his death in 1586.

What thus took place during Trubar's life was a condensation of Slovenian history. Both the Reformation and the general political situation provided him with a relatively favorable foundation, at least in the short term, on which to carry out his far-reaching initiatives. Archduke Charles, who ruled the provinces of Inner Austria between 1564 and 1590, was forced to make considerable concessions to the Protestants' demands, like his father Ferdinand I, and despite his dour Catholic upbringing. This was principally because many army commanders in the Military Frontier embraced and fervently defended Lutheranism. Particularly outstanding among them were two men from Turjak (Ger. Auersperg), the fearless Baron Herbard VIII and the Carniolan hereditary marshal Andrej of Turjak, the most distinguished victor of the great battle at Sisak on June 22, 1593. Trubar dedicated his translation of the final segment of the New Testament in 1577 to Andrej and several other patricians. Slovenian Protestant preachers, who also spoke fluent German, often accompanied the Christian troops heading to face the Turkish invaders, and Charles was compelled to allow the Styrian nobility their freedom of conscience and faith so as not to jeopardize the balance between religious and political affairs in his provinces. This was done in 1572 through the settlement known as the Pacification of Graz. Later, in 1578, he granted the same rights to the feudal lords in Carniola and Carinthia and the burghers of Graz, Judenburg, Klagenfurt, and Ljubljana (through the so-called Pacification of Bruck).[237] The ruler's oral promise was merely noted rather than presented as a solemn charter. Nevertheless, the Lutheran aristocracy interpreted it very loosely to fit their own purposes, extending the promise to their subjects and enabling Protestantism to reach part of the Slovenian peasantry (Slovenian serfs had turned to Lutheranism, particularly on Catholic Church estates). Since during the Counter-Reformation the Bishop of Ljubljana, Tomaž Hren, wrote that 40,000 souls had been converted in northern Carniola alone,[238] it can be concluded that the Lutherans spread their teachings successfully enough, despite the aversion of the provincial duke.

With both moral and material support from the Inner Austrian and particularly the Carniolan nobility, Trubar was able to perform major work

---

237  Čepič and Nećak, *Zgodovina Slovencev*, 292.
238  Grdina, *Od Brižinskih spomenikov do razsvetljenstva*, 218, 219.

which inspired his adherents among the clergy and laity until the suppression of Protestantism in the Inner Austrian provinces between 1598 and 1628. The Slovenian reformer, however, was well aware of the utter cultural backwardness of his countrymen, and thus aimed not only to spread Protestantism but also to establish a high-quality education system in his native country. In the Church Ordinance, he explicitly called for education in reading and writing that would be accessible to all of his countrymen, irrespective of their social standing:

> Not a province, a city, or a common can survive without schools, scholars, and men of knowledge, and are even less able to attend to and pursue the cares of the world and of the spirit. This any man of common sense can understand. For not only men of faith and God but also prudent pagans held, everywhere in their cities and provinces, schools and universities, and educated men in them. [...] [E]ach preacher and parson ought to have a schoolmaster or parish clerk [...] in his parish to teach these young servants and maids, noble and peasant children alike, how to read and write Slovene, as well as lecture them on the Catechism through this brief instruction.[239]

The first Protestant school was thus established in Ljubljana in 1563, and largely took on the nature of a grammar school. Its work mainly involved Philip Melanchthon's student Adam Bohorič from Lower Styria and Philipp Nikodemus Frischlin, who came to Carniola from the University of Tübingen. Bohorič received acclaim for compiling the first Slovene grammar, *Arcticae horulae* (written in Latin and printed at the same time as Jurij Dalmatin's translation of the Bible in Wittenberg; it was even reprinted twice in the 18th century). Frischlin became renowned as the author of scripts for school plays and as a Latinist poet. Frischlin's German countryman and student, Hieronymus Megiser, taught at the Protestant grammar school in Klagenfurt, which was established in 1553, and introduced the Slovene language into lexicography (*Dictionarium quatuor linguarum*, 1592; *Thesaurus Polyglottus*, 1603). He also published a book on the history of Carinthia under his own name, while the book had actually been written by pastor Michael Gotthard Christalnick (*Annales Carinthiae*, 1612).[240] The most important university attended by Protestant students from the Inner Austrian provinces in the 16th century was Tübingen. Before the Counter-Reformation, this university was attended by as many as 300 young men from Carniola, Carinthia, Styria, and Gorizia.[241] Slovenia's intellectual wealth increased in 1575 with the establishment of the first printing works in Ljubljana; however,

---

239   Rupel, *Slovenski protestantski pisci*, 173, 175.
240   Čepič and Nećak, *Zgodovina Slovencev*, 265, 297.
241   Štih and Simoniti, *Slovenska zgodovina do razsvetljenstva*, 197.

after intending to publish Dalmatin's translation of the Bible, Archduke Charles had the printer, Janž Mandelc, banished in 1582. The Catholics, however, failed to found any noteworthy intellectual centers until the establishment of Jesuit schools in the Inner Austrian cities of Graz, Ljubljana, Klagenfurt, Trieste, and Gorizia, and therefore had to seek education in Italy or Vienna: a major reason why the "brain drain" from the Slovenian provinces, so characteristic of the Late Middle Ages, lasted well into the 16th century. One of those who created his work abroad was Jacobus Gallus (1550–1591), the most distinguished composer ever born in Carniola. His numerous masses and motets, jewels of the Late Renaissance art of polyphony, aspired to glorify the Catholic understanding of the world (even though his opus also included the hymn "O herre Gott," dedicated to the Protestants).

**Figure 29.** Composer Jacobus Gallus Carniolus.

# From Counter-Reformation Rigor to Baroque Exuberance

The rapid rise of Protestantism across the Slovenian territory in the mid-16th century did not turn into a lasting Lutheran triumph. The provinces of Inner Austria were in many respects vital to the survival of the Roman Catholic Church in Europe: had the Lutherans gained access to the northern Adriatic through Trieste and Koper, the Catholic world would have been split into the Romance west and the Slavic–Hungarian east. Owing to its exposed position close to the Turkish border and the Orthodox world, the east could easily have fallen under further Protestant expansion.

Moreover, the Lutherans could also have pushed toward Italy after securing their position on the Slovenian territory. As early as December 1563, Trubar had written to the former Styrian Governor Ivan Ungnad, who financed the Bible Institute in Urach, claiming that publishing Luther's Church Postil would strike a serious blow against the Pope in the Apennine Peninsula.[242] In Trubar's opinion, the easiest way for Protestant beliefs to enter Italy, guarded by the Alps, was through Carniola and Gorizia. With the only pro-Reformation Habsburg Maximilian II (1564–1576)—to whom Trubar sent several letters—occupying the throne of the Holy Roman Empire, the situation was anything but promising for the Catholics. Yet as long as the Inner Austrian provinces remained in the hands of Archduke Charles, they could still hope that they would eventually prevail over the Lutherans. The position of the Turks on the southeastern border during Charles' reign was not precarious enough to allow for a more far-reaching Catholic restoration, but the Inner Austrian court in Graz was certainly preparing for it: episcopal positions were filled by increasingly persistent opponents of the reconciliation policy (Janez Tavčar in Ljubljana, 1580–1597; Georg III Stobäus von Palmburg in St. Andrä, 1584–1618). The early introduction of the Gregorian calendar to the Inner Austrian provinces in 1583 was another meaningful symbolic gesture. Furthermore, at that time some seigneuries and towns already witnessed the first expulsions of Protestants.

In the late 16th century, Styria, Carniola, and Carinthia saw the arrival of the Jesuits, who had already been held in high esteem by the Bishop of Ljubljana, Urbanus Textor. The members of the new order soon raised the education

---

242  Rajhman, *Pisma Primoža Trubarja*, 173.

level of previously often unlearned Catholic priests in the countryside. Unlike older monastic orders, burdened by the severe crisis of monasticism (many had disintegrated during the turbulent times; even the venerable Žiče Carthusian monastery struggled to survive), the Jesuits rapidly expanded their activities, establishing a "Latin school" in the Inner Austrian capital of Graz as early as in 1573 and a university college in 1585. Jesuit educational institutions were also opened in other cities. Colleges of higher education were established in Ljubljana (1597), Klagenfurt (1604), Gorizia (1615), and Trieste (1619). In Slovenian Styria, a grammar school with a similar educational program was set up by parish priests at Ruše (1644), and a century had to pass before it was replaced by its Jesuit counterpart in Maribor (1758). These institutions served as meeting places for young men of various backgrounds (the school at Ruše was attended by nearly 7,000 pupils; around 600 of them came from noble families; as many as 8 became bishops, and another 17 prelates).[243] Initially, the Jesuits had to demonstrate their virtues in an "open competition" against their Protestant rivals, which led to a further rise in the quality of their instruction.

In 1574, Catholics started publishing books in Slovene too, meaning that the "battle for souls" erupted in the very area previously under complete Protestant control. For a long time, however, their achievements in writing, translation, and publishing remained fairly modest in comparison with the Evangelicals. In the early 17th century, Catholic translations of Bible passages relied openly on those made by the Protestants. This not only considerably increased the quality of their publications, but also enabled the continuity of Slovene literature, whose development was therefore no longer influenced by religious belief. The Capuchins arrived in the Slovenian territory almost at the same time as the Jesuits. In the 17th century, they opened monasteries in virtually every major town. They developed a prolific preaching practice and the spread of their rhetorical prose vastly contributed to the refinement of Slovene literature.[244] Moreover, they made important contributions to the development of liturgical drama. In the early 18th century, they would perform the Passion Play in Škofja Loka with 278 costumed characters, based on a script by the Capuchin friar Romuald of Štandrež. Reportedly, the Slovene-language Passion of Christ was also performed in the town of Tržič and in Carinthia; other types of religious drama were staged in Ljubljana and Ruše.[245]

---

243   Franc Kovačič, *Slovenska Štajerska in Prekmurje. Zgodovinski opis*, Ljubljana: Slovenska matica, 1926, 280.
244   Štih and Simoniti, *Slovenska zgodovina do razsvetljenstva*, 212.
245   Gspan, *Zgodovina slovenskega slovstva I*, 305–308.

Despite their mutual antagonisms, Lutherans and Catholics could occasionally find common ground on crucial issues almost until the end of the 16th century. One such case involved the Croatian–Slovenian peasant uprising in late January 1573, which was soon stifled by the joint response from the Lutheran and "papist" nobility. Even the peasant rebels who aimed at establishing their own authority in Carniola, Lower Styria, and the provinces along the Turkish border by setting up some kind of imperial regency did not divide their feudal adversaries according to denomination. At that time, the entire aristocracy, regardless of their confessional affiliation, found itself in an unenviable position, as the rebels' carefully planned westward, southward, and eastward march from the Croatian and Slovenian plains of the Sotla basin was initially as effective as the movements of the nobility's troops and mercenaries. The Turkish threat had seemingly imbued the life of all social strata with military logic.

The excellent peasant organization, which owed much to Ilija Gregorić, unequivocally prompted an immediate and concerted response from feudalists, who were able to close ranks in times of trouble, regardless of religious differences and provincial borders. After a few fierce battles that lasted from February 5 to 9, 1573, the rebels were utterly routed (at Krško, Kerestinec, Šempeter pod Svetimi gorami, and Stubiške Toplice). Shortly afterwards, their leaders suffered brutal and degrading punishments. On February 15, Ambrož Gubec (also known as Matija Gubec), considered the rebels' most important "ringleader," was first crowned with a red-hot iron crown, then dragged along the streets of Zagreb, tortured with red-hot iron pincers, and finally quartered.[246] Two years later, Primož Trubar, then a renowned preacher in Württemberg, who had always remained keen to keep abreast of developments in his homeland, stressed that pious peasants:

> are obliged to endure cruelty and injustice in silence, complain with their families about their plight and poverty, and the cruelty and injustice to God, and above all restrain themselves from starting an insurrection or seeking vengeance, like the Magyars had done in 1508, the Carniolans in 1515, the Lower Carniolans in 1573, and Styrians in 1528—only in the end to be gruesomely slain, slaughtered, hanged and impaled on spikes.[247]

On the other hand, the Slovenian reformer was disquieted by "wicked lords" in a very different way than Luther had been, and offered a theological explanation of injustices which the lords inflicted upon the common man. His last catechism of

---

246 Grafenauer, *Boj za staro pravdo*, 276–309.
247 Rupel, *Slovenski protestantski pisci*, 243.

1575, intended primarily for the clergy, thus read: "The reason, known to every man, is that for all the sins of mankind, but most importantly for heathenism, false Divine Service, perverted masses and foul iniquities, God has sent the Turk and wicked lords, both ecclesiastical and lay, to exert their sway over us for so many years..."[248]

However, since such a standpoint by the Protestant Reformed Church was of little help to peasants, a new surge in "unofficial" popular piety did not come as a surprise. Once again the Slovenian provinces became home to *štiftarstvo*,[249] a slap in the face of Catholics and Evangelicals alike. This religious sect, which was most vigorous between 1583 and 1585, caused the authorities in Gorizia, Carniola, and Carinthia much anxiety: to the nobility it represented not only a form of heresy but also of concealed rebellion. Measures taken against the so-called štiftarji were therefore prompt and effective, and by the 17th century, larger štiftar communities could only be found in Styria (the Slovenske gorice region).[250]

The rapid and effective suppression of various forms of popular dissent signaled the consolidation of princely power in the territory between the Alps and the Adriatic. This was also a time when the construction of Karlovac and several other fortified settlements in Croatia along the Turkish border alleviated the threat of a full-scale Turkish incursion into Carniola and Styria. Between 1591 and 1593, the Bosnian beglerbeg Hasan Pasha Predojević tried to occupy Sisak three times, yet in vain. At the end of June 1593, he encamped outside the fortified town at the confluence of the Kolpa and Sava rivers for the last time. His 12,000-strong force was approached by the Habsburg army, less than half its size, and on June 22 a fierce battle ensued which ended in a crushing defeat of the sultan's army. Hasan Pasha and around 8,000 Turkish soldiers were buried in the battlefield or washed away by the Kolpa, while the Christian forces suffered no more than 50 casualties.

The Battle of Sisak, in which the Protestant commander of the Croatian Military Frontier, Andrej of Turjak, played a key role, led to the decline of Turkish power on land (the Ottoman navy had already been seriously depleted at the Battle of Lepanto in 1571). With increasing use of firearms and artillery, the Europeans

---

248  Ibid., 245.
249  A sectarian movement of peasant reformers who strove to build and establish new monasteries and churches; the Slovene term derives from the German word Stift, which denotes an institution such as a monastery or a church (transl. note).
250  Štih and Simoniti, *Slovenska zgodovina do razsvetljenstva*, 208, 209.

began to assert their supremacy over the Turkish traditionalist strategies and tactics, and from that point on, it took the Habsburgs much less military power to repel incursions from the east. The Bosnian beglerbeg's devastating defeat at Sisak ignited the Long Turkish War, which raged between the Habsburgs and the sultan until 1606. Both sides recorded a number of victories and defeats, especially in the Hungarian battlefields north of the Drava.

With the end of the Long Turkish War, the Inner Austrian provinces were at long last made safe from Turkish conquest. The Battle of Sisak, the great triumph of Christian arms, was the crucial turning point, and unleashed a wave of enthusiasm throughout Central Europe and Italy. Moreover, at the end of the 16th century, this victory enabled the archducal court in Graz to endow the provinces under its control with the right to freedom of religion: thus, the formula *Cuius regio, eius religio*,[251] summing up the provisions of the Peace of Augsburg between the Catholics and Lutherans, eventually started to apply in the Habsburg lands. Because the Habsburgs refused to silence Central European religious disputes through inquisition (as advocated by the Papal Nuncio in Graz), the Bishop of Lavant, Georg III Stobäus von Palmburg, designed a detailed plan for the restoration of Catholicism, giving the Habsburgs a free hand to pursue their domestic and foreign interests.[252]

Stobäus' plan, based on princely power, was realized by the so-called Reformation commissions. The Bishop of Seckau, Martin Brenner, led the persecution of Styrian and Carinthian Protestants, and his Ljubljana equivalent Tomaž Hren did the same in Carniola. The first to be expelled from the Inner Austrian provinces were Protestant preachers and teachers, including Johannes Kepler, who—along with other Lutheran teachers from Graz—sought temporary refuge in the Slovenian region of Prekmurje beyond the Styrian-Hungarian border. Scores were then settled with the peasants and the middle class. Restorers of Catholicism combed through the Inner Austrian provinces, burnt Protestant books (in Ljubljana alone, 11 carts of "heretical" literature were burned in 1600 and 1601), mocked and imprisoned their readers, and razed Lutheran sanctuaries and cemeteries to the ground. The few townspeople who withstood the pressure were driven from their homes.[253] Protestants could only continue to form a community in a few remote areas in Carinthia; around Villach, there was also a small number of Slovenes.

---

251 Whose the region, his the religion (transl. note).
252 Grdina, *Od Brižinskih spomenikov do razsvetljenstva*, 208, 209.
253 Ibid., 214–220.

The actions of the Catholic Reformation commissions caused untold cultural and material damage. Religious emigration resulted in not only the "brain drain," but also the loss of great entrepreneurial masters. Bishop Tomaž Hren, who in principle was very fond of the Slovene language (even though he himself preferred composing his verses in Latin), was the only Counter-Reformation cleric to advocate the use of Protestant books in educating Catholic priests. He even wrote twice to Rome, in 1602 and 1621, and was eventually given consent from the Pope's office for such use. It should also be pointed out that Hren fostered long-term publishing plans and tried to set up a printing office in Ljubljana, but never succeeded. Abroad, he was only able to publish the Slovene lectionary (in a surprisingly huge edition of 3,000 copies) and a small catechism.[254] Given that the Ljubljana Diocese administered parishes in Carniola, Carinthia, and Styria, he decided to use the standard Slovene language as introduced by Primož Trubar and his followers, to overcome the differences between the dialects spoken in individual Slovenian provinces.

Slovenian Evangelical preachers, whose publishing activities flourished until 1595, now sought assistance from their noble patrons, since the wave of re-Catholicization would only reach the next generation of the aristocracy. Under the increasing pressure of the Counter-Reformation through the princely offices, which enhanced Ferdinand II's power and successfully enforced the principle of monarchial absolutism, Evangelical preachers left the country and mostly migrated to German parts of the Holy Roman Empire. However, just before their expulsion, a number of them began to embrace the radical views most notably advocated by the preacher Janž Znojilšek, who drew on Matthias Flacius' theological interpretation in translating Luther's Catechism.[255]

Prekmurje, where Protestantism most probably took root slightly later than in the Inner Austrian provinces, experienced the spread of Lutheranism and even Calvinism. The Slovenian territory under the Crown of St. Stephen preserved its religious heterogeneity for a long time to come. The reason lay in the political necessity felt by the Habsburg monarchs to negotiate a religious treaty with western Hungary north of the Drava, which took painstaking effort to keep under their control. The sovereigns of the Holy Roman Empire feared that the Turks might take advantage of the dissatisfaction of the Protestant nobility during a ruthless

---

254　Gspan, *Zgodovina slovenskega slovstva I*, 278–280.
255　Rupel, *Slovenski protestantski pisci*, 43, 44. Later, Znojilšek's descendants moved to Sweden; one of his famous descendants was the distinguished poet and diplomat Carl Johan Gustaf, Count Snoilsky, born in the 19th century.

campaign to restore Catholicism. Feudal lords in the Hungarian crown lands (except Croatia) thus retained the freedom of conscience (or rather confession), while the serfs were obliged to be of the same religion as their masters. As a result, western and northern Hungary boasted a diverse religious landscape. The Treaty of Vienna, signed in 1606, which also applied to Prekmurje, might not have provided an ideal solution to the coexistence of various Christian churches, but still managed to prevent part of the Slovenian territory from forcible reconversion. Supported by German Pietists and Protestants from Bratislava, the several thousand Lutherans active in the area between the Rába River, the town of Lendava (Hun. Alsólendva), and the Styrian border proved to be relatively avid book publishers in the ensuing century. Štefan Küzmič even prepared a new translation of the complete New Testament for his fellow countrymen. Published in 1771 in Halle and reprinted in the late 19th century, the translation turned out to be the greatest achievement of Slovene literature from Prekmurje until the end of World War I.[256] The much weaker Calvinist communities in Prekmurje, on the other hand, could not boast similar accomplishments.

The restoration of Catholicism in the Inner Austrian provinces peaked at the turn of the 17th century, but at first circumvented the nobility completely. During the Long Turkish War, the court in Graz was not yet prepared to risk too much dissatisfaction from the local aristocracy, which continued to be the reservoir of senior and middle-ranking officers commanding the troops along the volatile eastern border of the Habsburg Empire. Similarly, the Inner Austrian aristocracy was fully aware of its advantageous position and remained corporately loyal to its Habsburg rulers even after they became embroiled in feuds with Protestants in Hungary, Upper and Lower Austria (1604–1608), and Bohemia (1604–1608; 1618). It is therefore possible to conclude that the partial stabilization of the situation promoted by the increasingly absolutist princely authorities suited the followers of all confessions, despite its Catholic connotation. Interestingly, however, not even the alarming reports on a remarkable number of impious or even highly suspicious members of the Catholic Church issued by the clergy upon their visitations in the first two decades of the 17th century could cause serious concern, let alone prompt swift action.[257]

---

256 Grdina, *Od Brižinskih spomenikov do razsvetljenstva*, 61.
257 Bogo Grafenauer, *Doba začasne obnovitve fevdalnega reda pod okriljem absolutne vlade vladarja ter nastajanja velikih premoženj od protireformacije do srede XVIII. stoletja*, vol. 4 of *Zgodovina slovenskega naroda*, Ljubljana: Kmečka knjiga, 1961, 22, 24.

If a more stable situation at home and abroad failed to give a major boost to the ambition to achieve full restoration of Catholicism in Carniola, Carinthia, and Styria, then the Thirty Years' War did precisely that. On August 1, 1628, Ferdinand II reconsolidated his personal authority over individual Habsburg lands in Central Europe and ordered the Inner Austrian Protestant nobility to join the Catholic Church or else leave the country within a year.

On the other hand, the advocates of the unyielding Counter-Reformation offensive dictated by the imperial court were elated by the military triumphs resounding from German provinces, as the Catholic troops had just encamped along the Baltic shores.

Even though a vast part of the nobility reconciled itself to the imposed change of religion during the reign of Ferdinand II, emigration was far from insignificant: some 100 noblemen and their families emigrated from Carniola, more than 150 from Carinthia, and at least 250 from Styria,[258] which, at least temporarily, drastically affected the manners and bearing of the aristocratic elite in the Inner Austrian provinces. For a long time thereafter, any idea of mounting serious opposition to princely absolutism was unimaginable. The nobility became totally dependent on the emperor, and advancement in his offices was regarded as the pinnacle of career success. It is little wonder that the aristocracy readily joined the imperial forces (from the Military Frontier) in crushing the continual and extensive peasant uprisings, the largest of which engulfed major parts of Slovenian Styria and Carniola in 1635. Consequently, rebellious serfs started to oppose all forms of authority, increasingly perceiving them as a uniform structure of oppression. In some places, they were acting very decidedly against the clergy.[259]

Peasant hostility toward representatives of lay and ecclesiastical lords continued into the next century and peaked again during the early spring of 1713 in a major insurrection sweeping across the Tolmin area, the central Soča basin, and the Karst region. The enraged rebels occupied Gorizia as the capital of provincial administration (and even pillaged and destroyed the house of the merciless tax collector Jakob Bandel). They protested against taxes and called for unification with Carniola, claiming that "His Imperial Majesty is no more than their servant." This time, too, pacification was brought about by the firm hand of the Habsburg authorities. When the Inner Austrian Treasury mobilized troops from the Military Frontier in Croatia and German soldiers from the hinterland,

---

258 Ibid., 33.
259 Grafenauer, *Kmečki upori na Slovenskem*, 299.

the insurrection soon lost momentum. In the end, 11 leaders were publicly beheaded in Gorizia and another 150 imprisoned in April 1714.[260]

As late as one generation after the emigration of aristocratic Protestant families (who possessed so much capital that in 1631 the Inner Austrian authorities resolved to prohibit its outflow), the nobility finally started to question the government and its absolutist tendencies. However, under Ferdinand II and his successors, provincial assemblies were degraded to representative bodies, deprived of the power to articulate views that would clearly run against the aspirations of the imperial court. Aristocratic opposition was therefore left with no other alternative than to go underground.

Isolated Inner Austrian critics of absolutism were given well-founded hopes to find companions, allies, and even supporters in the western Hungarian crown lands, where widespread dissatisfaction with the Habsburg rulers smoldered. Many Hungarian magnates could not forgive the Viennese court for disregarding their interests. After Raimondo Montecuccoli inflicted a crippling defeat on the Turks in the Battle of St. Gotthard (Hun. Szentgotthárd) on August 1, 1664— ending a one-year war between the Ottomans and the Habsburgs that had started over the selection of a new Transylvanian prince—the imperial forces did not head east, and missed the perfect opportunity to drive the Ottoman invaders out of Hungary for good. As a result, magnates from the Hungarian crown lands who aspired to increase their power by conquering the territory beyond the eastern border, decided to take an organized stand against the Habsburgs. The Hungarian conspirators against the government of Leopold I were initially led by the Hungarian poet and Croatian viceroy Miklós Zrínyi and, after his death in 1664, by his brother Peter. Between 1667 and 1668, they were joined by the wealthy Count Hans Erasmus Tattenbach, whose family estate had its center in Slovenian Styria, and by the notoriously capricious Governor of Gorizia, Karl Thurn.

The aristocratic rebels, who hoped to seize the provinces of Inner Austria and the western lands of the Crown of St. Stephen, were counting on the agreement with Turkey and the Venetian Republic, and on support from France, but their far-reaching and ambitious plans came to nothing. Louis XIV did not find the malcontents trustworthy enough, while Venice was losing its superpower status after the very last battle (1615–1617) with its increasingly powerful eastern neighbor, embarrassed by the Habsburgs' demonstration of power and strength in pillaging Istria and the lower course of the Soča River (moreover, in a subsequent

---

260   Ibid., 315–327.

war against the Turks between 1646 and 1669, La Serenissima could not even defend Crete). In March 1670, when the mobilization of the imperial forces clearly signaled that the Viennese court had gotten wind of the conspiracy, Count Tattenbach, who was to become the ruler of Styria or at least its southern part, provided his serfs with arms. However, battle plans were eventually abandoned because Péter Zrínyi became the Croatian viceroy in the spring of 1668 and was reluctant to wage an armed rebellion against the emperor. Tattenbach, who except for his political ideas was a fairly typical representative of the 17th-century hedonistic nobility, was beheaded in Graz on December 1, 1671.[261] Just before that, Zrínyi and his brother-in-law and accomplice Ferenc Kristóf Frangepán had faced the same punishment in Wiener Neustadt. If the rebels had prevailed over Leopold I, Frangepán, who was also accused of inciting hostility toward the Germans, would most likely have become the prince of Carniola.[262] However, the failed conspiracy strengthened the power of the court in Vienna, and the Habsburgs confiscated the Zrínyi and Frangepán estates and consolidated their position in the lands of the Crown of St. Stephen. Fully absorbed in his absolutist ideals, the emperor ascribed less and less importance to the aspirations of the nobility on the periphery and rather relied on German-speaking regions in the central Danubian basin and the Eastern Alps for his power.

After the threat of a Turkish conquest had been dealt with, the growing princely absolutism greatly increased certainty and predictability in all spheres of life, creating a positive environment for the economic consolidation of the Slovenian territory. Nevertheless, the growing financial needs of the Habsburg monarchs, who were being drawn into ever fiercer conflicts with France in the Rhineland and the Mediterranean basin, certainly precluded many from enjoying the improved conditions, although major changes in the geo-strategic position of Carniola, Carinthia, Styria, Gorizia, Istria, and Trieste at least brought great relief to the middle class and encouraged their entrepreneurial spirit. The Slovenian territory was no longer in the immediate rear of the battlefield as it had been in the late 16th century: in accordance with the Treaty of Madrid between the Venetians and the Habsburgs of September 26, 1617, it was incorporated into a zone designated as the central military reservoir of the imperial court in Vienna. The territories between the central Danubian basin in the north and the

---

261 Kovačič, *Slovenska Štajerska in Prekmurje*, 257–260.
262 Ferenc Kristóf Frangepán was also an important man of letters. Just before the conspiracy was revealed, he had started translating what was then Molière's latest comedy *Georges Dandin* into Slovene.

Adriatic in the south therefore witnessed only a few brief and inconsequential military operations until the Napoleonic Wars. In the late 17th century and early 18th century, Prekmurje became home to groups of Hungarian militants called *kruci*, who invaded eastern Styria on several occasions—first when they fought against the Turks and later when they took part in the uprising under Ferenc II Rákóczy[263] (the maternal grandson of Peter Zrínyi) against the reign of Leopold I in Hungary. Their plundering expeditions were most often warded off by the inhabitants of the Mura region.[264]

Despite a few tensions here and there, the general situation in the region between the northern Adriatic and the Eastern Alps continued to improve. Calmer circumstances were best reflected in changes in the aristocratic lifestyle: the nobility no longer lived in old castles built on remote hilltops, but started to build country mansions surrounded by magnificent parks and town palaces. Religious and secular buildings were increasingly modeled on Italian Baroque architecture, which became popular not only in the Littoral but also in the interior, particularly in Carniola and Styria.

Still, the preservation of strict feudal principles throughout the 17th century prevented the lower strata of the predominantly rural population from enjoying the benefits of greater prosperity. As a result, banditry remained firmly rooted in certain parts of the Slovenian provinces and caused the authorities great concern. In the late 17th century, Carniola, and occasionally its neighboring provinces, were terrorized by a sizeable group of bandits headed by Anže Košir, also known as the Infernal Rascal,[265] and it took the authorities quite some time to restrain it. Despite severe punishments, banditry survived until the abolition of feudalism in 1848, and in some places for even longer. Since it mainly targeted regional trade between the port of Trieste and its wider hinterland rather than the local economy, it sometimes attracted sympathies and even tacit support from the wider strata of the poor population. This became especially evident in the era of French Illyria (1809–1813), when outlaws were quick to take advantage of an almost complete lack of local knowledge of the French administration. Some bandits were even credited with a sense of justice. Therefore, in the 19th century,

---

263  It is not entirely clear whether *kruci*, who invaded the Slovenian part of Styria, were actually connected with Rákóczy or whether they only pretended to be his warriors. Traces of their invasions from Prekmurje to Styria have persevered in Slovene folk tradition.
264  Kovačič, *Slovenska Štajerska in Prekmurje*, 260–262.
265  Štih and Simoniti, *Slovenska zgodovina do razsvetljenstva*, 251.

some regarded Franc Guzaj as the "terror of Slovenian Styria," and others hailed him as the man who robbed from the rich to give to the poor.

The removed threat of Turkish conquest and a calmer situation on the Venetian border made it possible for the first far-reaching plans to facilitate regional economic activities to be conceived under Leopold I. In 1678, the emperor decided that Ljubljana should become the main repository for goods from the Habsburg lands destined for export to America, and Trieste should be made the main port of export. The northern Adriatic hinterland looked forward to even better prospects after the unsuccessful Turkish siege of Vienna in 1683. The ensuing military operations that ended with the Peace of Sremski Karlovci in January 1699 delivered to the Habsburgs almost the entire fertile Pannonian Plain. Owing to the drastic border revisions to the Ottomans' obvious detriment, security in Carniola, Carinthia, Styria, Gorizia, Istria, and Trieste increased and considerably facilitated trade with the eastern part of the central Danubian basin.

In light of such change, the Slovenian provinces became attractive to entrepreneurial immigrants in the mid-17th century. One of them was Jakob Schell, who moved to Ljubljana from Tyrol and developed incredibly successful trading activities. He quickly became rich by doing business with provincial assemblies, the imperial army, and individual feudal lords who entrusted him with the administration of their estates or advocacy of their interests in complex financial arrangements, so that they could fully commit themselves to their formal duties. By the end of the 17th century, Schell had been ennobled, and became a generous patron to prove worthy of his new status: he supported ecclesiastic and educational institutions (he was the founding father of the Ursuline monastery in Ljubljana, and in charge of Carniola's first girls' school), and made sure that all buildings erected under his patronage reflected the then fashionable Baroque tendencies. Many other merchants became businessmen as well and, as a rule, were raised to the rank of nobility. The most prominent merchant in the 18th century was Michelangelo Zois, who had come to Ljubljana from the Bergamo region and in the course of time became the richest businessman in Carniola. His main sources of income were his ironworks and trade in iron.[266]

The rapidly growing importance of non-agricultural activities in the 17th century was also reflected in the development of the mercury mine in Idrija. The mine was the second largest of its kind in the world and was a valuable asset to the economic vitality of the Habsburg court, which gained complete control of it as early as 1575 (after around 85 years of exploitation). Owing to its outstanding

---

266   Ibid., 238, 239.

economic importance, Idrija was granted administrative independence (1607), a market charter and, in the 18th century, a town charter. Idrija was also the first town in the Slovenian provinces to grow into an economic center based exclusively on a non-agrarian economic activity. The operation of the large mine hastened the early development of educational institutions (specializing in technology, land surveying, metallurgy, and chemistry).

With its economic power gradually increasing, the Slovenian territory strengthened its ties with other lands and opened its cultural arms to foreign influences, most notably in architecture, sculpture, painting, and music. The most successful Slovenian artist of the mid-17th century was composer Janez Krstnik Dolar, a Kamnik-born Jesuit. His masses and psalms made him famous in Vienna and Bohemia. Upon his death in February 1673, he was even praised for his music by Emperor Leopold I, himself a good composer. On the other hand, not a single Slovene book was published between 1615 and 1672. For a long time, the clergy's need for the written word was satisfied by copies from the enormous edition of Hren's lectionary and by the increasing volume of manuscripts. These manuscripts testified to a rapid restoration of the cult of saints and the veneration of the Virgin Mary, which had both suffered a severe blow during the rise of Protestantism. Manuscripts that have been preserved until the present day also reveal the prevailing rigorous interpretations of faith and piety (priests, for example, vigorously opposed folk poetry and its special treatment of the Christian spiritual tradition).[267]

A major change in literary production came in 1672 with the polymath Janez Ludvik Schönleben's republication of the Slovene lectionary. Schönleben held a doctorate in theology from the University of Padua and had a very keen interest in researching the history of Carniola. On his recommendation and at the request of the provincial assembly, Johann Baptist Mayr founded a printing house in Ljubljana in 1678, which soon turned into an important cultural center. Immediately upon arriving in Carniola, Mayr published a catalogue listing 2,500 titles available for sale. Before long, he also undertook to print Schönleben's most important work, *Carniolia antiqua et nova I* (1681), which, though unfinished, provided a major stimulus for research of the Slovenian past. Between 1707 and 1709, Mayr's successors even published the first newspaper in Slovenian territory (in German). Although it changed hands later on, the printing house operated until the mid-20th century.[268]

---

267  Grdina, *Od Brižinskih spomenikov do razsvetljenstva*, 72.
268  See Branko Reisp, *Kranjski polihistor Janez Vajkard Valvasor*, Ljubljana: Mladinska knjiga, 1983, 28–30.

The possibility of publishing Slovene books in their homeland proved advantageous to numerous ecclesiastical writers, who quickly equipped the inhabitants of the Slovenian provinces with the most common reading matter—prayer books, hymnbooks, catechisms, and meditative essays. The most ambitious and industrious author was Matija Kastelec, a canon from Novo Mesto, who by the end of the 17th century had prepared the first Catholic translation of the Bible and a Latin-Slovene dictionary, even though unfortunately neither was published. Slovene religious literature therefore reached its peak with a collection of five voluminous books of sermons written by the Capuchin friar Janez Svetokriški, and published under the Latin title *Sacrum promptuarium* between 1691 and 1707, partly in Venice and partly in Ljubljana. The collection contained nearly 2,900 pages.[269] Complemented by sermons by the Capuchin friar Rogerij of Ljubljana and the Jesuit orator Jernej Basar, who had distinguished himself at home and abroad, Svetokriški's Holy Handbook crucially contributed to the affirmation of the Slovene language. All three leading Slovene orators applied stylistic Baroque principles to their texts.[270]

The increasing number of books published in Slovene led Slovenian speakers to gradually introduce their mother tongue into personal correspondence, wills, inventories, and similar documents that had previously been written mostly in German or Latin. In the 17th century, the sharply defined medieval boundaries for the use of particular languages dissolved much more drastically than they had during the Protestant era, at least in some spheres of life. Moreover, authors writing in Slovene began to appear in all the provinces between the Adriatic and the Eastern Alps. All these developments called for the republication of Adam Bohorič's first Slovene grammar, which was eventually accomplished in 1715 by the Capuchin friar Janez Adam Gaiger from Novo Mesto.

On the other hand, secular literature, which had a very narrow circle of readers in the Slovenian provinces, mainly restricted to the aristocratic elite, continued to be published in German and Latin. In 1659, Adam Sebastian Siezenheim, a Carniolan provincial assembly official, had his book on education, *Speculum generosae juventutis*, published in Munich, while Franz Wiz, Baron Wizenstein, the ennobled son of Ljubljana's mayor and judge, had his adaptations of Italian Baroque novels printed in Nuremberg. Two additional writers who attracted readers' attention at the time were the aristocratic alchemist Johann Frederick von Rein, who lived in Strmol Castle (he donated one of his manuscripts

---

269   Grdina, *Od Brižinskih spomenikov do razsvetljenstva*, 69.
270   Gspan, *Zgodovina slovenskega slovstva I*, 284–293.

to Emperor Leopold I, who extended financial support to his work), and the physician Johann Baptist Ganser, from Novo Mesto, who studied women's diseases. The Jesuit Martin Bauer from Solkan wrote the *History of Noricum and Friuli*. Though unpublished, his Latin manuscript at least presented later authorities on the history of the Slovenian territory with a valuable collection of material.

However, most of those authors did not surpass the average quality of Central European literary production, increasingly defined by the reception of Italian belletristic and scientific works, while the 16th-century Slovenian territory and neighboring provinces predominantly followed trends from the German north. Gradually, the area encompassing Gorizia in the west, Istria in the south, and Carinthia, Styria, and Austria in the north and east, developed a unique symbiosis of influences from different sources, assimilated and adapted to local needs and possibilities.

If Janez Svetokriški, whose opus displayed a remarkable sense of both serious and humorous instructiveness, stood pre-eminent among the 17th-century ecclesiastical writers for his eloquence and thematic breadth, the polymath Janez Vajkard Valvasor deserves a special mention as the most distinguished secular writer. Born in 1641 in Ljubljana to a family originally from the Italian town of Bergamo, he received a good basic education at the Jesuit school in his hometown. His travels through German and Italian provinces, North Africa, France, and Switzerland made him well acquainted with the situation in the European West and the Mediterranean basin. In a relatively short period, he set up an impressive library holding around 10,000 books and 8,000 graphic prints (including those by Dürer and Callot), as well as collections of mathematical and cartographic instruments, minerals, fossils, antiquities, and old coins.[271] Although he took part in successful military campaigns during two Habsburg wars against the Ottomans (1663–1664 and 1683) and served as a Swiss Guard at the French royal court, he believed his mission in life was to study topics and phenomena related to his homeland. In order to present his land (and the neighboring regions) to his learned contemporaries as accurately as possible, he embarked on thorough research. He established contacts with the British Royal Society, which worthily awarded him a membership in December 1687 for his description of the disappearing Lake Cerknica and its "wonders." As a talented technical designer, he conceived a tunnel beneath the Ljubelj Pass at the

---

271 Reisp, *Kranjski polihistor Janez Vajkard Valvasor*, 104–108.

Carniolan-Carinthian provincial border (which was constructed as late as the 20th century).[272]

In 1678, Valvasor, who considered himself a baron (most probably through his mother's noble birth), founded a copperplate engraving press and a printing shop at his castle, Bogenšperk, and employed a number of designers, printers, and other workers from various provinces. In addition to three maps, a Passion booklet, scenes from Ovid's Metamorphoses, and horrible images of human death (partly modeled on Holbein's Dance of Death), he published several collections of books depicting the castles and towns of Carniola and Carinthia. He found inspiration and a role model in the outstanding copper engraver and publisher Matthäus Merian, who had produced and published topographies of various provinces of the Holy Roman Empire. Interestingly, the best works by Valvasor and his artisans display the same quality. The Bogenšperk graphic collections were invaluable for the Slovenian provinces and their neighboring regions.

Valvasor intended to complete Schönleben's unfinished history of his native country (above), but soon expanded the initial plan significantly by preparing a most accurate, all-encompassing description of contemporary Carniola, which included landscapes, towns, castles, language, customs, and outstanding personalities. Coverage of early Carniolan history and the editorship was entrusted to Erasmus Francisci, a Nuremberg polymath, who also saw to it that Valvasor's chef-d'oeuvre was written in fluent and appropriately embellished German. The masterpiece entitled *Die Ehre des Herzogthums Crain* (The Glory of the Duchy of Carniola) was truly impressive, both by 17th-century criteria and those of any later period. Published in 1689 in Nuremberg, this four-volume collection totaled more than 3,500 pages. *The Glory of the Duchy of Carniola* exhibited its author's unprecedented, fervent love for his homeland. Valvasor regarded the Slovene language as the Carniolans' mother tongue (as was also evident in his correspondence with the British Royal Society) and credited the Protestants with its affirmation. In 17th-century Central Europe, marked by the Catholic ecclesia triumphans, such a stance was anything but self-evident. However, under the influence of German humanist historians, Valvasor considered inhabitants of Carniola, much like other Slavs, a branch of the Germanic peoples.[273]

Valvasor drove the major project of his life onwards with great haste and completed it in four years (partly thanks to his previously produced

---

272  Ibid., 157.
273  Ibid., 219–223.

topographies). However, the publication of such a comprehensive work cost him a sum beyond his means. As a good manager—he had already bought Bogenšperk Castle as a young man—he strived to prevent his financial ruin, but in vain. In 1689, he was forced to start selling his property, including the library and graphics collection. The Carniolan provincial assembly, which had offered him only modest support for his encyclopedic work on their homeland, refused to provide him with financial assistance when his masterpiece was published by turning down the offer to purchase his library. Eventually, Valvasor was compelled to sell off his castle as well. He died in 1693 in the town of Krško, where he had bought a townhouse after his large-scale publishing plans had turned out to be financially disastrous. Valvasor's monumental work soon became a standard item in Carniolan aristocratic libraries. However, the utter economic ruin of this inquisitive author, with his homeland's interests very much at heart, indicated clearly that, even though material conditions in the Slovenian territory had improved significantly, the cultural level of the local aristocratic elite had remained considerably low. Regrettably, the Styrian nobility showed no willingness to further similar scientific achievements (and remained satisfied with Georg Matthäus Vischer's modest *Topography of Styria*, completed in 1696). Only the decline of the traditional feudal system finally released creative powers and enabled more ambitious ideas and visions to be realized. It was little wonder that later on, in the 18th and 19th centuries, Valvasor's homeland treatise became an important source of literary inspiration and a valuable historical description of the high Baroque.

# Scholars, Officials, and Patriots Changing the World

Despite Valvasor's bankruptcy, the publication of *The Glory of the Duchy of Carniola* was a unique portent of times to come. In the mid-15th century, the humanist Enea Silvio Piccolomini, who became acquainted with the Slovenian territory as a member of the Habsburg court and Bishop of Trieste before his elevation to Bishop of Rome, considered the regions alongside the northeastern borders of Italy to be barbaric. Similar sentiments persisted during the rapid rise of Protestantism. In early August 1565, Primož Trubar wrote to Adam Bohorič about his homeland's pervasive disdain for education, yet he believed that the establishment of a sufficient network of schools would bring an end to such crudeness.[274] His plans failed for two reasons: the colossal cost of the defense against the Turks, which rendered it impossible to significantly improve most of the population's standard of living and therefore education, and the expulsion of Protestants at the turn of the 17th century. Nevertheless, the belief remained that the Slovenian provinces were not destined to linger on the periphery of culture and civilization, but were able to articulate their ideas and priorities and establish their own centers. The idea that the Slovenian provinces were condemned to provinciality, which for a long time had simply been accepted as fact, began to be reconsidered. Although Valvasor's attempt to found a graphic institute failed, his masterpiece, *The Glory*, testified to the intellectual ambition which would not wane in future generations.

## Scholars

At the turn of the 18th century, the Slovenian provinces increasingly felt powerful influences from the Apennine Peninsula. The rapid spread of the contemporary Baroque style, which left its mark on architecture, painting, sculpture, and music, had accelerated the adoption of other trends from Italy. Accordingly, an academy (Academia Palladiana) was established in Koper quite early; its most distinguished member, a local named Santorio Santorio,[275] began

---

274 Gspan, *Zgodovina slovenskega slovstva I*, 211; Primož Trubar, *Cerkovna ordninga. Slowenische Kirchenordnung*, Munich: R. Trofenik, 1973, 78, 79.
275 Mirko Dražen Grmek, *Santorio Santorio i njegovi aparati i instrumenti*, Zagreb: Institut za medicinska istraživanja Jugoslavenske akademije, 1952, 9.

to introduce precise instrumental measurement procedures into medicine. With time, similar associations began to appear in nearby towns (Piran, Gorizia) and deeper in the hinterland (Ljubljana). The academies served as meeting places of the scholarly elite and encouraged special forms of selective sociability; only rarely did they leave visible traces as promoters of science and art. The very presence of educated people in the relatively small towns between the coasts of the northern Adriatic and the peaks of the Eastern Alps, who felt the need to organize themselves, testified to the great mental changes beneath the surface of 17th-century society. Until then, only church or religious associations had been common in the Slovenian territory. Some accepted only the chosen into their ranks: membership in the Ljubljana-based Noble Society of St. Dismas, established in 1688, was restricted to noblemen and distinguished men of science. The increasing importance of education was not only in achieving career success, but also in bringing individuals public acclaim.

The Academia operosorum (the Academy of Hard-Working Gentlemen), founded in Ljubljana in 1693 on the initiative of the historian and lawyer Janez Gregor Dolničar, and under the leadership of the cathedral provost Janez Krstnik Prešeren, was introduced to the public on December 13, 1701. Its members presented themselves to the society "amidst the sounds of trumpets and drums and to the symphonic unison of selected music."[276] The association, initially consisting of 23 prominent men (12 lawyers, six theologians, and five physicians), later accepted 25 new members, among them the President of the Rome-based Academy of Arcadia, Canon Giovanni Maria Crescimbeni;[277] it operated for about a quarter of a century. Its members were also represented in several Italian academies (Rome, Bologna, Forli, Venice, and Foligno). Most important for the Slovenian environment were the work of the physician Marko Grbec, an important proponent of medical prevention, and the research of the historian Janez Gregor Dolničar, who strove to continue the work of his uncle, Johann Ludwig Schönleben, and of Janez Vajkard Valvasor.

Before its decline, the Academia operosorum contributed to the foundation of several related institutions. The most significant was the Academia philharmonicorum, founded in 1701, the second oldest musical society in Central Europe (the oldest was a similar institution in St. Gallen, Switzerland). It not only sparked the production of more music, but also numbered quite a few

---

276 Primož Simoniti, "Spremna beseda," in: *Akademske čebele ljubljanskih operozov*, Ljubljana: SAZU, 1988, 80.
277 Gspan, *Zgodovina slovenskega slovstva I*, 299, 300.

composers among its ranks. Music in the Slovenian provinces was flourishing at that time; in Koper, a prolific composer, Antonio Tarsia,[278] created a number of sacral compositions, and oratorios and operas began to be performed in Ljubljana. Thereafter, Slovenian composers were no longer entirely dependent on foreign individuals and institutions for favored artistic talent—although exceptional talents such as Giuseppe Tartini, born in Piran in April 1692, were able to win recognition only abroad. The opera Il Tamerlano by Giuseppe Clemente Bonomi, a member of an immigrant family from Italy, was staged in Ljubljana in 1732.[279] Even earlier, oratorios composed by Johann Berthold von Höffer, one of the forefathers of the Ljubljana-based Academia philharmonicorum, and Mihael Omerza, the cathedral provost, were premiered.[280]

The Ljubljana branch of the Roman Academy of Arcadia, founded in 1709 as Academia Emonia, is also a noteworthy institution. By then, the Carniolan capital had already established itself as a regionally important center of Baroque culture. It was precisely through Ljubljana that the new style, which attracted determined advocates from the nobility, as well as ecclesiastical and lay intelligentsia, began to spread northwards and eastwards. The academy remained in the Slovenian territory until the late 18th century, boasting the membership of well-versed painters, sculptors, and architects. Alongside foreign masters (Andrea Pozzo, Giulio Quaglio, Francesco Robba, Johann Martin (Kremserschmidt) Schmidt), locals and completely naturalized immigrants (Fran Jelovšek, Fortunat Bergant, Anton Cebej, Valentin Metzinger) gradually became established as its driving forces.

In addition to scholars who met and formed partnerships in academies, some other individuals rose to prominence in and beyond the Slovenian territory. Baron Franz Albrecht Pelzhoffer, notorious for his difficult character that hampered his ambition to become a member of various elite associations, caused immense irritation among his contemporaries as a social philosopher. Some of Pelzhoffer's work was even banned in the hereditary lands of the Habsburg Monarchy at the beginning of the 18th century. Ironically, this speaks to the importance and originality of this unorthodox author, who was constantly at odds with his contemporaries (the baron also strove to protect his treatises against censorship

---

278  During this time, musical life in Koper already had a rich tradition; at the turn of the 17th century, the Tuscany-born composer Gabriello Puliti was active in Koper and other Istrian towns.
279  Jože Sivec, *Opera skozi stoletja*, Ljubljana: DZS, 1976, 57.
280  Dragotin Cvetko, *Slovenska glasba v evropskem prostoru*, Ljubljana: Slovenska matica, 1991, 148, 157, 158, 160–162.

by dedicating them to General Eugene of Savoy, Spanish King Charles III, and Emperor Joseph I). His works were published in great German cultural centers, and he remained a promoter of politics as a means of public welfare rather than a value in itself.[281] An even more impressive career was attained by Gregorius Carbonarius de Biseneg from Naklo, who traveled to Russia at the turn of the 18th century and became the personal doctor of Peter the Great.[282]

This however does not imply that the uptake of new ideas was always as rapid. Despite the increasing number of educated people, the conceptions that eventually overthrew the old inherited notions were rather slow to establish themselves, and even modern science (which, like the late medieval humanism, disregarded state boundaries) could not yet boast of a rapid expansion: for instance, precise instrumental measurement, as advocated by Santorio Santorio long before, only began to be widely introduced to practical medicine in the 18th century. At the time, the Enlightenment philosophy spreading from large Western European centers and particular courts toward the heart of the old continent ignited the hitherto inconceivable progress in scientific observation and comprehension of the world.

At the turn of the 18th century, even the most scholarly of elite circles between the Alps and the Adriatic were not acquainted with these concepts. The new empirical views were still hardly being disentangled from the traditional ones in the minds of their most eminent representatives, so that Valvasor's *Glory* paid considerable attention to the phenomenon of witchcraft in Carniola.[283] A generation later, Janez Jurij Hočevar, extremely important in the establishment of Ljubljana's Academia philharmonicorum, tried a radical method to suppress the deeply rooted work of the Devil across the Ribnica area. This austere *juris utriusque doctor*, who had learnt from Valvasor's book that Carniola, except eastern Inner Carniola, had already been considerably cleansed of witches, initiated ruthless operations ending in the cruel torture and death of several accused women in 1701.

Hočevar, a member of Academia operosorum with the meaningful nickname of "Candidus" (The Pure), was just one of many persecutors of witches at the turn of the 18th century. The major witch trials that peaked at that time in Carniola

---

281 Evgenij Vasilevič Spektorskij, *Zgodovina socialne filozofije I*, Ljubljana: Slovenska matica, 1932, 220, 221.
282 See Marjan Drnovšek, *Nakljanec Gregor Voglar (1651–1717), zdravnik v Rusiji*, Naklo: Občina, 2002.
283 Janez Vajkard Valvasor, *Slava vojvodine Kranjske*, Ljubljana: Mladinska knjiga, 1978, 183, 184.

(Škofja Loka, Ljubljana, Ribnica, Bočkovo pri Ložu, Kočevska, Krško), Carinthia (Pliberk, Ženek, Humberk, Rožek), and Styria (Ormož, Ptuj, Ljutomer, Maribor, Hrastovec, Radgona) led to a few hundred executions between the Late Middle Ages and 1746, when the last prosecution in the Slovenian territory of men supposedly possessed by the Devil took place in Gornja Radgona.[284] Most of them were simple rural folk, acquainted only with the basics of Church teachings. Members of the already universally subordinate serf class were completely helpless before legally and theologically educated judges. Despite its irrationality, the impulsive persecution of witches, which claimed some 400 lives in the late 17th and early 18th centuries alone,[285] can be regarded as a particular means of intimidating the perpetually restless serfs. This insight is supported by the specific geography of the witch trials, which did not include the Littoral regions where feudalism took on a slightly different form from the system in the continental hinterland. It may also be concluded that belief in witches among Slovenes did not have indigenous roots, but rather originated from the German north.

## Politics and Economic Development

The great paradoxes of the Baroque age were not only manifested in the coexistence of great cultural and scholarly effervescence with a stupendous expansion of superstition, but also in the increasingly pronounced economic and political crisis of the Habsburg lands in Central Europe, which were given the common name of the Austrian Monarchy in 1711. The failed plans to expand overseas trade under Emperor Leopold I and the questionable results of the old nobility's entrepreneurial activity in the 17th and 18th centuries (the provincial assembly's cloth factory in Ljubljana performed poorly for a long time and was eventually sold to a private entrepreneur in 1747[286]) exposed the impassable obstacles of the traditional system and mentality. However, the reform plans of Joseph I's reign could not express all the advantages of modernization, due to the fierce struggle for the Spanish succession. Military reforms made the most impact, and were a court priority thanks to the situation in the field. Military conscription was introduced in 1705 and provided the Habsburgs with a regular influx of soldiers into their operational and rearguard units. In the Slovenian

---

284  Štih and Simoniti, *Slovenska zgodovina do razsvetljenstva*, 251, 253.
285  Ibid., 252.
286  Milko Kos, *Zgodovina Slovencev*, Ljubljana: Slovenska matica, 1979, 367.

territory, between 500 and 2,500 men a year could be put into uniform.[287] The armies and mercenary troops of the provincial assemblies disappeared.

After 1711, when Charles VI ascended the throne in Vienna, the Austrian Monarchy, engaged in battles in the west, north, and southeast, was initially at its largest (c. 725,000).[288] But its foreign policy entanglements, which led to several wars in the Iberian, Italian, German, Balkan, and Polish theaters, soon proved highly problematic, if not completely misguided. Many territorial gains from the early 18th-century peace treaties were soon lost, and even the weakened Turkey saw success in the war against Charles VI between 1736 and 1739, forcing the border of the Habsburg Monarchy northwards and westwards from central Serbia and Wallachia (the 1718 frontier) back to the Sava and Danube rivers and the Southern Carpathians. This was a clear indication of a serious crisis in the Habsburg state. To make matters worse, Russia's rising profile in European politics meant that the court of Charles VI could no longer maintain the illusion that its future role as "administrator" of the Ottoman Empire's "insolvency" was secure. In the early 18th century, when Peter the Great's reforms had not yet borne the expected fruits, the Austrian Monarchy indeed seemed predestined to become the heir to "the sick man on the Bosporus."

Charles VI, originally the Viennese claimant to the Spanish throne, who received the imperial crown after the death of his brother Joseph I, could not even devote his full attention to the slowly evolving lands after the War of Spanish Succession was over. The European balance of power forced the Habsburgs to give up their enormous properties on the Iberian Peninsula and in their colonies, and the dynasty was also confronted with the burning issue of succession. Charles VI's lack of a male heir constantly forced the Viennese court into solving problems of a purely political nature. The Habsburg dynasty eventually managed to secure the possibility of succession in the female line through several agreements and concessions inside and outside the monarchy, while enforcing the principle of the Habsburg dominions' indivisibility (the "Pragmatic Sanction" of 1713).[289] However, the theoretical solution to the problem could not significantly affect a thorough reorganization of life in the

---

287  Čepič and Nećak, *Zgodovina Slovencev*, 332.
288  Ibid., 329, 330.
289  Bogo Grafenauer, *Začetki slovenskega narodnega prebujenja v obdobju manufakture in začetkov industrijske proizvodnje ter razkroja fevdalnih organizacijskih oblik med sredo XVIII. in sredo XIX. stoletja*, vol. 5 of *Zgodovina slovenskega naroda*, Ljubljana: Kmečka knjiga, 1974, 5.

Austrian Monarchy. Even if the leadership had realized that reorganization was at least to some extent necessary, its preoccupation with diplomatic problems remained. Therefore, it is little wonder that plans to expand trade with overseas territories, which should have laid a most solid foundation for welfare according to the prevalent mercantilist doctrine, also failed during Charles' reign. There was simply not enough capital for the realization of such ambitious economic ideas, and the Austrian Monarchy could not even take advantage of trade privileges stemming from the Treaty of Passarowitz (present-day Požarevac), concluded with Turkey at the end of the 1716–1718 war.[290]

Nonetheless, the mercantilist policy of the Viennese court was very important for the Slovenian territory at an exposed strategic junction of transport routes linking Central Europe and the Mediterranean. The anticipated expansion of trade stimulated the reconstruction of the road system, which would usually have only been overhauled for non-economic reasons, such as Emperor Leopold I's visit to Carniola in 1660. Bridge building was a particularly significant step forward and severely limited the role of less efficient ferries at river crossings.[291] In Carinthia alone, investments in road construction and maintenance rose from a negligible 326 florins in the mid-17th century to nearly 10,000 florins per year between 1718 and 1729.[292] While responsibility for roads and transport had previously been borne by individual provinces, it now became a state matter as well.

In 1717, Charles VI declared freedom of navigation in the Adriatic, despite Venetian protests. Since his interests were in line with the aspirations of other great European powers, he was able to enforce his will. Two years later, he went even further and declared Trieste and Rijeka free ports.[293] Although trade did not flourish as originally envisaged, it nevertheless witnessed considerable growth: around 3,500 tons of salt were transported from the Adriatic coast into the hinterland every year, and the timber trade was another important industry for the Littoral. As in previous centuries, livestock, iron industry products, and linen were being exported from the Slovenian provinces to Italy. Sericulture, first introduced in Gorizia in the 17th century, became an important industrial sector, as did some other crafts (for instance, sieve manufacturing, gunmaking, weaving) in the hinterland.

---

290 Čepič and Nećak, *Zgodovina Slovencev*, 330.
291 Ibid., 341, 342.
292 Ibid., 341, 344.
293 Ibid., 330.

The guild system, which was under state patronage since 1732, greatly hindered the expansion of handicraft and entrepreneurship, at least in cities and market towns.[294] Rural handicrafts were able to develop more freely, yet the limitations imposed by the traditional feudal fetters prevented it from contributing significantly to economic modernization. The only appreciable impact could be seen in the rural trade developing in the hinterland of the free ports, especially Trieste, despite restrictive legal regulation (the 1737 patent).[295] The mercantile mentality only managed to loosen the existing system to a limited extent, since economically and structurally most of the Austrian Monarchy lagged considerably behind the Western European countries. Individual peasants in the Karst region and Inner Carniola became real small businessmen through their involvement in trade and transportation.

## The Theresian Reforms

The death of Charles VI in 1740 meant the extinction of the male line of the Habsburgs and ultimately revealed the Austrian Monarchy's insufficient synchronicity with the European state of affairs. Prussia, which was small in size but governed according to what were then modern principles, snatched most of Silesia from Maria Theresa after her accession to the throne. King Frederick the Great's new administration promptly managed to squeeze twice as many taxes out of the conquered land than the Habsburg sovereigns had, without economically destroying the country. The bellicose Hohenzollern could therefore use the income from a much smaller territory than the one controlled by the Viennese court to maintain an army that could compete with Austria's in every respect. Maria Theresa was no longer able to close her eyes to the need for a universal modernization of the state, since insistence on the existing system would, despite the Pragmatic Sanction, sooner or later raise the issue of her survival.

Count Friedrich Wilhelm von Haugwitz, entrusted with the reorganization of state administration in the Habsburg part of Silesia, immediately undertook to adopt the Prussian system. The royal plenipotentiary and a native of Silesia soon took the principled view that political and financial administration had to be assigned to a single office rather than to the existing multitude of administrative bodies. He also believed that it was imperative to reassess property and assign tax collection to the national authority.[296] Haugwitz believed that the predicament

---

294   Ibid., 341.
295   Ibid., 344.
296   Grafenauer, *Začetki slovenskega narodnega prebujenja*, 11, 12.

of the Austrian Monarchy could be resolved by a bureaucratic apparatus tied directly to the monarch, since the excessively complex administrative structure allowed for grave abuses (especially when levies were settled partly in money and partly in kind). The old nobility in individual provinces, which had already lost its political power under Ferdinand II, still controlled much of the financial and trade flows. It thus maintained its elite position that was otherwise threatened by prevalent absolutism along with the nobility's decreasing role in the mercenary- and later conscription-based military. Since the old nobility, unlike the ennobled middle class, had a mentality rooted in land-holding economics, they did not see a role for themselves in the emerging trade system; instead, they tried to maximize their benefits from their service and functions in different provinces.

The situation was particularly difficult in Carniola and Carinthia. Local aristocrats accused one another of machinations or corruption, thus attracting the attention of the court, which, encouraged by the success of reforms in Silesia, was already leaning toward the introduction of novelties throughout the Austrian Monarchy (with the exception of Hungary).[297] In 1747, Haugwitz journeyed to Ljubljana as the prince's investigative commissioner and began an examination of the management of the provincial assemblies. He discovered that Carniola, which had to pay around 100,000 florins in tax annually, was heavily encumbered with debts, which amounted to around 2,800,000 florins. It very soon turned out that the situation in Carinthia was even worse: the provincial debts totaled about 4,000,000 florins. Haugwitz justly blamed the irresponsible nobility for the unbearable situation, since it did not feel obliged to contribute to the payment of taxes; the court thus found a welcome argument for introducing a new administrative system on the southwestern margin of the Austrian Monarchy.[298] Thereafter, the rhythm of life was increasingly dictated by state officials, who rapidly established themselves as the skeleton of provincial assemblies' governments, which were entirely dependent on the will of the Viennese court.

Noblemen who did not adjust to these new circumstances found themselves in evident disgrace; the Carinthian provincial assembly even lost its right to participate in the setting of taxation levels for 20 years (1749–1769) as a result of

---

297 Owing to the great distress in which she found herself after the outbreak of war with Prussia and its allies, Maria Theresa solemnly approved the Hungarian provincial assemblies' existing privileges in 1741.
298 Victor Lucien Tapié, *Marija Terezija. Od baroka do razsvetljenstva*, transl. Vital Klabus, Maribor: Založba Obzorja, 1991, 103–105.

their complaints. In Carniola and Styria, where the new regime was introduced in 1748, aristocrats were more reasonable and cooperated with the executors of the sovereign's will in the field, albeit only after an explicit request from the court to "voluntarily" comply with the court's wishes.[299]

Major Theresian reforms, introduced as a necessary measure after the Austrian Monarchy's defeats by the Prussians, soon started to affect the way of life, encompassing more and more territory and interfering with more and more spheres. The resolution of financial problems through the newly established bureaucratic apparatus and the (at least rudimentary) modern structure of government resulted in the army becoming larger and increasingly better equipped; the army's very existence demanded an organized infrastructure and economic stability or prosperity in the hinterland. Education, too, became problematic, since the new trends in economic management and the intensification of production in traditional branches of the industry called for ever more knowledge. In fact, the reform initiatives in the Austrian Monarchy were not connected to the principles of the Enlightenment, but in practice their execution resembled the endeavors of rationalist "philosophers" who saw themselves as a principled opposition to the "Christians." Maria Theresa and a great majority of her officials were in no way opposed to religious dogmas, but the urge to establish a modern school system in the Austrian Monarchy, and aspirations to resolve legal and administrative issues surrounding the Church forced the court to interfere in spheres which had previously belonged to the ecclesiastical domain. The reforms concerned everything and everyone and demanded a change of mentality from each and every individual. Their implementation marked people's lives differently, but nevertheless represented the population's first encounter—and confrontation—with the evolving modern state administration.

The Theresian reforms not only established a new relationship between the highest authority and the provinces, but also brought about radical changes to the ruling system. Uniform legislation began to be enforced for the entire non-Hungarian part of the Austrian Monarchy. In 1768 and 1769, a uniform Penal Code (*Constitutio Criminalis Theresiana*) was introduced in the territory from the Adriatic in the south to the Bohemian crown lands in the north. Initially, it still provided for torture, but the empress rescinded this part in 1776.[300] Ancient boundaries between individual territories drawn by the nobility began to change

---

299  Grafenauer, *Začetki slovenskega narodnega prebujenja*, 21.
300  Štih and Simoniti, *Slovenska zgodovina do razsvetljenstva*, 253.

## The Theresian Reforms

in accordance with the needs of the court's policy. In 1748, the Mercantile Province of the Littoral was established, which was not territorially united and comprised Aquileia, Trieste, Rijeka, Bakar, and Kraljevica.[301] The Habsburg ports on the Adriatic became an instrument of state economic policy through uniform administration. Carniola, which had previously linked the coast and hinterland, began to acquire an explicitly continental character, a process which concluded after the decline of Napoleon's Illyria.

The formerly unified provinces were divided into *kresije* (Ger. *Kreis*),[302] which became units of state administration. Cities in which officials of the crown settled began to take on the character of regional centers, although many contained less than 10,000 inhabitants. The commissioners in charge of kresije administrations were responsible for the relations between the state and the unit's population. Although seigneurs, burghers, market town settlers, and serfs did not have the same obligations and rights, they were subjected to the same administrative authority.

In Carniola and Carinthia, the Slovenes lived in all three kresije (with centers in Ljubljana, Novo Mesto, and Postojna; and in Villach, Klagenfurt, and Völkermarkt), and in Styria in two (Leibnitz, or later Maribor, and Celje). The Mercantile Province of the Littoral had its center in Trieste. The territorially small Gorizia with Gradisca, originally attached to Carniola, obtained its own special government in 1754.[303]

After the new structure of authority had been set up and stabilized, the administrative reforms, intended from the outset to increase state efficiency and thus to enhance its battlefield strength, continued at an undiminished pace. The first census was carried out in 1754. Its results were unreliable, since people avoided the count, realizing that it would serve either for taxation or for military service. Only the numbering of the houses, ordered by a patent of March 10, 1770, brought more precise data with the new census of 1771. This census allows one to calculate that in the late 18th century approximately 750,000 inhabitants lived in the territory of the present-day Republic of Slovenia, with the total number of Slovenes amounting to about 900,000. The number of inhabitants did not rise substantially; on average, birth rates averaged between 35 and 39 per thousand, while the death rate was slightly lower. There were marked differences among individual provinces. The predominantly mountainous Carinthia

---

301   Čepič and Nećak, *Zgodovina Slovencev*, 354.
302   Administrative units roughly corresponding to districts (transl. note).
303   Vasilij Melik, "Slovenci v času Marije Terezije," in: Tapié, *Marija Terezija*, 364.

had considerably lower birth and mortality rates (28.5 and 28 per thousand, respectively) compared to the regions to its south.[304]

Conscription after the census proved that people's fears were justified. Between 1771 and 1773, a new conscription system was introduced, providing the Austrian Monarchy's armed forces with many more soldiers than the previous system had produced: Count Haugwitz's plan, devised long before, for a peacetime standing army of 108,000 men (requiring an annual sum of 14 million florins) began to acquire solid foundations.[305] The new recruitment system relied on delegated recruiting authorities. But a more efficient recruitment of soldiers, who could only be exempted from permanent service on grounds of serious disability, obviously had a darker side. In Carinthia, it was calculated that conscription had cost the province nearly 39,000 men between 1771 and 1790, almost one seventh of the population according to the 1771 census.[306] Conscription mainly affected the poorest classes, since it excluded the nobility, clergy, officials, physicians, lawyers, artisans, city merchants, peasants with large and medium-sized properties, and members of some other "vital" occupations (their firstborns or even entire families were sometimes exempted as well). Exemptions were also granted to the inhabitants of certain places, such as the rapidly evolving Trieste, whose population increased from around 7,000 to 28,000 in the late 18th century.[307]

Administrative and military reforms became irreversible due to an obsessive idea in court circles that a retaliatory war against Prussia was necessary. The reforms would remain unfinished without new legal arrangements for trade, the regulation of relations between feudal lords and serfs, or without interference with the Church and education, since the modernization of particular sectors alone could not produce the desired results. The economy needed a major incentive, because the traditional feudal system no longer guaranteed further economic growth. Internal customs borders were abolished gradually (except for the borders with Hungarian provinces); toward 1770, physiocratic views emphasizing the value of individual economic initiative were gaining ground. However, the role of state authorities was irreplaceable in projects demanding a considerable concentration of financial resources. The drainage of the Ljubljana

---

304 Grafenauer, *Začetki slovenskega narodnega prebujenja*, 29.
305 Tapié, *Marija Terezija*, 107.
306 Grafenauer, *Začetki slovenskega narodnega prebujenja*, 30.
307 Ibid., 29, 30, 74. Trieste itself, without its immediate hinterland, recorded an even more dramatic rise, from 4,000 to 21,000 inhabitants.

Marshes, discussed as early as the 16th century, only became possible after Maria Theresa developed enthusiasm for it. Control over the works, which mainly involved constructing a canal between the Golovec and Castle hills in Ljubljana, was entrusted to the Vienna-born Jesuit scholar Gabriel Gruber, who helped establish the Carniolan capital as an important center of exact sciences and engineering (one of his students was the renowned mathematician and engineer Jurij Vega). Even though the distinguished builder did not finish the canal that he had designed, his plan was realized to the last detail under the leadership of Vincenc Struppi.[308]

The emergence of the physiocratic doctrine, based on the idea that wealth was essentially derived from land, stimulated the founding of agricultural societies in individual provinces (in Carinthia in 1764, in Styria and Gorizia in 1765, and in Carniola in 1767). Their members were primarily officials and seigneurs inspired by the state's new economic policy. However, the introduction of new crops (corn, potatoes, and clover), land cultivation practices (orderly crop rotation; a greater role for legumes), and methods of house or farm construction spanned several decades, well into the early 19th century.[309] In non-agricultural branches, the main novelty was the rapid establishment of coal as a source of energy, reputed to be "dragon's blood." It was excavated almost across the entire Slovenian territory, from the Littoral to Styria.[310] The rise of coal mining testified to the growth in economic activity, since the traditional iron industry (already starting to feel the consequences of the increasingly exploited iron ore deposits) remained attached to charcoal.

The everyday life of the overwhelming majority of the population was mostly affected by labor service restrictions. The state introduced them with great caution, since outbursts of discontent among either the peasantry or feudal lords could have proven lethal to the economic stability of the predominantly agrarian Austrian Monarchy. In 1778, labor service was limited to three days a week in Styria and Carinthia, while remaining slightly longer in Carniola even after the regulation of 1782.[311] Tenure (known as the *colonatio*) in the Littoral remained untouched, although its abolition was under consideration. The cadaster produced between 1748 and 1755 tried to eliminate various anomalies

---

308 Albert Struna, *Naši znameniti tehniki*, Ljubljana: Zveza inženirjev in tehnikov Slovenije, 1966, 42–45.
309 Grafenauer, *Začetki slovenskega narodnega prebujenja*, 44–47.
310 Ibid., 60.
311 Čepič and Nećak, *Zgodovina Slovencev*, 360.

in previously accumulated levies or charges and define them in relation to taxable persons' real economic strength.

The authorities' stabilization projects also involved an attempt to reduce or abolish the institution of temporary tenure (lease) and promote hereditary tenure (lease in perpetuity), which gave serfs much more security and limited the despotic power of feudal lords. Despite a principled arrangement of this issue (in 1766 or 1772 in Carinthia, in 1788 elsewhere), in some places the old relationships endured right up until the 1848 revolution.[312]

The state also seriously interfered with the population's daily rhythm through its education reforms, which from the outset greatly affected the Catholic Church. The Church had largely monopolized education since the decline of Protestantism until Charles VI's time, but under Maria Theresa the Austrian Monarchy started taking various measures which systematically limited the influence of foreign centers of power on the functioning of the Catholic Church within its borders. For this reason, the Habsburg part of the Aquileia Patriarchate, dissolved in 1751, passed under the authority of the newly established Archdiocese of Gorizia; it was forbidden to issue Papal Orders and Pastoral Epistles without the secular authority's permission, while priests could only attend domestic seminaries and universities and could no longer send money abroad. For the same reason, the state did not and could not avoid taking control over education. The process had begun during Charles VI's reign in response to the ossified educational system, but was greatly accelerated after 1740, with new goals set. The functioning of the state under the new circumstances required people to be able to follow modernization initiatives, leading to the conclusion that each man must master a particular field of knowledge. The need to fill the increasing number of clerical posts and intensify economic activities made the establishment of the new school system inevitable. To make matters worse, the serious disputes of the Portuguese, French, and Spanish crowns with the successors of St. Ignatius brought the Pope to suppress the Jesuit order, which hitherto had had the main say in educating the subjects of the Viennese sovereigns.

In 1770, the Austrian Monarchy declared schools to be "politicum," a matter of the state, providing the basis for further state interventions in education. Mandatory education in the western part of the Austrian Monarchy was introduced in 1774. For the Hungarian crown lands, i.e., for the territory where the Prekmurje Slovenes lived, the same measure was adopted slightly later, in 1777. Four-year schools were introduced (as preparatory institutions for

---

312   Ibid., 360, 361.

grammar schools) and main schools were founded in larger towns, while trivial schools (one-class schools) were set up in rural areas for children aged between 6 and 12.[313] Teachers were supposed to have pedagogical education, but in the beginning the practice greatly lagged behind the theory. The most far-reaching goals of the Theresian reform of the education system were achieved gradually; in fact, mandatory education did not begin to function properly until around 1900. In the Slovenian territory, the greatest difficulties with regular school attendance were experienced in northwestern Istria and Prekmurje, i.e., in the provinces which in the late 18th century were not central to educational modernization because they belonged to the Venetian Republic and Hungary, respectively.

Although the Theresian reforms occasionally caused great dissatisfaction among the nobility, they did not generate universal opposition anywhere. Even though the measures were introduced by the state center, over time they attracted fervent advocates even in very different parts of the Austrian Monarchy. Individuals who felt intellectually or economically limited by the traditional system became ardent supporters and even executors of the Viennese court's measures. In the Slovenian territory, a typical representative of pro-reform scholars was Blaž Kumerdej, who had studied philosophy and law in Vienna (at the same time helping his fellow countryman Anton Janša to write the famous *Razprava o rojenju čebel* (A Treatise on Swarms of Bees)). On his own initiative, in 1773, he prepared and submitted to the authorities *Domoljubni načrt, kako bi se dalo kranjsko prebivalstvo najuspešneje poučevati v pisanju in branju* (A Patriotic Plan on How to Teach Carniolans to Write and Read with the Greatest Success). He was of the opinion that the Slovene language should be used in schools first, and that the teaching of other languages, especially German, should begin later. Although the Viennese authorities failed to observe his proposal, Kumerdej was appointed headmaster of the Ljubljana normal school (Ger. *Normalschule*), which he transformed into the best educational institution of its kind in the Austrian Monarchy. During the reign of Emperor Joseph II, he rose even higher and became the district school supervisor in Celje (1786–1792). Initially, Kumerdej managed to make the authorities assign Slovene, rather than exclusively German (which was spreading throughout the entire western half of the Habsburg state), a certain role in the educational process. Later, however, his contagious enthusiasm for educating simple compatriots surmounted many a bureaucratic restriction. Kumerdej, who researched his mother tongue from a scientific perspective (grammar, vocabulary), also worked tirelessly on

---

313 Čepič and Nećak, *Zgodovina Slovencev*, 375, 376.

translating textbooks into Slovene. His views were also embraced by Count Janez Nepomuk Jakob Edling, who played an important role in introducing Slovene to elementary schools (especially in Carniola).[314] The initiatives of individual state officials in practice significantly modified the policy or aspirations of the Viennese court.

Kumerdej managed to combine the fervor of Theresian reform with the consciousness of Slovenian individuality that was steadily growing due to systematic pressures of Germanization. The instability of the traditional system made individual and community identities more salient. Even the polymath Janez Žiga Valentin Popovič, born in Arclin near Celje and pursued a thrilling career as a private scientist between 1753 and 1766, before occupying the new Chair of German at the University of Vienna, at the time greatly emphasized the connectedness and linguistic unity of Carniolans and inhabitants of the neighboring provinces, even though he later spent most of his time abroad. Popovič believed that his fellow countrymen and South Slavs were victims of history and their geographical position, forced to sacrifice their own progress for the defense of the Western world against the Ottoman conquerors. Popovič's work later became an important inspiration to both Slovenian and German scientists who drew material and ideas for their research from his published and unpublished papers.[315]

Of course, even more important for the future was the work of writers who, like Kumerdej, maintained daily contacts with their homeland. Remarkable for its volume and diversity was the literary work of the barefoot Augustinian monk Marko Pohlin, who published *Kraynska grammatika* (A Carniolan grammar) in 1768. He equipped his grammar, written in German, with an enthusiastic foreword in which he called upon his fellow countrymen not to be ashamed of their mother tongue. Although his work was initially intended for Carniolans alone, the industrious Pohlin eventually subscribed to the trans-provincial or all-Slovenian concept (probably under the influence of Popovič, with whom he was in contact after 1775, when Popovič was in Vienna), which was also supported by the critics of his often idiosyncratic linguistic conceptions—especially the Jesuit Ožbalt Gutsman in Carinthia. In his Slovene grammar, which was first published in Klagenfurt in 1777 (and afterwards reprinted five times), and in other works, Gutsman firmly advocated the cultural unity of his fellow countrymen divided among many provinces.[316]

---

314   Gspan, *Zgodovina slovenskega slovstva I*, 368–370.
315   Ibid., 352, 353.
316   Ibid., 353–361.

The opus of Pohlin and Gutsman inspired passionate writing of many other authors. In addition to religious writings, works appeared in the Slovene language that tried to improve the common people's daily life with various kinds of advice. Literature also emerged that was committed to the then modern aesthetic standards of the 18th century. In 1776, Pohlin published a physiocratically-oriented "textbook of life," entitled *Kmetam za potrebo inu pomoč* (To peasants for their need and help) (after Zacharias Becker). He also simultaneously translated the literary letters of Christian Fürchtegott Gellert. Characteristically, he added a chapter on metrics to the second edition of his grammar in 1783. Alongside the somewhat aesthetically conservative Pohlin, around whom younger patriotic writers began to gather, Fr. Janez Damascen Dev, editor of a poetry almanac, *Pisanice od lepeh umetnost* (A collection of fine literature; between 1779 and 1781), was also literarily active in Ljubljana. The almanac included several stylistic trends, from Baroque (a text for a short opera) and Rococo (epigrams and lyrics) to pre-Romanticism (a poem about Bürger's famous ballad, "Lenore").[317] Slovenian literature, which (except for folk literature) until then had been predominantly religious, thus took on a secular dimension. Dev and the Pohlin circle, to which the first openly patriotic Slovenian poet and journalist, Valentin Vodnik, belonged, were crucial in synchronizing what had been one-sided, predominantly religious, literary developments in the territory between the Eastern Alps and the Adriatic with the developments in Europe's "republic of spirits." The task of setting to music the libretto of a short opera published in *Pisanice*, which was undertaken in 1780 or 1782 by Jakob Frančišek Zupan, a composer based in Kamnik and Komenda, demonstrated that the changes had permeated music as well as literature.[318] Before Dev, Jurij Japelj, who translated Metastasio's frequently composed libretto *Artaserse*, had endeavored to establish the Slovene language on the opera stage. This ambitious priest also translated individual works by Alexander Pope, Jean Racine, Gellert, and the "Germanic Plato," Moses Mendelssohn (of Jewish descent) into his mother tongue. He later became the chief translator of the first Catholic edition of the Bible into Slovene (1784–1802) and died as the appointed (but not yet invested) Bishop of Trieste.[319]

---

317 See Janko Kos, *Primerjalna zgodovina slovenske literature*, Ljubljana: Znanstveni inštitut Filozofske fakultete, Partizanska knjiga, 1987, 15–24.
318 Cvetko, *Slovenska glasba v evropskem prostoru*, 182–184.
319 Gspan, *Zgodovina slovenskega slovstva I*, 371, 372.

## Joseph II's Enlightenment Inclinations

The death of Maria Theresa at the end of November 1780 caused much sincere sorrow among the nascent Slovenian intellectual elite. In honor of the late empress, perhaps the only person worthy of the title "Great" among the Habsburgs, two elegies by Dev and Vodnik were published in *Pisanice*. Her successor, Joseph II, at first enjoyed a reputation as the executor of her reform policy. He unequivocally achieved great popularity as the emperor who strove to improve the conditions of his subjects (Dev praised him as "our Titus" as early as 1779). Some were very fond of his Enlightenment ideological orientation that visibly distinguished him from his mother. In early April 1781, Anton Tomaž Linhart exclaimed in the recently renovated Academia operosorum in Ljubljana: "We have freedom of thought and Joseph on the throne!"[320]

Yet the new ruler's reforms in the 1780s soon proved to have a very different frame of reference from their Theresian precursors, going beyond intensification or radicalization to new objectives. Pragmatism was giving way to clear ideological stances, since Joseph II did not hide his adherence to Enlightenment principles. In 1782, the emperor abolished serfdom, and in 1789, he did the same with labor service through tax and land register regulation. He also put an end to peasants paying duties in kind. Joseph's emancipation measures were accompanied by the preparation of a new land register replacing the Theresian one (it goes without saying, of course, that the new measuring and accounting methods under state control were not accepted enthusiastically, since more accurate data usually led to higher duties).

The emperor's administrative reforms made the territories of several provinces subject to Gubernia (Styria, Carinthia, and Carniola fell under the administration in Graz, the Littoral under that in Trieste) and abolished provincial assembly committees,[321] greatly diminishing the powers of the traditional elites. Even wider discontent was caused when German was declared the official language of the entire Austrian Monarchy in 1784; this measure could not satisfy all state officials or advocates of Enlightenment philosophy. Universal centralization also significantly affected smaller centers. In the Josephine period, Ljubljana temporarily lost higher education studies of philosophy and theology (which

---

320   Alfonz Gspan, *Anton Tomaž Linhart. Zbrano delo*, Ljubljana: DZS, 1950, 326.
321   A major administrative unit in the Habsburg Monarchy (transl. note).

had survived the suppression of Jesuit colleges).[322] The Carniolan capital also lost its appeal for the enterprising Gabriel Gruber. In 1785, he discreetly set off for Russia, where the Jesuit colleges continued to function under the auspices of the highest authorities.

The emperor's wish that the entire Austrian Monarchy be managed according to uniform principles led in practice to an attempt to create a sort of German-speaking "Habsburg nation." However, his policy soon ran into insurmountable obstacles. In Hungary, the emperor's centralization tendencies provoked rebellious sentiments, expressed in the failure to settle fiscal charges and the obstruction of the new land survey, and in Belgium the general discontent even escalated into a full-blown insurrection. Before his death, the revolutionary reformer was therefore forced to recognize that his attempts to transform the Habsburg Monarchy into a unified state had failed. In December 1789, Joseph II promised the abolition of several measures he himself had introduced. In Hungary, where he would not crown himself ruler, he even intended to reestablish the Theresian system.[323]

Figure 30. Dramatist, poet, and historian Anton Tomaž Linhart.

---

322 Grafenauer, *Začetki slovenskega narodnega prebujenja*, 97, 98. During the reign of Joseph II, the only new subject to be studied in Ljubljana was medicine; philosophy was restored in 1788, three years after its abolition.
323 Hans Magenschab, *Jožef II. Revolucionar po Božji milosti*, transl. Vital Klabus, Maribor: založba Obzorja, 1984, 348–353.

However, Joseph provoked an even greater outcry among the population by systematically intervening in the domains of religion and the Church. The early restrictions on celebrating many saints' cults were not particularly welcomed, since religious festivals were a way of alleviating the serfs' arduous daily existence. Further state interventions in this sphere incited loud indignation. Although the Edict of Toleration, issued in 1781, did not essentially challenge the leading role of Catholicism, it nevertheless had great symbolic value, allowing the state to demonstrate its supreme authority in religious matters as well, whereas in practice it proved valuable only for religious minorities—Protestants (in the Slovenian territory in Prekmurje and Carinthia) and Jews (Trieste, Gorizia, and Prekmurje).[324] Many Catholics of traditional upbringing disapproved of the policy. The dissolution of the monasteries of contemplative orders, which Joseph believed did not carry out useful work for the community, additionally fanned discontent. By suppressing monastic orders (especially the Cistercian order in Stična and the Carthusian order in Žiče and Bistra), Josephinism inflicted enormous cultural damage to the Slovenian territory—in certain areas, this damage was certainly more devastating than the consequences of communist policies in the 20th century—since many ancient manuscripts and other treasures were removed from the culture they represented. The estates of dissolved monasteries began to deteriorate in every way: in many places, this led to a decline in the sound administration of land.

At the time of Joseph's universal transformation of the empire, the Catholic Church as a whole could only be satisfied with the more reasonable delineation of diocesan boundaries, which were redrawn to correspond with state administration units[325] (even though individual archbishops who lost their competence over certain territories understandably opposed this imperial intervention), and with the establishment of new parishes to enable priests regular contact with their entire congregation. However, new parishes were established on the assumption that future generations of clergymen, educated in general theological seminaries (a seminary for Styria, Carniola, and Carinthia was established in Graz), would follow the state policy. The Church was expected to surrender completely to the will of the Viennese court, which at the height of the new reformist fervor in

---

324  Grafenauer, *Začetki slovenskega narodnega prebujenja*, 87, 88. Orthodox Christians (who also lived in Trieste as immigrants) had already been granted freedom of worship in the Theresian period.

325  The Diocese of Lavant now (until 1859) encompassed the territory of the former Völkermarkt (Slov. Velikovec) district in East Carinthia and the Celje kresija in southern Styria; to its north stretched the Diocese of Seckau with its see in Graz.

1784 decreed that the dead should be buried in sacks so as not to waste precious wood.[326]

Only a few priests in the Austrian Monarchy supported Joseph's course of action without self-serving motives, the most prominent being the Prince-Bishop of Ljubljana, Karel Janez Herberstein. This was another reason why his diocese increased considerably when the boundaries were redrawn (it covered almost the whole of Carniola). For some time, Ljubljana even became a metropolitan see or archdiocese. But the frequent rearrangement of ecclesiastical boundaries, status alterations, and relocations of individual diocesan sees (especially in the Littoral) showed the instability of Joseph's ecclesiastical policy: it failed to convince most clergy and vast masses of lay believers of its good intentions. The unpopularity of the emperor's religious and ecclesiastical measures was soon demonstrated by the enthusiastic reception of Pope Pius VI, who in 1782 journeyed to Vienna through the Slovenian provinces to dissuade Joseph II from further reformist impulses (though in vain). Simple people under the influence of priests who opposed the emperor or monks banished from their monasteries often interpreted the Viennese authorities' measures as an outright attack on their traditional faith. Even though the emperor himself had yet to become the target of open criticism, the executors of his will in the field were looked down upon as "vipers" who called "God's punishment" upon themselves. The "enlightened world" that the emperor tried to establish in both secular and religious spheres became a synonym for disorder and hard life, since taxes represented an increasingly excruciating burden. The commoners simply did not consider it appropriate for everyone to "master multiplication like the Ten Commandments."[327]

Joseph II, who died in 1790 during the poorly conducted Austrian (Austro-Russian) war against Turkey and the rapidly radicalizing French Revolution, did not strengthen the state with his reforms; on the contrary, he thoroughly shook its foundations. As a ruler by the grace of God, he did not find it necessary to take into account the sentiments of his subjects. Instead, without any sense of the boundaries of the possible, he tried to enforce his will, which he considered to be an expression of the noblest volition in man. When the disgruntled views of the population became rather important, as the Austrian Monarchy's foreign policy

---

326 Magenschab, *Jožef II*, 278.
327 The simple folk's perception of the Josephine reforms was preserved in the anonymous poem *Pesem od tega rezsvetleniga sveta*, produced after Joseph's death. Most likely written by a priest, the poem spread among peasants in the countryside. See Alfonz Gspan, *Cvetnik slovenskega umetnega pesništva do srede XIX. stoletja*, vol. 1, Ljubljana: Slovenska matica, 1978, 324–327 and 379–382.

collapsed—with ineffectual wars in the Balkans (1788–1791) and rivalry with Prussia for Bavaria (1778–1779; 1785)—he realized that all his enterprises had failed. His brother, Leopold II, was forced to end the war with Turkey as quickly as possible and at almost any cost so as to concentrate on consolidating the state and the ever-greater problems for the Viennese court posed by the French Revolution. The reform period was concluded and the most radical Josephine measures abolished. Dissatisfaction with the fundamental political changes, driven by the fear of a return to the most rigid feudalism, and escalated into peasant uprisings throughout the Slovenian territory (especially in the kresije of Celje and Maribor), which were quickly quelled by the authorities.[328]

Within a short period of time, until Leopold's death in March 1792, the Austrian Monarchy completed its internal stabilization on foundations which maintained the achievements of the Theresian period in their entirety. A number of Joseph II's measures also remained in force, although more determined strides toward modernization had proved to cause instability. It was impossible to break the confines of the contemporary times and mentality. Above all, Josephinism had not only revealed an orientation toward the future, but had also sworn by absolutism, and did not consider tolerance a universal principle. This was precisely why it caused dissatisfaction among many people who were originally in favor of reform. Proper anti-Josephine coalitions ranging from principled conservatives and local aristocrats to moderate advocates of the Enlightenment were established in individual provinces in opposition to the measures of the imperial bureaucracy. Conservatives disapproved of the reforms' increasingly radical turn toward modernization (especially when it came to religion); nobles were dismayed by the loss of traditional privileges and their diminished role in individual provinces; and the pro-Enlightenment moderates opposed Viennese centralism. They found a common denominator in the emerging spirit of nations that began to spread in the context of the late-Enlightenment and pre-Romantic philosophies, both of which emphasized in different ways the importance of the natural life embodied in the common people.

It was mostly the pro-modernization circles that strengthened their position in the Slovenian territory during Joseph's reign: a unique expression of this was the relatively favorable coverage of the early phases of the French Revolution in the Ljubljana press, and Freemasonry (with lodges in Ljubljana, Klagenfurt, Trieste, and Maribor) also began to spread.[329] All of them gradually became very critical of the court's reformist efforts.

---

328   Grafenauer, *Začetki slovenskega narodnega prebujenja*, 85.
329   Peter Vodopivec, *Od Pohlinove slovnice do samostojne države. Slovenska zgodovina od konca 18. stoletja do konca 20. stoletja*, Ljubljana: Modrijan, 2006, 13.

This trend was characterized by Anton Tomaž Linhart, a universally ambitious intellectual who attended the lectures of the enlightened theorist Joseph von Sonnenfels in Vienna and became an admirer of Montesquieu.[330] He tried to combine the Italian spirit and German culture in his early poetic work (even the imperial poet laureate, Pietro Metastasio, praised his euphonic and melodious skill in Slovene[331]), and was enthusiastic about freedom of thought and the enlightened ruler in Vienna. Upon returning to Ljubljana, he became the most active member of the renovated Academia operosorum (1781). Yet contact with reality soon "cured" him of an exaggerated enthusiasm for the fruits of Josephine progress. Although he was a successful state official who saw to the establishment of numerous new schools in Upper Carniola (an amazing 27 schools between 1786 and 1790), he began to point out its problematic tendencies in his light-hearted comedy, *Županova Micka* (Micka, the Mayor's Daughter), which successfully premiered in Ljubljana in 1789, and the comedy *Ta veseli dan ali Matiček se ženi* (This Happy Day, or Matiček Gets Married). The latter, published in 1790 but not performed until 1848, was one of the most autonomous adaptations of Beaumarchais' *The Marriage of Figaro*.[332]

In Linhart's works, German swindlers who come into contact with Slovenian peasants and the German language enforced by the bureaucracy cause confusion and complications, but common sense ultimately prevails. Both plays very suggestively represent the imagery and ideas of patriotic enlightened men; the author chose to use the dramatic form because the masses of illiterate people could only come into contact with high-level literature and its ideas through theater (Ljubljana opened the modern "House of Thalia" in 1765[333]). Moreover, between 1788 and 1791, Linhart, who had also admired Shakespeare in his youth, even wrote a German Sturm und Drang tragedy, *Miss Jenny Love*, and published a historiographical study, *An Attempt at the History of Carniola and Other Lands of the South Slavs of Austria*, exemplary for its time. Although he only succeeded in conducting a critical examination of the period until the reign of Charlemagne, the work's trans-provincial concept, which no longer took into consideration the existing boundaries of the Habsburg crown lands, but

---

330 Gspan, *Anton Tomaž Linhart. Zbrano delo*, 278.
331 Gspan, *Zgodovina slovenskega slovstva I*, 392.
332 *Ta veseli dan ali Matiček se ženi* may also have been influenced by da Ponte's libretto for Mozart's opera *The Marriage of Figaro*. Linhart envisaged his play to have three musical numbers, composed by the Ljubljana-based composer Janez Krstnik Novak. Novak entitled his work—which was both stylistically and expressively modeled upon Mozart—*Figaro*. See Cvetko, *Slovenska glasba v evropskem prostoru*, 221–224.
333 Vodopivec, *Od Pohlinove slovnice do samostojne države*, 18.

was conceived as a description of people's lives along with their material and spiritual culture, represented an enormous advance in Slovenian self-perception. Linhart's great historiographical essay, written in German, was soon reprinted in Nuremberg in 1796, which in and of itself testifies to the quality of the essay. For a long time thereafter, no study of comparable quality was published about the territory between the Eastern Alps and the Adriatic.

As was common at the time, after the death of Joseph II, Linhart prepared for the new sovereign a memorandum of the Carniolan provincial assembly that presented the views of the local nobility. The drawing up of this document did not mean that he had abandoned his rather radical Enlightenment views (according to him, the Bishop of Ljubljana Herberstein was a "mitered madman"),[334] but was merely an attempt to achieve consensus in the anti-centralist coalition on the Slovenian soil. He considered the provincial assemblies an important factor in resisting the (self-)will of the imperial administration. However, the nobility soon completely adapted itself to Viennese politics. A fair bit more productive for the future was Linhart's notion that the national structure of the Habsburg Monarchy made it predominantly a Slavic state. Although the concept of Austro-Slavism, developed in the 19th century by Jernej Kopitar, never prevailed, it nevertheless had important cultural significance. It was thanks to this idea that Vienna became the most important center for Slavic studies, embodied by none other than Franc Miklošič, Linhart's compatriot from Styria. The purely political significance of Austro-Slavism was incomparably smaller.

The activities of Anton Tomaž Linhart and other enlightened Slovenes were made possible in many ways by the patron Žiga Zois, who combined cosmopolitanism and patriotism as successfully as his affinity for humanistic disciplines with his interest in mineralogy, geology, botany, and zoology (he was, however, not a very successful entrepreneur and could not pride himself on such brilliant successes as his father, Michelangelo). Zois was particularly interested in natural sciences, which grew remarkably in the Slovenian territory in the late 18th century (Giovanni Antonio Scopolli, who worked as a physician in Idrija and corresponded with Linnaeus, catalogued the Carniolan flora; his successor Balthasar Hacquet was renowned as the first explorer of the Julian Alps).[335] Zois, who was himself a poet (he translated Bürger's "Lenore" into Slovene) and gradually accumulated a voluminous library in his palace, might have already supported Dev's *Pisanice od lepeh umetnost* financially, but as a mentor he definitely steered the course of Slovenian literature after 1781. His

---

334   Gspan, *Anton Tomaž Linhart. Zbrano delo*, 287.
335   Struna, *Naši znameniti tehniki*, 49–54, 177–180.

parlor was a gathering place for Slovenian illuminati, although some more conservative individuals could also be found there. Jernej Kopitar, later an eminent and influential Viennese Slavist, took his first steps into the world of science precisely in Zois' palace, and in the early 19th century the Zois circle ensured that Ljubljana periodically held theatrical performances in Slovene[336] (among the actors was Franc Pollini, a relative of Zois and later the benefactor of the composer Vincenzo Bellini).

**Figure 31.** Mentor and reformer baron Žiga Zois.

---

336   Cvetko, *Slovenska glasba v evropskem prostoru*, 224.

The Ljubljana illuminati were above all oriented toward science and literature, but they did not particularly distinguish themselves in philosophy. In this respect, they could not measure up to the Klagenfurt-based circle of Baron Franz Paul Herbert, an adherent of Kant. No real opportunities existed for the spread of Enlightenment philosophy in the Slovenian territory, since higher education in smaller centers had suffered severe blows during the Josephine reforms. Individual Slovenes nonetheless became prominent philosophers, among them Franc Samuel Karpe, an ardent admirer of Leibniz and Wolff (but a critic of Kant) and a lecturer at the universities of Olomouc, Brno, and Vienna. Karpe conceived a "philosophy without surnames."[337] Like the foremost Slovenian mathematician Jurij Vega, whose 7-place and 10-place logarithm tables established his reputation among contemporaries and future generations, as well as earning him membership in several European academies of science,[338] he had not (previously) had the opportunity to universally develop his talent in his homeland.

---

337   Frane Jerman, *Slovenska modroslovna pamet*, Ljubljana: Prešernova družba, 1987, 40–46.
338   Struna, *Naši znameniti tehniki*, 214–219.

Modernization and National Emancipation

# French Rule

In 1792 the Habsburg crown passed to the untried and conservative emperor Francis II. His imposition of stringent censorship and political pressure as well as his persecution of defenders of the Enlightenment and sympathizers with French revolutionary ideas hampered the reformist impetus in the Habsburg Monarchy. However, it did not completely paralyze the adherents of new ideological movements, who continued to meet in secret societies and Masonic lodges. One border regiment officer, Baron Siegfried von Taufferer, not only embraced the ideas of the French Revolution, but also tried to use them to bring change to the monarchy. He established direct contact with the revolutionary government in Paris and organized volunteer companies of Austrian prisoners of war to fight for the French, whose assistance and revolutionary experience he considered instrumental in replacing the absolutist system with a government formed on the principles of the French Revolution. He also drew up a plan for the secession of the provinces with Slovenian and Croatian population and their incorporation into the French Empire. In 1795, he was captured by the Austrians and executed in Vienna.

Still, the Slovene-speaking population not only experienced French influence ideologically and spiritually, but also directly. In 1797, French troops nipped at the heels of the defeated Austrians, who were retreating across northern Italy and along the Soča River toward Carinthia. Gorizia fell first on March 20, after the French army had crossed the Soča, followed within the next three days by Trieste and Idrija, "where two million francs' worth of quicksilver and cinnabar fell into French hands as a war trophy. Besides, the inhabitants of Idrija, as much as those of any other town, were obliged to supply French soldiers with bread, yeast, eggs, lard, smoked meat, clothing and shoes."[339] Postojna came next, whence the people fled in fear for their lives, so it was "no wonder that the sight of the Austrian troops tamely withdrawing across the Sava River also instilled a sense of terror in the inhabitants of Ljubljana, who dreaded the infamous subversives and the Jacobins."[340] The French commander General Jean-Baptiste Bernadotte addressed the people of Ljubljana with proclamations—including one in Slovene—offering reassurances that the French would honor the manners and customs of all and that there was no need to fear either for the chastity of

---

339    Josip Mal, *Zgodovina slovenskega naroda*, Celje: Mohorjeva družba, 1993, 19.
340    Ibid., 19.

their women or for the preservation of their religion. To this he added that any French soldier to be found looting would be sentenced to death.

After two months of French occupation, when Ljubljana also received a visit from Napoleon, the French retreated from Slovene-speaking territories. Under the Treaty of Campo Formio in the fall of 1797, Austria was awarded Venice and the possessions east of the Adige River to compensate for the loss of its territories in northern Italy. Since Austria was also given the Slovenian inhabited Venetian Slovenia[341] under this treaty, Slovenian populated territories were briefly merged under the same crown. Eight years later, in 1805, another war culminated in a French victory, compelling Austria to relinquish all the territories acquired under the previous truce. The Austrian government instructed the population to deal calmly with the French. Yet the memory of the second French occupation, which lasted two months, was more bitter still, since this time the French not only seized the entire Austrian state property, but also impoverished the population with their exorbitant taxes and demands for army equipment.

Learning from other European states' experience with Napoleon's army, even the Austrian sovereign ultimately realized that expensive mercenary troops alone would not guarantee a successful defense and that additional armed contingents had to be raised. On the initiative of Archduke John, the emperor issued a patent on June 9, 1808, ordering all Austrian provinces to establish a National Guard with men between the ages of 45 and 60 who were capable of bearing arms and who were not serving in the regular army. Fervent preparations were underway for a new battle with Napoleon. The authorities sought to stir up the population with anti-French propaganda, and the National Guard poets promoted Austrian patriotism. Nevertheless, the French army entered Slovenian populated territories in May 1809 for the third time—only for its measures to grow even more severe, combined with even more exhausting requisitions, taxes, and war contributions, while the burden of army supplies, again, fell on the provinces. In such circumstances, the population's discontent turned into popular uprisings, which boiled over in October in Lower and Inner Carniola and in several clashes between the French army and the National Guard. The latter should have formally ceased to exist immediately after the arrival of Napoleon, who had dissolved it, supposedly to preclude the Emperor of Austria from arming even women against him.[342] The tumultuous events therefore only added to French pressure.

---

341 A mountainous region in the northeastern part of the province of Friuli (Italy) that borders the Friulian plain and extends toward the Adriatic Sea, situated in the hinterland of the city of Cividale. The Slovene-speaking region was historically named Schiavonia, and later Slavia Veneta (Slov. Beneška Slovenija).
342 Ibid., 52.

The situation eventually calmed after the Austrian defeat at Wagram, which was followed by the Treaty of Schönbrunn (near Vienna). Under this treaty, Austria ceded to France the territories of Western Carinthia, Carniola, and the Littoral with Gorizia, Trieste, and Austrian Istria, as well as the Croatian provinces south of the Sava. Napoleon established the Illyrian Provinces in this territory with a special decree dated October 14, 1809, which further included Venetian Istria, Dalmatia, and Boka Kotorska. Eastern Tyrol was incorporated one year later. From its source to its mouth, the Soča represented the western, "natural"[343] border of the Illyrian Provinces with Napoleon's Italian Kingdom; the border with the Austrian Empire coincided with that of the Klagenfurt district in the east and the provincial border between Carniola and Styria in the north. Ljubljana was the capital of the Illyrian Provinces. The territory encompassing 55,000 km$^2$ of land and 1.5 million inhabitants has never been nationally, politically, or economically homogeneous. The Illyrian Provinces were created primarily for economic and military reasons, e.g., to prevent Austria from having an outlet to the sea and to deny Great Britain access to all European ports. At the same time, their territory constituted a vital link to the East for France.

The disruption of the traditional economic flows between the central part of the Habsburg Monarchy and the Adriatic, which had been the driving force of the rapid economic development of the Austrian hereditary lands since Trieste and Rijeka were proclaimed free ports, plunged the Illyrian Provinces into a major economic crisis. The rural population was bereft of additional sources of income (carting, coastal shipping, rural crafts), and urban areas were facing equally grim prospects. Under international law, the Illyrian Provinces were part of the French Empire, with certain unique features in contrast to Napoleon's other territorial gains. However, although their territory was under the French flag and the imperial coat of arms, and to a large extent subject to French legislation, their inhabitants were Illyrian citizens. Provincial offices, of which some bore French and others Illyrian names, were responsible to Paris ministries. The administrative organization of the Illyrian Provinces did not follow the example of the French departments. Instead, they consisted of provinces, which were subdivided into administrative units (districts, cantons, and communes) and headed by the governor, usually a military strongman. The official language was French, in which the *Télégraphe Officiel*, the official gazette of the Illyrian Provinces, was published. German and Italian were auxiliary languages, while

---

343 Janez Šumrada, "Poglavitne poteze napoleonske politike v Ilirskih provincah," *Zgodovinski časopis*, 1–2 (2007): 76.

Slovene was still unable to find its way into public offices due to deficiencies in legal and administrative terminology and a poorly developed civil service. The French era brought about significant changes to administration and legislation. Its modern administrative system based on communes has been preserved in memory as the *mairie* (mayoralty) with an appointed mayor. This period saw the introduction of equality before the law, universal military service for all citizens, civil marriage, a streamlined and equitable system of taxation, and the abolition of tax privileges. Judicial administration was placed under state control, guilds and patrimonial courts were dissolved, and seigneuries were stripped of their legal functions.

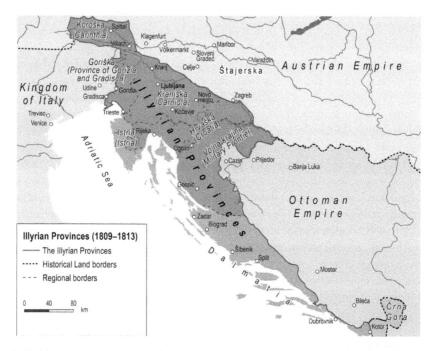

**Map 4.** Ilyrian Provinces (1809–1813).

In its effort to modernize the school system, the French administration passed a number of reforms that eliminated Church supervision. These reforms were introduced in 1810 by Marshal Marmont, the first Governor-General of the Illyrian Provinces and the former Governor of Dalmatia. The Austrian school system, which comprised three categories of elementary schools, was replaced by a uniform four-year elementary education. The network of lower grammar schools was expanded, and higher grammar schools (lyceums) were founded in Trieste, Gorizia, and Koper. Ljubljana also opened a trade school. The central school, also established in Ljubljana, was renamed the academy in 1810 and became the first university on territories with Slovenian population. It consisted of five departments: theology (which attracted most students), philosophy, law, medicine, and initially also engineering. The overall number of students ranged between 200 and 300. Lectures were given in French, Italian, and Latin. As established by Peter Vodopivec, Ljubljana became an important education center, hosting more than 1,000 pupils.[344]

Figure 32. Napoleon's Illyria Memorial, French Revolution Square, Ljubljana.

---

344   Vodopivec, *Od Pohlinove slovnice do samostojne države*, 27.

The costs of maintaining the education system were borne solely by the Provinces, while lower grammar schools and elementary schools were even placed under the financial administration of communes. Since many were incapable of shouldering such a burden, many schools were closed; in some places, their number even declined compared to the Austrian era. As an authority on the national and linguistic situation in the Provinces, Governor-General Marmont envisaged the use of the vernacular, "Illyrian," in schools. This naturally met with a sympathetic response from several Slovenian adherents of the Enlightenment movement, because it also gave the Slovene language an opportunity to find its place in textbooks. Elementary and lower grammar school textbooks, albeit in very limited numbers, were written by the Slovenian scholar and poet Valentin Vodnik, and Slovene became the language of instruction. In the opinion of French officials, including the experienced Marmont, Illyrian was a common Slavic language. However, they were eventually convinced by the unremitting demonstrations by the linguist Jernej Kopitar and Valentin Vodnik that the population of the Provinces spoke two separate languages. During this period, Kopitar and Vodnik also embraced the idea of the unity of the Slovene language and a common national name, "Slovenes."

From an economic perspective, the Illyrian Provinces closed off traditional transport and trade routes. The customs blockade had an adverse impact on the development of ports in Trieste and Istria. The French projected the construction of an east–west road connection, but failed to complete it fully. All financial funds, whether intended for roads, the army, or the administration, were to be collected from the Provinces, through direct taxation of their population. The French system of taxation became widely known as *fronki* and left resentful memories for a long time as well as aroused indignation. Similarly, the French measures prompted exasperation from the peasants, who were eager to finally experience civil equality before the law, having been emancipated from serfdom with their personal service and bondage abolished. Furthermore, contrary to expectations, the French did not abandon feudalism. The status of Illyrian peasants was not on a par with that of French peasants. Lords who owned the land, which was cultivated by the peasants, could recover debts by imposing duties and various forms of service. The discontent among the peasantry therefore grew into peasant uprisings, while poor material conditions of large segments of the population were also reflected in increasing banditry and emigration.

The Illyrian Provinces found sympathizers and supporters among Slovenian enlighteners in particular. The Illyrian name of the newly-established Provinces filled them with high hopes for national and linguistic development, even though Napoleon's introduction of the name was inspired by his narcissistic flirting with the heroic ancient past rather than any awareness of the ethnic affiliations of the occupied provinces. However, the Slovenes, like all other Balkan Slavs, recognized in the name the pre-Roman Illyrian origins of the Slavs, including the South Slavs and the Illyrian-speaking population of Dalmatia and Bosnia. Particularly interesting was the attitude of the three founders of the Slovenian Enlightenment toward the French: the wealthiest Carniolan Baron Žiga Zois, Valentin Vodnik, and Jernej Kopitar.[345] Žiga Zois was loyal to the French, but nevertheless reserved and nostalgic about the bygone Austrian times. Kopitar "awaited" the French in Vienna, urging Zois to arrange a French translation of his grammar from the new masters, but when he was offered a post at the Imperial Court Library in Vienna, he divested himself of all interest in cooperating with the Illyrian administration. Valentin Vodnik was initially extremely ill-disposed toward the French authorities. However, the introduction of the "local" language both as a subject and a language of instruction in the new school system enabled him to publish his Slovene grammar *Pismenost ali gramatika za prve šole* (Literacy or Grammar for Elementary Schools), and his repugnance gradually turned into acceptance. Moreover, his ode "Ilirija oživljena" (Illyria Revived) made him the harbinger of Napoleon's revived ancient Slavic Illyria, but found very few followers among Slovenian enlighteners. In his document *Nota über die Pismenost ali Gramatika* (1811), Vodnik unequivocally expressed his linguistic, political, and cultural views.[346] In his opinion, Slovene and Čakavian Croatian[347] were two dialects originating from the same Illyrian language and should have their own grammars (which had already been accomplished) and dictionaries. Once these

---

345   See Šumrada, "Poglavitne poteze," 81–83.
346   Ibid., 82.
347   Together with Štokavian and Kajkavian, Čakavian represents one of the three main dialects of the Croatian language. The Čakavian dialect provided the basis for the first Croatian language used in public life. Today, Čakavian is spoken by around 12 percent of Croats in the northeastern Adriatic (Istria, Kvarner, Adriatic islands lying north of the Pelješac Peninsula), as well as in part of the Lika area, Gacka, and Gorski Kotar. The name of the dialect stems from the interrogative pronoun "what" (*ča* or *ca*). The Štokavian dialect also formed the basis for the standardized languages of Serbs, Bosnians, and Montenegrins. This name, too, derives from the interrogative pronoun "what" (*što* or *šta*). The dialect is divided according to two principles: into New Štokavian and Old Štokavian dialects, and with regard to the changes in the Old Slavic phoneme *jat*. The Kajkavian dialect comprises the northwestern and central parts of the Croatian language territory; it is spoken by approximately one third of Croats and, like the other two dialects, its name derives from the interrogative pronoun "what" (*kaj*).

"conditions" were met, a common literary language would be artificially devised, introduced in schools, and used as a language of communication by intellectuals, while the people would continue to speak the same language as before. Illyrian thus would become the administrative and cultural language of Illyria. However, after the first Governor-General, Marmont, retired from his position, the Illyrian strong administration denounced this policy, partly because of the economic crisis, but above all because of the administration's aversion to such plans.

Anti-French sentiment among the population started to proliferate, spreading from disgruntled peasantry, tradesmen embittered by the abolition of guilds, and citizens crippled by taxes, to the clergy, which had lost its authority over the school system. The division of power between the state and the Church, the proclamation of religious equality, the institution of civil marriage, and the transfer of the responsibilities for registering births, marriages, and deaths to the civil administration caused much distrust among a vast portion of the clergy, and subsequently among the Catholic Slovene-speaking population. Moreover, the return of the Jews to Carniola and Carinthia did not win the new rulers many sympathies either, except from an insignificant number of intellectuals, officials, and well-established merchants. On the other hand, the French era saw the formation of many Masonic lodges; the Ljubljana lodge, which had less than 50 members (and only 15 of them locals), operated in 1812–1813. The Slovenes in Venetian Slovenia were the victims of utter contempt for their indigenous heritage on the part of the French. They not only dismantled the autonomous judicial system granted to them as guardians of the border with the Venetian Republic, but also destroyed all visible symbols of their unique culture (tables, trees).[348] These measures were not repealed even after the French had finally withdrawn from the Slovenian populated territories.

Given the brief duration of the French rule, and the possibility that Napoleon did not envisage the Provinces as a permanent formation, but was waiting for the first suitable opportunity to exchange them for other territories, many French measures failed to have their expected or desired effect.[349] Nevertheless, the

---

[348] Stone tables under linden trees were a meeting place for the heads of villages. Composing a so-called bank (community of villages), such assemblies were held to discuss and pass judgments on common matters. Village heads were elected by members of the village community and a grand mayor was appointed by the bank. According to Sergij Vilfan, this was a long-standing Old Slovenian tradition that survived into the 19th century (*Pravna zgodovina Slovencev*, Ljubljana: Slovenska matica, 1961, 333).

[349] Vodopivec, *Od Pohlinove slovnice do samostojne države*, 25.

elevation of Slovene into the language of instruction was a very important phase in the development and self-consciousness of the Slovenian national movement. The Illyrian Provinces ceased to exist in fall of 1813, after Napoleon had suffered a crushing defeat against Russia. Austrian troops started returning to the territory of the former Illyrian Provinces. After 25 years of war, the Vienna Congress (1814–1815) not only restored the Illyrian Provinces and other territories to Austria, but also sought to bring a lasting peace to Europe. The Illyrian Provinces nominally remained in existence within a new unit, the Illyrian Kingdom (1816), although it only encompassed Carniola, Carinthia and Gorizia, Trieste with its immediate hinterlands, Gradisca, Cividale del Friuli and its environs, and Istria, as well as Rijeka and Civil Croatia (until 1822), south of the Sava River. Administratively, the kingdom was of no significance, being divided into the Trieste and Ljubljana gubernia, which were under direct authority of the central administration in Vienna. Thus, Ljubljana also retained control over the whole and undivided territory of Carinthia until 1848. Styria, which had first fallen outside the borders of the Illyrian Provinces, now remained outside the boundaries of the Illyrian Kingdom and belonged to the gubernium in Graz, which largely contributed to its singular position and further development. Archduke John, the brother of Emperor Francis, lived in Graz from 1809 onwards and thereafter oversaw economic and cultural progress in Styria, making his influence felt within the broader Slovenian populated territories as well.[350] The provinces and provincial assemblies were vested with no authority. For a long time the name of the Illyrian Kingdom referred to the Littoral, which was known as the Austrian-Illyrian Littoral. At the end of the 19th century, the Illyrian name also appeared in the name of the town Ilirska Bistrica.

Although the Austrian imperial court was very disinclined to any major change, it abided by a number of measures from the Napoleonic era after Austria reoccupied the Provinces. Guilds remained abolished, as did patrimonial justice and the right of lords to collect taxes. The territories of the former Illyrian Provinces saw the introduction of the Austrian Civil Code, which was passed in 1811 and whose enlightened-rationalist basis made it one of the most advanced civil codes of the time.[351] The French administrative borders remained unchanged as did the counties, now renamed to main municipalities. County administration again passed into the hands of the so-called delegated authorities, who were required to possess proper qualifications. As these matters gradually came under the

---

350 Stane Granda, "Predmarčno obdobje," in: Janez Cvirn (ed.), *Slovenska kronika XIX. stoletja*, Ljubljana: Nova revija, 2001, 115.
351 Vodopivec, *Od Pohlinove slovnice do samostojne države*, 30.

jurisdiction of local commissariats due to costs, this transition also signaled the beginning of the gradual nationalization of public administration. The Austrian authorities, recognizing the peasants' ever-decreasing obligations to their lords as a potential threat to the monarchy's feudal system, warned the peasantry to continue with their urbarium tax payments and labor service, and above all to pay any outstanding fees due to their lords. Otherwise, the requirement for peasants to pay taxes to the state also reduced their obligations to their lords to one fifth of the previous amount. The education system was restored to its previous condition: grammar schools and the Ljubljana Academy were dissolved and the former Austrian grammar schools and lyceums were reopened, with instruction pursuant to Austrian school legislation. The organization of the Church was also returned to its former state (the French had dissolved the Diocese of Koper in 1810). An Illyrian unit was established in 1830, when the Archbishop of Gorizia became the Metropolitan of Illyria, presiding over the dioceses of Trieste, Koper, and Ljubljana. This ecclesiastical province remained in existence until World War I. The Illyrian Kingdom had its own coat of arms (a ship), and its name remained part of the imperial title. The changes in civil and criminal legislation reinvested the registration of births, marriages, and deaths and the performance of wedding ceremonies in the clergy. Civil marriages contracted in the Napoleonic period required further recognition from the clergy.

# The Pre-March Era, the Time of Non-freedom

The pre-March era (1815-1848), also known as the Metternich age after the Austrian Chancellor Clemens Metternich, appeared to be a period of peace. Metternich's main concerns were the continued existence and consolidation of Austria and the continued legitimacy of sovereigns, which he promoted through the Holy Alliance—a coalition created in 1815 by the Russian Tsar Alexander I, the Prussian Emperor William III, and the Austrian Emperor Francis I. This coalition in turn became a symbol of reaction, police control, stringent censorship, and fanatical persecution of progressive ideas. The state saw every rebellion, and every revolutionary or democratic movement emerging in Europe, as a threat. Even though this period has been characterized as an era of cancellarial absolutism, this observation nonetheless did Metternich a disservice, especially in light of his intense international involvement. Furthermore, Austria's domestic affairs were also profoundly influenced by imperial court circles, whose power increased even further after Emperor Ferdinand (1835-1848) acceded to the throne.[352]

Although the Viennese authorities acquiesced to it only grudgingly, the profound and radical change was no longer possible to avoid altogether. The Industrial Revolution and the onset of the modern industrial era, which brought the advent of the railway, the steam engine, and steam-powered factories, as well as England's supremacy in economy and trade, eventually stirred the Austrian Empire into action, too. The most powerful harbinger of the modern times was the railway, which entered the Slovenian populated territories with the Vienna-Trieste line of the Southern Railway. From then on, economic development was measured in railway kilometers and the total power of steam engines.[353] Trieste was one of the most advanced Slovenian cities to benefit from the advantages of the "iron road," while the railway's construction and operation also gave the economic development of other Slovenian towns fresh impetus. Nevertheless, the effects of the Industrial Revolution were not always positive, as several industries, such as carting and transportation on the Sava River, started to yield considerably smaller profits or were even brought to a complete standstill.

Non-agricultural industries were very slow to embrace change. The first steam engine was put into operation at the Ljubljana sugar refinery in 1835,

---

352 Granda, "Predmarčno obdobje," 113.
353 Ibid., 115.

and the number of mechanized manufacturing establishments increased only gradually. The iron industry was the most important sector beside agriculture. Iron ore came from the Carnic Alps, the Pokljuka and Jelovica plateaus, the Suha krajina region, and deposits around Novo Mesto. The ore was worked in smaller ironworks such as iron foundries, iron workshops, and forges, where production was still based on traditional methods and charcoal. Iron companies that had amalgamated smaller ironworks were in the hands of lords—the Zois, Ruard, and Auersperg families in Carniola, and the Eggers in Slovenian Carinthia. Some companies, however, were even quicker to modernize their processing and manufacturing facilities: most notably, in the pre-March era, the Prevalje Ironworks, owned by the Rosthorn brothers (English immigrants), was the first factory in the monarchy to specialize in the production of railway tracks. Other non-agricultural sectors, too, provided an important source of income—for instance, the mercury mine in Idrija, which, despite producing less than in the last decades of the 18th century, employed more than 500 miners. Lead production and coal extraction also increased.

The greatest boom in the craft sector was experienced by the textile industry, which was precisely where technical progress and the modern mechanized steam-powered factories advanced furthest, as modern mechanized cotton mills (in Ajdovščina, Ljubljana, and Prebold) gradually replaced the production of cotton at home and by hand. The cloth industry in Upper Carniola, the linen industry in Škofja Loka, and the silk industry in Gorizia bore the heaviest brunt of these changes and survived only in a few countryside towns. The glass, paper, and sugar industries underwent progressive modernization, while sugar cane processing and shoemaking forged ahead. But in spite of these advances, the development of the non-agricultural sector as a whole remained primitive and pre-industrial, causing the Slovenian populated provinces to lag far behind the most developed parts of the monarchy. One of these provinces was Styria, which largely owed its unique position to the keen interest of its prominent resident—the emperor's enlightened brother, Archduke John. It was on his initiative that the Association for the Support and Promotion of Industry and Artisanship in Inner Austria was established in Graz in 1837, with two subsidiaries in Ljubljana and Klagenfurt. In firm belief that knowledge and will were the central driving forces of economic progress, the association took it as its principal task to promulgate information on technical and industrial novelties and to encourage the entrepreneurial spirit. The idea of promoting awareness aroused much enthusiasm, but only among the scanty circle of its members, which included enterprising lords, iron foundry owners, manorial stewards, well-established artisans, and high officials. The association also found supporters in Carniola and Carinthia, although they were no more successful in fulfilling their ambitions and aspirations toward modernization than were their counterparts in Styria.

Agriculture underwent significant change in this period, with the gradual introduction of physiocratic principles and methods of cultivation, such as a three- or four-year crop rotation without fallow periods. The production of potatoes ("the bread of the poor"), corn, and forage crops increased. Owing to larger quantities of clover and other forage, the predominantly free-grazing cattle were moved into barns. Modernization in the production of fruit and industrial plants was slow, but the increased concern for forestry soon had noticeable effects. New crops, especially potatoes and corn, required more farm labor, being more demanding than cereals. However, besides the burden of tending the fields, peasants were further oppressed by feudal obligations, especially labor service. The Slovenian populated rural landscape was therefore undergoing transformation as well, but not as rapidly or drastically as major Slovenian cities. New opportunities also came with new transport connections, which some were able to skillfully take advantage of to sell their products, while others were not. This, in addition to the divisibility of land plots (a policy introduced by the French and maintained by the Austrians), only contributed to the further stratification of the Slovenian countryside. While the number of small farmers and cottagers was rising, there were fewer and fewer wealthier farmers or big landowners. Non-agricultural activities (e.g., blacksmithing, carrier transport, and handicrafts) remained important sources of income for the peasants, but the modernization of transport and the hesitant introduction of the manufacturing industry persistently hindered their extra earnings. Modernization was seen most clearly in the introduction of individual technical improvements and tools, but did not interfere with traditional methods and organization of work.

The pre-March era was also the time when banks began to cautiously enter the economy in Slovene-speaking provinces. In 1820, Kranjska hranilnica was established in Ljubljana as the monarchy's second savings bank. Favorable conditions for the development of financial businesses were most certainly created by the growing capital inflow from intermediary commerce between Vienna and the Adriatic. But like everywhere else, the savings banks invested in real estate rather than the expansion of businesses, and simultaneously made savings deposits soar. New savings banks were founded in Klagenfurt, Trieste, and Gorizia, followed by modern insurance companies. Supplementary fire insurance spread from Graz and Vienna, while "basic" insurance first emerged in Trieste after the Lloyd Steamship Company was founded and connections were established between Trieste and other ports in 1836. Most investors in new technologies came from the Western Europe and the United States: in 1818, the first steamship in Trieste was owned by an American; the first working steam mill in Trieste a year later was owned by a Frenchman; and the first steam engine installed at the Ljubljana sugar refinery (Cukrarna) in 1835 was owned by a Briton (William Moline), as was the one obtained by the Ljubljana mechanized spinning mill four years later. The Prebold spinning mill acquired its first steam engine in 1844. The Slovenian investor Franc Černe became the owner of a steam mill in Kočevje in 1845. Although the Habsburg Monarchy as a whole was quite slow and cautious in equipping itself with steam machinery, the Slovenian populated territories faced additional obstacles.[354]

Domestic entrepreneurs gradually ascended to notable positions as lords and owners of foundries or mines, having at their disposal considerable capital; later, they were joined by enterprising merchants and artisans from rural backgrounds who successfully ventured into business. Janez Kalister, an uneducated but remarkably shrewd local merchant, earned his place among the most successful and wealthiest Slovenes through commerce, tax leasing, and speculations. The founder of the entrepreneurial ascendancy of the Kozler merchant family, which had become rich at the end of the French wars by selling fruit and purchasing real estate, was a peasant's son called Ivan. The father of Fidelis Trpinc, one of the most prosperous entrepreneurs of the time, was a small rural shopkeeper who first made his way into business by leasing bridge toll and excise taxes, and advanced by trading in agricultural products. Their most distinguished trait was their great mobility, as their businesses covered the territories of the entire monarchy, while the then Slovenian business world measured its success by the size of estates or mansions and commercial premises in Trieste.[355] The railway was an important acquisition for the life in Slovenian populated provinces; the authorities first informed the public about the arrival of the "iron road" in 1836, announcing the construction of the Vienna–Trieste line. Expectations were high, and future plans were ambitiously drawn up at the very beginning of construction in 1838. Agricultural societies, however, wondered what the new transport connection had in store for towns it would serve and what kind of economic drought would follow in its wake. They were aware that a brighter economic future lay in modernized agriculture and a modernized iron industry, but they still failed to address the antiquated and unprofitable organization of the

---

354  Jože Šorn, *Začetki industrije na Slovenskem*, Maribor: Založba Obzorja, 1984, 200–201.
355  Granda, "Predmarčno obdobje," 116.

agricultural economy seriously. The railway first reached Celje in 1846, and the extension to Ljubljana opened in 1849.

In 1844, the state introduced a new system of taxation built on stable tax, calculated on the basis of a stable cadaster. Previously, tax obligations such as land and direct taxes had been undergoing constant and drastic change. To ensure an appropriate tax basis thus required taking accurate measurements of the entire territory and determining the net yield of arable lands, part of which was then envisaged to constitute land tax. It took the state more than 10 years to complete the cadaster, while "[t]he Franziscean cadastral survey, creating the basis for the determination of land tax, became the greatest achievement of the pre-March regime alongside the construction of the railway system."[356] However, even the new taxation system failed to alleviate conditions everywhere. Quite to the contrary, the amount of land tax payable by the peasants around Celje and the district of Novo Mesto increased, and provoked mutinous sentiments. Mindful that a possible revolt would be brutally suppressed, the Carniolan nobility, headed by Anton Aleksander Auersperg (Anastasius Grün), drew up a special memorandum containing statistical data, which in 1849 provided some taxpayers with tax relief.

Relatively favorable economic conditions promoted rapid population growth, as well as an improvement in health and sanitary conditions. Much like widespread famines, several infectious diseases were becoming less and less prevalent, even though their causes remained largely unknown. New diseases were emerging, though: in 1831, the monarchy suffered the first cholera epidemic, which Carniola at first effectively warded off by erecting a sanitary cordon along the Carniolan–Croatian border. By 1836, the epidemic had also struck Carniola, but the mortality rate was lower, due especially to a decline in child mortality. The size of rural families grew. The population of the Slovene-speaking provinces increased from 838,000 to 1,077,000 between 1818 and 1846—by 28.5 percent.[357] In some areas, a higher number of survivors even resulted in overpopulation. Farms were unable to sustain entire families, so people migrated to towns. At the same time, an intense stratification of the rural population was taking place, accelerated by the commercialization of agriculture and the law on the division of inherited land. Ljubljana and Trieste recorded the most significant population growth: in 1817, Ljubljana had 9,900 inhabitants, and by 1847 it had 17,000, an increase of 75 percent. This trend was even stronger in Trieste: between 1815

---

356 Ibid., 117.
357 Janez Cvirn and Andrej Studen, *Zgodovina 3*, Ljubljana: DZS, 2007, 30.

and 1841, the population increased from 36,000 to almost 68,000 (an estimated 86.4 percent).

Trieste was becoming the largest Slovenian city, although it did not have a majority Slovenian population; Ljubljana, however, was rather slow to obtain the status of a Slovenian metropolis. Like Trieste and Graz, it remained the capital of its gubernium, and maintained authority over two provinces (Carniola and Carinthia). Other towns were growing as well; owing to a mild climate, Gorizia developed into a holiday destination for elderly retirees, but its overall progress lagged behind Ljubljana. Comprehensive change gave the social structure new dimensions, as class differences began to fade. The nobility was gradually losing its leading social role; conversely, the middle class no longer accepted the absolutist regime unconditionally but treated it with criticism and aspired to gain an active role in society, rising in power and influence. Members of the middle class found security and stability in a new lifestyle and artistic expression, Biedermeier; they largely withdrew to their homes and restricted their expressions of political or other opinions to private circles. These conditions subsequently led to the expansion of middle-class social life.

In 1821, Ljubljana hosted the second in a series of international diplomatic congresses of the Holy Alliance and was the center of European politics during the four months. The Holy Alliance met to discuss the revolutionary disturbances at the end of the post-Napoleonic period. The congress was as much a delight for diplomats and the crowned heads of Europe as it was for the population of Ljubljana, with a multitude of dances, theatrical performances, fireworks, and parades. From January 10 until May 22, 1821, Ljubljana was both a residence and a meeting place for the representatives of the Holy Alliance. The congress was attended by Tsar Alexander I of Russia, John Kapodistrias (the Russian prime minister and secretary of state), Emperor Francis I of Austria, King Ferdinand IV of Naples, and Duke Francis IV of Modena, as well as some 500 ministers and representatives from France, Great Britain, Prussia, and individual Italian states. The initiator and moderator of the congress was the Austrian Chancellor, Prince Metternich. The reason for the Holy Alliance to hold this congress was the constitutional revolution that had erupted in the Kingdom of Naples in June 1820. The absolutist rulers of the European Great Powers, Russia, Austria, and Prussia, decided to take austere measures against the rebels. They offered assistance to Francis, King of Naples, and following 45-day consultations (from January 12 until February 26) sent the Austrian army to crush the revolutionary forces in Naples. In the eyes of its participants, the Alliance's triumph was complete, as the revolutionaries were defeated as early as March 20. This marked the end of the first phase of the congress. During the second phase, the European rulers discussed (and suppressed) a new revolution in Piedmont, and also touched upon the insurrections in Moldova and Wallachia against Turkish hegemony, which they only condemned in principle. The public was regularly informed of the political developments through articles in *Laibacher Zeitung*. The Holy Alliance accomplished its objective in Ljubljana and secured the sovereigns' absolutist authority.

In addition to the distinguished guests, Ljubljana also received a number of domestic and foreign dignitaries, and anyone seeking an opportunity to pursue their own personal advantage. The city intensified security controls and undertook a multitude of renovation works. The paving of squares and streets, the leveling of Capuchin Square (Kapučinski trg, known today as Congress Square— Kongresni trg) for parade ceremonies, the installation of sewage and lighting systems, the renovation of the theater and the dance hall, and the extension of postal services were completed before the congress was opened. The guests not only brought politics to Ljubljana, but also proved to be generous tourists who participated in numerous social events. Every day there were military parades, concerts by the Philharmonic Society, dances, masquerades, fireworks, theatrical and operatic performances by domestic and foreign artists, boat trips on the Ljubljanica River, hunting expeditions, and masses. Ljubljana and its 20,000 inhabitants lent themselves to the splendors of Europe. The memory of the occasion remained vivid long after the congress had ended; even today, the names Kongresni trg (Congress Square), Cesta dveh cesarjev (Two Emperors Street), and the Pri ruskem carju (Russian Tsar Inn) bear witness to the gathering of the Holy Alliance.

Even though the period was marked by absolutism, police pressure and rigorous censorship did not impede cultural progress, much less discourage it. On the contrary, it was precisely the cultural sphere, besides compulsory schooling and the growing significance of education, which laid the foundations for the Slovenian national movement and provided the basis for strengthening modern national consciousness. In the Slovenian populated territories, every seventh child attended school in 1810 and every third in 1847, but the percentage of schoolchildren varied from one province to another. In the Podravje region, for instance, attendance was almost 100 percent,[358] while it was recorded at its lowest (less than 20 percent) in Lower and Inner Carniola. After the departure of the French, the Austrian authorities restored the Austrian system of compulsory

---

358   Cvirn and Studen, *Zgodovina 3*, 32.

elementary (*Trivialschulen*), general (*Hauptschulen*) and normal elementary (*Normalschulen*) schools.

The Slovene-speaking territories had the highest number of compulsory elementary schools, only 16 general and normal elementary schools, and 9 grammar schools (Ljubljana, Novo Mesto, Celje, Maribor, Klagenfurt, Gorizia, Koper, Trieste, and Idrija). Lyceums in Ljubljana, Klagenfurt, and Gorizia provided higher education studies, which were restricted to two years of philosophy as a prerequisite for the further pursuit of a full-time university degree or four-year theology studies. Ljubljana and Klagenfurt also provided studies in medico-surgical sciences. Owing to the authorities' reluctance to extend philosophy studies and introduce the study of law at the Ljubljana lyceum, lyceums complemented their curricula with elective subjects offering instruction in a wide variety of subjects: agriculture, education, botany, Italian, and within theological studies Hebrew, Armenian, and Arabic. At the lyceum in Klagenfurt, agriculture was taught by a Slovene, Matija Ahacel, while the midwifery school remained in operation even after the abolition of "medicine" as a subject. The most ambitious Slovenian students, however, had to seek further education in Vienna. The University of Graz was renovated in 1827, although it had only two faculties, theology and law. This has increased opportunities to pursue higher education for many Slovenian students, particularly those from Styria, for whom Graz was closer and less expensive than Vienna. In 1812, the lyceum in Graz founded a chair in the Slovene language, allowing Slovenian students in Graz to form the first Slovenian cultural societies, which among other things importantly influenced the creation of the "United Slovenia" program. In 1811, the visionary Archduke John endowed Graz with the so-called Joanneum,[359] the first foundation for the establishment of the technical school in Graz and the mining school in Leoben, the two closest technical colleges for Slovenian students. The school of maritime trade in Trieste and the trade school in Ljubljana (founded by a merchant, Jakob Marn) had a reputation as prestigious vocational schools.

The situation and development of the school system also helps to understand the position of the Slovene language within it. German was the language of instruction in elementary schools in cities and major towns, Italian in the Littoral and Gorizia, and Slovene in rural elementary schools throughout most of the Slovene-speaking territories. The case was somewhat different in Styria, where German schools formed a majority, whereas schools in the Carinthian countryside were exclusively German. On the other hand, the use of Slovene

---

359  Today FH Joanneum, University of Applied Sciences (transl. note).

increased in so-called Sunday schools that offered remedial instruction, which enabled compulsory schooling to expand in 1816 and include rural children. German and Latin were the exclusive languages of instruction in grammar schools and lyceums. Following the example of the lyceum in Graz, the second chair in Slovene was established in the Ljubljana lyceum in 1817, which enabled Slovene to become a subject of instruction in lyceums. The primary aim of both chairs was to provide Slovene education to non-Slovene-speaking officials and priests. Slovene made its way into other lyceums and grammar schools at a slow but steady pace: the grammar school and lyceum in Gorizia, for instance, embraced it as late as 1848, while it was taught in the midwifery school and the theological seminary in Klagenfurt by none other than Anton Martin Slomšek, the future Bishop of Lavant.

An important role in the national movement, which concentrated primarily on establishing the Slovene language in public life, was played by the Slovenian clergy, which was mindful of the fact that religion could not exist without a perfected vernacular. The priesthood then represented the majority of Slovenian scholarship. Not only were priests independent from the state, but they also had the capacity to engage actively in the Slovenian Enlightenment during the pre-March era, when there were few other Slovenian scholars. They found inspiration and guidance in humanities and the Enlightenment, and in the belief that culture stood as the pillar of national development—even though most inhabitants of the provinces with Slovenian population did not define themselves in terms of national identity, but rather by their affiliation with Slovene-speaking provinces and local communities, and, undoubtedly, with the imperial family and the monarchy it ruled. The German, Italian, and Hungarian cultures, which had converged in the Slovenian populated territories, were embodied in cultivated noble and middle-class society, whereas the Slovenian culture was most remarkable for its popular customs and folk songs. The tide of national consciousness that swept across the European nations in the pre-March era also reached the members of the more progressive cultures in the Slovene-speaking territories; but it was a very slow process. During its course, "[t]he sentiment of national identity [...] was not inevitably related to the consciousness of one's origin and mother tongue, for many Slovenes declared themselves as Germans on account of their traditional affiliation with the German cultural world."[360] Oscillations or double cultural and national affiliations were completely natural. In everyday life, different cultures and language practices thus coexisted in the

---

360   Vodopivec, *Od Pohlinove slovnice do samostojne države*, 46.

absence of major national tensions. Every potential change in the traditional cultural and linguistic hierarchy was nonetheless viewed with apprehension by the Austrian government, which tried to prevent clear national definitions.

Still, it was no longer possible to stop the surge of the Slovenian national movement, which gathered momentum from a growing number of Slovene books and, of course, supporters. The terms "Slovenec" (a Slovene), "slovenski" (Slovenian), and "Slovenija" (Slovenia), referring to the territory inhabited by Slovenes, which had already been used in Valentin Vodnik's poem "Ilirija zveličana" (Illyria redeemed), written in 1816 (but not published until 1859), were gradually replacing the names of the provinces, particularly Carniola. People of various views and professions began expressing their national identity with greater clarity, which was also reflected in their attitude toward language and culture. Moreover, the changes in the perception of language and national affiliation were clearly reflected in Slovenian literature of the period, demonstrating a gradual formation of a Slovenian national identity.[361] A vast majority of them were priests who recognized the simple enlightenment of the people as the guiding principle of their labors to assert the written Slovene language through religious texts, entertaining and instructional books, and agricultural manuals. Thus, "[i]n 1828 [...], the parish priest Franc Veriti, a native of Horjul, made a surprisingly strong impact on the Slovenian literary milieu,"[362] by starting the four-volume hagiographic collection *Življenje svetnikov* (The Life of Saints), which was a roaring success on the Slovenian book market. The priests brought simple verse to the people using themes from the folk tradition, and blended folk motifs with impassioned national and lyrical narratives. The second mission assumed by the priesthood was to persuade the state authorities of the necessity to introduce Slovene into education and public life.

Another group of Slovenian scholars, better versed, more liberal, and yet numerically weaker, deemed the simplicity of Slovenian literary expression unsatisfactory. Its most distinguished members were Slovenian scholars close to the linguist Matija Čop and the Slovenian Romantic poet France Prešeren (1800–1849). It was Čop and Prešeren who put the Slovene language on the European literary map and raised it to the highest level of literary art.[363] This

---

361 Štih et. al., *Temelji slovenstva*, 176.
362 Igor Grdina, "Življenje svetnikov—ena prvih uspešnic," in: Cvirn, *Slovenska kronika XIX. stoletja*, 172.
363 Igor Grdina, "Življenje ječa, čas v nji rabelj hudi," in: Cvirn, *Slovenska kronika XIX. stoletja*, 186.

was a time when the idea of a periodical publication to disseminate Slovenian literature came to life in the four volumes of the *Kranjska čbelica* (Carniolan Bee) almanac (1830–1834). The almanac received ardent support from the circle, which also used it to publish their poetry. However, not all shared Čop's enthusiasm, least of all the Jansenists, who viewed the almanac as a culmination of the cultural and political work of Slovenian secular circles and the Slovenian Romantics. The most important member of this last group was France Prešeren, an eloquent author who praised the Slovenian national liberation and "the unity of all Slavs"[364] in his poem "Zdravljica" (A Toast). This was one of the reasons why censors claimed this poem could be understood as pan-Slavic.

In 1846, Prešeren issued the collection of poems *Poezije* (*Poems*; dated 1847), in which he

> combined his previous outstanding poetical work into a beautiful new piece of art, wherein they all breathe in the same harmonious rhythm and sensation, as if born out of the same central experience. *Poezije* reflect […] the highest ascendancy of our spirit; no Slovene has been capable of attaining such artistic originality until the present day nor given so humanly consistent and artistically perfected testimonies about himself as Prešeren did in his *Poezije* in 1846,

wrote Anton Slodnjak with rapturous praise in the 1952 edition of *Poezije*.[365] Prešeren was the first to "transcend the boundaries of didactic rhyming and constituted poetry as an autonomous art, evaluated according to the aesthetic criteria of developed literatures,"[366] while according to the writer and politician Josip Stritar, who edited a new edition of *Poezije* in 1866, Prešeren rendered the same service to the Slovenes as Shakespeare to the English, Racine to the French, Dante to the Italians, Goethe to the Germans, Pushkin to the Russians, and Mickiewicz to the Poles. During that period, Prešeren rose to the pinnacle of Slovenian culture and became a national myth, owing to his "historic contribution to cultivating and thereby asserting the Slov[ene] language and also to his demands for the equality of the Slovenes with other nations, based on the idea of Slov[enian] state independence."[367]

---

364 Mal, *Zgodovina slovenskega naroda*, 336.
365 France Prešeren, *Poezije*, ed. with intr. by Anton Slodnjak, Ljubljana: Slovenski knjižni zavod, 1952, xxxii.
366 Boris Paternu, "Prešeren France," in: *Enciklopedija Slovenije*, vol. 9, Ljubljana: Mladinska knjiga, 1995, 313.
367 Ibid., 313.

**Figure 33.** Poet France Prešeren.

Censorship shaped Slovenian literary creativity and was unfavorable to *Kranjska čbelica*. In July 1835, the aspirations to make it a long-standing Slovenian literary gazette came to nothing, not least because of the death of Matija Čop, its greatest patron and for many a "giant of scholarship" and "hope of humanity."[368] In 1824,

---

368  Igor Grdina, "Mlad umre, kdor je bogovom drag," in: Cvirn, *Slovenska kronika XIX. Stoletja*.

censors prevented the publication of *Slavijan*, a local history and literary gazette which Franc Ksaver Andrioli, Ignacij Holzapfel, and Janez Cigler (a parish priest from Višnja Gora) had struggled to establish. Cigler, an active associate of *Kranjska čbelica*, concentrated his endeavors on spreading Slovenian tales instead. In 1836, he published the first Slovenian tale *Sreča v nesreči. Podučenje starim in mladim, revnim in bogatim* (A Blessing in Disguise: Instructing Young and Old, and Rich and Poor), thus laying the foundations for the future development of Slovenian evening tales. *Sreča v nesreči* was printed in 1,500 copies by Jožef Blaznik, who became the owner of a printing house in Ljubljana in 1829 and provided various customers with both German and Slovene texts. Blaznik provided generous support to the Slovenian authors of the period, and printed fundamental Slovenian literary and review works: from *Kranjska čbelica* to Prešeren's *Krst pri Savici* (The Baptism at the Savica) and *Poezije*, and from Vodnik's poetry and Fran Levstik's prose to gazettes appearing in Carniola.

The silence imposed by censorship on the periodical press was broken in 1838 with the appearance of a local history review and a humorist gazette of art, literature, theater, and social life, as the subtitle of the new periodical *Carniola* indicated. Its preoccupation with the native country was evident from its essays on the motherland. It also published literary, historical, ethnographic, and geographic records, entertained readers with anecdotes and crosswords, and above all explored the Slovenian past, which it further entwined with folk literature. Folk songs, on the other hand, appeared in German translations. During the time when Slovenian periodicals were banned from circulation, *Carniola* published Prešeren's poems in German. However, six years after *Carniola* first appeared, it was precisely its Slovenian patriotic orientation that eventually brought its circulation to an end, in spite of its unequivocal loyalty to the Austrian regime. The termination of *Carniola* ushered in a tumultuous period of emergent Slovenian journalism, starting with the appearance of *Kmetijske in rokodelske novice* (Agricultural and Handicraft News) in 1843, the result of Blaznik's years-long negotiations with the police and censorship authorities.

Blaznik's aim was to publish a gazette that would avoid conveying any political or religious content. Rather, it would disseminate general knowledge, especially in economics and agriculture, and entertain its readers with stories, travelers' journals, poems, and historical accounts. Blaznik's idea met with a sympathetic response from the Carniolan Agricultural Society and the Inner Austrian Industrial and Artisans Association. When the veterinarian Janez Bleiweis, who was appointed Secretary of the Carniolan Agricultural Society in 1842, prepared a detailed program defining the gazette as a guide exclusively devoted to improving economic practices, the Viennese censorship office granted permission to publish the weekly under the patronage of the Carniolan Agricultural Society. Even though the editor of *Kmetijske in rokodelske novice* faithfully abided by the above principles, which were the prerequisite for its circulation, he became the central figure of the Slovenian national movement with the weekly as his herald. Bleiweis prudently avoided any delicate political and social issues, continually expressing his loyalty to the monarchy and the Viennese imperial court, "[but] in the same breath, [*Novice*] overtly fostered the awareness of the Slovenian national affiliation and the love for the Slovene language,"[369] thus also drawing readers' attention to subjects relating to history, literature, geography, language, and the alphabet. The weekly played a central role in the introduction of a common Slovene alphabet, which was a turbulent process and subjected to different national and ideological influences.

An important service to the cause of the common Slovene alphabet was rendered by the Illyrian movement and its founder Ljudevit Gaj, who argued in his *Kratka osnova horvatsko-slavenskog pravopisanja* (Brief Basics of Croatian-Slavonic Orthography) for the orthographic unification of the Slavs on the basis of the Czech alphabet. In the shadow of the pan-Slavic ideas of Ján Kollár, who advocated the concept of four great Slavic languages (Russian, Polish, Czech, and Illyrian), Gaj recognized the opportunity to develop a common "Illyrian" literary language, drawing on the idea that the Illyrians were the ancestors of the South Slavs. Having fulfilled the important task of promoting national enlightenment and endeavors to achieve linguistic unification, the movement was excellently received in Croatia. However, since it was largely perceived as a threat to Metternich's policy, the Illyrian name was banned in 1843 and the movement was confined to Croatia. Its merits were nonetheless viewed sympathetically in the Slovenian populated territories, which became embroiled in an outright war of alphabets during the 1830s, sparked by aspirations for a new alphabet where every sound would have its own letter. The most fervent proponents of alphabet reform were the Styrian Peter Dajnko and the Carniolan Franc Serafin Metelko. In 1824 and 1825, they not only published their respective alphabet proposals—the Dajnko alphabet[370] and the Metelko

---

369  Vodopivec, *Od Pohlinove slovnice do samostojne države*, 49.
370  The Dajnko alphabet or *dajnčica* was a new system for writing sibilants, fricatives, and affricates. While sibilants were given the same letters (*c, s, z*) that Gaj's later alphabet would use, special symbols were devised for fricatives and affricates. Dajnko also introduced symbols for *nj* and the Eastern Styrian vowel *ü* (which ceased to be used after 1829). *Dajnčica* was in use between 1824 and 1839.

alphabet[371]—in grammar books, but used them in their textbooks and religious texts. Slovene orthography therefore began to differ by provinces, and differences also arose among supporters of the various grammars.

When Čop and Prešeren eventually decided to join the so-called "battle of the letters," Čop revealed his literary and cultural program, which saw the future of Slovenian literature in a highly aesthetic literary production at a European standard. Prešeren implemented the program more than persuasively, and his poetic grandeur ultimately helped Čop to persuade the scientific court commission to ban the Metelko alphabet from schools. They thus paved the way for the introduction of Gaj's alphabet[372] as the common Slovenian alphabet; the largest contribution came from Bleiweis' *Novice*, or rather from their patron's decision to embrace Gaj's alphabet as Slovenian. The ideological tenets of Gaj's Illyrian movement also resounded across the provinces in which the Slovenes lived. However, their advocacy of a new language system under the banner of national and linguistic unification meant that Slovene would be "degraded" to a folk language used for simple instructional literature, while Illyrian would become the language of the scholarly elite. Since the recognition to raise Slovenian culture to a European level had strengthened, Illyrism could not find a wide group of supporters, but it nevertheless attracted a number of visible Slovenian adherents, predominantly from Styria. Their most notable representatives were the poet Stanko Vraz, who for a brief period was joined by Franc Miklošič and Davorin Trstenjak, Carinthians like Urban Jarnik (in his old age) and Matija Majar, and a number of Carniolans.

From the very beginning, *Novice* was charged with a specific mission, of which Bleiweis must have been well aware. Although he opposed profound social and political change, he nevertheless appreciated the significance of gradual modernization of conditions in the Slovenian populated territories, particularly the modernization of agriculture. However, *Novice* reached out to all Slovenes across the provinces where they lived, even though it principally offered useful information to peasants and was predominantly read by the clergy and educated laity. During Emperor Ferdinand's visit to Ljubljana in September 1844, *Novice* published an ode by Jovan Vesel Koseski entitled *Slovenija presvitlemu,*

---

371 *Metelčica* was in use between 1825 and 1833. Its author, Metelko, devised it by complementing the Latin alphabet with Cyrillic letters in order to accord every sound its specific symbol.
372 The Slovenian version of Gaj's Latin alphabet or *gajica*, composed of 25 letters, followed the Czech example in using the graphemes for affricate *č* and fricatives *š*, and *ž*.

*premilostljivemu gospodu in cesarju Ferdinandu pervemu, ob veselim dohodu njih veličanstva v Ljubljano* (Slovenia to the Most Illustrious, Most Gracious Lord and Emperor, Ferdinand I, on His Majesty's Happy Arrival in Ljubljana). This new term—Slovenia—for the territory inhabited by Slovenians, which Koseski also adopted as the title for his ode, was widely celebrated. Even though the ode evinced profound loyalty to the emperor, it was a unique embodiment of the Slovenian national idea.

Awareness of Slovenian national affiliation thus strengthened in the pre-March era, even though the centers of the Slovenian national movement sprang up sporadically in individual cities that attracted Slovenian intellectuals. Still, the Slovenian population identified with other forms of collective affiliation long into the second half of the 19th century.[373] Apart from Carniola and the so-called Ljubljana group, much agitation was engendered by Slovenes in Carinthia, including the priest Urban Jarnik, who had already drawn attention to the contraction of the Slovene-speaking territories in 1826. His efforts to incorporate Carinthian Slovenes into a pan-Slovenian national development were continued and expanded by Matija Majar Ziljski, the ideological father of the March program "United Slovenia." Majar's adherent and teacher was Anton Martin Slomšek, who recognized the importance of providing language education for theology students and introducing the mother tongue to ecclesiastical circles.

Graz was another important center of the Slovenian national movement, where the main initiative rested with Jožef Muršec Živkov. He brought together Styrian Slovenes, pupils, and students in Graz, and strove with his followers to spread pan-Slovenian consciousness in Styria. Vienna, too, saw the formation of a group of patriotic Slovenian students, who had gathered around the renowned Slavic scholar Franc Miklošič and simultaneously benefited from their intimate acquaintance with the cultural, political, and ideological pulse of the monarchy's capital. Other centers of the Slovenian national movement were in the districts of Trieste and Gorizia, where its inception was catalyzed, among other things, by the Italian national movement. The time had come for the emergence of numerous national movements throughout Europe.

---

373  Štih et al., *Temelji slovenstva*, 185.

# "The Year of Freedom," the 1848 Revolution, and United Slovenia

> 1848 is a deep and clearly visible fissure that has cut the 19th century in two. It is an axis around which the fate of the nations orbits, a barrier separating the spirits, classes and states and putting a definite end to the era of manorial feudalism. However, it is also the source and fount of new powers that have been called by this remarkable year to the forefront of public life. The idea of nationality will no longer be driven into a corner, much less let itself be completely erased.[374]

Political and economic turmoil, demands for expanded nationality rights, and the ever-louder voices proclaiming the urgency of dismantling the feudal social system first erupted in Paris in February 1848 and reverberated across much of Europe. When they reached Vienna on March 13, 1848, they first knocked the unpopular Chancellor Metternich off his position and in the ensuing days forced Emperor Ferdinand I to promise that he would pass a new constitution, convoke the Parliament, abolish censorship, reinstate freedom of the press, association, and speech, and permit the establishment of National Guards. This way, the Slovenian national activists "were given an opportunity to publicly formulate Slovenian national demands and expand Slovenian political programs for the first time in history."[375] Although the 1848 revolutions were typically mainly middle class, they differed from state to state. It is also generally believed that throughout a vast portion of Europe the revolutionary ferment marked the Springtime of Nations, which made their demands that year with unequivocal clarity.

In the Habsburg Monarchy, however, the revolution followed a separate and distinct course from that in the rest of Europe, "because it had grown from its own roots and faced problems that were altogether different from those confronted elsewhere."[376] Three days after the revolution broke out in the capital, a traveler brought the news to Ljubljana; it blazed across the city, causing a great furor. Distinguished citizens of Ljubljana and representatives of the local authorities hailed the emperor's decisions at the theater, while the rest of the population of Ljubljana and its environs, particularly workers and students, gathered in the streets, demonstrated, rioted, destroyed a statue of Metternich, attacked excise

---

374 Mal, *Zgodovina slovenskega naroda*, 635.
375 Štih et al., *Temelji slovenstva*, 177.
376 Stane Granda, "Revolucionarno leto 1848 in Slovenci," in: Cvirn, *Slovenska kronika XIX. stoletja*, 306.

offices, and drove Ljubljana's mayor to flee. Following the Viennese example, the citizens of Ljubljana set up a National Guard to restore public order in the city and protect private property, but that did not completely relieve the city from fear and disquiet. More than anything else, the citizens rejoiced at the emperor's promises of a new constitution, the freedom of the press, and the right of association, whereas other, more moderate Slovenian scholars (including Bleiweis) decided to wait patiently for the changes the new legislation would bring.

In the countryside, too, discontent was brewing, as the calls to abolish absolutism and the remnants of feudalism had reached the rural population. The peasantry perceived the emperor's promises as a possible guarantee of their emancipation, when the land they cultivated would finally become their own. Therefore they decided to stop paying fees and refused to do labor service or pay taxes. The brunt of their wrath was directed at the lords; the peasants vented their rage against mansions and castles, and plundered feudal estates. They even rose up against the clergy and a number of parish priests. The largest peasant uprising took place at Ig (on the outskirts of Ljubljana), where on March 21 enraged peasants pillaged and burnt down the castle. The wave of insurrections surged until mid-April, except in Carinthia, where it did not recede until May. But it was most extensive in Carniola, where the authorities proclaimed a state of emergency, set up a summary court, and imposed collective responsibility for the damage suffered by rural municipalities. Townspeople and most of the clergy, their liberal orientation notwithstanding, showed no sympathy for the peasants' discontent either; national awakeners condemned violence and called for obedience and restraint. This lack of understanding, though, would have precluded the Slovenian national movement from grounding itself in the wider society if it had supported the peasantry.[377] The rebellious peasants were eventually silenced by the proclamation of their emancipation in the early fall of 1848.

## "United Slovenia"

The initial stage of the revolution revealed no signs of major national divisions within the Slovenian populated territories. Disputes were solved by compromises, as in the case of the Ljubljana Castle tower, where they eventually resolved to fly both the Frankfurter German flag and the Carniolan colors of white, blue, and red, which later became the colors of the Slovenian national flag. The exhilaration

---

377   Vodopivec, *Od Pohlinove slovnice do samostojne države*, 52.

of freedom and the belief that the nations of the monarchy could thenceforth settle their relations in peace still bound the Germans and the Slovenes together at that time. In mid-April, a delegation from the expanded Carniolan provincial assembly, including Bleiweis, the editor of *Novice*, visited the capital to declare their support for the imperial court. However, Viennese Slovenes like Franc Miklošič, Peter Kozler, and the lawyer Matija Dolenc persuaded Bleiweis to appeal to Archduke John on the question of unifying all Slovenes into one province. The archduke shrewdly sidestepped the appeal by inquiring whether Bleiweis could demonstrate pan-Slovenian support for this demand. Bleiweis then realized that, as he wrote, the Slovenes should express their common will by issuing municipal proclamations.[378] At the same time, they also contemplated the idea of a Greater Germany that would also incorporate Austria, perceiving such a constitutional arrangement as self-evident, while completely disregarding its consequences for the existence of the monarchy or the Slovenian nation.[379] Yet when the "well-informed"[380] Slovenian scholars recognized that it was imperative for the Slovenes to formulate their national demands following the Czech and Croatian examples, Slovenian–German relations started to show obvious signs of strain.

The task of preparing the national program was undertaken by Matija Majar, the Cathedral Vicar of Klagenfurt. In late March 1848, Majar explained the significance of the revolution in the monarchy to his fellow countrymen:

> There has never been a time so ripe for all Slovenes for as long as the Sun has shone, and only God knows whether there will ever come a time like this again. Now we can behold the prosperity of our nation, now our liberated nation can join other free nations—an equal and free partner to all.[381]

Initially, Majar restricted his demands to the use of Slovene in public offices and schools; but having sent his memorandum to Ljudevit Gaj in April, Gaj encouraged him to expand his claims. For the Slovenes as a separate nation, he then further envisioned a special Sabor[382] following the example of the Sabor of the Banovina of Croatia, demanded the recognition of equal status for Slovene,

---

378 Stane Granda, *Prva odločitev Slovencev za Slovenijo*, Ljubljana: Nova revija, 1999, 562.
379 Vodopivec, *Od Pohlinove slovnice do samostojne države*, 53.
380 Granda, *Prva odločitev Slovencev*, 562.
381 "Slava bogu v višavah in na zemlji mir ljudem dobrega serca," *Kmetijske in rokodelske novice*, March 29, 1848. Quoted from Granda, "Revolucionarno leto 1848," 306.
382 The name of the Croatian Parliament. In Croatian the word *sabor* means "assembly," "meeting," or "congress" (transl. note).

and endeavored to establish closer ties with the Croats. Separate petitions were also drawn up by Slovenes in Graz surrounding the local Slavic scholar and professor Josip Muršec, and by Slovenian scholars and students in Vienna. Their memoranda urged the protection of the Slovenian nationality, demanded equality for the Slovene language, and called on the Slovenes to lobby the emperor for the unification of the Slovene-speaking territories, hitherto divided among different crown lands, into a single Slovenian province. The program was finalized when the Slovenija Society was established in Vienna on April 20, 1848, with Franc Miklošič as its president.

The program that became known as "United Slovenia" was based on Majar's ideas and the petition of the Slovenes in Graz. It demanded the foundation of the Kingdom of Slovenia with its own provincial diet that would be part of the Habsburg Empire, but not part of the German Empire. The authors—the Slovenian national activists—demanded equality of the Slovene language with German and the introduction of Slovene into schools and public administration. The bilingual petition was printed in several thousand copies with an accompanying leaflet, entitled *Kaj bomo Slovenci cesarja prosili* (What Shall We, the Slovenes, Ask of the Emperor), and circulated for signatures. Similar petitions were presented by the Slovenian associations that have sprung up in Slovenian cities (Graz, Vienna, Ljubljana, Klagenfurt, and Gorizia). Slovenes in Hungary and Venetian Slovenia were also included within the borders of Matija Majar's "United Slovenia." The program did away with the old provincial boundaries and argued for the reorganization of the Habsburg Monarchy. The leaders of the Slovenian national movement recognized their future in a federal Austria where the Slavs would undeniably constitute a majority population, assume all rights as its citizens, and above all enjoy administrative, political, and economic independence. But the vigorous demands failed to attract wide support from Slovenian townspeople or scholars. Even Bleiweis, who had in principle accepted the program or at least not rejected it entirely,[383] warned that the Slovenes did not yet possess enough "spiritual strength" to achieve their ambitious plan. The consciousness of provinces was still too firm for the old and well-established provincial boundaries to be relinquished in favor of new, unknown borders of a future Slovenia. However, this could not discourage the persistent and successful agitation for the "United Slovenia" program, which peaked with the signing of the petition.

---

383   Vodopivec, *Od Pohlinove slovnice do samostojne države*, 54.

The petition in support of "United Slovenia" was signed by nationally minded scholars and peasants alike, and by mid-May 1848, Styrian Slovenes had gathered more than one thousand signatures. The Viennese Slovenian circle received encouraging support throughout the Slovene-speaking territories, except from Carniola and its capital Ljubljana, where it evoked only a lukewarm response—although on April 25, the Slovenija Society was also founded in Ljubljana. On May 12, the Ljubljana society was visited by a delegation from Vienna's Slovenija Society headed by Miklošič, which is when the gathering of signatures started throughout the Slovenian populated territories. National life, also characterized by an increasing awareness of the Slovenian national program, primarily continued to run its course within national societies in Trieste, Gorizia, Klagenfurt, and Celje. Furthermore, the promotion and dissemination of the ideas of "United Slovenia" became a mission of the new Slovenian newspapers that appeared in late 1848. Ljubljana issued the first political newspaper, *Slovenija*, Celje launched the weekly, *Celske slovenske novine* (Slovenian News from Celje), and Trieste saw the emergence of the monthly, *Slavjanski rodoljub* (The Slavic Patriot). Ecclesiastical circles in Ljubljana gave further impetus to the publishing efforts with their newspaper *Slovenski cerkveni časopis* (Slovenian Ecclesiastical Newspaper).

"United Slovenia" thus became a fusion of all Slovenian national programs and national activists of the time, which also unanimously and firmly rejected the possibility of Slovenia's incorporation into a united Germany. It was nonetheless necessary to turn this position into reality. Following the Czech example, the Slovenes therefore boycotted the elections to the Frankfurt Parliament and urged their supporters to do the same, reminding them that they were Slovenes and not Germans. They found guidance in the Czech František Palacký, who opposed Greater Germany in his letter dated April 11, 1848. Faithful to the Austro-Slavic spirit, they firmly believed that the Austrian Slavs could only have a future in a reconstructed monarchy that could protect them from German, Russian, and Italian hegemony, and they contested Polish demands for a restoration of the kingdom.

Again, Slovenian scholars had conflicting views about the elections to the Frankfurt National Assembly, with those in opposition in the minority. Therefore, both the sympathizers with the revolution, who saw the elections as an opportunity to develop democracy, and fervent supporters of the Viennese imperial court supported the elections. Their close ally, the poet Count Anton Aleksander Auersperg, tried to convince the Slovenes to participate. Bleiweis, who had succumbed to a fear of possible Slovenian-German tensions, urged his readers to support the elections, arguing that no one could ever deprive the

Slovenes of their national consciousness.[384] Several Slovenian liberal scholars declared themselves in favor of the elections, feeling that the all-German Parliament granted them a possibility to assert their liberal and democratic values, while a potential federation of German states might allow the Austrian and Slovenian federal units to obtain equal status at a national level. Arguments for and against participating in the elections thus elicited very different responses and results. The Slovenija Society in Vienna issued a bill urging people to boycott the elections. Its calls also reached parts of the Slovenian countryside, where it exposed the threat of higher taxes and conscription into the German army in particular. It emphasized that only the imperial court, the emperor, and the Vienna Parliament would enable the abolition of feudalism and the settlement of other issues. Opposition to the elections has therefore received some favorable response, both in Carniola (which nevertheless elected Count Auersperg) and in Lower Styria. The elections to the Frankfurt Assembly were the first modern elections for the population of the German federation, including the Slovenes, even though they took place at a very turbulent period in the lives of its citizens.

The anti-election agitation ran concurrently with the Slovenian declaration for Slovenia and the circulation of the petition. Many peasants' names appeared on the list of signatories as well, since the signing had extended to the Slovenian countryside. The principal driving forces of the Slovenian national movement nevertheless continued their work outside Ljubljana, in Graz and Vienna, where they also tried to strengthen their relations with the Croats and the Czechs. Issuing a call, to which the sole Slovenian signatory was Miklošič, the Czechs convened a pan-Slavic Congress in Prague on May 1, 1848, as a counterweight to the pan-German Congress in Frankfurt. The Prague Congress was attended by 300 people, including around 40 South Slavic representatives. Four Slovenes presented the "United Slovenia" program, demanding unification of the Slovenian populated provinces into the Kingdom of Slovenia with its capital in Ljubljana, equal status for the Slovene language and its introduction into public administration, schools, and courts, and the establishment of a university in Ljubljana. "United Slovenia" was successfully incorporated into the final memorandum of the Prague Congress, which principally aimed to reorganize the monarchy into a federal state. The Prague Congress was concluded in late June, when the Austrian army suppressed the insurrection in Prague. During the same time, the speaker of the Slovenija Society in Graz participated in a Croatian Sabor session and called for Slovenian–Croatian reciprocity, which, primarily

---

384  Ibid., 56.

recognized as a basis for forging alliances, was positively received across the provinces with Slovenian population. Meanwhile, a completely unrelated peasant movement was established in Lower Carniola and along the Slovenian–Croatian border in Styria, which pressed for the incorporation of local border areas into Croatia. The movement was inspired by the steps taken in the Hungarian part of the monarchy, where feudal obligations had been abolished at the very outbreak of the revolution, with some of the compensation to be paid by the government.

In May, after the emperor and the imperial court resolved to impose a new constitution, Vienna was struck by another rebellion that forced the emperor to retreat to Innsbruck. The situation calmed somewhat in early June, when preparations for the elections to the Vienna Parliament began as a first step toward fulfilling the May rebels' demands. The knowledge that the new constitution would be deliberated by the elected Parliament also led to a change in the attitude of the Bleiweis circle towards the current conditions and, above all, to a consensus on unequivocal support for the demands of "United Slovenia."

The elections to the Vienna Parliament took place in mid-June 1848. The right to vote was accorded to all men over the age of 24, but with numerous restrictions based on the principle of independence; these particularly affected the rural constituencies, since the vote was granted only to peasants who were engaged in handicrafts or owned their own land. The situation was effectively a repetition of the Frankfurt elections, as the turnout was again low in Carniola and Lower Styria. Elsewhere, however, the peasantry grew much more aware of the importance of the Vienna Parliament and voted either for peasant candidates or scholars. The decision to vote rested primarily on trust, which neither priests nor castle lords enjoyed at that time; accordingly, they were not elected parliamentary representatives for the Slovene-speaking territories. A majority of votes thus went to liberal-minded people of pro-Slovenian inclinations, while those who were less than supportive of the Slovenes slipped into the minority. Nevertheless, a lack of unity within the national movement and among elected deputies in the Vienna Parliament remained a constant in Slovenian politics of the period. When the Vienna Parliament opened on July 7, 1848, Slovenian deputies failed to propose a clearly defined common program, but rather presented an incoherent farrago of individual, personal views. By sheer coincidence, they were almost unanimous on one of the most important issues under parliamentary consideration, the law about the emancipation of peasants, which brought them close to what was then the left wing of the Parliament.[385]

---

385   Ibid., 59.

The deputies' opinions still clashed over the issue of compensation: left-wing deputies sought to impose the least possible burden on the peasantry, while some deputies, including the Slovenian ones, actually opposed it outright, because it violated the principles of civic equality ensured by the French Revolution. The motion for compensation with partial land redemption still won an eventual majority, and was enacted by the Law on the Abolition of Serfdom promulgated on September 7, 1848. One part of the lords' compensation was to be paid by the state, the other by the peasants. Even though the law did not fully meet the demands of the peasants, it at least assuaged them enough to stop their involvement in further violent struggles.

The deputies also had differing views on the reorganization of the Habsburg Monarchy into a federation of national units as argued by the Slovenian lawyer Matija Kavčič: "The nations make up the foundation stone of the Austrian state. I have the honor of being a Slav."[386] The constitution therefore became the central point of dispute among the Viennese deputies and the discussion dragged on until the outbreak of yet another revolution in Vienna, this time in support of the unyielding Hungary. In the aftermath of the spring revolutions, Hungary secured itself considerable autonomy and tried to benefit from the current situation by achieving the independence of historical Hungary, but refused to recognize the national rights of other nations (the Croats, the Slovaks), which naturally precipitated resistance from both the Slovaks and the Croats. This led the Croatian Ban Josip Jelačić to terminate all contact with the Hungarian government, which removed him from his position—only for the emperor to reinstate him by early September in return for his much needed assistance in suppressing the Hungarian revolution. Jelačić's army first helped quell the revolution in Vienna and then proceeded to Hungary, which declared independence on April 11, 1849, but returned to Habsburg authority after the ultimate defeat of the Hungarian forces with the help of Russian military intervention. The Slovenian and other Slavic deputies in the Vienna Parliament based their attitude toward the October insurrection on the belief that only the Habsburg Monarchy could give them national rights. In the hope that their loyalty to the emperor and the government's decisions would be rewarded with imperial recognition of national equality, they thus expressed their support to them and left the tumultuous Vienna. Meanwhile, the imperial court consolidated its power; a new government was installed under Prince Felix of Schwarzenberg, and on December 2, 1848, the 18-year-old Franz Joseph ascended the imperial throne.

---

386   Ibid., 59.

The Vienna Parliament reconvened in late November 1848 in Kroměříž, Moravia, and was marked by countless proposals for the reconstruction of the monarchy. A German liberal deputy, Ludwig von Löhner, devised a plan providing for the equality of nations and the reorganization of the western part of the empire into a union of five national groups, the Germans, the Czechs, the Poles, the Italians, and the Slovenes (save Trieste and Slovenian Istria). This plan appealed to some Slovenian deputies, or at least was not rejected by them, unlike the plan submitted by the lawyer Matija Kavčič, who envisaged a reorganization of the monarchy into fourteen national and historical units under which the Slovenian unit would include Trieste and the Littoral but not Carinthia. The prevailing view in the Parliament was that of the so-called German center, favoring the preservation of the old crown lands, divided into districts enjoying limited local autonomy, while their boundaries were to correspond with ethnic ones as closely as possible. In spring 1849, the Constitutional Committee drafted a constitution and submitted it to the Parliament. At the same time, however, preparations were already being made by the imperial court to impose the octroyed constitution (composed by the Interior Minister, Count Franz Stadion) on March 4, 1849, and to dissolve the Parliament on March 7. The new constitution dealt with the equality of nations while simultaneously conferring wide-ranging powers on the sovereign and the government. The suppression of the revolutions in Hungary and Italy gave the ruling circles a fresh start, and their willingness to compromise gradually waned.

These events took the top Slovenian politicians by surprise. The conservative faction hoped for the octroyed constitution to remain in force, especially because it referred to the Illyrian Kingdom established in 1816, which in their opinion (later proved as unfounded) signaled the Viennese government's readiness to establish a separate Illyrian unit that would also incorporate the united Slovenian territory. The more liberal faction, however, was vehemently disappointed, noting that the new administrative arrangement abolished gubernia but preserved the territorial division into provinces. The octroyed constitution, furthermore, provided for the formation of districts that would correspond with ethnic borders, but it only delivered on its commitment in Styria. The demands of "United Slovenia" for equal status for the Slovene language were widely disregarded even after the passage of the constitution, even though Slovene had won official recognition through the government's decision to translate the Austrian Civil Code into the so-called languages of the provinces (*Landessprachen*), because similar provisions should also have been made for the legislation of the provinces themselves. Thus, Ljubljana obtained a Slovene law gazette, *Ljubljanski časnik* (Ljubljana Gazette), which led to the official recognition of the Slovene name for the Slovenian nation and its language; this, in turn, also contributed to the gradual formation of Slovene legal terminology and to the standardization of a common literary language. However, the abolition of the octroyed constitution in 1851 also eliminated the possibility of using the Slovene language in public administration. Elementary instruction in the mother tongue took place only in the countryside, and even there it was restricted to nationally homogeneous areas, while German was the exclusive medium of secondary instruction. Following grammar school reforms, Slovene became a subject of instruction in some places.

Shortly before the repeal of the octroyed constitution, government circles undertook to restore absolutism by eliminating the freedoms and political rights of the citizens, which led to a drop in political engagement. Press freedom was abandoned altogether, and Slovenian newspapers disappeared, except *Novice*, which returned to offering advice on economic development. The Catholic weekly *Slovenski cerkveni časopis* was renamed *Zgodnja danica* (Early Morning Star). Ministers answered to the emperor alone and the so-called Sylvester Patent of December 31, 1851, repealed the octroyed constitution. The period of neo-absolutism began, during which authority relied once more on the army, the police, the civil service, and the Church.

## In the Shadow of New Absolutism

The failed revolution, tightened censorship, silenced newspapers, and the ever-decreasing number of political societies brought to a brief standstill the Slovenian national movement that "became limited to a small circle of individuals still harboring the Slovene sentiment."[387] Under Matija Majar's instructions from 1851, Slovenian national and political efforts concentrated on culture and, above all, on literary creation.[388] Slovenian political societies, like the Slovenija Society in Graz, were reformed into reading and cultural societies, enabled by the imperial patent in 1852. Prolific literary production inspired renewed attempts to establish literary gazettes, the longest-lasting being the Klagenfurt monthly *Učiteljski tovariš* (The Teacher's Companion). From 1851 until the present day, the Mohorjeva družba publishing house, established in Klagenfurt in the same year, has made a particularly valuable contribution to the spread of Slovenian literature.

---

387 Štih et al., *Temelji slovenstva*, 191.
388 Ibid., 68.

Despite its explicit literary and cultural focus, the Slovenian national movement began to split into three groups as early as the 1850s, even though Slovenian politicians' viewpoints were much more diversified. The group surrounding Bleiweis, who had never wavered from his deep commitment to the Slovenian movement—and yet at the same time remained indulgent of and loyal to the Austrian authorities—took up his view that people needed more practical education. Another group gathered around Slomšek, the Bishop of Lavant, and Luka Jeran, the editor of *Zgodnja danica*. Slomšek made an outstanding contribution to raising awareness of the Slovene language; he wrote Slovene religious and educational texts, energetically advocated the authority of God's laws, and opposed the revolution, industrial capitalism, and liberalism. He found great support in Jeran as a defender of clerical views. Both groups continued to prevail on the Slovenian national and political stage during the age of absolutism associated with the Minister of the Interior, Alexander Bach,[389] since their circle of supporters was wide and included the clergy, which provided the most guidance to literary and publishing activities. Both *Novice* and *Zgodnja danica* printed articles with religious, educational, and patriotic content, and *Novice* also devoted space to more contemporary, original Slovenian literary works. These also occupied a central position in the literary magazine *Slovenski glasnik* (Slovenian Herald), first published in 1858 by the Slovenian Carinthian agitator Anton Janežič.

The third group, whose most visible representatives were the writer Fran Levstik and the poet Simon Jenko, was more liberally oriented. Most of its members were secondary school pupils and students, fervent readers and admirers of Prešeren, who drew their inspiration from Slavic and Slovenian slogans of the March Revolution. Their poetry and prose were moving progressively closer to Realism. Their stories had an expressly national and political character, and found readers both among the educated and the simple people. In 1858, Fran Levstik wrote his famous tale *Martin Krpan*, which remains compulsory reading for Slovenian elementary schoolchildren. This was the story of a doughty Carniolan smuggler Martin Krpan, who saved the emperor and the Viennese nobility from the Turkish giant Brdavs and was rewarded with ingratitude. At the time, *Martin Krpan* was overshadowed by the story of Ferdo Kočevar-Žavčanin, *Mlinarjev Janez* (1849), which dealt with the rise of a Slovenian Styrian village. It took the story of the Counts of Celje as its historical background and was a literary sensation of the time.

---

389  The period was hence termed "Bach's Absolutism" (transl. note).

Scientific production was also on the rise, spurred on by historical societies, including the Carniola Historical Society. Their efforts concentrated on investigating and recording histories of individual provinces, albeit exclusively in German. In the late 1850s, *Novice* started printing essays of dubious historical accuracy penned by Davorin Trstenjak, an ardent proponent of Slovenian autochthonism and the Slovenes' millennial presence in the territory between the Alps and the Adriatic. Even though his essays did not produce much response, they "revealed [...] Slovenian feelings of powerlessness and insecurity under absolutism and renewed German pressure."[390] German pressure was also manifested in the Austrian authorities' distrust of any public expression of Slovenian national affiliation. The unveiling of a monument to Prešeren in 1852 at the Kranj cemetery, with Bleiweis as the official speaker, was therefore the only public ceremony in commitment to the national initiative until 1858. In 1853, censors prevented the publication of a map of Slovenian provinces by the cartographer Peter Kozler, which determined the borders of United Slovenia and was to bear the title *Map of the Slovenian Land and Regions*, for "violating the lawful union of Austrian lands."[391] The end of the decade marked the end of absolutism. Its decline in the Slovene-speaking territories was already heralded by the solemn ceremony held in Ljubljana in 1858 to commemorate the hundredth anniversary of the birth of the poet Valentin Vodnik.

The linguistic boundaries of the territory, which was densely populated by Slovenes, remained largely unchanged during the mid-19th century. The western linguistic border ran from the Gulf of Trieste to the Tagliamento River, along the edge of the Friulian plain and the Karst Plateau. Slovenes inhabited the Resia Valley, from where the border ascended through Kanin and Pontebba and then descended to the Kanal Valley, crossing the Carnic Alps toward the Gail River in Carinthia. The Slovenian–German border crossed the Gail Valley east of Hermagor, then turned at Villach, crossed the Ossiacher Tauern range, Zollfeld, and the Sau Alps, and continued until the area north of Diex, whence it turned toward the Drava River and crossed Styria eastwards. This part of the boundary ran north of the present-day Slovenia–Austria state border. The Prekmurje region remained part of Hungary, where Slovenes also inhabited the area stretching to the Rába River. The Sotla River, the Gorjanci range, and the Kolpa River separated the Slovenian populated territory from Hungary and Croatia. In Istria, the Slovenian and Croatian population inhabited both sides of the

---

390  Vodopivec, *Od Pohlinove slovnice do samostojne države*, 70.
391  Štih et al., *Temelji slovenstva*, 195.

present-day Slovenia–Croatia border. The Slovene-speaking territories then had around 1,300,000 inhabitants, 88.9 percent of whom were Slovenes (1,150,000). Provincial borders were drawn in such a way that Slovenes formed a majority only in Carniola and Gorizia; the German element remained very strong in the Kočevje area, while German settlers also lived in the Kanal Valley and Bela Peč. Germans generally inhabited cities and their neighborhoods. Littoral towns maintained a Romance majority; 50 percent of Gorizia's inhabitants were of Italian or Friulian origin and the city also had a sizeable German population, whereas the surroundings were mostly Slovenian. The population of Trieste was ethnically extremely diverse, comprising not only Italians, Slovenes, Croats, and Germans, but also Jews, Greeks, Serbs, and Armenians.

Figure 34. Peter Kozler's Map of Slovenia, first printed in 1853.

Bach's administrative reforms entailed, among other things, the administrative division of the Slovenian populated territories in 1849, which remained in effect until the end of the monarchy in 1918. While the old historical boundaries still separated historical crown lands, the old *kresije* (pre-March state administrative units) were abolished and replaced by district commissions that exerted authority over municipalities, the smallest administrative units. The division of Styria into three regions, Upper, Central, and Lower Styria, was "congruent" with ethnic borders, but Slovenes formed a majority only in the Maribor (Lower Styrian) region. The Littoral administrative unit was divided into the regions of Istria and Gorizia, while the city of Trieste and its environs were accorded special status and their own constitutions. The judicial system was composed of judicial districts distributed across individual district commissions.

Administrative reforms were followed by Church reorganization in Styria and Carinthia. In 1859, the predominantly Slovenian Diocese of Lavant ceded its territory in Carinthia to the Diocese of Gurk, and received parishes within the "Maribor" region from the Bishop of Graz (Seckau). The then Bishop of Lavant, Anton Martin Slomšek, moved the diocesan seat from the marginal St. Andrä in Carinthia to Maribor, which thus became the fifth diocesan seat in the Slovenian populated territories, besides Trieste, Ljubljana, Gorizia, and Klagenfurt. Parts of the provinces with the Slovenian population also belonged to the dioceses of Graz (Seckau), Zagreb, and Szombathely. Protestants living in the Prekmurje and Rába areas had their own church organization.

Population growth in the Slovenian populated territories during the mid-19th century lagged behind the European (including Austro-Hungarian) demographic average, and continued to do so until the outbreak of World War I. The reasons were manifold, but stemmed largely from changing economic conditions and increased emigration. The 1850s were therefore also a groundbreaking time for the economy in the provinces with Slovenian population. They saw the advent of a modern market economy, encouraged by the construction of the Southern Railway Vienna–Trieste line in 1857, the emancipation of peasants, the introduction of a free market, and a more open trade policy. Slovenian peasants too had to adopt market practices, because their new, "post-emancipation" obligations (taxes and levies) were now to be paid in money, which obliged them to sell their crops. Because the emancipation process dragged on into the mid-1850s, peasants were able to market their produce well and the countryside did not lack money. Artisans had to contend with harsh competition from industrial products. The arrival of the railway led to a decline in carting, compelling peasants to find other sources of income. Nevertheless, before the completion of the railway, when Ljubljana (then still a rail terminus) had grown into a buzzing trade and commercial center, carting and railway construction offered abundant opportunities for profit. A favorable economic wind sweeping across Western Europe also reached the Habsburg Monarchy. Institutions promoting trade and commercial development, chambers of trade and commerce, were permeated by optimism and confidence in great prospects for the Slovenian populated provinces.

However, the economic boom of the early 1850s started to wind down before the end of the decade. The frequent mismanagement of money proved fatal; there were no savings and money was spent extremely unwisely on cheap industrial products and, even worse, on excessive alcohol consumption.[392] The misery was further aggravated by the crippling burden inflicted on the countryside by the estimated compensation for the emancipation of peasants. Of a total of 290 million florins calculated to reimburse the Austrian half of the monarchy, Slovenian peasants had to pay 20 million. Moreover, the question of so-called servitude rights was yet to be solved, and the issue of pasture and forest exploitation remained unsettled until the end of the 1870s, further increasing anxiety among peasants. Small farmers and cottagers found themselves in the worst situation, as vast forests and arable lands remained the property of big landowners even after emancipation, while municipal pastures were distributed among individual farmers. Rural trade was in decline as well, and the largely Slovenian populated countryside was thrust into a period of excessive peasant borrowing. However, agricultural societies and Bleiweis' *Novice*, led by a firm belief that the modernization of agriculture was essential to ensure a brighter and more successful future for the Slovenian population, were even more distrustful of the industrial development of the Slovene-speaking provinces. Only Trieste experienced the desired economic progress: as a Southern Railway terminus, it obtained an arsenal, a large shipyard, and a ship-repairing facility belonging to the Austrian Lloyd, the largest steam-navigation company in the Mediterranean. Elsewhere in the Slovenian populated territories, economic progress mostly took the form of setting up small trade and manufacture companies. Contemporaries sought the main reasons for sluggish economic growth and, according to *Novice*, the main problem lay in poor education. The rural population still often resorted to superstition to explain their misfortunes, attributing it to various natural phenomena. But the reasons were much deeper-seated and stemmed from the lack of domestic capital as well as the lack of interest of foreign investors in the Slovenian market.

---

392   Ibid., 65.

The general education system remained more or less the same, namely, far from satisfactory. This was largely the result of the obsolete school system, which hadn't changed since the pre-revolutionary period. On the other hand, drastic changes were made in secondary and higher education with a major reorganization of grammar schools in 1849. The existing six-year grammar schools were transformed into eight-year general schools and remained as such until the fall of the monarchy in 1918. School reorganization also brought about changes in curriculum content: the number of natural science courses increased and the teaching of Latin declined in favor of German, which became the only language of instruction in grammar schools during neo-absolutism. A secondary school diploma (*matura*) was introduced for the completion of secondary education and simultaneously served as the qualification for university admission. Teaching in secondary schools now required a university degree and additional special examinations. The school reform led to the mushrooming of new schools of mathematics and natural sciences, the so-called secondary modern schools. Lyceums, which had previously provided a transition between secondary school and university, were abolished. However, the abolition of lyceums in Ljubljana, Gorizia, and Klagenfurt also threatened the total collapse of higher education as a whole in the Slovene-speaking territories, with the exception of theology in these three cities and Maribor. The Concordat between the Austro-Hungarian Monarchy and the Vatican in 1855 granted the Church more authority over schools and teachers and also represented an important gain for the Slovene language and its use in schools. This step was especially welcomed by the nationally-minded Slovenian clergy, which struggled to promote Slovene as the language of instruction.

By the end of the tumultuous 1850s, absolutist bonds were beginning to break: "[m]uch like the contingencies of war had brought absolutism to power in 1849, the defeats in 1859 clearly demonstrated what a poor service the arrogant bureaucracy had rendered [...]."[393] The Slovenes were entering a period of the highest ascendancy in their history.[394]

---

393 Mal, *Zgodovina slovenskega naroda*, 964.
394 Vasilij Melik, "Ustavna doba in Slovenci," in: Cvirn, *Slovenska kronika XIX. stoletja*, 13.

# The Slovenes in the Constitutional Era

Military defeat in the war against France brought a dramatic reversal of fortune to the Habsburg Monarchy when, entering the 1860s, it lost Lombardy and thus unwillingly created the perfect opportunity for the unification of Italy under Piedmontese rule. The Habsburg Monarchy found itself in the uncomfortable position of deciding what to do next, because Italian unification also represented a triumph of national tendencies and liberal ideas, and conversely a defeat of legitimism and the Church. The state suffered a series of military and economic disasters, and public debt soared, since the European economic boom had barely touched the Austrian provinces. The emperor eventually had to decide to impose a new constitution. After signing the Treaty of Villafranca between Austria and France, he drew up the Laxenburg Manifesto in July 1859, in which he promised to make long overdue legislative improvements. In late August 1859, he deposed the central figure of the absolutist regime, Alexander Bach, and formed a new government under Agenor Gołuchowski. The imperial cabinet announced an expansion of the autonomy of provinces and the restoration of representations of the provincial assemblies in the crown lands. In the ensuing months, the emperor introduced various language concessions to accommodate the non-German historical nations of Austria. Other than that, the imperial court in Vienna did not envision any drastic political reforms; quite to the contrary, at the new government's session on August 25, 1859, the emperor even urged decisive action against demands for a new constitutional system. However, amid general discontent and facing a growing crisis in Hungary, he was eventually forced to retreat from his unbending stance. On March 5, 1860, the emperor issued a special imperial patent to reinforce the Imperial Council (provided for by the 1849 octroyed constitution and appointed as an advisory body in 1851) with new associate members.

The composition of the reinforced Imperial Council made it a pale substitute for the Parliament with its House of Lords and House of Deputies. Besides lifelong members, it also had 38 provincial representatives holding a six-year mandate; each provincial representation appointed three candidates—but since there were no provincial diets, representatives were appointed by the emperor himself. Each province with Slovene-speaking population, and Gorizia and Istria combined, was represented by one member, none of whom was a Slovene. The reinforced Imperial Council, which convened between May 31 and September

28, 1860, had no legislative power and restricted itself to discussing the budget, financial reports, important draft laws, and proposals submitted by provincial representations. During its session, the Imperial Council also addressed major issues concerning internal state organization and submitted its opinions to the emperor. In mid-September 1860, discussion thus focused on the future organization of the empire. Representatives of the conservative nobility, who formed a majority, called for the enhanced autonomy of historical provinces; they advocated major constitutional amendments to accommodate individual nations that would thus be given an opportunity to actively participate in the development of the state. The minority, on the other hand, urged the establishment of a modern, centralist state built on liberal foundations. In the end, the majority opinion prevailed and the Imperial Council passed a proposal that would reorganize the Habsburg Monarchy on the basis of historical (traditional) federalism. It was approved by the emperor after due reflection. Following a decade of absolutist centralism, the Imperial Council's conclusions were also greeted with satisfaction by the Slovenes. The government took three weeks after the end of the session to determine the form of a future constitution, but the emperor eventually made this decision alone, opting for the principle of historical federalism.

**Map 5.** Austro-Hungarian Empire (1867–1918).

On October 20, 1860, the emperor issued a manifesto and the so-called October Diploma, which regulated internal constitutional relations and so definitively brought absolutism to an end. The October Diploma reflected the majority view on the reinforced Imperial Council that the best solution for the state lay in the recognition of the specific historico-political circumstances of individual kingdoms and provinces, in the equality of nations, and in the enhanced administrative and legislative autonomy of the provinces. The October Diploma distributed legislative power among the emperor, provincial diets, and the Parliament (*Reichsrat*),[395] which would consist of land representatives and replace the previous reinforced Imperial Council. It would have legislative power and competence over financial matters. Besides the Parliament for the entire Monarchy, the October Diploma also envisaged a parliament for all non-Hungarian provinces. The constitutions of the lands of the Crown of St. Stephen were restored and four additional manifestos were drawn up for the rest, including Styria and Carinthia.

Predictably, the amended internal constitutional system, and especially its undemocratic conception, met with opposition from both German centralists, who criticized the pronounced role of provincial diets and land autonomies, and proponents of enhanced democratization. Under the new state organization, provincial diets would consist of representatives of the clergy, nobility and big landowners, chambers of trade and commerce, cities, and other local communities. In cities, the vote was limited to municipal representation, whereas in the countryside it was only granted to mayors and one member from each Municipal Council. In such composition, provincial diets barely differed from the previous provincial assemblies in the feudal period. Furthermore, because such an electoral system was not based on the actual population, it clearly prioritized the Germans in national terms as well as big landowners and the middle class in terms of social position. Therefore, even in Carniola, where the Slovenian majority was the largest, the curia of large landowners remained in German hands until the end of the monarchy.

But before the October Diploma could even be enacted, the emperor had to change ministers. The chancellorship was passed to Anton Schmerling, who declared the so-called February Patent on February 26, 1861. The patent was in fact a law issued by the emperor that contained the emperor's manifesto, the

---

395 To avoid ambiguity, the Slovenes referred to it as the National Assembly, since the literal Slovene translation of its German name, *Reichsrat*, denotes "Imperial Council."

Basic Law on the Parliament, which was soon amended because it was initially intended for the entire Habsburg Monarchy and as such was never enforced.

The Parliament was divided into the House of Lords and the House of Deputies. The members of the House of Lords were appointed by the emperor and included two Slovenes: Franc Miklošič and, after his death, the castle lord Oton Detela. Sitting in the House of Lords were also archdukes, bishops, and heads of certain noble families. The 343 members of the House of Deputies were elected by provincial diets and from 1873 onwards by direct popular vote. Immediately after its constitution, the Parliament split into two camps over the division of jurisdiction between the Parliament and provincial diets. The left wing and the center (the so-called centralists) demanded more power and authority for the central Parliament and central authorities, whereas the right wing (the so-called federalists) called for the strengthened autonomy of historical provinces. The first camp represented the stipulations of the February Patent in particular, and the second embraced the October Diploma. The centralists formed the government, and the federalists the opposition. The struggle for centralism was emblematic of many liberal states in Europe.

The February Patent provided individual provinces of the Austrian half of the monarchy with provincial constitutions containing their respective provincial statutes and election rules of the Provincial Diet, which remained in force until the end of the monarchy. This marked the creation of a two-tier system of governance: on the one hand, the central government and Parliament, which delegated the administration of the provinces to state authorities on behalf of the Viennese government and its ministries, and on the other hand the provinces and provincial diets with wide-ranging jurisdiction.

In Styria and the Littoral (Trieste, Gorizia, and Istria), the Viennese government was represented by the imperial (later imperial-royal) stadtholder (*Statthalter*). In smaller provinces like Carniola and Carinthia, it was represented by the president of the province (*Landespräsident*). The president of the province had his seat in the provincial palace and was responsible directly to the government in Vienna, whose decisions he carried out faithfully. Presidents of provinces and stadtholders represented the high, predominantly non-Slovenian nobility; the sole exception was Carniola, which in 1880–1892 had a president of Slovenian nationality, Andrej Winkler. Provincial stadtholders and presidents had to learn the language of the province over which they presided. In Carniola this was Slovene, whereas in Carinthia and Styria (and also in the Littoral and Istria), where the Slovenes formed a minority population, it was German (or

Italian).³⁹⁶ Even the emperor himself cultivated his linguistic education by learning Hungarian, Czech, and other languages that he used during his visits to individual provinces.

The main provincial authority, however, rested in the Provincial Diet and governor of the province (*Landeshauptmann*), appointed by the emperor from the majority party in the Provincial Diet. In Carniola, the provincial governor was most often a Slovene, whereas in provinces with a minority Slovenian population he was usually a German or Italian. The emperor also appointed the deputy provincial governor, who was usually a member of the minority party in the Provincial Diet. The provincial governor presided over the provincial committee, which carried out its business as an autonomous government, i.e., the executive body of the Provincial Diet, and was entrusted with autonomous provincial administration. The provincial committee was responsible to the Provincial Diet; all elected members of the provincial committee had to be residents of the province's capital.

The electoral system of 1861 introduced five electoral classes or curiae: virilists (bishops and university chancellors, acting as ex-officio deputies), large landowners, chambers of trade and commerce, cities, and rural communities. Each curia had a predetermined number of deputies. Such a composition allowed certain social classes to elect a greater number of deputies than others, because the vote was weighted according to the amount of taxes paid or property owned. This right to vote was a liberal principle which did not recognize the universal suffrage demanded by the democrats, who argued that everyone ought to have the right to vote—just as everyone, whether they paid taxes or not, had to serve in the army. The vote under the Austrian electoral system was extremely limited, but still less so than in Hungary, Italy, or Belgium. In the 1870s, it was granted to 7–8 percent of the population. Therefore, for example, a handful of landowners had as much as a quarter of the deputies (between 100 and 200 in the Slovene-speaking provinces), while 80 percent of the rural population only had 40 percent of the deputies. In the Slovenian populated territories, all bishops were virilists. In the city and rural curiae, the vote was generally held by members of the first two classes of qualified voters under municipal law. Representatives of the city curia were elected by direct vote, while those of the rural curia were elected through electors appointed by each rural community. The resulting composition of provincial diets therefore showed obvious traces

---

396 Janez Cvirn, *Trdnjavski trikotnik. Politična orientacija Nemcev na Spodnjem Štajerskem (1861–1914)*, Maribor: Založba Obzorja, 1997.

of the provincial assemblies. The same curial system applied to both provincial diets and the Parliament, because election by curiae and an unequal distribution of deputies among individual provinces were the main objectives of Schmerling's electoral geometry, which sought to maintain German dominance even in non-German areas with a minority German population.

The provinces were free to manage their often valuable possessions as they chose, and had the right to state tax allowances and various taxes. Under general Austrian law, the provinces also had a number of powers to settle local, school, and Church issues and deal with matters of local interest, e.g., charity funds, hospitals, cultural and public works, and industrial facilities. Provincial autonomy, however, proved much smaller in practice than in principle. By definition, it also included the elementary school system, pursuant to the law that obliged all autonomous provinces to maintain their respective school systems, appoint teachers, and found new schools. Moreover, because teachers' salaries fell under the purview of provinces, they varied considerably from one province to another, so that, for instance, a teacher in Carniola was paid less than his peer in Styria. The law also stipulated that schools were to be supervised by school councils, which were set up for each administration level and consisted of representatives of the state, municipality, county, province, the Church, teachers, and parents. Each province passed a school act that provided for the composition of school councils.

The year of 1861, when the February Patent was issued and caused much indignation across the provinces with Slovenian population, was also marked by the first elections to the provincial diets. After serious consideration, Bleiweis realized that there were too few patriots to warrant the success of the Slovenian national movement. Slovenian expectations mainly stemmed from the belief that the national language was an essential prerequisite for national development. This was also articulated in a petition addressed to Chancellor Schmerling that demanded the introduction of exclusively Slovene-language elementary schools, the guarantee of equality for the Slovene language in secondary and high schools, public offices, and legislation, and the unification of all provinces with the Slovene-speaking population under one rule. By April 1861, 20,000 people had signed the petition.[397]

The firm belief in linguistic rights as fundamental national rights also found its expression in the political contest before the first general elections to the provincial diets in 1861, especially in appeals to vote for men who would uphold

---

397  Mal, *Zgodovina slovenskega naroda*, 937.

the freedom endowed by the emperor, but also fight fervently for the interests of their nation and their (linguistic) rights. Before the first Provincial Diet elections, there were no significant shifts in the Slovenian attitude toward the Germans or Italians. However, the impotence of Slovenian politics at the time was fully revealed in poor election results for the Slovenian national camp. Its complete lack of organization was, furthermore, most eloquently demonstrated by Bleiweis' election in no less than three constituencies, so that he had to resign from two. Two reasons for this were that the pre-election campaign had been left in the hands of individual nationally-minded Slovenes in the field, and that voters' opinions were more strongly influenced by acquaintances than by political determinations.

The elections moreover revealed two concurrent struggles: one between liberal and conservative or clerical views, and another for and against Slovene language rights. The conservative forces were sorely beaten, losing all the rural votes and having only one priest elected for the entire Slovenian populated territories. The Slovenian party was well organized in Styria as well as in Ljubljana, whose new mayor was a popular official from 1848, Miha Ambrož. In his commitment to maintain mutual trust, brotherhood, and the traditions of 1848, Ambrož endeavored to govern the city by finding common ground between Slovenes and Germans. He even delivered speeches in both languages, which also differed in substance. In general, it may be concluded that the Slovenian national movement was defeated at the 1861 elections. Only 13 of 36 deputies declared themselves as Slovenes in the Provincial Diet of Carniola, and 7 of 21 in the Provincial Diet of Gorizia, whereas national declarations in other provincial diets had not yet been made. Consequently, the Slovenes had a very marginal voice in the Parliament, in which their views were represented by three unaffiliated deputies—two from Carniola and one from Gorizia. What is more, during the first Parliament sessions, the Slovenes discovered with disappointment that all the emperor's promises to protect the rights of all nations had been fraudulent and deceitful. The State Minister, Schmerling, flatly rejected the Slovenian petition and all the linguistic demands of the Slovenian deputies.

After constitutional life had been restored, the Slovenian national movement continued to gather around the prudent and pragmatic *Novice* editor Bleiweis, who insisted that nationality and freedom were inseparable rights and that Slovene ought to be introduced in schools and in the bureaucracy. New and more radical clerical views began to emerge nonetheless, embodied by the *Zgodnja danica* editor Luka Jeran, who placed religion above nation and freedom, to the disapproval of the clergy itself. Even though the central concern of the Slovenian politicians at the time was a federalist transformation of the monarchy, they were not even able to put up a principled opposition to the government. The Slovenian national-political movement still showed outward unity, but it had already been split into two (nationally) conflicting political camps—the Slovenian and the German. Bleiweis divided the adversaries of the Slovenian movement (who, he warned, also included Slovenes) into three groups, based on their attitude toward the Slovene language. Members of the first group, public officials and artisans, spoke Slovene at home but German in public, partly through a desire to improve their reputation. Members of the second group closely depended on the first group and were in principle not anti-Slovene. The third group consisted of members of German middle and official classes who energetically opposed the Slovenian movement and described the Slovenian language demands as indicators of separatism, treachery, Pan-Slavism, and maybe even the aspiration to create a South Slavic kingdom under Nicholas of Montenegro.[398]

Crossing over to the opposite national camp, as witnessed by the Slovenes at the very onset of the constitutional era in 1861, caused an outcry. Dragotin Dežman, a scholar previously devoted to the Slovenian cause, found himself at odds with the Slovenian political leadership of the time and swung from the Slovenian national camp to the German side. He was particularly opposed to forging alliances with the Croats and imitating the Czech example. After having "deserted" the Slovenian side, he fervently advocated German views and earned himself pejorative names like "renegade" or "national traitor" with his extremely degrading judgment that the Slovenes ought to be humbled by their cultural and economic existence, which they largely owed to the Germans, whose culture was their sole guarantee of progress. However, his observation that Slovene could still not prevail over German in public and cultural life was applauded by many of the then predominantly bilingual middle class, which simultaneously empathized and identified with both the Slovenian and German cultures. Some may have deemed Slovenian national politics too narrow, while others may have considered them too conservative or totally unacceptable for their unscrupulous flirtation with South Slavic policies. Therefore the problem of "defection" was obviously multi-faceted and stemmed from economic, social, even personal, and above all political reasons. A scornful designation for a Slovene who took the German side in the Slovenian-German dispute became *nemškutar* (meaning "Germaniser"). Josip Jurčič characterized them as follows:

> Would you agree, my fellow Slovenian men and countrymen, that it is a foul bird that defiles or even abandons its own nest; that he who bites the hand that feeds him, ends up

---

398   Vodopivec, *Od Pohlinove slovnice do samostojne države*, 76.

licking the boot that kicks him; that he who is ashamed of his mother and his father, who had loved and nourished him, is a disgraceful son, of a heart of stone and unworthy of trust and respect. And behold, such a bird, such a person, such a son is a nemškutar. He scorns his compatriots, feels contempt for the speech and the language of his own race, and is ashamed to be a Slovene, straining to the top of his bent to become a German and consorts with the Germans, our adversaries.[399]

The outwardly displayed internal unity of the Slovenian "national" party started to crack under the emerging rifts; besides the radical clericals, younger liberally inclined Slovenian politicians (Simon Jenko, Josip Stritar, Josip Jurčič, and Josip Vošnjak) started successfully asserting their position in literary and broader cultural spheres. In Styria, the leading role in promoting national politics was in the hands of extreme liberalists (Janko Sernec and Ferdinand Domenkuš), whose liberal orientation and ideology impregnated the entire Slovenian national movement. A German official, Mihael Herman, was one of the first Styrian Slovenian leaders and an advocate of Slovenian interests. After the first elections, Styrian Slovenes were the first to organize their political agitation, since in this period Styria possessed a much greater political flexibility, which resulted from, among other things, better financial and organizational capabilities. It was precisely in Styria that the idea of the Slovenian Society (Slovenska matica), established in 1864 in Ljubljana, first took shape.

During the four year of Schmerling government, the successes of the Slovenian national movement in asserting the Slovene language continued to be modest. While elementary schools in Carniola and Gorizia noted improvements, German remained the dominant language of instruction in secondary schools throughout the provinces with Slovenian population, as well as in courts and public administration.

## Reading Societies, *Tabori*, and the Popularization of National Ideas

During the constitutional era, the Slovenes confined their political and cultural efforts to reading societies, since constitutional life in Austria was making very slow progress and did not grant enough political freedom for establishing proper political societies. Reading societies therefore provided a perfect niche for many cultural and political activists. The first reading society was founded in January 1861 in Trieste as Slovanska čitalnica, the Slavic Reading Society, also

---

399   Josip Jurčič, *Zbrano delo*, vol. 11, Ljubljana: DZS, 1984, 74.

intended to serve other South Slavs in Trieste. Fran Levstik was appointed its secretary. During the first year of operation, the Slavic Reading Society had as many as 230 members. That same year, reading societies began to spring up in Maribor, Ljubljana, and Celje. After that, their number rose each year; by 1864 there were 12, and after the adoption of the December Constitution in 1867 their number continued to increase until it reached 57 in 1869. Reading societies were established in cities, market towns, and minor administration centers, in the Littoral, to some extent in Inner Carniola, and in villages. In 1869, reading societies had around 4,000 members. They continued to emerge in and after the 1870s, but not with the same intensity as before. Their "golden age" may be placed in the 1860s, when they played a crucial political role.

Reading societies, whose origins can be found in pre-March reading clubs, provided a meeting place where people of the same views would organize so-called *besede* (words) at various locations, but most often in taverns. These were evenings with entertainment, recitations, singing, concerts, speeches, and lectures, permeated with patriotic content. Many had their own choirs and singing schools; they collected domestic and foreign newspapers and in some places even founded proper libraries. In cities and market towns, reading societies became the centers of social life. Even though they varied in accessibility, reading societies were, in general, middle-class institutions (scholars, respectable tradesmen). They provided a gathering place for nationally-minded intelligentsia and the privileged Slovenian elite in cities and market towns, and for well-off rural people, artisans, and small merchants in the countryside. However, reading societies did not encompass the largest segments of the population, except in the Littoral, where they were also popular among peasants. As early as 1863, Styrian reading societies organized outdoor festivities, bringing together large crowds to celebrate specific events and providing a wide array of performances to attract as many people as possible. Organizationally, this form of assembly was becoming increasingly similar to the later *tabori* meetings,[400] but, unlike them, it had no political connotations.

Initially, reading societies were also frequented by Germans, which led to a language problem, since the "official" language in reading societies was Slovene, while the women, in particular, communicated in German. The position of Slovene in reading societies was eventually improved through active involvement

---

[400] *Tabor* is a Slovene term usually denoting a camp or encampment; however, in reference to the period 1868–1871, it stands for a mass outdoor political gathering (transl. note).

by the generation of post-1848 grammar schools, where Slovene had become the language of instruction and the Slovene writing system had been fully established. The only criterion for cultural (and especially literary) production was national fervor and patriotic enthusiasm, while literary criticism was at the time virtually unknown to the Slovenes. In 1869, the central Slovenian reading society in Ljubljana had 300 members. Gradually becoming a center of Slovenian cultural and political life, Ljubljana was evolving into a long desired Slovenian capital, a modern city with gas lighting. Owing to the unjust election rules of Schmerling's patent, as Josip Mal called the election legislation in understandable frustration,[401] Germans had a majority in Ljubljana's City Council until 1869, but the city itself was run by Slovenian mayors.

The growing vigor of the Slovenian national movement was also reflected in the revived establishment of societies. The only society that had persisted since the 1850s was the Mohorjeva družba publishing house, established in 1851 on the basis of legislation from the revolutionary year of 1848. This legislation remained formally in effect until the age of reaction, when it was nearly impossible to set up any society, especially a political one, until 1867. Since the idea of organized dissemination of Slovene books among the population was fathered by Slomšek, the reorganized Mohorjeva družba became a semi-ecclesiastical organization whose popularity rose during the 1860s and peaked on the eve of World War I. The books from Mohorjeva družba were ordered by readers throughout the Slovenian populated territories, irrespective of their political beliefs. Their content was well balanced and structured, comprising a calendar with articles, Slovenian vespers (a collection of tales), church books, and various instructional books (in economics and popular science). The period of 1874–1891 also saw a five-volume edition of Josip Stare's *Obča zgodovina za slovensko ljudstvo* (General History for the Slovenian People), written in vibrant, intelligible, and beautiful language, and whose first edition sold out in a very short time. Before World War I, there were plans to publish a new edition of world history in ten volumes, to which selected authors would contribute their descriptions of individual historical periods. Eventually, Anton Sovre's *Stari Grki* (The Ancient Greeks, 1939) was the only volume actually published, as other publications were overtaken by the outbreak of war, after which a project of such scale could no longer be completed. Between World War I and World War II, *Zgodovina slovenskega naroda* (The History of the Slovenian Nation) was published, initiated by Josip Gruden (who wrote the first 1,000 pages, up to Maria Theresa

---

401   Mal, *Zgodovina slovenskega naroda*, 972.

and Joseph II). After Gruden's death, Mohorjeva družba found his successor in the Provincial Museum director Josip Mal, who contributed a complete segment covering the period from the February Revolution to World War II.

In 1864, Mohorjeva družba was joined by the first Slovenian scientific and literary society, Slovenska matica, whose task was to publish academic books, original Slovenian and translated literature, for a more refined readership. The name "matica" was adopted from its sister societies (Matica srpska in Novi Sad and Ilirska matica in Zagreb). The fundamental mission of Slovenska matica, which still operates today, was the promotion of the Slovenian scientific press. The foremost imperative of this arduous task was to assert Slovene as the language of science and construct an appropriate terminology. Slovenska matica published a vast volume of books and the first scientific journal, *Letopis—Zbornik Matice Slovenske* (Yearbook—The Slovenska Matica Collection), featuring Slovene-language discussions in various scientific fields. In the early stages, most texts were in fact translations from other languages, but the selection of Slovenian scientific papers continued to grow.

In close relation with reading societies, and pursuing the same mission, theater lovers established the Drama Society in Ljubljana in 1866. Following the Czech example, and in line with the 19th-century European trends which marked the beginning and development of sporting life, Ljubljana saw the founding in 1863 of the gymnastic society Južni sokol (Southern Falcon), which fostered the love of country and people among the youth by organizing competitions and trips to the countryside. Južni sokol was a liberal gymnastic organization, while the Catholics established their own, Orel (named after another bird, the eagle), in 1906. The two organizations remained separate until the introduction of the royal dictatorship in 1929, when they united into a single pan-Yugoslav association. The Slovenian Alpine Club (Slovensko planinsko društvo), founded in 1893 and still active today, also promoted the national spirit, as well as physical exercise through exploring the alpine, mountainous and karst worlds of Slovenia.

Furthermore, with the introduction of the 1867 Constitution, which further expanded freedom of association, the first workers' educational societies began to emerge by the end of the 1860s. Favorable conditions also facilitated the establishment of political societies, whose membership was strictly male. New societies sprang up throughout the Slovenian populated territories, but were confined to the boundaries of each respective province, despite their all-Slovenian names, such as Slovenija in Carniola (for the protection of national rights), Sloga (Concord) in the Littoral, and Edinost (Unity) in Trieste. This was above all a reflection of the contemporary state of Slovenian politics at the time, which was led differently in each province. Numerous minor societies were also springing up in individual cities and market towns and crucially contributed to Slovenian cultural and national development.

This was also a period of brisk literary activity, which mainly followed the post-Romantic orientation combined with early elements of Realism, whose outstanding representatives were Fran Levstik, the poet Simon Jenko, and Josip Jurčič, the author of the first Slovenian novel *Deseti brat* (The Tenth Brother, 1866). In 1866, Ljubljana's reading society staged the first operetta, *Tičnik* (Birdcage), composed by the most important representative of Slovenian Romantic music, Benjamin Ipavec.

In the 1860s, the newspaper publishing blossomed amid awakening political passions, which prompted the need for a special political newspaper for the Slovenes. Bleiweis' *Novice* not only had a modest circulation but appeared only once per week, while its editor insisted on maintaining the antiquated and ineffectual concept of writing in a clear and concise fashion. In these new times, the Slovenes needed a political newspaper that would specialize exclusively in political issues. A group of liberally oriented Slovenian intellectuals and politicians, the so-called Young Slovenes, made three attempts at launching a political newspaper. In 1863, the well-off castle lord and poet Miroslav Vilhar embarked on the publication of a twice-weekly gazette, *Naprej* (Onward), and appointed Fran Levstik as its chief editor and author. Even though *Naprej* did not mar the unity of the Slovenian political camp or directly contradict the Old Slovenes, it represented a step forward with its sound and realistic judgment of the situation in the Slovene-speaking crown lands, with its principled stance toward matters of national interest, and its call for improvements in the organization of and commitment to the national movement. *Naprej* unequivocally advocated "United Slovenia" and the forging of ties with other South Slavs, but rejected the proposals to adopt Croatian as the principal scientific language, stressing that precisely scientific work was of central importance to the Slovene language. Crippled by numerous lawsuits brought against it, the gazette remained in circulation for no more than six months. It should be borne in mind that this was still a time when the press was not yet free, and editors were often brought before the court on charges of publishing articles that incited hatred between the two nations. The gazette's owner, Vilhar, was even imprisoned and fined for printing an article calling for the abolition of the old provincial borders and the unification of the nations that shared the same language.

The Bleiweis circle, however, was not so fond of the idea of a Slovene political language, but rather published a gazette that would present the Slovenian perspective in German. Levstik and many Slovenes outside Carniola energetically opposed this concept, and the ensuing discussions about what kind of a political newspaper the Slovenes ought to have eventually split the Slovenian political leadership into Young Slovenes and Old Slovenes, who had acquired their name after the example of the "Old Czechs." Between 1865 and 1870, the Old Slovenes thus published *Triglav*, styled the "newspaper for homeland interests," which presented the Slovenian point of view; even though it was essentially more Carniolan than pan-Slovenian. Meanwhile, Andrej Einspieler, a Carinthian Slovenian priest, a conservative, and an ardent supporter of the rights of the Catholic Church, launched the newspaper *Slovenec* (The Slovenian) in Klagenfurt, which lasted no more than two years. Its principal aim was to instruct and educate the Slovenian people in matters of religious, political, national, and economic interest. Moreover, in 1868 a new political newspaper, *Slovenski narod* (Slovenian Nation), was established in Maribor, which initially appeared several days per week and ultimately became the first Slovenian daily. *Slovenski narod* became the first permanent Slovenian newspaper and played an important role as the herald of the liberal camp until the 1920s, when it came to represent the views of only one Liberal faction. In 1920, the liberal camp launched a new newspaper, *Jutro* (Morning), while *Slovenski narod* denounced its political orientation and continued to be published until the end of World War II. With time, each province started issuing weeklies as their special journals. *Soča* (1871–1915) was most renowned in Gorizia, *Mir* (Peace, 1882–1920) in Carinthia, *Slovenski gospodar* (Slovenian Master, 1867–1941) in Styria, and *Edinost* (Unity, 1876–1928) in Trieste. This period also saw the emergence of humor magazines, such as Levstik's *Pavliha* (Jester) in 1870, as well as *Osa* and *Brencelj* (both named after insects: the wasp and the horsefly, respectively).

The mid-1860s also saw major shifts in the broader Habsburg political arena. Schmerling's government tried to lead the monarchy in a centralist spirit, but could not garner Hungarian and Croatian support for the October Diploma and even less for the February Patent and the elections of the Parliament. The opposition was also gaining ground in the Bohemian lands, where the liberals and the old conservative nobility stood against the system under the February Patent. Insisting on the principles of Czech historical law, they sharply rejected centralism and called for enhanced local autonomy. Czech politics became radicalized, and from 1863 onwards Czech deputies no longer participated in parliamentary sessions in Vienna. Opposition to centralism was also voiced by other Slavic nations. Schmerling's policy of waiting even became a source of frustration for German liberals, who demanded democratic constitutional amendments, the expansion of political rights and freedoms, and the revocation of the 1855 Concordat with the Catholic Church.

After Chancellor Otto von Bismarck came to power in Prussia and Schmerling was no longer capable of administering state affairs, the monarchy's deteriorating relations with Prussia and fear of its disintegration called for immediate internal consolidation. The total failure of Schmerling's policy brought down his government in the summer of 1865. Count Richard Belcredi became the new prime minister, and for a brief period the federalist and conservative nobility took the helm of the state. The September Patent of 1865 retained the Basic Law on State Representation, based on the argument that it was first necessary to reach an agreement with the Hungarian Parliament and the Croatian Sabor, which had hitherto refused to cooperate. In the eyes of German liberals, the retention of the February Patent was an obvious attack on the constitutional rights and parliamentarianism; conservatives and clericals greeted it with satisfaction, while the leaders of non-German nations welcomed the September Patent as a guarantee of the future equality of the nations, recognition of the historical rights of the provinces, and an end to centralist pressure.

Belcredi's arrival filled the Slovenes with hope of a brighter future for their nation, especially after he issued an encyclical demanding that officials in ethnically mixed provinces respect the language of the local population, and expressed a commitment to abide by the Slovenes' wishes. Slovenian political involvement became more intense and daring; the resurfacing ideas of "United Slovenia" were now to be harmonized with the principles of historical law, which naturally constituted an imperative for the reconstruction of the monarchy. The initiator and steadfast supporter of the monarchy's reconstruction, Andrej Einspieler, had envisioned the creation of a special group composed of the Inner Austrian provinces, Styria, Carinthia, Carniola, and the Littoral, but excluding Prekmurje and Venetian Slovenia. The so-called Inner Austria would have a common parliament and government, and would simultaneously form a federation with the rest of the Austrian provinces. Einspieler opened the door wide to collaborators from the broader Slovene-speaking territories, including the Young Slovenes and Levstik, whose articles in *Slovenec* dealt with the most pressing issues of Slovenian politics.

At a meeting of Slovenian politicians in Maribor in 1865, the Young Slovenes called for a reconstruction of the former Inner Austria, whose boundaries would encompass all Slovenes, including those from Venetian Slovenia and Prekmurje. According to the so-called Maribor Program, each province would have its own diet, consisting of deputies of all nations living there who would be elected on the basis of nationality curiae. The common affairs of the Inner Austrian group would fall under the competence of the joint diet. They thus envisaged the unification of Slovenes without breaching the principle of indivisibility of provinces advocated by the Carinthian and Styrian Germans. However, since the Slovenes would have made up about 43 percent of the Inner Austrian population, the problem of equality still remained to be solved. One way or another, the plan was of no interest to the government circles and conservative Germans, while the Program even engendered sharp opposition and criticism among the Slovenes. The Maribor Program was therefore only a brief episode in the Slovenes' struggle to fulfill their national objectives. The Slovenian movement reverted to the original "United Slovenia" program and was, furthermore, stimulated by the tumultuous events of 1866 and 1867: Austria's defeat by Prussia, resulting in the unification of Germany, and Austria's defeat by Italy in the summer of 1866, carrying with it the loss of Venetian Slovenia. Following the example of kingdoms constituting the new united Italy, a plebiscite was also held in Venetian Slovenia, at which the population, including about 27,000 Slovenes, voted for their annexation to the Kingdom of Italy. This marked the beginning of the process of Italianization. The new Italian rulers assured the local population that Italianization would not be forcibly imposed, but would be spread by teaching the language and culture of the predominant Italian civilization. Italianization was therefore mainly conducted in education, agriculture, and literature. Such an approach, in the Italians' opinion, was a most effective strategy to strengthen their borders. What is more, if the whole population inhabiting the mountainous regions in the Province of Udine and the Resia Valley were to receive Italian instruction on Italian culture, Italian would become the sole language used by the next generation.

Another sobering moment for the Slovenian political leadership came with the Austro-Hungarian Agreement. Parallel to the Slavic (particularly the Czechs') demands for a federalist reconstruction of the monarchy, advocates of a dualistic reorganization of the state were gaining ground among the leading circles, and completely prevailed in the wake of Austria's military defeats in 1866. After protracted negotiations with Hungarian liberals, the government ultimately opted for a dualist reconstruction of the Habsburg Monarchy in early 1867. The Austro-Hungarian Agreement declared the union of two sovereign states within the same monarchy, and the internal organization of each part was determined by a special constitution. Three areas of government were designated as concerning both parts of the newly established Austro-Hungarian Monarchy: foreign affairs, military affairs, and (partly) finance. Individual matters fell under the purview of new joint ministries that had their names prefixed with abbreviations which were also incorporated into professional titles or personal names, giving them a completely new meaning. These abbreviations were: k. k., denoting imperial-royal (*kaiserlich-königlich*); k. u. k., denoting imperial and royal (*kaiserlich und königlich*); and k., denoting royal (*königlich*). Emperor Franz Joseph ruled as Emperor of Austria, but only as King of Hungary. The common Austro-Hungarian army was under German command. Furthermore, each state unit also formed its own provincial defense regiments (called *Honvéd* in Hungary and Croatia, and *Landwehr* in the Austrian half of the monarchy) as part of provincial autonomy. In each defense unit, the language of command was a "vernacular" (German, Hungarian, and Croatian). Ljubljana was accorded its own provincial defense barracks, built with financial support from the province and the city, in the early 20th century.

The Slavs, who had aspired and hoped to achieve a federation of nations, regarded dualism as their greatest misfortune. The Czech leader František Palacký declared the day on which dualism was proclaimed as the birthday of Pan-Slavism in its least desired form. In a similar vein, Bleiweis described the Austrian decision in favor of dualism as playing with fire and the greatest catastrophe that would one day befall Austria. The introduction of dualism also inflicted a physical loss upon the Slovenes, because Hungary assumed control over Prekmurje and consequently severed the contacts of 45,000 Slovenes from Prekmurje with their motherland. Dualism cut the Slavic nations of the monarchy into two halves and separated natural allies from each other.

Despite all that, the crucial years of 1866 and 1867 also brought some positive changes to the Slovenes. The new and more liberal legislation granted them more freedom, including the freedom to pursue political life.

The Provincial Diet elections in January 1867, in which the Slovenes took part as an organized political body for the first time, brought the Slovenian national movement its first major victory. This triumph, furthermore, owed much to the profound commitment demonstrated by a vast majority of the rural population in Gorizia, Carniola, and Styria to the political cause in their unequivocal expression of Slovenian national consciousness. Namely, all Slovenian populated rural constituencies in these three provinces elected Slovenian deputies. The Slovenes also triumphed in Ljubljana, securing themselves a majority of seats in the Carniolan Diet; in Gorizia, the number of Slovenian deputies almost equaled that of Italians; and in Lower Styria, Slovenian deputies emerged as the strongest force in the rural curia.

In February 1867, the new prime minister, Friederich Ferdinand von Beust, dissolved the mostly pro-federalist provincial diets, including that of Carniola. But this did not preclude the Slovenes from triumphing again in the ensuing March elections. Slightly more frustrating was the percentage of Slovenian representatives in the Vienna Parliament. They were thus presented with three options: to follow the path of principled abstention like the Czechs did; to form a principled opposition in the Parliament; or to offer opportunistic tactical opposition. The Slovenian deputies settled for the third path. In May 1867, they participated in the parliamentary session in Vienna and voted for a dualist reconstruction of the monarchy, which provoked an outburst of indignation from the Slovenian public. The Slovenian representatives in the Vienna Parliament justified their decision by citing the government's promise to expand provincial autonomy and grant the Slovene language more rights in schools and public administration. However, the only promise kept by the government was the approval of an Upper Carniolan railway line between Ljubljana and Tarvisio. The concession was won by the deputy Lovro Toman, who then sold it for a fortune to a construction company. Toman's act turned into a political scandal that produced furious reactions, as the Slovenian deputies found themselves overwhelmed by accusations of casting corrupt votes in favor of dualism.

From yet another perspective, the year 1867 ushered in a period of growth and of deepening ideological antagonisms between Young Slovenes (liberals, Slovenian national activists) and Old Slovenes (Catholic conservatives), signaling the first major shift toward the clericalization of one part of Slovenian politics. One reason for this was the ever-fiercer struggle throughout the state against the Concordat and ecclesiastical standpoints: the German liberal camp saw the Church as the main obstacle to the development of a constitutional state and the rule of law, while the Church felt threatened by the introduction of civil marriage and of state control over the school system, which undermined the fundamental principles of the Concordat. In such circumstances, the clergy was compelled to strengthen its political commitment.

Most of the Slovenian clergy was nationally conscious. Moreover, as the central organizational body in the Slovenian rural camp, it wielded considerable influence and control over the Slovenian national movement, thus forcing the national movement's leadership to acquiesce to its demands for the unification by defending Catholic principles and Church influence. Another cause of considerable friction within the Slovenian national movement was the political tactics of the Slovenian deputies in the Parliament: one side advocated real, opportunistic, or compromise-driven politics, while the other side agitated for national policies drawing on "United Slovenia." However, given the deteriorating political situation, even the Slovenian liberal camp came to the conclusion that the defense of Catholic principles and the Church's influence on public life was imperative; as such, these were eventually incorporated into the Slovenian national program. The March elections of 1867 gave rise to the principle "Everything for Faith, Home, and the Emperor." The Slovenian deputies in the Parliament established a common club with clerical Tyroleans as a sign of resolute opposition to the Germanization policies pursued by German liberals. In July 1867, Luka Svetec was the first Slovenian state deputy to raise his voice in support of the Concordat.

The decision to endorse dualism also struck a heavy blow to the Slovenes because it allowed the Germans and the Hungarians to make mutual arrangements to the detriment of other nations in the monarchy. Disappointed by unfavorable internal constitutional developments and in fear of German pressure, Slovenian national leaders conceived the idea of forging stronger ties with other Slavic nations. As early as 1868, a number of Slovenian dignitaries asked the Croatian Ban whether it would be possible to rely on Hungarian support if they decided to join the lands of the Crown of St. Stephen. Budapest's negative response naturally dictated the same response from the Ban. The strong resentment toward the German-Hungarian division of power triggered even more pronounced pan-Slavic expressions of dissatisfaction, for instance through the defiant attendance of the Czechs, Slovenes, Croats, and Galician Ruthenians at an ethnographic exhibition in Moscow in spring 1867, as a clear indication that all their hopes were with Russia.[402] "Majar stood pre-eminent among Slovenian pilgrims to Russia," Mal wrote, further explaining that:

> [t]his was truly a time when they would turn with great love and sympathy toward their eastern matuška.[403] At a meeting held in the fall of 1869, Slovenian pupils expressed their "firm belief that, in the interest of high literature, it is absolutely imperative for us to learn Russian." This was, furthermore, a time when all true Slovenes espoused the slogan "better Russian than Prussian,"[404]

in the event of the monarchy disintegrating or becoming threatened by Prussian domination. But in seeking allies that would ensure or guarantee the Slovenes their national existence, the Young Slovenes also learned a few sobering lessons that called for serious reconsideration. They were primarily stunned by the realization that the Russians cared far less for the Slovenes than they did for Orthodox Slavs, to the extent that the Russians actually sought to form an alliance with the Prussians. The second sobering lesson came at the Ljubljana (or Yugoslav) Congress in December 1879, attended by Serbs, Slovenes, and Croats, who reiterated the need for the unification of South Slavic nations. On that occasion, the Vojvodinian Serb Svetozar Miletić stated that it was utterly absurd to forge alliances with the Slovenes or the Czechs, because should war

---

402 Mal, *Zgodovina slovenskega naroda*, 976.
403 A Russian term of endearment for a mother (transl. note).
404 Mal, *Zgodovina slovenskega naroda*, 976.

break out, they would surely be annexed to Germany. The sole tangible result of this South Slavic meeting was the six-month circulation of the newspaper *Südslawische Zeitung* (South-Slavic Newspaper).

From another point of view, dualism did grant greater political freedom by recognizing equality before the law as well as freedom of speech and association, by passing the Law on Criminal Procedure, and by ensuring the victory of several parliamentary principles that brought about the new constitution (the December Constitution) and installed the first middleclass government. Article 19 of the new constitution guaranteed the equality of nations and languages. But this right applied only to the so-called historical nations, while the rest were even denied the opportunity to receive high school education in their mother tongue. Also, Slovenes succeeded in establishing only one state grammar school (in Gorizia), while equality in academic education was even further from becoming a reality.[405] Irrespective of the declared freedoms, however, the same constitution "[e]nsured [...] nothing short of individual periods of absolutist rule"[406] by authorizing the Ministerial Council to issue laws in case the Parliament were dissolved.

This liberal regime lasted until 1879, except for an interlude between April 1870 and October 1871, when the coalition of conservatives and federalists briefly held power. During this period, Austrian governments were not proper partisan governments, since they consisted of the emperor's closest agents and high nobility and would, due to their more conservative leaning, constitute the "right wing" in a liberal government. Domestic political life in the monarchy continued in the shadow of international political events that also had an impact on the Slovenian national-political situation. The most important events of this period were the Prussian–Austrian War (1870–1871), Prussia's victory, and the consequent establishment of a unified Germany (January 18, 1871) under the first German Emperor William I. Imperial Germany soon grew into the most powerful state on the continent. This also meant Prussia's triumph in the German question and Austria's ultimate exclusion from Germany.

When the Old Slovenes' camp and its deputies in the Parliament (Janez Bleiweis, Etbin Costa, Lovro Toman, and Luka Svetec) opted for the aforementioned policy of opportunism, Slovenian politics became radicalized. The demand for a resolute opposition to the government was spearheaded by the Young Slovenes Fran Levstik, Valentin Zarnik, and Josip Vošnjak, who mostly came from more

---

405 Igor Grdina, *Slovenci med tradicijo in perspektivo. Politični mozaik 1861–1918*, Ljubljana: Študentska založba, 2003, 31–32.
406 Ibid., 32.

liberal Styria. They presented their views to the public in *Slovenski narod*, which became a daily after it moved to Ljubljana in 1872. On the one hand, *Slovenski narod* supported a strong Austria that would protect the Slovenes against Italian and Prussian tendencies, but at the same time called for a free and federalist state, thus expressing its strong anti-dualism. *Slovenski narod* propagated the "United Slovenia" program and promoted cooperation between the Slavic and South Slavic peoples. Yet in certain respects, it also turned against the principles of European liberalism, for example by strongly advocating an unbreakable bond between the Slovenian nation and the Catholic Church. On economic questions, it again defended the domestic market by rejecting free commerce and trade. It was "typical" of the then Slovenian liberal camp "to continually seek a balance between liberal principles and domestic reality."[407]

Another front in propagating national-political views focused on attracting large masses and assumed the form of national rallies. These were enabled by liberal regulations granting more freedom of assembly and became the basis for the organization of outdoor assemblies, the so-called *tabori* meetings, which reached beyond the spatial and selective confines of reading societies. Thus the European tradition of mass meetings, also fairly well known in Austria, became rooted in Slovenian populated lands. Moreover, as the Czech experience (the Hussites' gathering at the Biblical Mount of Tabor in 1434) shows, the Slovenian political leadership embraced the tradition in both form and name, although *tabori* also had a long-established significance in Slovenian tradition (the word denoted fortifications, e.g., *tabor* churches). In Styria, the so-called "words under the open sky" were organized from 1863 onwards and attracted mass audiences. The very first *tabor* was held at Ljutomer in Styria, on August 9, 1868. Over the next three years, 18 *tabori* occurred and the largest was held at Vižmarje near Ljubljana, with 30,000 people.

On average, *tabori* attracted crowds of 5,000–6,000. Accompanied by the sounds of bands and clad in national costume, crowds would come on foot or horseback, in decorated wagons, or by trains offering cheap fares. All *tabori* had similar programs: they were convened on Sunday afternoons and featured a number of speakers, including national activists or deputies. The centerpiece of every *tabor* was the "United Slovenia" program, and other events were all related to it in some way, such as demands to introduce Slovene in schools, offices, churches, and courts, and to establish new schools and a Slovenian university.[408]

---

407   Vodopivec, *Od Pohlinove slovnice do samostojne države*, 84.
408   Ibid., 84.

Speakers would then turn to issues of local or economic importance. The official part of a *tabor* was concluded by passing resolutions and appealing to the authorities to endorse them, and followed by an entertainment program. Two of the most popular speakers at *tabori* were Valentin Zarnik and the priest Božidar Raič. The opponents of the Young Slovenes maintained their distance from these outdoor gatherings, or rather joined them only after the *tabor* movement had reached its full swing. *Tabori* filled the Slovenes with enthusiasm, exciting both the urban and the rural population and boosting the sympathies among common people for "United Slovenia." They therefore represented a unique manifestation of the national program, and the demand for "United Slovenia" moved from the *tabori* to the Styrian and Carniolan provincial diets. Mihael Hermann thus gave an explicit warning in the Styrian Provincial Diet that foreigners held complete sway over the Slovenian populated territories, whereas the Slovenes were no better than servants in their own house. The Provincial Diet of Gorizia addressed the government in 1869 with the question whether it was willing to concede to the Slovenian wishes contained in "United Slovenia," and the Carniolan Provincial Diet passed a similar resolution no sooner than 1870.

Liberal legislation in 1869 introduced the Law on Elementary Education. The Church lost its control over the school system it had enjoyed through the Concordat. Education was no longer at the discretion of local authorities, and became the responsibility of provinces and the state. The previous six years of compulsory school attendance were increased to eight. Substantial changes were also made to the curriculum: in addition to the four basic subjects (reading, writing, arithmetic, and religious instruction), children also received important instruction in natural sciences, geography, history, geometry, singing, and physical exercise. The new law was aimed at ensuring equal educational opportunities, improving the quality of instruction, and eliminating school fees, from which only the poorest children had hitherto been exempt. In accordance with new legislation, elementary education was placed under the supervision of provincial school councils that, in cooperation with municipalities, also decided on the language of instruction. Slovene thus entered elementary schools in Carniola, the Slovenian part of Gorizia, the neighborhood of Trieste, and several Styrian municipalities.

Since the costly administration of the school reform caused the state serious difficulties, less developed provinces (Carniola, Gorizia, and Istria) were offered the opportunity to adapt the duration of schooling to their conditions. Istria introduced six-year compulsory school attendance with two extra years of compulsory evening classes. In Carniola, eight-year school attendance was introduced in major cities and market towns, and the countryside retained six-year schooling. Meeting the new provisions on universal compulsory school attendance, the new modern elementary school included all children aged between 6 and 12 or 14. Further steps were taken in the 1870s to provide for craft and trade education, which was also provided in extra-curricular courses. Illiteracy in provinces with Slovenian population had been steadily decreasing, so that on the eve of World War I, the Slovenes scored high on the literacy scale, only lagging behind the Germans, the Czechs, and the Italians. There were still differences among individual provinces: the illiteracy rate in Carniola, Styria, Trieste, and Gorizia was 11–15 percent; while in Carinthia it amounted to 23 percent due to Germanization, and the situation in Istria was even worse.

The 1860s may also be regarded as a period of cultural struggle that partly arose from the liberal–clerical antagonism that had permeated Austria since 1867. The liberal majority in the Parliament tried to push through a series of laws to minimize the role of the clergy in the state as much as possible (the Law on Education, the organization of monasteries, the Law on Catholic and Heterodox Marriage), thus sparking off fierce polemics. Similar tensions also arose in the Slovenian populated provinces, causing ever-deeper divisions within the Slovenian national movement. The *tabor* movement began to dwindle; it was banned during the Franco–Prussian war in 1870 and was also banned later for promoting pan-Slavic and Yugoslav ideas. Yet the Slovenian national movement still retained a unanimous stance toward the German camp, which declared itself liberal, anti-clerical, and constitutional (*Verfassungstreu*). In order to demonstrate their firm commitment to their principles, Germans established the Constitutional Association (*Verfassungsverein*) in Ljubljana in 1868 and began publishing a newspaper, *Laibacher Tagblatt*. Their example was soon followed by Germans in Celje and Maribor.

German pressure increased at the same time as the fissure within the Slovenian national movement deepened. To fend off liberal demands, which were materialized above all in liberal legislation, the clergy from the Austrian half of the monarchy began to organize itself into a Catholic party that first appeared at the elections in 1870, achieving notable success in the countryside while the liberals prevailed in urban areas. This wave also reached Slovene-speaking territories, where Catholic societies were springing up and bringing together Catholics of all national affiliations. The vigor of the conservative Catholic course was reflected in the appearance of new newspapers: Maribor's *Slovenski gospodar* emerged in 1871 and Ljubljana's Catholic *Slovenec* began to be published in 1872. Interesting and vibrant newspaper publishing activity also took place in Gorizia, where the Catholic newspaper *Glas* (Voice) was started in 1872. The Catholic faction of the political society Soča, established in 1868, found the homonymous newspaper too liberal and founded a new newspaper less than a year later. The atmosphere was saturated with controversy between Gorizia's Old and Young Slovenes and generated a new "Slovenian national-political society on a religious basis" in 1873 that became known as *Gorica*. *Glas* became the central newspaper of the new political society, and retained its position until *Gorica* was terminated as a sign of reconciliation in 1876, when *Soča* became the newspaper of the unified society of Gorizia Slovenes, symbolically named Sloga. Gorizia's long-standing publishing tradition was unfortunately cut short by the annexation of the Littoral to the Kingdom of Italy after World War I. In 1949, a new Catholic newspaper *Katoliški glas* (Catholic Voice) was issued in Gorizia as the successor of *Slovenski Primorec* (1945–1948, published in Gorizia under Zone A of Julian March). It remained in circulation until 1995, when it merged with the Trieste *Novi list* (New Paper) into a new Catholic weekly, *Novi glas* (New Voice).

The Slovenes disapproved of this forced ideological division: because it came too early, it precluded them from waging a unified struggle, and it adversely affected the Slovenian national development as a whole. Most Slovenian clergy agitated for the Slovenian cause and kept a vigilant eye on the position of the Slovene language in schools. University professors, however, were predominantly liberal. Under new legislation, teachers were raised to reputable positions, enabling the Liberal Party to extend its majority support to the countryside. Thus, the Slovene-speaking provinces too saw the formation of two political camps, Catholic and liberal, while Bleiweis' group of conservatives warned that it was absolutely vital for the Slovenian national movement to preserve its unity in the national struggle. In 1872, German liberals took control of the Austrian part of the state, while the Austrian conservative forces, who had founded the Constitutional Party (*Reichspartei*) in the meantime, reiterated the demand to restore the historical autonomy of provinces. They advocated the equality of nations and languages and stressed the need to reinstate the Church's former control over education and public life. This, however, was in complete contradiction to the principles advocated by Young Slovenes, who recognized the Old Slovenes' nod to the Austrian Constitutional Party as a reversal of the obsolete historical concept of provincial autonomy arrangement. The committee elections at the general assembly of Slovenska matica in 1872, won by the conservative Old Slovene side, also showed that the Slovenian national movement was on the verge of an ideological rupture. Furthermore, antagonisms became a prominent feature of the Carniolan Provincial Diet and the pre-election period in 1873, when writs were issued for new elections to the Vienna Parliament. As the first direct elections, they reflected an important internal political reform, which represented a significant step toward the democratization of political life, but also required more effective political organization. The creation of new political constituencies gave the Slovenes an opportunity to elect their representatives to the Viennese Parliament from all Slovene-speaking provinces—Lower Styria, Gorizia, Istria, Trieste, and even Carinthia. However, the era of German liberal predominance was also a time of intense pressures on Slovenian officials to demonstrate their loyalty to their superiors—even though the vast number of German officials in the Slovenian populated territories had itself already secured a large percentage of votes for the German party.

The eventual election results were very even and, with merely eight Slovenian elected deputies, extremely unfavorable for the entire Slovenian national movement. The elections also showed that the Slovenian liberal camp remained weak in urban areas, where a majority of votes were given to the German Liberal Party, but had much more success in the countryside, where the support of liberal-minded teachers and well-to-do farmers proved crucial. A similar incapability to pursue a unified policy could be observed among Slovenian representatives in the Viennese Parliament, where the Catholic and liberal deputies formed separate clubs and did not re-affiliate until the mid-1870s. The Slovenian Catholic camp recognized its main leader in Count Karl Hohenwart, who in a way became the key protagonist of Slovenian politics (even their club was named after him) while simultaneously holding most of the German Catholic conservatives together. The two contesting Slovenian political camps fought a dirty and personal campaign against each other; they eventually brought their quarrels to their respective newspapers, which were soon reduced to petty propagandist brochures. Moreover, their vicious disputes made both camps each other's worst enemies, since they would rather cast their votes for Germans than for any member of the opposite Slovenian camp. This situation filled Slovenian voters with indignation and frustration. But once the ideological divisions in the Austrian half of the state finally started to subside in early 1876, and especially when the liberals joined the Hohenwart Club in the same year, unity also came to the Slovenian populated territories.

This unity did not mean that the Slovenes united into a single political party, but rather that all camps, while retaining their autonomy, committed themselves to mutual political cooperation. According to Grdina, the "[u]tilitarianism of [the Old Slovenes] and the fervent activity of [the Young Slovenes] found common ground in national-political pragmatism," and governmental pressure was much too great for the Slovenes to remain at each other's throats forever.[409] Solemn

---

409   Grdina, *Slovenci med tradicijo in perspektivo*, 66.

events, such as the 70th anniversary of the "father of the nation," Bleiweis, loudly celebrated by torchlight in November 1878, heralded the unification of the Slovenian national movement. Moreover, the movement now also welcomed the active participation of Slovenian women, because:

> [a]n old Slovenian saying goes: The house is not built upon the ground, but upon a woman. And today this old Slovenian saying has a broader meaning [...]: Slovenia is not only built upon Slovenian men, but also upon Slovenian women.[410]

The alliance between the united Slovenian camp and German conservatives represented a step away from the national cause and the stipulations of "United Slovenia." The Slovenian deputies joined forces in Parliament and at elections, and continued to do so until the early 1890s, even though both camps also continued to follow their respective paths independently. At the turn of the 1860s and 1870s, new political forces emerged that mainly focused on the social question, or rather on the question of workers. This period saw the foundation of supranational workers' educational societies that initially banded together tradesmen and artisans. Applications for membership came from Slovenian and German liberals as well as Catholics, since both sides fully recognized the urgency of the workers' problem and the implications of potential outbursts of dissatisfaction. In the 1870s, the first workers' strikes took place across the Slovenian populated territories much as they did everywhere else, and political connections were established with the Austrian workers' movement. The vivid memory of the crushed Paris Commune raised general awareness about the need for a more systematic social policy.

The development of Slovenian politics varied from one province to another, and was in many respects determined by economic conditions and the position of the Slovenian intelligentsia and the middle class. The Styrian Slovenian intelligentsia was politically very active and liberal-minded, but too weak to win a majority of votes in elections. The situation in Carniola appeared somewhat better, but the fierce electoral competition proved fatal for the Slovenian movement, which was defeated at the Provincial Diet elections in 1877 after losing support from the urban curia. In Carinthia, the Slovenian movement was progressing slowly, due to unfavorable ethnic conditions and a small body of lay Slovenian intelligentsia. Slovenes in the Littoral, Trieste, Gorizia, and Istria faced two strong adversaries, Italians and Austrian state officials. Even though

---

410 Anton Bezenšek, *Svečanost o priliki sedemdesetletnice Dr. Janeza Bleiweisa*, Zagreb: Uredništvo "Jugoslavenskog stenografa," 1879, 54, quoted in Grdina, *Slovenci med tradicijo in perspektivo*, 67.

the Italian movement itself branched off into two directions—irredentist and pro-Austrian—it remained unanimously anti-Slovenian and anti-Slavic. The conditions appeared most favorable for Slovenes in Gorizia, where they made up 75 percent of the population, and even though that did not warrant their majority in the Provincial Diet, it did not give Italian deputies complete predominance. Slovene was therefore equal to Italian in the Provincial Diet of Gorizia, while Slovenian politicians also reaped strong support from the Slovenian press and Slovenian societies. Political unanimity, which reached the central Slovenian areas by the late 1870s, had already produced important national-political results in Gorizia a year or so earlier. In Trieste, any successful political effort was rendered almost inconceivable due to low Slovenian representation in the City Council, but the economically revitalized Slovenian liberal middle class had nevertheless gained enough political power and know-how in 1874 to unite into the Slovenian political society Edinost. Edinost gradually grew into an all-Littoral Slovenian political society and published a newspaper of the same name that remained in continuous circulation until 1928, when it was banned by the Fascist regime. The struggle against Italian domination in Istria compelled the Slovenes to join forces with the Croats.

Slovenes in Hungary lived in principle under the protection of a special nationality law, which allowed the use of non-Hungarian languages in public life. However, the law was not implemented in the Slovenian populated area between the Rába and Mura rivers. For the few elementary schools that provided instruction in Slovene, the 1879 law demanded obligatory Hungarian lessons, further increasing Hungarian pressure. The Mura River, the major natural obstacle between the Hungarian Slovenes and their motherland, was most successfully surmounted by Mohorjeva družba, whose books significantly contributed to preserving the Slovene language among the Prekmurje Slovenes.

The Slovenes in Venetian Slovenia, on the other hand, lived under completely different political and ethnic circumstances. They populated eleven municipalities, six of which were ethnically mixed. Due to adverse economic conditions, this region was perpetually marked by mass emigration, most notably from Slovenian communities in mountainous areas. From a formal perspective, Slovenes in Venetian Slovenia were Italian citizens possessing equal rights before the law with the exception of equal nationality rights, because the Italian Kingdom legally only recognized the French ethnic minority in Aosta Valley and subjected all other ethnic groups to assimilation. Such an attitude mostly stemmed from the general belief that the (Italian) state ought to be based on language and cultural unity. The contemporary Friulian journalist Pacifico Valussi wrote that, although the Slavs in San Pietro al Natisone were clearly not of

Italian origin, they nevertheless identified themselves with Italian culture. From 1869 onwards, Italian was declared the mandatory language of instruction, while Slovene was preserved, albeit in a dialect form, by a Slovene catechism produced in 1869, Slovene sermons, and Slovene books published by Mohorjeva družba. Yet, despite the unfavorable ethnic conditions, the Slovenes in Venetian Slovenia formed a thriving scientific and cultural community, imbued with national and patriotic fervor. Except for professional workers' associations, all Slovenian societies in Venetian Slovenia were distinguished by their roles in strengthening and promoting Slovenian professional, cultural, scientific, and artistic pursuits. Glasbena matica systematically fostered and promoted Slovenian folk and classical music, Slovenska matica devoted its efforts to developing science and Slovene scientific terminology, and Mohorjeva družba promulgated a reading culture among Slovenes with its exceptional literary and publishing endeavors.

The central literary personality of the period was Josip Stritar, a writer and poet, publisher of the literary magazine *Zvon* (The Bell), and a committed believer in a distinctive artistic philosophy according to which "art was the only way out of the conflict between man's expectations and the realities of life."[411] His view was shared by his collaborators Fran Levstik and Josip Jurčič, and a younger generation of Slovenian literary authors started to come to the fore, its most noteworthy representatives being Ivan Tavčar, Janko Kersnik, both approaching the form of a true bourgeois novel,[412] and the poet Simon Gregorčič. If prose was predominantly realist, poetry was still essentially post-Romantic. Yet, despite the high level of literary creativity, the time was not ripe for liberal thinking; simplicity and an uncritical atmosphere, dictated by cultural and political circles of a conservative bent, left more room for entertaining and artistically undemanding creations and events. Music was still largely influenced by Romanticism. This was also a golden age of operetta in the shape of Benjamin Ipavec's[413] *Tičnik* (Birdcage) and Anton Foerster's *Gorenjski slavček* (The Nightingale of Upper Carniola—adapted from the work of the Slovenian poet and writer Luiza Pesjakova). Whereas Ipavec was the most outstanding composer of the period, simple melodies and patriotic compositions also gained popularity for Davorin Jenko and Anton Hajdrih. Slovenian painting was marked by a shift

---

411 Vodopivec, *Od Pohlinove slovnice do samostojne države*, 92.
412 Štih et al., *Temelji slovenstva*, 212.
413 Igor Grdina, *Ipavci. Zgodovina slovenske meščanske dinastije*, Ljubljana: Založba ZRC, 2002.

toward academic realism of the modern period embodied in the brothers Janez and Jurij Šubic.[414]

Economically, however, the Slovenian populated provinces were experiencing turbulent times. On May 1, 1873, Vienna solemnly opened a world economic exhibition to showcase the economic progress of the Habsburg Monarchy. Yet this progress was illusory: the very next day the Vienna stock exchange crashed, numerous banks and joint stock companies were ruined, and many businessmen lost all their property. While the crisis most severely affected the most industrially developed parts of the monarchy, it barely impacted the predominantly agricultural Slovenian populated provinces. The agricultural situation had been bleak for a long time as a result of the emancipation of the peasants, the slow modernization of agriculture, the loss of income from non-agricultural activities, the decline in rural handicrafts and carting, and the lower prices of agricultural products due to foreign competition. The position of farmers had further deteriorated with the entry of Russian and American corn onto the European market, which caused corn prices to plummet. At the same time, the development of the agricultural economy was affected by the vicinity of cities and major towns, which enabled resourceful farmers to successfully market their produce. However, the law on the emancipation of the peasants most severely affected the Slovenian countryside, where farmers paid one third of the acquired land value within 20 years, resulting in ever-increasing debt. The 1868 law allowed free trade in farming land, which subsequently ruined small and middle-sized farms, because farmers could only pay their debts by selling their property, whereas large estates were growing larger.

This process was most evident in Styria, Prekmurje, and Carinthia. What is more, even geographical conditions themselves were extremely unfavorable to agriculture in the Slovenian populated territories, since vast portions of land were unsuitable for agricultural production. This in turn led to a growing fragmentation of farming land and a decreasing agricultural population. An attempt to solve the Slovenian agricultural crisis was the establishment of a cooperative movement, stimulated by the adoption of the law on cooperatives in 1873, which drew on social reformers' ideas of self-assistance and made a crucial contribution to improving the situation of farmers.

Credit cooperatives were the most important cooperative societies at the end of the century. In the Slovene-speaking territories, the most notable were the Raiffeisen cooperatives and savings banks introduced by Janez Evangelist Krek.

---

414  Vodopivec, *Od Pohlinove slovnice do samostojne države*, 93.

Credit cooperatives accumulated a vast amount of small savings and distributed them as affordable loans among farmers and artisans. The savings banks' assets grew, the number of members increased, and savings deposits soared. This halted the deterioration of agricultural holdings and facilitated their modernization, and thus significantly contributed to improving the situation of agricultural holdings in general. Before agricultural or savings banks were founded, farmers had usually taken loans from rural usurers on terms set by them. Thus, on the one hand, moneylending became an important source of Slovenian capital, and on the other, it proved to be a ruinous business for Slovenian farmers.[415] This was clearly illustrated by the example of Carniola, where 14 percent of farming estates were sold at auction between 1868 and 1893. The agony of Slovenian farmers manifested itself in various ways; some joined the army of the emperor's brother Archduke Maximilian on his military expedition to Mexico in 1867, and a vast majority emigrated to Western Europe and America. This exodus could not even be dammed by Bleiweis, who was fundamentally more in favor of emigration to Slavic countries.

Farming, livestock breeding, and forestry constituted the main sources of agricultural income, and the provinces with Slovenian population even ranked among livestock- and timber-exporting areas. Agricultural production in these provinces largely remained within the confines of provincial borders, with the exception of winemaking, which always boasted a wide audience of consumers: Styrian wines found enthusiastic tasters in Graz, wines from Vipava and the Karst region were sold to Trieste and the coastal hinterland, while those from Lower Carniola suited less demanding consumers in Upper Carniola.[416]

Mining and forestry developed in close mutual correlation and were placed under the same ministry for technological and ownership reasons. The Slovenian mining industry was based on coal, iron, lead, zinc, and mercury, with coalmining as the most important sector. The unique position and value of mineral wealth gave miners special rights and obligations, as well as their own mining legislation arranged by mining regulations (the General Mining Act of 1854). The Trbovlje Mining Company, founded in Vienna in 1872, consolidated all the most prominent coal mines in the Slovene-speaking territories (80 percent share) until 1904.

Systematic industrialization in the Slovenian populated provinces dated back to the second half of the 19th century, beginning with the 1859 Patent on the

---

415   Ibid., 96–97.
416   Granda, "Od razcveta v streznitev," in: Cvirn, *Slovenska kronika XIX. stoletja*, 129.

Peasants' Emancipation and Crafts Act, which ended the privilege-granting process, set the conditions for the development of non-agricultural industries, and thus paved the way for the gradual transition from an agricultural to a non-agricultural economy. Besides the mining and coal industries, the most remarkable progress in this period was experienced by the iron industry with the establishment of four major ironworks (Ravne na Koroškem, Jesenice, Štore, and Prevalje). In 1869, three businessmen from Ljubljana, members of a commercial company that owned the Ljubljana steam mill, founded the Carniolan Industrial Company in association with the leadership of the indebted enterprise Karl Zois & Sons, which owned several ironworks in Upper Carniola. By merging Zois' works, Viktor Ruard's mines, and works in Upper Carniola, the Carniolan Industrial Company established a modern iron industry network. By the end of the century it had expanded its activity to Servola near Trieste, where it built the first smelting furnace in 1897 and put it into operation with the assistance of ironworkers from Jesenice.

Metallurgy was a key agent of industrialization in this period, and a state zinc processing plant was founded in Celje in 1873. The glass industry also took major steps forward; the Slovenian timber industry was fragmented (its primary sector was timber processing, whereas the manufacture of finished products was rather modest); the paper and leather industries experienced major change (the modernization, specialization, and expansion of paper mills were accelerated by inflows of foreign capital, and the paper industry was concentrated around Ljubljana). The chemical industry established its first factories after 1869; the textile industry underwent minor modifications; the food industry made advances in food, cereal, oilseed, and hop processing. Large clusters of individual industrial plants were especially concentrated along railway lines, between Maribor and Trieste, and between Ljubljana and Jesenice (forming the so-called industrial epsilon). This gave rise to the development of new industrial centers (Celje, the Zagorje–Trbovlje–Hrastnik triangle, Ljubljana, Trieste, and Jesenice). Parallel to industrial development, the percentage of population engaged in handicrafts and industry rapidly increased, from 7 percent in 1869 to 10.4 percent in 1910.

Another important contribution to economic and general development was the expansion of financial institutions, especially savings banks. Agricultural cooperatives also began to emerge. The first attempts at setting up a Slovenian bank occurred in the second half of the 19th century, until it finally bore fruit at the end of the century with the founding of the Ljubljanska kreditna banka, fully supported by Czech finance. By the outbreak of war, it built a network of affiliates throughout the Slovenian populated territories (Klagenfurt, Trieste, Gorizia, and

Celje). In 1905, Jadranska banka was founded in Trieste, and Kranjska kreditna banka was founded in Ljubljana at the end of the decade. The initial advances in the Slovenian banking sector coincided with the development of the Slovenian insurance business.

A major change, if not a genuine revolution, was also underway in transportation. The greatest achievement was the construction of the Vienna–Trieste line of the Southern Railway. In Upper Carniola, the line connecting Ljubljana and Tarvisio was built in 1870; in 1873 the Ljubljana–Tarvisio line was extended to Villach; the Tarvisio–Pontebba section was built in 1879, the Pivka–Rijeka section in 1872, and the section between Divača and Pula, the main Austrian naval port, in 1879. The Bohinj line connected Jesenice and Gorizia in 1906. At the beginning of the 20th century, the major challenge to the railway came with the introduction of the automobile. Roads in the Slovenian populated territories were categorized as state or main commercial roads, provincial, county, and municipal roads. The main commercial road was the one connecting Vienna and Trieste. The development of the railway network was further encouraged by the construction of the Suez Canal in 1869. By 1872, the first transport and economic plans were made to integrate the entire territory into a whole, but the central Viennese government didn't show the slightest interest.[417]

The Suez Canal could also have crucially contributed to the progress of the Trieste hinterland. The city of Trieste itself, which was becoming an important port, had every opportunity to prosper, but expectations that it would become the Austrian Manchester proved unrealistic.

The beginnings of tourism in the Slovenian populated territories date back to the last two decades of the 19th century, with spas being the fastest growing branch of tourism, especially in Lower Styria (Dobrna, Laško, Rimske Toplice, Rogaška). Bled was among the most famous tourist resorts and spas before World War I, and has retained its renown as a prestigious tourist destination to the present day.

Slovenian middle-class representatives, whose fundamental economic activity still resembled that of the rural rather than the urban population, improved their material position in the late 19th century. The middle class consisted of a few large businessmen and industrial entrepreneurs, and a slightly higher number of tradesmen, city merchants, and well-off farmers who had recently migrated to cities. The social elite of every city was formed by lawyers, notaries, physicians, and professors; but it was the scholars who were most involved in the national struggle.

---

417   Vodopivec, *Od Pohlinove slovnice do samostojne države*, 101.

Reading Societies, *Tabori*, and the Popularization ... 347

**Figure 35.** Ljubljana, latter half of the 18th century.

# Unity and National Existence

The summer of 1879 introduced significant changes into the Austrian internal political arena. The key catalyst was the so-called Eastern Crisis, which erupted in 1875 with the Bosnian insurrection against the Turks. This largest and most powerful insurrection to date aroused attention in the whole of Europe, as well as among the Slovenes, whose sympathies lay strongly with any liberation struggle against Turkish hegemony. Many Slovenian towns therefore started gathering aid for the rebels, and many Slovenian volunteers even set out for Bosnia to join their struggle. Austria had to take a stand and Russia called on Austria to join the war against Turkey, after which they would partition the Turkish territories in Europe among themselves. The idea of a war against Turkey was extremely popular among the Slavic peoples of Austria-Hungary, but condemned by German liberals and Hungarian parties, who found their greatest apprehension in a powerful Russia and the Slavs in an expanded Austria. The workers' movement headed by Marx and Engels sided with Turkey. Engels even praised the Turkish peasant as a harbinger of progress, but this was mainly because he did not tolerate Russian and Slavic aspirations. This resulted in a bizarre situation in which a majority in both the Austrian and Hungarian parliaments opposed the intervention. The government tried to find a neutral solution, whereas the concerns of the Foreign Minister, Gyula Andrássy, largely revolved around territorial acquisitions for Austria-Hungary. Russia therefore entered the war against Turkey alone and advanced as far as Constantinople.

Russia's invasion was eventually stopped by British intervention and followed by the Treaty of San Stefano, which Europe refused to recognize. The government then successfully persuaded the Viennese Parliament to make arrangements for the occupation. While Young Slovenes hoped that a new strong Slavic state would rise from the territory of European Turkey, Old Slovenes expressed support for the occupation of Bosnia and Herzegovina. The Congress of Berlin held in 1878 placed Bosnia and Herzegovina under the administration of Austria-Hungary; Old Slovenes even proposed a unification of both provinces with the Slovenian-speaking and Croatian territories into a South Slavic unit of Illyria. Between 9,000 and 10,000 Slovenian soldiers took part in the occupation of Bosnia and Herzegovina. The military operation was completed in less than three months and, judging by its statistics, was the Austro-Hungarian army's greatest military enterprise in the 19th century.

After the Congress of Berlin, the emperor dissolved the entire German liberal government and formed a new one under Count Eduard Taaffe. Taaffe was the president of Austria's government, which consisted of a right-wing coalition of Slavic nations and German Catholic deputies. The new government's task was to settle political disputes and establish internal balance in the monarchy. By this time, the distribution of power in the Parliament had changed completely after the Czechs abandoned their futile policy of abstention and returned to their parliamentary seats. On their return, they delivered a principled statement that referred to their historical rights and stressed that they did not recognize the Parliament but would nevertheless take part in its proceedings. Both sides, the German liberals and the right wing, were equally strong. The German Liberal Party retained a great number of deputies, since the electoral legislation continued to prioritize cities and large landowners. The question of a majority nonetheless remained the cause of constant friction, with each side striving to win smaller groups. Taaffe's essentially independent government relied upon the parliamentary coalition, but had to make constant concessions to one or the other side scrambling for their interests. In response to such pressure, the government threatened to resign and surrender its power to the German liberals, which was precisely the maneuver that secured it the longest mandate. However, Taaffe's government had no intention of changing constitution or solving the national question on the basis of some general principles; the only thing that had kept it in power so long was abiding by the principle of "from hand to mouth," as Fran Zwitter called the government's policy of concessions.[418] Although the era of Taaffe's rule did see some decline in tensions across the Austrian part of the monarchy, this improvement did not come without a price: "Politics, which had hitherto revolved around principled questions (the struggle between centralism and federalism, the conflict between German liberals and the Catholic Church), had turned into a matter of sheer 'pragmatism.' [...] [T]he path to success could solely be paved with unheroic steps that led to notable changes only after a longer period of time."[419] The Slovenes were therefore not overenthusiastic about the government, but settled for the scraps from Taaffe's table. That said, the period of Taaffe's government brought a major positive change for Carniolan Slovenes by recognizing Carniola as a Slovenian province and simultaneously signaling the government's waning support for the German party in Carniola, which still

---

418   Fran Zwitter, *Nacionalni problemi v Habsburški monarhiji*, Ljubljana: Slovenska matica, 1962, 163.
419   Grdina, *Slovenci med tradicijo in perspektivo*, 80.

managed to retain important economic positions. Its daily *Laibacher Zeitung* was reduced to a weekly publication. In 1879 the Carniolan German party participated in the elections for the last time, without success. In 1882, it lost its majority in the City Council, and a year later its majority in the Provincial Diet. The provincial presidency passed to a Slovene, Andrej Winkler, an accomplished official from Gorizia who was extremely popular with the people, without ranking among the great patriots.

Winkler rendered a remarkable service to the Slovenes by quietly asserting Slovene as an official language; at the same time, he remained loyal to the government, skillfully navigating between the two sides and never acting to the detriment of Slovenian interests. Over a few years, Winkler discreetly reorganized the Carniolan administration and reinforced it with Slovenian personnel. This was most evident in the growing and eventually predominant share of Slovenes occupying important positions, such as headmasters of grammar schools, school supervisors, or court presidents. During Winkler's provincial presidency, Slovene also began to be used in official communications. Yet, throughout this time, Winkler successfully safeguarded the image of an unbiased leader. Another testament to his adaptability was his decision to declare one half of his family as German and the other as Slovenian in a population census. The German party celebrated his retirement as a victory of its own. Carniola thus had its first and last provincial president between 1880 and 1892. Even before Taaffe's mandate had expired, government policy and pressure from German liberals forced Winkler to resign.

The situation of the Styrian and Carinthian Slovenes did not change so much. They constituted a minority in both provinces, whereas Germans held not only political power but also significant economic power, which further strengthened their position despite the government's policy. The situation of Slovene, which was already trickling into public life in Carniola, was much more frustrating in Styria and Carinthia, where most officials were Germans. The national existence of Carinthian Slovenes, who were economically and socially weak, was especially threatened, another reason why Slovenian urban communities abstained from participating in elections. In these provinces, the effect of government policy was the opposite of that in Carniola, since Styrian and Carinthian Germans were putting ever-stronger pressure on the government, which they believed was promoting Slovenian national development.

The introduction of Slovene into the bureaucracy and courts was gradual and, as mentioned above, most successful in Carniola. However, even though it was possible in theory to communicate in any language in the Parliament, Slovenian deputies would not speak Slovene at all until they took their oaths in 1867. Even then, they continued to speak in German until World War I, using only rare sentences and paragraphs in Slovene to underline their demands for linguistic equality. Non-German speeches were neither translated nor transcribed. In provincial diets, Slovene was only granted equal status with German in Carniola and with Italian in Gorizia. Official translations of the Official Gazette were also published in Slovene, but only the original text in German was recognized as authentic. In addition, all provinces, including Slovenian, issued their own official gazettes in their respective languages. Public events were most often bilingual; in Carniola, they began to be conducted exclusively in Slovene only toward the end of the mayoralty of Ivan Hribar (1896–1910). Gorizia, too, had a bilingual character, whereas cities in Lower Styria were completely Germanized.

Although the introduction of the Slovene language and Slovene education were the fundamental points of the Slovenian national program of 1848, before World War I, Slovene only fully asserted itself in elementary education. The language of instruction was determined by provincial school councils; by 1914, all Carniolan elementary schools were Slovene, except in the Kočevje area, which had a strong German community, in Bela Peč, and in Ljubljana, which also had German elementary schools. In Slovene schools, German became a mandatory subject from the third grade onwards, and vice versa. Slovene education made great strides in the Littoral, much to the displeasure of the Italian majority in Gorizia and Trieste; the first Slovene-language school in Gorizia was only founded in 1895, whereas in Trieste these efforts remained fruitless. Before World War I, Slovene schools were also set up in the Lower Styrian countryside, while those in German populated cities and market towns were predominantly German. In Carinthia, Slovene schools emerged in Jezersko, Globasnitz, St. Michael, Zell, and St. Jakob im Rosental. The situation was most deplorable in Prekmurje and Venetian Slovenia, where there were no Slovene schools. The struggle for predominance over education became more organized after 1880, when the German School Association (Deutscher Schulverein) was founded in Vienna to establish and maintain German schools in Austrian provinces with ethnically mixed populations. In Italian populated provinces, the same task was undertaken by a similar organization, Pro Patria (For the Homeland), which was succeeded by the Lega Nazionale (National League). The leading promoter of Slovenian education efforts was the St. Cyril and Methodius Society (Družba sv. Cirila in Metoda), founded in 1885 in Ljubljana. Its objective was to facilitate the establishment of private kindergartens and schools; by 1914, it had 21 kindergartens and 8 elementary schools in the Littoral, Styria, and Carniola. In Gorizia, kindergartens were also founded under the political society Sloga, whose patronage over the education sphere was continued by the Šolski dom society in 1897.

The active and vibrant political life of the Taaffe era presented a major challenge for Slovenian political unity, which was greatly attenuated because compromises with the government made it difficult to take a principled stand thereafter. Both political groups, the clericals and the liberals, maintained relatively conservative social views and suited a society that only acknowledged a limited right to vote.

# In the Shackles of Political Parties

Political life at the beginning of the 1880s was characterized by generational change. Janez Bleiweis, the "father of the nation" and "a moderate conservative, but not a clerical,"[420] died in 1881 without a successor capable of restraining the growing divergence of views and political intolerance. Indeed, the Catholic camp proclaimed Luka Svetec his successor, but their "choice" turned out not to be effective. Instead, it was the canon Karel Klun, one of the first organizers of the Catholic camp in Carniola, who rose to prominence. Klun actually embodied what was considered to be a conservative Catholic faction. In harmony with developments within the Catholic Church as a whole, which became more actively involved in the social matters under Pope Leo XIII and adopted a more rigorous attitude toward liberal social and economic principles, the Slovenian Catholic camp began to strengthen its position. A Catholic printing house was established, and the paper *Slovenec* became a daily. The new Bishop of Ljubljana, Jakob Missia (consecrated in 1884), supported Klun unconditionally. His decision was fueled by a profound fear of the advancement of the social democratic party, which he identified as the gravest threat, so that he "allowed the rapid politicization of the Church network."[421]

A series of articles entitled *Dvanajst večerov* (Twelve Nights) by Anton Mahnič, a professor at the theological school in Gorizia, marked the beginning of what came to be known as the "division of souls" across the Slovene-speaking lands. The series listed the principles and demands of Catholic reform, which Mahnič reasserted and even radicalized in 1888 in the journal *Rimski katolik* (The Roman Catholic). For Mahnič, the only legitimate standard in literature and arts was Catholic morality. He strove to encourage renewed Christianization in all walks of life, demanding radical confession and behavior based on Catholic principles. He condemned the poems of Simon Gregorčič, a poet and priest, branding them as complete nonsense and contrary to religious tenets, and initiated a general criticism of Slovenian realism in literature. Slovenian realists responded hesitatingly for the sake of political unity, and it was not until 1888

---

420  Stane Granda, "Od razcveta v streznitev," in: Cvirn, *Slovenska kronika XIX. stoletja*, 129.
421  Grdina, *Slovenci med tradicijo in perspektivo*, 124.

that Mahnič encountered more decisive opposition, from Janko Kersnik in the literary journal *Ljubljanski zvon* (The Ljubljana Bell). The Catholic camp, for its part, launched its own literary journal, *Dom in svet* (Home and World, 1888–1944), edited by Frančišek Lampet.

Mahnič's intervention in the political situation in Slovenia had enduring consequences. His was a war against every form of liberalism, including Catholic liberalism, since for him religious principles were the compulsory guidelines in public and political life. Work for the good of the people, in his view, was impossible without a consensus on religious tenets. In other words, he called for the division of souls along the lines of principles. With Mahnič, the principled and combative Catholic trend striving for a reform of modern society according to Christian values, as encouraged by Pope Leo XIII and his encyclicals, began to gain a foothold in Slovenia. The division of souls provoked political divisions, beginning with the Gorizia region in 1889, when a radical Catholic group led by Mahnič and Josip Tonkli separated from the Sloga society. Most of the members remained with Sloga, which was led by the liberal Catholic Anton Gregorčič. The struggle between the two sides ended during the parliamentary elections in 1891, when Gregorčič won a victory over Tonkli and ushered in a new period of unity. In Carniola, unity dissolved when the Sloga leadership failed to nominate its joint candidates for the 1889 provincial elections.

In 1890, the Catholic Political Association (Katoliško politično društvo) was established in Ljubljana in the wake of the second Austrian Catholic Convention held in 1889, while the first Slovenian Catholic Convention in 1892 sparked off the foundation of a series of local political Catholic associations. In October 1895, before the provincial elections, these many associations came together under the umbrella of the Catholic National Party (Katoliška narodna stranka). Its official leader until his death in 1896 was Karel Klun, but the actual driving force behind it was Dr. Ivan Šusteršič, who took over in 1902.

On the liberal side, the most persistent protagonists were Josip Vošnjak and, working in his shadow, the writer Josip Jurčič, the editor of the journal *Slovenski narod*. However, low social esteem of men of letters prevented Jurčič from becoming the leading liberal politician. After Jurčič's death in 1881, the younger generation began to establish a foothold within the liberal camp. In 1883, after the victory of the Slovenian national camp in the Diet of Carniola, the delegates split over a disagreement provoked by the demand to invalidate the appointments of the three German representatives of the large landowners' curia. This was a split between the policy of gradualism and adaptation to circumstances, advocated by the so-called faction of "elastikarji," and the policy of concentrating on practical gains in national demands, advocated by the radical faction.[422] The most prominent influence among the radical liberals was the writer Ivan Tavčar, who argued that while the Slovenes and Slavs were entitled to making compromises, a compromise with the Germans was out of the question.[423] Tavčar had a staunch supporter in Ivan Hribar, who later became the mayor of Ljubljana (1886–1910). The consolidation of the Catholic camp forced the liberals to rise to the challenge. In 1891, they established the Slovenian Association (Slovensko društvo), and in 1894 the National Party (Narodna stranka), which was renamed the National Progressive Party (Narodna napredna stranka) in 1905. Its leaders were Karel Bleiweis, Ivan Hribar, and Ivan Tavčar.

The workers' movement also began to gain political influence during the 1880s. It positioned itself against the two traditional political camps, and evolved in response to the course of events in Central and Western Europe. The first workers' associations in Ljubljana and Maribor were established during the late 1860s. These were primarily trade associations, then education associations that originated in towns and spread to industrial and mining centers. The early workers' associations were not organized around common worldviews or national goals, but over time their unifying trait became an anti-bourgeois stance. Their viewpoints began to crystallize through debates on how to improve the workers' situation. One option was cooperatives, meaning self-help through pooling small sums of money, as advocated by liberals; the other was state aid, as advocated by the so-called "lassallovci," adherents of the German social and political activist Ferdinand Lassalle. Anarchist trends originating in the Austrian workers' movement during the 1870s also gained ground, but anarchists were persecuted for supporting terrorist tactics. The leader of the Ljubljana workers' association, Franc Železnikar, who appealed to his comrades with the slogan "dynamite and paraffin," also belonged to this group.[424] He and his associates were brought to trial in Klagenfurt and sentenced to several years in prison.

A modern labor movement took shape toward the end of the 1880s, and the first larger socialist organizations (social democratic parties pursuing Marxist ideology) emerged. The Austrian Social Democratic Party was established in 1888/89. Although there were some Slovenes among its founders, the party's success in the Slovenian populated provinces with a predominantly peasant population

---

422   The name "elastikarji" (elastics) originated in the Slovenian writer Janko Kersnik's explanation of gradualist policy, where he argued for the need "to be elastic."
423   Vodopivec, *Od Pohlinove slovnice do samostojne države*, 107.
424   Ibid., 109.

was relatively modest. The Yugoslav Social Democratic Party (Jugoslovanska socialdemokratska stranka, JSDS) was established in 1896, when the Austrian Social Democratic Party began to split into individual national parties. In Styria and Carinthia, the workers' organizing efforts had German national overtones, and membership in the Styrian and Carinthian social democratic parties implied affiliation with German culture. German leaders labeled the Slovenian labor movement as backward. Indeed, the Slovenian party in Carinthia was clerical, so the social democrats associated the Slovenian labor movement with clericalism. The social democratic party was also active in Celje and Maribor, while the Littoral region had two separate social democratic parties, one Slovenian and one Italian. Both had their seats in Trieste, one of the workers' centers at the time. In 1890, Labor Day was celebrated in Slovenia for the first time.

The Taaffe government, which actually enabled the development of political parties in Slovenia, began to crumble in 1890 and eventually collapsed over a domestic policy issue that dominated the last decade of the 19th century—the battle for universal suffrage. In 1893, Taaffe surprised the coalition partners with a proposal for electoral reform that envisaged the extension of voting rights, but was met with tough opposition from the largest parliamentary clubs (the German liberal left wing, Hohenwart's conservative club, and the Polish club). The main goal behind the proposal was to harness the political, national, and social radicalism of the new parties that were unyielding in their national demands and sought support from wider society. Taaffe eventually lost the emperor's support, and leadership of the government was entrusted to Alfred Windischgrätz. The new governmental coalition seemingly brought together the Germans and the Slovenes, so six Slovenian delegates left Hohenwart's club, while seven chose to remain. The Slovenian negotiating position improved, because now the Slovenian delegates were in a position to tip the balance on which the survival of the government depended, and independent delegates could directly present their demands.

The Slovenian delegates' common goal was the introduction of parallel classes in Slovene at the grammar school in Celje. Over the next two years, this became an important political issue whose implications were felt all over the monarchy. In 1888 the Taaffe government introduced the parallel classes in Maribor and, in exchange for their support in the Parliament, the Slovenian delegates were promised parallel classes in Celje as well. The promise was realized in 1895, but the Germans living in Celje, emboldened by support from German nationalist liberals, adamantly rejected the move, seeing it as the beginning of the Slovenization of Lower Styria. The liberals withdrew from the coalition and the Windischgrätz government resigned. The new cabinet was formed by Kazimir

Badeni, who first set about settling the German-Czech conflict by passing a resolution requiring German officials in Bohemia and Moravia to speak Czech. Once again, the Germans fiercely resisted.

The Germans' pugnacious campaign, directed at the Slovenes in Styria and Carinthia among others, was run by nationalist, anti-Slav, and anti-Semitic parties which had by then superseded the waning liberals. Particularly successful were parties backed up by big money, with good economic connections and the support of nationalist organizations. In the Slovene-speaking lands, their most reliable ally was the German School Association, which supported German private schools, played an important role in the Germanization process, and collected money for the protection of German national interests. A similar organization was the Südmark (South Province), which bought farms from bankrupt Slovenian peasants and sold them to Germans. German settlement, and various other methods developed by such organizations, led to a decrease in the proportion of the Slovenian population in Styria and Carinthia, although those Slovenes who declared themselves as Germans also helped this trend.[425] In the areas where the Germans were economically dominant, they exerted pressure on the economically dependent Slovenes. The Slovenian ethnic border in Carinthia was steadily pushed southward. One apparent sign that the Slovenes were losing the battle for national recognition was a rubric in the census form that read "language of communication" rather than "mother tongue."

The German–Slovene strife culminated in physical clashes, which occurred frequently especially in Celje during cultural shows and social events. In the spring of 1903 a protest meeting in Ljubljana, organized in support of a Croatian resistance movement, turned into an anti-German demonstration. However, the most serious national unrest began in 1908 in Ptuj, after the Germans failed to prevent the general meeting of the St. Cyril and Methodius Society through legal means. Emboldened by support from their compatriots elsewhere in their struggle for German national interests, and with the tacit consent of the municipal police, they attacked the participants at the meeting on September 13, causing a general brawl followed by numerous arrests. The reports of the events in Ptuj triggered demonstrations on both sides and in many towns across the monarchy. The worst conflicts occurred between September 18 and 20 in Ljubljana, provoking intervention by the military and ending in two deaths. The two victims' funerals turned into a massive anti-German manifestation. These events made it clear that:

---

425  Vodopivec, *Od Pohlinove slovnice do samostojne države*, 112.

the greatest threat to the national future of the Slovenes is posed by the Germans and their "Drang nach Süden," while the German–Slovene conflict is just one part of the "thousand-year-old" struggle between the Slavs and the Germans, which is currently approaching the decisive stage and must sooner or later end with Slovenian victory.[426]

In the Littoral, national relations were less tense. Italian irredentism and nationalism were on the rise, but so was the national awareness of the Slovenian middle class. Italian and Slovenian workers, meanwhile, demonstrated exemplary cooperation.

In the Austrian Parliament, the Slovenian delegates from both political camps and the Croats established their own club, the Slavic Christian People's Association (Slovanska krščanska narodna zveza). The German resistance resulted in parallel classes being abolished in Celje in 1897, but preserved in Maribor as a compromise solution. Badeni's language ordinance on the compulsory knowledge of Czech in Bohemia and Moravia provoked opposition from German nationalists, causing continuous unrest, street demonstrations, and the obstruction of Parliament, eventually bringing down the Badeni government. National strife continued, rendering Parliament usually inoperable. The successive governments, most of them caretaker governments, have had difficulty forming coalitions, and frequently had to resort to Article 14 of the Constitution, allowing the emperor to issue ordinances pertaining to parliamentary competencies if Parliament was prevented from sitting. The Slovenian delegates initially gave their support to certain governmental coalitions, but eventually joined the Czech obstructionists.

Slovenian politics were under strong German pressure and were plagued by factional divisions. The Catholic program envisaged schools fully based on religious principles, and the introduction of Slovene as the language of instruction in all regions inhabited by Slovenes. Liberals and Marxists were proclaimed as their adversaries. It was planned that the program would be implemented through political and educational associations. The Catholics were especially active in rural regions, where they established savings and loan societies and cooperatives for the sale and procurement of agricultural products, rapidly expanding their rural constituency. The new party gained a foothold on the political stage thanks to its astute populist leaders Ivan Šušteršič, Janez Evangelist Krek, Vinko Gregorčič, and Ignacij Žitnik.

Political life in the Gorizia region took a slightly different course. There, the pace was set by the prominent Catholic leader Anton Gregorčič, who found political partners among the Slovenian and Italian liberals, and resisted the clericalism

---

426  Ibid., 148.

that dominated in Carniola and perpetuated the Kulturkampf. In March 1900, Andrej Gabršček and Henrik Tuma established the National Progressive Party of Gorizia (Narodna napredna stranka na Goriškem). Gregorčič and his followers became linked with the adherents of a new current within the Christian Social movement led by Josip Pavlica. Political, organizational, and social activities were taken over by the Christian socialists. Toward the end of 1907, the long-established Sloga society was renamed the Slovenian People's Party of Gorizia (Slovenska ljudska stranka za Goriško).

Unity in Styria was also dissolving, and the protagonists of intellectual unity were most concerned over national rights. The antagonism between liberals and Catholics became particularly conspicuous after 1895, when the two camps each nominated their own candidates at the parliamentary elections. The parties eventually split during the parliamentary elections in 1907. At the end of 1906, the Styrian National Party (Narodna stranka za Štajersko) was founded under the leadership of Vekoslav Kukovec, and in 1907 the Slovenian Peasants' Association of Styria (Slovenska kmečka zveza za Štajersko) was founded under the leadership of Dr. Anton Korošec.

Divisions between political parties naturally led to separate social and professional organizations. The Sokol society was countered by Orel, while Slovenska matica obtained its opposite pole in Leonova družba, established in 1896. In 1900, Catholic teachers formed their own organization, Slomškova zveza.

Unity was only preserved where the greatest threat to the national cause existed. The leading movement in Carinthia, the Catholic Political and Economic Association of the Slovenes in Carinthia, was Catholic-oriented but harmonious. In Trieste, national issues were in the hands of the liberal society Edinost. The Slovenes in Istria still maintained close connections with the Croats in their struggle against the Italians: the Political Association of the Croats and the Slovenes in Istria (Politično društvo za Hrvate in Slovence v Istri) was established in 1902. New names appeared on the Slovenian political stage, and some would remain there for decades, shaping the political and national future.

The Catholic party's program called for engagement in political, economic, social, and educational work. "[In] all fields and among all classes a lively work has begun, such as cuts deep furrows into the public and spiritual life of the nation."[427]

However, not only furrows, but also scars were inflicted by a policy that caused the division of souls, although it was aware of the social and economic

---

427  Mal, *Zgodovina slovenskega naroda*, 1089.

problems and determined to resolve them. The lively work, as Josip Mal described it, was taken up by the "young squad of Mahnič's supporters,"[428] the Christian Socials whose agenda, after the second Catholic Convention in Vienna in 1889, was influenced by the Vienna circle (Karel Vogelsang) and Pope Leo XIII's *Rerum novarum* (1891), the encyclical on the condition of the working class. They advocated a radical reform of social and economic legislation for the benefit of peasants and workers, and established peasant cooperatives and trade and workers' associations providing material assistance. Under the influence of the younger generation, the Catholic party indeed lost some of its conservative character, but its conservative leaders continued to insist on a hierarchical class society, while refusing to model themselves after the Austrian Christian Social movement. After Klun's death in 1896, the influence of Mahnič's adherents increased. They were convinced that the Christian Social movement was the only correct alternative to the ominous, disliked, and most of all godless socialism. Dr. Janez Evangelist Krek, firmly convinced that just social legislation implied granting rights rather than giving alms, took control of the Christian Social movement, and "through his democratic views attracted all classes of society to him and his party, and had a particularly strong educational impact on the workers and students."[429]

Krek laid out the Christian Social movement's program in a book, *Črne bukve kmečkega stanu* (The Black Books of the Peasant Class), published in 1895. According to this program, society was class-based and rested on an agreement between workers and the owners of capital. Over time, he discarded the concept of a class society and sought solutions to social disparities within a modern parliamentary society, whereby cooperatives would be the main drive in transforming the capitalist order. His most loyal collaborator was Dr. Ignacij Žitnik, whose activity was concentrated in rural regions, where he established Catholic political and educational associations, while Dr. Ivan Šušteršič assisted him "with his tactical skillfulness and his talent for organization."[430] In the decades before World War I, Krek earned a reputation equal to that previously enjoyed by Bleiweis.[431] His persistent effort led to the establishment of numerous workers' associations and peasant cooperatives that eventually united to form the Slovenian Christian Social Association (Slovenska krščanskosocialna

---

428  Ibid., 1090.
429  Ibid., 1090.
430  Ibid., 1090.
431  Vodopivec, *Od Pohlinove slovnice do samostojne države*, 115.

zveza). Its main tasks included the education and the improvement of material conditions for workers. The comprehensive network of loan societies and cooperatives helped to stem the selling-off of Slovenian farms, the destruction of the peasantry, and emigration.

In 1909, Krek's initiative resulted in the establishment of a Catholic trade union organization, the Yugoslav Professional Association (Jugoslovanska strokovna zveza). Although he considered the possibility of founding a Christian social party, he remained with the Catholic party (known since 1905 as the Slovenian People's Party). Krek's activities covered various areas of public life: he was a member of the party leadership, a delegate to the Parliament and the Diet of Carniola, the editor of the daily *Slovenec*, and he also watched over Slovenian Catholic students. In 1909, the party reached beyond provincial borders and united all local parties with a Catholic program into the Pan-Slovenian People's Party (Vseslovenska ljudska stranka). However, its penetration into other Slovenian populated provinces did not proceed without obstacles. The Catholic politicians from Gorizia and Carinthia frequently opposed the Slovenian People's Party leadership, both on a personal and ideological level.

The pace of political life among the Slovenes had clearly been dictated by the Catholic camp since the turn of the century.[432] Its achievements could partly be attributed to the apathy of the liberal National Progressive Party, which displayed no particular gift for resolving economic and social issues, especially in rural areas, and did not "attach importance to the organization of the masses."[433] The liberals chose to rest on their laurels and, in the midst of conflicts and political clashes, "at the dawn of the new century [they] reminisced melancholically about the golden era of their political camp."[434] The Slovenian liberals relied on the urban middle class and the newly formed rural middle class. Many secular intellectuals, teachers, and students were also liberally oriented. In rural regions, liberals initially followed Krek's cooperative model, albeit unconvincingly. Although they were aware of their political opponents' superiority and appeal to voters, they began to reject Krek's system of cooperatives, which diverted many a peasant voter from their party. Their weakness was also demonstrated through their decision to enter into a coalition with large German landowners in Carniola, forcing them to avoid addressing national issues for the sake of good cooperation. Although this lack of principle earned them public condemnation,

---

432    Grdina, *Slovenci med tradicijo in perspektivo*, 165.
433    Mal, *Zgodovina slovenskega naroda*, 1090.
434    Grdina, *Slovenci med tradicijo in perspektivo*, 151.

the coalition did bring some benefits to Slovenian culture: the liberals were able to secure funds for the establishment of a Slovenian theater, a higher school for girls in Ljubljana, and a civil school in Postojna. The most prominent party members and its leaders were well-off people from both urban centers and rural areas. Students, once a mainstay of the party, became disappointed with its politics and lukewarm program, and deserted. Liberally inclined Slovenian students in Graz, Prague, and Vienna gathered around the academic newspaper *Omladina* (Youth) and began a search for ideological concepts that could be used to harness political Catholicism.

During the 1890s, Slovenian students in Prague, most notably Dragotin Lončar and Anton Dermota, embraced the philosophy of the Czech professor Tomáš Garrigue Masaryk, who criticized political Catholicism, liberalism, and Marxism, and demanded democracy and the socialization of culture and politics. In their *Poslanica slovenski mladimi* (Communication to the Slovenian Youth), issued in 1901, and in the newspaper *Naši zapiski* (Our Notes), Masaryk's followers (also known as realists) concentrated on promoting culture and education. In 1906, most of them joined the Yugoslav Social Democratic Party, to which they were attracted by its program for just and gradual political and social reform. Their ideas also inspired Albin Prepeluh, who advocated a special Slovenian concept of socialism; for him, the origin of the social and national crises was the pressing peasant issue. In 1902, Masaryk's adherents caused a split among the liberal Slovenian students in Vienna, resulting in the emergence of a radical-populist group led by Gregor Žerjal, which acted independently of the National Progressive Party. It demanded radical reform in all spheres, but envisioned its activity beginning with the intelligentsia and spreading to the common people only later. The group expounded its program in 1905 at the first gathering of the national-radical students in Trieste, but after 1909 their dependence on the National Progressive Party increased until the group eventually merged with it. In 1906, Vladimir Ravnihar and some others split from the group and established the short-lived Slovenian Economic Party (it disintegrated in 1908), calling for the economic independence of the Slovenes and collaboration with wider society.

The liberal party was the party of the upper social classes. It resisted the introduction of universal suffrage and sided with tradesmen in their opposition to consumer associations and loan societies. This cost them the support of the lower classes, which was transferred to the Catholic party. They adamantly opposed the Church and the Catholic political movement, with anti-clericalism at the core of their political activity, and accused the Church of abusing religion to achieve political goals. Their persistent denunciation of eminent figures from the Catholic camp, dubbed the "anti-priest campaign," should be attributed to the ambiguity of their ideological and political setup. The campaign met with some disapproval (e.g., from Ivan Hribar), but passions were already running high, and it was too late to calm them.

The Austrian political arena had meanwhile been embroiled in the struggle for electoral reform since the second half of the 1890s. In January 1896, the government reformed the electoral system by introducing a general *curia* that gave the vote to all men 24 years of age and over. One advantage of the extended franchise was that it accorded an active political role to ordinary people. Its negative side was that it did not automatically imply greater individual freedom, but rather gave more power to individual parties, in line with the trends elsewhere in Europe. The reform opened the door to Parliament for social democrats, but it was the German nationalist parties, backed by big capital, that proved most skillful in taking advantage of the new electoral system. The general curia, i.e., a qualified universal male suffrage, was also introduced in provincial diets. The next electoral landmark was the 1907 law on universal male suffrage passed by the government of Baron von Beck, which abolished voting in curiae. Large German landowners, Slovenian progressive delegates, and Czech Catholics, supported by social democrats, balked at the law. The first elections to the Parliament under the new system were held that same year. The Slovenes had 24 delegates in the Reichsrat, most of them from the Catholic camp; the number of delegates was roughly proportional to the Slovenian population in the Austrian part of the monarchy. The new electoral geometry was meant to placate national dissatisfaction, and it even raised hopes among the Slavs that Austrian politics would radically shift when confronted with a Slavic majority in the Parliament. A group of neo-Slavicists was formed within the Parliament under the Czech delegate Karel Kramař and the Slovenian delegate Dr. Ivan Hribar. This new Slavic movement attempted to bring together all the Slavic nations in the monarchy and beyond; they advocated close economic, political, and cultural links among the Slavs and desired the federalization of the monarchy. Regarding foreign affairs, their view was that Austria should lean on Russia.

The electoral reform of 1907, in short, opened the door to Parliament for the Social Democratic Party, but its Slovenian branch was still too weak to hope for success. Early in the 20th century, social democrats in the Slovenian ethnic provinces maintained a national platform, although they aimed to unite all the South Slavic social democrats in the monarchy; in fact, already in 1896 they founded the Yugoslav Social Democratic Party to unite Slovene workers with Serbs and Croats in Istria and Dalmatia. Unfortunately, the workers' movements in other lands were too weak, rendering this goal unachievable. They then concentrated on larger industrial and mining centers (Trieste, Gorizia, Idrija, Ljubljana, Jesenice, Zasavje, Celje, Maribor, and Klagenfurt) and on publishing the newspapers *Delavec* (The Worker), *Rdeči prapor* (The Red Standard), and *Naši zapiski*. In their work, they relied on professional organizations, strove to explain their principles and program at political gatherings, and actively supported strikes. The Slovenian workers in ethnically mixed provinces joined the Austrian and Italian social democratic parties; the Italian social democrats in particular established close cooperation with their Slovenian counterparts. In terms of goals, they modeled themselves on the Austrian social democrats. Their long-term goal was the elimination of the capitalist system, but also the democratization of political life in the monarchy, universal, equal, direct, and secret suffrage, a strict separation of the Church from the state, and free and non-religious education. Until the disintegration of the monarchy, the most prominent leaders of the party were Etbin and Anton Kristan, Albin Prepeluh, Melhior Čobal, and Josip Kopač.

At the turn of the century, the "United Slovenia" program occasionally figured in considerations of potential solutions to the Slovenian national issue, but it no longer represented the driving force. The Slovenian political parties exhausted themselves in the national struggle. Consequently, both the liberal and the Catholic parties only paid lip service to national autonomy and the equality of Slovene, which could only be achieved through the system of Slovenian educational and cultural institutions. The monarchy, for its part, contemplated possible ways to resolve the pressing, and eventually fateful, national issue. The options considered were the reorganization of the judicial system and voting in national curiae, which were meant to guarantee national protection. However, no change or progress was made in this respect regarding the Slovenes, although during the early 20th century a reform of the provincial diet and its structure was being negotiated in some parts of the country and carried out in Moravia and Galicia. The leading ideologist of the Slovenian social democratic movement, Etbin Kristan, proposed a unique approach to the national issue. In 1898, the Czech newspaper *Akademie* featured an article by Kristan proposing so-called "personal national autonomy." "Full equality is achieved when the members of one nation enjoy, regardless of their place of residence, equal rights regarding their inter-personal relations and relations with other nations. One precondition to achieve this is the unity of the nation, meaning cultural unity, given that territorial unity is inevitably excluded."[435] Clearly, in Kristan's view, the national

---

435 Quoted from Vasilij Melik, "Problemi slovenske družbe 1897–1914," in: Viktor Vrbnjak (ed.), *Slovenci 1848–1918. Razprave in članki*, Maribor: Litera, 2002, 604.

issue was exclusively cultural, but he did not give any suggestion as to how to achieve autonomy. Kristan presented this idea at the 1897 congress of the Austrian Social Democratic Party in Brno, but it was rejected. Even in Slovenia it never met with a positive response. The party's official standpoint was that Austria-Hungary should be transformed into a federation of nations.

After the split within the Slovenian national movement, the main concern of the Slovenian political parties in the Parliament became political alliances irrespective of national affiliations. However, the heated response of the Germans to Badeni's language ordinances and the collapse of his government in 1897 clearly showed that the strategy of seeking agreement with the Germans did not promise any benefit for the Slovenes. The Slovenian delegates continued to support the government out of caution, but at a meeting of all Slovenian delegates in July 1897 in Ljubljana, they resolved no longer to yield to the German parties but to respond with a pan-Slovenian convention. It was organized on September 14, 1897 in Ljubljana, with the Croat, Ruthenian, and Czech representatives as guest participants. On this occasion, Catholics and liberals declared their support for the integration of all Slovenian regions into one administrative unit with Ljubljana as its center, and for an extended franchise, social reform, and a united movement. Unfortunately, their reconciliation was short-lived. In 1899, the so-called Pentecost Program of the Austrian bourgeois (*bürgerlich*) parties made it clear that German political parties were ill-disposed toward the Czechs and Slovenes. From then on, the Slovenian delegates mixed only with other Slavic parliamentary groups. However, the Slovenian delegates were no longer united in a single club after 1901, because the differences between the clericals and liberals spread into the Parliament. Šusteršič succeeded in uniting the Slovenian, Serbian, Croatian, Czech, and Ruthenian delegates within the Slavic Society (Slovanska Jednota), which became the largest parliamentary group. It "established the reputation of the Slovenes and brought Šusteršič to prominence as an excellent politician, but did not bring any substantial achievement," according to Melik.[436]

Krek, too, was actively involved in a search for fruitful connections that would lead to the resolution of the Slovenian national issue. In 1890 he advocated a union with the South Slavs, and at the first Catholic Convention in 1892, he supported the proposal of Slovenian–Croatian mutual assistance. In the summer of 1898, he began to call for political integration with the Croatian parties, particularly the Croatian Party of Rights. In his view, the solution to the Slovenian

---

436   Ibid., 603.

problem lay in integration with the Croats based on Croatian state right. Such integration would exclude the Slovenes in Prekmurje, Venetian Slovenia, and the Littoral, so it was considered an inappropriate approach to the national issue. The convention and the so-called tripartite program were criticized by the liberals, and other Catholic politicians were no less skeptical. However, after the proclamation of the German mutual warranty, which prioritized German throughout Cisleithania except in Bohemia and Moravia, Slovenian–Croatian mutuality began to be emphasized. In 1899, Krek reaffirmed his allegiance to the Serbs and Croats when he refused to accept a seat on the board of the Austrian Christian Social Workers' Association.

The opinion of Slovenian politicians about an appropriate solution to the Slovenian national question crystallized, in short, through their allegiance to the Czechs, Slavs, and South Slavic nations, and the Yugoslav idea appealed to the largest number of them. "The Yugoslav idea was not a single idea, but many ideas greatly differing from one another, so that not one party was of one mind as to the meaning of this notion."[437] Ideas about a tripartite concept within Austria-Hungary were similarly divergent. One possibility was the monarchy's transformation from a dual into a tripartite state, with the south Slavic part as its third constituent, or into a federation of equal nations. The tripartite concept, in the narrow sense of the word, envisaged Croatia, Slavonia, Dalmatia, and Bosnia and Herzegovina as a third constituent, but for the Slovenes the only acceptable arrangement would also include the Slovene-speaking provinces. Until the Balkan Wars, the various tripartite concepts, which were all tied to the framework of the Austrian state, had also counted the Bulgarians among the Yugoslav group. Slovenian political parties had differing views on this issue. Catholic circles, for example, hoped for the victory of Catholicism if the tripartite concept was realized. Undoubtedly, Slovene interest in developments in Croatia was increasing alongside growing resistance there to the pressure exerted by the Hungarians. The Slovenes were in favor of the change of dynasty in Serbia, where the Karadjordjević dynasty took the throne in 1903, and they also devoted close attention to the Ilinden uprising in Macedonia. However, in 1905 the representatives of all Croatian parties adopted the Rijeka and Zadar resolutions, also signed by the Croatian Serbs, in which they urged the inclusion of Dalmatia in the Croatian kingdom. With a view to realizing this goal, they even sought allies among the Hungarian opposition and Italian irredenta. With the latter, they planned to reach an agreement on the border in the coastal region, under

---

437   Ibid., 603.

which Trieste, western Istria, and Gorizia would go to Italy. The Slovenian parties therefore found themselves without a reliable ally in their effort to find a solution to their problem, and that remained unchanged for another decade.

On October 6, 1908, Austria-Hungary proclaimed the annexation of Bosnia and Herzegovina. The move was not entirely unexpected, but did change Austrian policy in the Balkans. Bosnia and Herzegovina became a separate imperial province under the administration of both parts of the empire and a joint finance minister. The annexation inevitably gave an impetus to the tripartite ambitions of the Catholic and liberal camps, whereby the major initiator and the propagator of the tripartite concept was the Slovenian People's Party. The annexation of Bosnia and Herzegovina was greeted with approval as the first step toward the union of all South Slavic nations within one administrative unit under the Habsburg crown. Undoubtedly, at that time no one, apart from some rare individuals, imagined or desired the disintegration of Austria-Hungary, because it ensured protection against the territorial aspirations of the German Empire and the Kingdom of Italy. Most Slovenes firmly believed in and were loyal to the crown and the Habsburg dynasty until the pivotal period of World War I in 1916–1917. The appointment of Franz Ferdinand as the successor to the throne raised new hopes that the third, Yugoslav entity might become a reality, given his antipathy toward the Hungarians.

In the summer of 1909, Ivan Šušteršič sent a memorandum to the heir presumptive proposing a third unit consisting of the south Slavic countries: Croatia including Slavonia, the Serbian provinces located within the Hungarian part of the empire, Bosnia and Herzegovina, and the Croatian and Slovenian populated lands in the Austrian part, namely Dalmatia, the Littoral including Istria, Carniola, and the Slovenian parts of Styria and Carinthia. The social democratic views, on the other hand, were quite different, and were proclaimed in the Tivoli Resolution at their conference in November 1909. They condemned the annexation of Bosnia and Herzegovina as an imperialistic act and demanded "a new political concept of national autonomy, i.e., the transformation of Austria-Hungary in such a way that all nations living within the integrated economic territory regardless of historical borders would have their unity, independence and self-governance in all national and cultural matters guaranteed."[438] They also advocated closer ties with the South Slavs with whom they argued the Slovenes should form a united nation.

---

438  Vasilij Melik, "Načrti za reformo Avstro-Ogrske in Slovenci," in: Vrbnjak (ed.), *Slovenci 1848–1918*, 645.

Looking back at several decades of national development should have filled Slovenes with a sense of pride and satisfaction in the first decade of the 20th century; the reality was exactly the opposite, and pessimism was the order of the day. The most important aspirations, embodied in "United Slovenia," were still unfulfilled, the proportion of the Slovenian population was in decline because of a lower birth rate and German pressure, and the extension of the franchise had not brought the desired and expected results everywhere.[439]

The feeling was ever more intense that the Slovenes could not achieve their goals on their own and that their salvation lay in connecting with a bigger and more powerful ally. Advocates of South Slavic unity were convinced that the natural environment of the Slovene language was too small to enable a successful and multifarious development within literature and science: "In so thinking, some had in mind a truly close relationship that would enable the development of a common scientific terminology, the elimination of Germanisms and Turcisms, and an egalitarian amalgamation of languages."[440] Others, in contrast, believed that Slovene should have been dropped in scientific and intellectual communication and that Croatian or Serbo-Croatian should be used instead.

The adherents of revived Illyrism, or neo-Illyrians as they were known, referred to the chairman of Slovenska matica, the Slavicist Fran Ilešič, who argued that linguistic unity was the most obvious sign of national unity, much like the language itself was the most obvious sign of a nation. Most Slovenes rejected neo-Illyrism and advocated Slovenian national individuality. The liberal Mihajlo Rostohar was one of those who stood firm against neo-Illyrism, and another prominent figure who opposed it was Ivan Cankar. In his April 1913 lecture "Slovenci in Jugoslovani" (The Slovenes and the Yugoslavs), he drew a clear line between political and cultural Yugoslavism. He advocated the political unification of South Slavic nations into a federal Yugoslav republic, but fully rejected the cultural and linguistic elements in the Yugoslav concept, describing the Yugoslav nations as brothers by blood and cousins by language, much more separate in terms of culture than a peasant from Upper Carniola was separate from a peasant in Tyrol.

The ideas cherished by the revivalist movement Preporod (Rebirth) stood in stark contrast to all those concepts that envisaged the South Slavic nations as part of Austria-Hungary. They promulgated explicit anti-Austrian ideas and hoped for

---

[439] Vasilij Melik, "Slovenci v času Cankarjevega predavanja o jugoslovanstvu," in: Vrbnjak (ed.), *Slovenci 1848–1918*, 688–689.
[440] Ibid., 692–693.

the break-up of the monarchy. A clandestine organization of Rebirthers, which published the eponymous newspaper *Preporod* (1912/1913) and represented the students championing Yugoslavia, argued that the solution to the Yugoslav and Slovenian issues was the revolutionary path. They were convinced that the federalization of Austria-Hungary was not possible, because it could not be carried out peacefully, only by resorting to arms, which were in the hands of the Germans and Hungarians. They stressed that a new Yugoslav state could only be built on the ruins of Austria-Hungary. Preporod had few followers, and after the outbreak of World War I, the authorities staged what came to be known as a high treason trial against them. Most students were sentenced to short prison terms, while Ivan Endlicher and Janež Novak received longer sentences. Novak was thrown in jail for five years, while Ivan Endlicher died in a remand center on September 4, 1915.

The tripartite concepts were put on hold by the Balkan Wars (1912/1913), but the victories of the Serbs inspired enthusiasm and rekindled Yugoslav sentiments. All Slovenian political parties supported the political aspect of Yugoslavism, but their views on the potential allies, the Serbs and Croats, differed radically. In 1912, the Pan-Slovenian People's Party even merged with the Croatian Party of Rights and accepted its program unconditionally; their joint resolution stressed that the Slovenes and the Croats represented a national whole. Their expectations were unfounded.[441] One rare good result was closer cooperation within the Croatian-Slovenian club in the Parliament. The Slovenian liberals established contacts with the Croats, but they also pondered a partnership with the Serbs and Bulgarians and were enthusiastic about the newly crowned Karadjordjević family, and even about the Serbian Orthodox Church, which they saw as less aggressive and closer to believers. Yet they were not successful in establishing links and could only conclude with disappointment that the Croatian-Serbian coalition gave priority to the Croatian version of the tripartite concept that excluded the Slovenes. It is therefore possible to say that in the first decade of the 20th century, the ones most successful in making alliances were the social democrats.

The Yugoslav-oriented Slovenian leaders looking for a solution to the national issue were primarily guided by a wish to secure protection against the Germans and Italians. Under the influence of the Pan-Slovenian People's Party, the peasant population warmed to the union of the South Slavs living in Austria-Hungary within a Slovenian-Croatian state.

---

441   Vodopivec, *Od Pohlinove slovnice do samostojne države*, 149.

A battle for the democratization of the electoral system was also fought on the local level. In 1902 the Catholic National Party in the Diet of Carniola began a campaign for universal and equal suffrage, which would have resulted in fewer diet seats for the liberals. The parties only reached an agreement on electoral reform in August 1908. In Carinthia, the general curia was introduced in 1904 and then reformed in 1909. The electoral reform of the Carinthian Diet was introduced in 1902. Only after a second reform of the electoral system in 1909, which introduced electoral districts based on the nationality principle, did the position of Slovenes in the Diet of Carinthia and the Vienna Parliament improve somewhat. The electoral reform in the Gorizia region did not change the proportion of the Slovenes with respect to the Italians, who still outnumbered them. The situation in Istria and Trieste was similar. A common feature of electoral reform in provincial diets was the introduction of a general curia. However, since universal suffrage had not yet been introduced, the Slovenes in Istria, Trieste, Carinthia, and Styria could not attain proportionate representation in provincial diets.

Local electoral reform therefore failed to resolve national disputes in provinces, but it decisively changed the power relations among political forces, particularly in Carniola, where the (pan-)Slovenian People's Party had an absolute majority and governed on its own without difficulties, with the provincial governor also coming from its ranks. From 1908 to 1912, the governor was Fran Šuklje, who later changed sides and joined the Catholic camp. He was succeeded by Ivan Šusteršič. Thanks to its power, the Slovenian People's Party (Slovenska ljudska stranka) subordinated its local politics to party interests, and the earliest and most conspicuous result was the persecution of the liberals. On the other hand, it pursued an active economic policy, introducing measures to enhance agriculture, traffic, and electrification. It also established a local bank. Šusteršič, its leader and the provincial governor, strengthened his position. Dr. Anton Korošec took over the leadership of the Croatian-Slovenian parliamentary club in Vienna with support from Krek, becoming the third most influential party member and later an influential person on the Slovenian political stage.

On the eve of World War I, the Slovenian political arena was dominated by the Catholic movement. In the Littoral, the political balance of power tipped alternately between the liberals and Catholics. In Prekmurje and Porabje, political activity was focused on national cultural issues, unrelated to political life elsewhere in the Slovene-speaking provinces. There, Slovenian priests and secular intelligentsia concentrated their efforts on the maintenance and evolution of the Slovene language, historical heritage, folk customs, and tradition. Those who were more politically active relied on the Hungarian Catholic party.

Slovenian expatriates in the United States and Western Europe also began to organize themselves, as did Slovenian emigrants living in the mining regions of the German Empire. The largest organization to integrate all Slovenes living in the United States was the Slovenian National Benefit Society (Slovenska narodna podporna jednota), established in 1904.

## Demography

Until the break-up of the monarchy, the territory inhabited by the Slovenes was divided into four historical provinces (*dežele*): Carniola, Styria, Carinthia, and the Littoral (comprising the Gorizia region, the Gradisca d'Isonzo region, and Istria). During the second half of the 19th century, the northern ethnic border was pushed south. From the perspective of the national issue, larger towns were peculiar in that a significant part of their population was of German and Italian origin. The earliest official statistical data on national affiliation comes from the 1880 census, whereby nationality was determined on the basis of a disputable category, the language of communication. According to this unreliable system, 23 percent of the residents of Ljubljana were Germans in 1880, but only 15 percent in 1910. Large proportions of German-speaking citizens could be found not only in the three largest Styrian towns (Maribor, Ptuj, Celje), but also elsewhere. The towns along the edges of the Slovenian ethnic border with large Slovenian populations (Trieste, Gorizia, and Klagenfurt) presented a special case. For example, 35 percent (57,000) of the Trieste population were Slovenes according to their language of communication, so it would be justified to say that, in terms of national representation, Trieste was the largest Slovenian city. In Gorizia, this figure was 38 percent (11,000), and in Klagenfurt just 2 percent (600). Analysts of these statistics attempted to make corrections by combining the language of communication with birthplace (homeland rights). These results suggest that the majority of the population originated from Slovene-speaking crown lands. According to some calculations and taking into account the said corrections, in 1846, 88.9 percent of the population living within the Slovenian populated territories were Slovenes, with Germans and Italians as the next largest ethnic groups. In 1910, estimates based on the language of communication put the percentage of Slovenes at only 76.9 percent; for 12.1 percent of the population, the language of communication was German, and for 11 percent Italian.

Tracing demographic development only became possible after 1857, when the population census in Austria-Hungary became a legal requirement. The first census was taken in 1857 and it concentrated on the permanent home residence of "local inhabitants." Subsequent censuses (in 1869, 1880, 1890,

1900, and 1910) were modernized and sought to establish the "currently present population." In 1869, there were 1,447,241 people living within the Slovenian populated crown lands in the Austrian part of the monarchy, and 64,795 living in the Hungarian part (Prekmurje), yielding 1,512,036 in total. The population of the current territory of Slovenia was 1,193,563—20 percent smaller than the above figure. In 1910, there were 1,795,376 people living in the Slovene-speaking provinces in the Austrian part of the monarchy, and 90,670 living in the Hungarian part, a total of 1,886,046 people. In 1869, there were 46 settlements with more than 1,000 inhabitants in the Slovenian populated territories, and 19 settlements with more than 2,000 inhabitants; in 1910, the number of the former had risen to 67, and the latter to 30.[442]

During the 19th century, the territories with Slovenian population experienced urbanization similar to other lands. However, the relatively weak industrialization and resulting underdevelopment of the region meant that only a few larger centers developed there before World War I. The only large city was Trieste, whose population doubled between 1880 and 1910 (rising from 74,554 to 161,000), pushing the town's borders ever further outward. Among the towns with a population above 10,000 in the mid-19th century were Klagenfurt, Gorizia, and Ljubljana (26,284 inhabitants in 1880, and 41,727 in 1910), later joined by Maribor, Celje, and Ptuj.

Two important factors affected demographic development in the Slovenian populated territories during the second half of the 19th century. Much like elsewhere in Europe, the birth rate rose while the death rate stagnated or even fell. The Littoral had the highest birth rate, and Carinthia the lowest. During the decade before World War I, the birth rate in Carniola rose to the same level as the Littoral, while the Lower Styria lagged behind in this respect. Moreover, there were large differences between urban and rural areas. The other factor that decisively shaped the demographic picture of the Slovene-speaking provinces was massive emigration before World War I. According to the above estimates, and referring to the present territory of Slovenia, between 170,000 and 300,000 people emigrated between 1860 and 1914. The widely accepted belief today is that emigration, which in Carniola reached its peak during the first decade of the 20th century, absorbed half the population growth in the decade preceding World War I. Emigration also affected the age structure of the population, which

---

442 Jasna Fischer, "Slovensko narodno ozemlje in razvoj prebivalstva," in: Zdenko Čepič et al. (eds.), *Slovenska novejša zgodovina*, Ljubljana: Mladinska knjiga—Inštitut za novejšo zgodovino, 2005, 17–21.

until 1869 had been rather balanced, with only small differences between women and men as a consequence of the 19th-century wars. A special form of Slovenian emigration during the 19th century and the first half of the 20th century was the practice of Slovenian women from Littoral moving to Egypt, where they worked as housemaids. Popularly called "aleksandrinke," these women and girls were mainly motivated by the opportunity to earn extra income in an attempt to rescue their heavily indebted farms at home. There were between 5,000 and 7,000 Slovenian women and girls working in Alexandria (until 1960), with special organizations taking care of their well-being.[443] Consequently, population growth was slower than in comparable European environments.

## Economy

During the last decades of the 19th century and at the turn of the century, the economy of the Slovenian populated territories mainly depended on agriculture; in 1910, 54.5 percent of the population relied on some form of agriculture for their livelihood, and agricultural production accounted for more than two thirds of the total value of production.[444] The most widespread form of agriculture was subsistence farming; only corn was not produced in sufficient quantities. Stockbreeding and farming generated the highest income. Although emigration most severely affected agricultural regions, the shock was offset by a boost from the flourishing cooperatives, loan societies, and general education, even if the outcome was modest compared to other, more developed parts of the empire. Modernization also penetrated the rural regions, bringing with it new products and urban customs. The rural population came in contact with the latter through their work in factories or other employment in towns which they accepted in order to secure additional income. Peasant women were especially skillful in taking advantage of the new circumstances; the money they earned as washerwomen, cooks, and maids contributed to the family budget.

Crafts, too, underwent gradual modernization. Artisans' cooperatives made modernization and the transition to the industrial model of production easier and faster. Large crafts transformed into industries and stifled traditional crafts and trades, such as weaving, linen-making, cloth-making, silk-making, leatherwork, shoemaking, locksmithing, nail-making, carpentry, and logging. On the other hand, some handicrafts (such as dressmaking and tailoring) developed even

---

443   Kalc et al., *Doba velikih migracij na Slovenskem*, 58–59.
444   France Kresal, "Struktura slovenskega od 1851–1914," *Časopis za zgodovino in narodopisje*, 2 (2002): 108.

faster. As early as the mid-19th century, the development of commerce and crafts was safeguarded by trade associations (trade chambers), which devoted special attention to educational activities aimed at merchants and artisans, the establishment of new commercial and craft companies, and the promotion of new products. Industrial companies were small and lagged behind those in Bohemia and other developed parts of Austria-Hungary. The most important industries, both in terms of their output and the number of workers, were the timber industry, mining, the iron industry, and the metal industry.

The leaders in the iron industry were Kranjska industrijska družba (The Industrial Company of Carniola), the main supplier of pig iron and the company that controlled most of the metal industry in Carniola, and the Alpine Montangesellschaft in Styria. The textile companies were integrated within the "foreign-owned" Mautner's textile concern with its seat in Vienna. Most of the Slovenian paper industry was also part of a joint stock company (under the control of the Graz-based Leykam-Josefthal since 1870). Several modern industrial companies emerged, for example, Westen, the Celje-based manufacturer of enameled kitchenware (later renamed Emo Celje). The Slovenian populated provinces were self-sufficient in coal production, and coal mining doubled during the last 25 years of the monarchy. Lead, zinc, and mercury productions were also rising.

The largest industrial companies were in the hands of foreign capital, with their headquarters outside the Slovenian populated crown lands. Before the outbreak of World War I, the ratio of domestic to foreign capital was 1:7 (some sources put it at 1:8 or even 1:10).[445] Nevertheless, the financial sector was poorly developed (before the collapse of the Vienna stock exchange in 1873, there were 150 banks in the monarchy, but none of them in the Slovene-speaking provinces).[446] The banks only began to develop after 1900, and Ljubljanska banka was founded with the help of Czech capital, liberal cooperatives, and individual investments. The liberals also played a role in the establishment of Jadranska banka in Trieste in 1905. The Slovenian People's Party contributed to the establishment of Kranjska deželna banka in 1910 (until then, Catholic financial institutions had relied on capital provided by Krek's cooperatives, such as the Ljudska posojilnica loan bank and the Vzajemna insurance company). The third Slovenian joint stock bank was Ilirska banka, established in 1916. The most widespread financial institutions were the savings cooperatives and loan societies present in almost

---

445  Vodopivec, *Od Pohlinove slovnice do samostojne države*, 144.
446  Kresal, "Struktura slovenskega gospodarstva," 120.

every Slovenian village. These provided loans on favorable terms, helping peasants to weather crises, accelerating the modernization of rural regions, and encouraging the development of industry, trade, and commerce through larger investments.

Industrial development led to an increase in the number of workers, which in turn gave rise to more resolute labor movements and strikes calling for higher wages, regulation, and the improvement of working conditions. Trieste had the largest number of industrial workers, around 41,000; within the territory of present-day Slovenia, the total number of industrial workers was approximately 86,000.[447]

Factories were the first to take steps toward electrification, while street lighting and municipal facilities (such as water pumps and trams) were electrified somewhat later. However, in mining and certain other industries, electrically powered machines were only introduced after World War I, and electricity reached rural regions at about the same time. In some cases, electrification fully kept pace with European trends: for instance, the steam mill in Maribor acquired its first light bulb in 1883, four years after its invention. Planned electrification began in 1909; ambitious plans were drawn up in 1912 for the construction of eleven power plants on the Sava River, but the work was delayed by war. The first power plant, Završnica, was nevertheless built in 1915, and the transmission network was completed the following year. The large power plant Fala on the Drava, which supplied electricity to the town of Maribor, was constructed between 1912 and 1918.[448]

Much as elsewhere, life in Slovenian towns was affected by the urbanization, which was the prime mover behind the modern history of civilization. It shaped social norms and mentality, and consequently the life of 20th-century urban and industrial societies. Slovenian towns were plagued by a housing crisis in the wake of migration to urban centers, and by worsening supply and traffic problems. The culture of housing underwent rapid change. Wooden houses in rural regions were replaced with brick and stone houses. Town dwellers began to devote attention to both privacy and appearance: drawing rooms became luxurious, bedrooms turned into intimate spaces, while bathrooms were still a rarity (before 1914, only 6.1 percent of Ljubljana households had bathrooms). Flush toilets only entered the Slovenian urban households at the turn of the 20th century. The lower classes lived in poorly furnished apartments with too few

---

447  Vodopivec, *Od Pohlinove slovnice do samostojne države*, 146.
448  Kresal, "Struktura slovenskega gospodarstva," 112.

beds and too many residents squeezed into damp and moldy rooms. The streets of 19th-century Slovenian towns were muddy and overwhelmed by the stench of cesspits. Hygiene gradually improved over time, mainly in the wake of cholera epidemics (in 1836, 1849, 1850, 1855, 1866, and 1886).[449] The construction of mains water supplies and sewage systems made the largest contribution to the improvement of hygiene (Kamnik obtained them in 1888, Ljubljana in 1890, Maribor and Škofja Loka in 1902, Celje in 1908, and Kranj in 1911).

Urbanization also changed dressing habits. The use of linen in dressmaking was in decline, and new materials called for new approaches in design and brought more relaxed dressing styles. Traditional costumes became confined to folklore. Diet and dining culture changed, too. By the mid-19th century, hunger had been eliminated even among the lowest social classes, although everyone's diet continued to be affected by economic fluctuations. Cookbooks and cookery classes for young girls became a must in urban households. The serving of better dishes was mainly an outward sign of higher social status. The main components that distinguished the diet of higher classes from that of the majority population were meat and diversity. The majority population lived on a meatless diet composed of simple dishes, while the diet of the working class was more meager still.

At the turn of the century, the construction of the railway network was completed. It took 70 years and the total length of railway tracks was more than 1,700 km. An interesting comparison can be made based on France Kresal's information on the total length of the road network within the Slovenian populated crown lands: there were 3,750 km of national and provincial macadam roads.[450]

## To Seek and Find Justice in One's Own Language

Social life and culture gained momentum. Slovenian artists were on a par with their counterparts elsewhere in Europe and were establishing much needed and important connections. However, the official position of the Slovene language was improving only gradually. The Slovenian and Czech delegates to the Parliament strove for the equality of all nations within Austria-Hungary when it came to shorthand, for a purely practical reason: speeches written in shorthand could not be confiscated by censors. However, it was not until 1907 that non-German

---

449  For more on the cholera epidemics, see Katarina Keber, *Čas kolere. Epidemije kolere na Kranjskem v 19. stoletju*, Ljubljana: Založba ZRC, 2007.
450  Kresal, "Struktura slovenskega gospodarstva," 112.

delegates were allowed to request a separate shorthand writer.[451] From 1882, the use of Slovene in Carniolan, Lower Styrian and Slovene-Carinthian administration buildings was extended to clients who expressly requested it, but German and Italian was still more prevalent.[452] In 1898, two Slovenian delegates proposed a law to regulate the language issue once and for all. According to this proposal, every citizen would be "entitled to seek and find justice in his own language, with all competent bodies, even if located outside the territory of his own language. All commonly used languages within the state would be equivalent and on an equal footing in all public and internal proceedings taken by both the state and provincial authorities."[453] Mal assessed this as the "most far-reaching" demand ever for the recognition of linguistic equality within the monarchy, adding that "over the next few years, it became obscured amid the parliamentary unrest."[454]

At the provincial level, the status of Slovene was equal to that of German only in Carniola. Unfortunately, even in this most Slovenian of all provinces, Slovene had a hard time establishing itself in public life. Winkler's successor, Baron Hein, argued that the administrative business in Carniola should be as German as possible. Bishop Anton Bonaventura Jeglič categorically defied him when he ordered the diocesan ordinary's office to use Slovene exclusively in its correspondence with the provincial government. Provincial Governor Hein did not yield. In 1899, he ordered the annulment of the municipal resolution that street names should be written in Slovene only, although such a decision was in the domain of the municipal authorities. The Diet of Carniola ordered him to apologize for belittling the majority language.

However, a brighter prospect was in view for Slovene: from the end of 1905, the records of chamber sessions were written and read in Slovene. A proposal by the delegate Ivan Hribar required that all written correspondence between the provincial board and state offices and autonomous authorities in Carniola be exclusively in Slovene. Hribar also demanded that the streets in Ljubljana be named after Slavs of merit, and furthermore that the street names be exclusively in Slovene. In 1897, Slovene-German street signs were erected with a longer text in Slovene; they became entirely Slovene from 1908 onwards.[455] In other provinces, German and Italian continued to be the languages of administration. In 1905, the

---

451 Mal, *Zgodovina slovenskega naroda*, 1173.
452 Štih et al., *Temelji slovenstva*, 222.
453 Mal, *Zgodovina slovenskega naroda*, 1173–1174.
454 Ibid., 1174.
455 Štih et al., *Temelji slovenstva*, 223.

Diet of Carinthia instituted German as the exclusive language of communication, although the local board also accepted and processed appeals in Slovene, while invariably using German in its communication with Slovenian municipalities. In Styria, the language of public administration was not a bone of contention. What was more disputable was the obstinacy of municipal authorities in Maribor, Ptuj, and Celje in refusing to take into account the equality of languages, so the state court called on them to observe bilingualism on official signs and seals used by political and financial authorities. In the Styrian Diet, Slovenian delegates could use Slovene.

No less painstaking was the penetration of Slovene into schools. The state entrusted the supervision of elementary education and the language of instruction to provincial authorities, retaining control only over secondary and higher education. The practical effects varied from province to province: "Of course, the consequences of this local self-governance in education were felt by the Slovenes wherever their bread was sliced by the Germans or the Italians, who did not allow them to have schools in their national language, especially not in border regions or in towns."[456] The establishment and maintenance of schools in border regions was the task of the St. Cyril and Methodius Society, which attempted in this way to resist the denationalizing influence of the Germans and Italians. The authorities of Carinthia adhered to bilingual schools. There were 82 of them in total, and Slovene was replaced with German within the first few years of their operation; there were only three Slovene-language elementary schools in Carinthia. In Carniola, public elementary schools were Slovene schools (285 altogether), except within several German language enclaves in the Kočevsko region, in Bela Peč, and in Ljubljana. In the Littoral, the Slovenes obtained their first public elementary school in 1895, but in Trieste such wishes and aspirations were never realized. In Lower Styria, the German middle class bitterly opposed public schools in Slovene, but the St. Cyril and Methodius Society nevertheless managed to establish two private elementary schools. The Slovenes living in Venetian Slovenia and Hungary had no alternative but to attend Italian/Hungarian schools. Awareness of the necessity of elementary education varied from place to place. In Carinthia, 80 percent of children aged 6–12 attended elementary school. Their percentage in Lower Styria was also high, approximately 82 percent, while elsewhere this figure ranged from 72 to 76 percent.[457] However, the percentage of children in elementary schools did not reflect the illiteracy

---

456 Ibid., 1180.
457 Vodopivec, *Od Pohlinove slovnice do samostojne države*, 129.

rate: in Styria, Carniola, Trieste, and the Gorizia region, this figure was between 11 and 15 percent, while in Carinthia it was as high as 23 percent.

The determination of the language of instruction in grammar schools was the domain of the education minister. In Carniola, two state grammar schools were exclusively German, and four were German-Slovene. After 1908, Slovene also became the language of instruction in higher grades, although not evenly across all provinces. The introduction of the "notorious" parallel classes in Celje and Maribor was a painful process, while "the grammar school in Ptuj was not even allowed to have Slovenian departments lest it ruin its German character."[458] The three grammar schools in Carinthia were exclusively German. In Carniola, on the other hand, grammar schools had already been transformed during the Taaffe period so that in some schools the language of instruction in lower grades was Slovene; after 1908, this was extended to higher grades. The first exclusively Slovene grammar school was the diocesan grammar school in Šentvid nad Ljubljano, established in 1905.

The grammar school in Gorizia was the only state school in which the language of instruction was exclusively Slovene. It was established in 1913 by dividing the previously German-only school into Slovene, German, and Italian schools. In all provinces with a Slovenian population, Slovene was an optional subject, including in non-Slovene grammar schools. Regarding vocational schools, two secondary schools (*realka*) in Trieste were Italian, and others were German. Only in Idrija was there a Slovene-German *realka*. *Realka* in Gorizia offered religious classes in Slovene in the lower two grades. Postojna had a Slovene secondary school, as did Žalec (since 1913); the women's and men's teacher training schools in Gorizia and Škofja Loka were Slovene-only, while in Ljubljana the languages of instruction in teacher training schools were Slovene and German.

Slovenian students studied in Vienna, Graz, and Prague, and in the decade before World War I also in Kraków. There were 300–400 Slovenian students enrolled in Austrian universities and higher schools during the 1870s and 1880s; around the year 1900, there were already some 650, and just before World War I their number had risen to more than 930 (two-thirds of the total number of Slovenian students studied in Vienna). Until the turn of the century, the greatest number of Slovenian students had been enrolled in theology courses; afterwards, their number was exceeded by law students. The number of students studying technical science was also growing. The first Slovenian women enrolled at the University of Vienna in the early 20th century.

---

458  Mal, *Zgodovina slovenskega naroda*, 1181.

Although the Slovenian intelligentsia came from non-Slovene-language schools, the "national idea was already so strong that it was capable of attracting an increasing number of workers to work in culture. The struggle and competition with the economically, politically, and culturally stronger opponent forced the Slovenes to organize themselves resolutely and to rely on their own work and initiative."[459] Awareness of the shared spiritual world and culture was growing, a trend reflected in the flourishing of education societies, libraries, choirs, bands, theater groups, and gymnastic and other associations. These institutions were mainly active in rural regions, since peasants, as the largest class, represented an important mainstay of the Slovenian national character and were instrumental in reinforcing national awareness.

Increased literacy and education ushered in a new leisure time activity—reading. It soon became the most popular and the most widespread way of spending leisure time. There were five daily newspapers in Slovene: the Catholic daily, *Slovenec*, the liberal daily, *Slovenski narod*, the socialist, *Zarja* (Dawn), the independent liberal, *Dan* (Day), and the liberal daily, *Edinost* (Unity), published in Trieste; plus several papers offering entertaining content. The most influential literary organization of the time, or book organization as they were called, was Mohorjeva družba, which had about 71,000 members in 1895 and 91,000 by the end of World War I. Its network spread to the regions beyond the borders of the Austrian part of the empire, namely Hungary and Venetian Slovenia. The easiest way for scientific texts to reach readers was through various collections of scientific papers. The main mission was still carried out by Slovenska matica. Educational content was disseminated by the Slovenska šolska matica (from 1900), while scientific societies and societies for the promotion of knowledge about the homeland published their own publications. The Slovanska Library disseminated knowledge about the literature of Slavic nations. Since publishing involved financial risks, literary societies and organizations were the main publishers, with individual entrepreneurs only joining in toward the end of the century. The queen of art forms among the Slovenes was literature. The priests Simon Gregorčič and Anton Aškerc were the most prominent poets of the older generation. The former gained a reputation as a writer of ballads and romances, which, thanks to "the relaxed resonance of his language and his good feel for rhythm,"[460] were widely popular and some even became part of Slovenian folk culture. Anton Aškerc devoted himself exclusively to epic poems, ballads, and romances.

---

459 Ibid., 1183.
460 Ibid., 1079.

During the second half of the 1890s, a new trend in literature, named *moderna* (modernity) for its break with the Slovene literary tradition by the Slovenian writers Fran Govekar and Ivan Cankar, began to acquire a reputation. This was a very diverse literary trend in terms of style. Among the poets, the most prominent representatives of modernity were Dragotin Kette and Josip Murn—Aleksandrov, two very talented poets who unfortunately died young at the turn of the century. No less renowned was Oton Župančič, who "introduced so much of the stylistic, linguistic and mental richness based on folk songs into Slovenian poetry and literature."[461] Župančič was also an accomplished translator of many preeminent works of world literature. Together with Cankar, the poet, playwright, and author, these writers brought Slovenian literature closer to European trends for the first time since France Prešeren. Mal has summarized the basic features of Cankar's work:

> He is a great artist of a shorter narrative in a lyrical form, of novel and short story; he created a completely new and unique narrative style. In these writings, much like in his plays and satires, he castigated social injustice and, among other things, unmasked with reformist enthusiasm the insincerity and selfishness of political patriotic society. His deeply psychological descriptions also earned him fame and reputation among great European nations, which included several works by Cankar in their literature.[462]

The protagonists of modernity in literature advocated the autonomy of art, and strove above all for sincerity in their artistic confessions. Although their knowledge about contemporary literary trends came from Vienna, Slovenian writers modeled themselves after Western European authors. They found inspiration in various contemporary trends, such as decadence, impressionism, and symbolism, with an admixture of realism, naturalism, futurism, and expressionism. Their artistic style was vehemently opposed by the Church. This hostility went so far that the then Bishop of Ljubljana, Anton Bonaventura Jeglič, purchased the majority of Cankar's published collection of poems, *Erotika* (1899), and had it burnt. But the major influence of the Slovenian *moderna* could no longer be suppressed. On the contrary, Cankar set an example that earned him supporters even among the Catholics.

Traditional literary styles also maintained their continuity, and some incorporated certain elements of modern trends. Fran Saleški Finžgar was renowned for his close observation of popular and social development, which he portrayed in the traditional style. He made his reputation with his historical

---

461  Ibid., 1185.
462  Ibid., 1185.

novel *Pod svobodnim soncem* (Under the Free Sun), written in the style of Henrik Sienkiewicz.

Other art forms also combined old and new elements, in some cases exceeding traditional and established ways. Vocal, and particularly choral music was giving way to instrumental music. The Slovenian musical magazine *Novi akordi* (New Chords, 1901) offered systematic music criticism and published original Slovenian music. Glasbena matica continued to be the central Slovenian musical institution, joined by the Slovenian Philharmonic in 1908. With the opening of the new Provincial Theater in Ljubljana in 1892, operas also found their place on the stage. The first Slovenian opera performed on the domestic stage was *Teharski plemiči* (The Nobles of Teharje) by Benjamin Ipavec. Other Slovenian operatic composers who rose to fame included Viktor Parma, Anton Foerster, and Risto Savin. The author of the first Slovenian ballet, *Možiček* (Jumping Jack, 1901), was Josip Ipavec, Benjamin Ipavec's nephew. Trieste also boasted an opera ensemble, and operettas were staged in Maribor. Choral music and singing retained their traditional place in Slovenian music and in people's minds: in 1913, the Association of Slovenian Choral Societies (Zveza slovenskih pevskih društev) had more than 300 member choirs.

The Provincial Theater provided a stage for theatrical art that was primarily cultivated and guided by the Dramatic Association (Dramatično društvo), under whose auspices the first professional actors received their training. They staged classical works of internationally renowned playwrights (Shakespeare, Goethe, Schiller, Ibsen) and Slovenian playwrights (Josip Vošnjak, Anton Aškerc, Anton Funtek, Fran Govekar, and Ivan Cankar). Amateur theater groups catered for the less demanding theatrical audience that enjoyed merry folk plays; performances were also staged in rural regions. In 1905, Karol Grossman, a lawyer from Ljutomer, entered the international world of cinematography with two short documentaries, the first Slovenian films ever: *Odhod od maše v Ljutomeru* (Dismissal from mass in Ljutomer) and *Sejem v Ljutomeru* (The Ljutomer Fair).

During the last decades of the 19th century, visual arts in Slovenia ceased to follow blindly the canon of religious-romantic painting and searched for an authentic course. Jožef Petkovšek and the brothers Janez and Jurij Šubic influenced the expression of Slovenian artists, but could not make a profound impact on the development of Slovenian painting because they worked abroad and their early deaths cut short their careers. In contrast, Anton Ažbe was a tremendous influence. He became the teacher and friend of the young generation of Slovenian painters, but cultural narrow-mindedness in Carniola drove him away, to Munich. The most eminent representatives of the generation of painters who moved out of the studios and worked in nature were Ivan Grohar, Rihard

Jakopič, Matija Jama, and Matej Sternen. Drawing on impressionism, they made color the basic medium of their expression. Although their popularity at home came only after they had won acclaim in Vienna, they found inspiration in Slovenian art, local landscapes, and local people. In 1900, this artistic group established the Slovenian Arts Society (Slovensko umetniško društvo) and organized the first group exhibition in Ljubljana in the same year, followed by exhibitions in other European centers. In 1901, Jakopič established the first exhibition pavilion in Tivoli Park in Ljubljana.

The architects Jože Plečnik, Maks Fabiani, and Ivan Jager followed the path of Slovenian painters. However, at the time only Fabiani became famous at home, when the restoration of Ljubljana after the 1895 earthquake gave him an opportunity to draw up an urban plan and design several buildings. They found work and fame beyond borders of present-day Slovenia: Plečnik and Fabiani co-created Vienna, and Plečnik also left his mark in Prague before he returned to Ljubljana, while Jager worked in China and in the USA.

The fundamental mission of the Slovenian intelligentsia continued to be the promotion of Slovene in public life, with great success in various areas of culture. Slovene became a fully formed language. It was also used in science; the first studies in the history of art were published, and professional terminology was established. Slovene was used at public events, and an increasing number of professionals wrote in Slovene. The Slovenian intelligentsia could now receive elementary education in Slovene, although most cultural treasures came through German. A very popular format were the so-called "reklamke," low-priced German translations of works ranging from ancient classics to contemporaries, named after the German publisher Reclam.

Academic institutions were few in number and academic work was mainly performed by individuals working in museums or for numerous professional associations and their publications. Slovenska matica and its yearbook continued to be the driving force. In 1889, Slovenian lawyers established their own association, Pravnik (Lawyer), and launched a publication with the same name. In 1891, the Museum Association of Carniola launched the journal *Izvestje* (Annual Report), and later *Carniola* (in 1910). These two papers were the main Slovenian publications dealing with historical and natural sciences, and they soon extended their activities to cover all the Slovenian populated provinces. In 1903, the Historical Society (Zgodovinsko društvo) was established in Maribor. Its publication, *Časopis za zgodovino in narodopisje* (Review for History and Ethnography), still exists today. The last decades of the 19th century saw the publication of many professional treatises: *Stavbinski slogi* (Architectural Styles) by Janez Flis appeared in 1885; *Uvod v modroslovje* (An Introduction to Philosophy,

1887) by Frančišek Lampe laid down the foundations of Slovene philosophical terminology; *Temelji vremenoznanstva* (The Basics of Meteorology) by Simon Šubic was published in 1900; *Gradivo za zgodovino Slovencev* (Material for the History of the Slovenes) by Franc Kos began to be published in 1902. Between 1874 and 1891, Mohorjeva družba published the series *Obča zgodovina za slovensko ljudstvo* (General History for the Slovenian People), written by professor Josip Stare. Between 1893 and 1895, Maks Pleteršnik published a Slovene-German dictionary, and in 1894 Karel Glaser completed *Zgodovina slovenskega slovstva* (A History of Slovenian Literature); Karel Štrekelj published the collection *Slovenske ljudske pesmi* (Slovenian Folk Songs) in 1895. Josip Gruden and Josip Mal put the history of Slovenia into words in a comprehensive book that ventured beyond provincial borders and beyond the traditional bounds of historiographical description. The part written by Mal in particular offers a complete picture of life across the Slovene-speaking lands. A growing number of Slovenian intellectuals were cementing their professional reputation, contributing to various academic and scientific disciplines across both the monarchy and Europe. Among them were the physicist Nace Klemenčič, the mathematician Josip Plemelj, the law historian Vladimir Levec, and the renowned physicist Jožef Stefan. The latter later moved to Vienna, "left the national ranks" and became an Austrian scientist both in speech and in writing.

National political and social life became better organized and more diversified. The old forms of political and social life, such as meetings, excursions organized by gymnastic or choral societies, and literary reading events, were replaced by new ones. The most widespread were firefighting, educational, sporting, and choral associations, as well as humanitarian and national defense societies. The initial centers of their activities were reading rooms, but over time national halls took their place and became the hubs of social activities. The first national hall was built in Ljubljana (1896), followed by Celje (1897), Maribor (1899), and finally Trieste and Gorizia (1904). They hosted performances, housed libraries and coffeehouses, and provided opportunities for establishing social and business contacts.

Flourishing of sport culture led to the emergence of gymnastic societies. The two main societies, Sokol and Orel, were politically tinged. Apart from physical exercises, they also offered their members spiritual training in the form of lectures and various courses, and even exercises in rhetoric. Together with educational societies, they also took part in theatrical plays. Although the preoccupation with fitness and appearance was indeed more a concern of the well-off class and intellectuals, some estimates suggest that before World War I, one fifth of the active population occasionally engaged in physical exercise.[463] The most popular recreational activity of the time was mountaineering (and it continues to be so today). In 1893, the Slovenes left the German-oriented Austrian mountaineer society and established the Slovenian Mountaineering Association in Ljubljana. It immediately contended with the Austrian organization over mountain huts. This so-called Slovenian–German war for mountain huts culminated in 1895 when the priest Jakob Aljaž from Dovje bought the summit of the highest Slovenian mountain, Triglav, for 5 florins and erected a tower there as a symbol of the Slovenian nation. Other sport associations included the Cycling Association and the Slovenian football club, Hermes (1906); the Slovenian chess player Milan Vidmar earned an international reputation.

Everyday life of Slovenian towns was much like that of other towns within the monarchy. The towns were predominantly German in appearance (or Italian in the coastal areas), reflecting the bilingual and binational population structure. Among them, Ljubljana underwent the most radical changes in urban planning, which were in a way accidental and forced by circumstances. In 1895, it was hit by a massive earthquake, after which it developed into a modern capital with new Art Nouveau palaces. A modern hospital was built in 1895, the first movie was screened a year later, and the first telephones arrived in 1897. Power lines were installed the following year, and in 1901 Ljubljana introduced electric streetcars. The first Slovenian street signs were first installed in 1908.

---

463   Vodopivec, *Od Pohlinove slovnice do samostojne države*, 136.

# The Other Side of History: Herstory

The hallmarks of Slovenian women entering political and cultural life were recorded in their national affiliation and involvement in national activities. As early as 1809, the *Blejke* (brave women from Bled) prevented the French from removing the treasures from the church on Bled Island.[464] Yet their first great opportunities came with the Springtime of Nations; in the chronicles of this era, their names can be found in records of social unrest, proto-elections, newspaper publishing, and political publications. This turbulent period gave reputable middle-class women, who had been traditionally engaged in charity work up until that point, more opportunities for involvement in social events. They were engaged in flag making for the National Guard and various other associations, as well as took part in charity lotteries, donating their hand-made products or personal items. On the other hand, women who did not belong to the social elite of the time, who were primarily widows, tradeswomen, and farm owners, were also active contributors. In village communities, widowed women were formally equal to men, as the widow status carried the same rights and obligations enjoyed by members of the opposite sex. In the general cultural development of rural regions, the role of peasant women was particularly exceptional, since reading and writing lessons were mainly a woman's domain. Women sent children to school and frequently provided their meals.

During the first half of the 19th century, *kazina* societies (the *kazina* being a place for social events at the time) played an especially important role in the social and cultural development of the Slovenian population. These were part of the urban culture and through them the social role of women became more conspicuous. During the pre-March era, Slovenian women were entrusted with an important nation-awakening role, which was reflected in their effort to instill a love for Slovenian in their children and to provide them with schooling.

During the revolutionary years of 1848 and 1849, Slovenian women were able to publicly express their national affiliation. Jožefina Oblak from Graz, for example, chose to fly the Slovenian flag, enraging her German neighbors. Others expressed their national sentiment by encouraging and supporting their husbands and sons in their endeavors toward the preservation of the Slovenian

---

464  Stane Granda, "Ženske in revolucija 1848 na Slovenskem," in: Nataša Budna Kodrič and Aleksandra Serše (eds.), *Splošno žensko društvo 1901–1945*, Ljubljana: Arhiv Republike Slovenije, 2003, 7.

nation. Nationally aware and active Slovenian women could also be found among the signatories of the "United Slovenia" petition. During the revolutionary years, women throughout the Slovenian populated lands became involved in public political matters in newspaper articles.

In the summer of 1848, the Slovenian newspaper organization in Celje that published *Celjske Slovenske Novine* (Slovenian News from Celje), later renamed *Slovenske Novine* in 1849, was joined by the first female Slovenian poet, Fani Hausmann. Her poem, "Vojaka izhod" (The Soldier's Farewell) enchanted Slovenian populists.[465] After Hausmann paved the way, early female writers and artists could present their work to reading circles and drawing rooms.

During the 1860s and 1870s, this path was determinedly taken by Marija Murnik Horak, an activist and organizer of charitable, educational, and women's societies. This was also the period when Slovenian politicians and writers began to publicly express their views on women's role in society. In 1871, Radoslav Razlag delivered a lecture in Ljubljana entitled "On the Autonomy of the Female Gender" and acknowledged that women were allowed to engage in social, national, and political activities, but only if the nation was in danger. In other circumstances, he argued, their active role was within the family. During the 1880s, Fran Celestin suggested that the status of women was an indicator of society's culture and that men should guide the women's movement toward the good of the nation.

However, some didn't approve of the nationally-inspired activities by women; some thought that such engagement was undermining traditional relations within society and the family. Anton Mahnič was among those who strongly resisted it and tried to convince readers that the women's liberation movement harbored the seeds of corruption and decadence as the consequences of capitalism and liberalism. On top of that, he viewed women as unequal by nature, by value, and by rights. Andrej Gabršček strongly resisted this line of thought and responded to Mahnič by manipulating his thesis that women were at a lower stage of development, saying that, if this were true, they should be given more opportunities for education.[466] The admittance of women into public life was not only disputed by the Catholic camp, but also by some liberal Slovenian

---

465   Mira Delavec, "Fani Hausmann," in: Alenka Šelih et al. (eds.), *Pozabljena polovica. Portreti žensk 19. in 20. stoletja na Slovenskem*, Ljubljana: Založba Tuma, 2007, 31.
466   Nataša Budna Kodrič and Aleksandra Serše, "Žensko gibanje na Slovenskem do druge svetovne vojne," in: Kodrič and Serše (eds.), *Splošno žensko društvo 1901–1945*, 18–19.

politicians. An interesting viewpoint was that of Ivan Tavčar, whose wife, Franja, was versatile and active, while he himself relentlessly opposed giving political rights to women.[467]

Women were forbidden from entering political societies by the Societies Act until 1867, and another two decades passed before a tangible victory was achieved in 1892, when Marija Murnik Horak initiated the establishment of a women's branch of the Society of St. Cyril and St. Methodius in Ljubljana, modeled on those in Trieste and Gorizia. Perhaps it was the multi-ethnic environment in the Gorizia crown land and Trieste that provided the inspiration to raise a new nationally aware generation, which, in turn, brought up the question of women's education and possible ways of eliminating the obstacles which stood in the way of equality in education. In 1900, the supplement of the Trieste paper *Edinost* (Unity) with the title *Slovenka* (Slovenian Woman) became an independent journal and the first women's newspaper in the Slovenian language. Its mission was to educate female Slovenian readers. In the meantime, the first female Slovenian poets and writers gained recognition, writing for the literary journal, *Ljubljanski zvon* (The Ljubljana Bell), and the Klagenfurt-based *Kres: Slovanski svet* (Bonfire: The Slavic World) introduced a women's column in the 1890s.

It was the economically independent and educated women who expressed resolute demands for political equality, since their good material standing enabled them to discard male patronage, as instituted by the General Civil Code in 1811. The roads to equality were paved through societies as one of the most firmly established and recognized forms of public activity. The recognition that working women (such as teachers, administrative workers, or postal clerks) were paid less than men for equal work, and that this was a consequence not only of the capitalist system but also of gender discrimination, led to the establishment of the Slovenian Women Teachers' Association (Društvo slovenskih učiteljic) in 1898. Its primary task was the fight against gender discrimination within the teaching profession. In 1900, they were also joined by postal workers. Their struggle shook the social and political relations of the time, entrenched in several hundred years of oppression, so it became a political fact that took on new meaning over the next few years, through the struggle for universal suffrage.

The entry of Slovenian women into political life provoked mixed responses. The Catholic camp attempted to isolate educated women and cancel out their social effect by establishing societies that represented a new form of supervision

---

467   Marta Verginella, "Mesto žensk pod steklenim stropom," in: Kodrič and Serše (eds.), *Splošno žensko društvo 1901–1945*, iii.

over women,[468] for example the Catholic Society of Women (Katoliško društvo za delavke), the Christian Women's Society (Krščanska ženska zveza, established in 1900), and St. Mary's societies. In 1910, the Slovenian People's Party supported the Ljubljana municipal ordinance giving female taxpayers and teachers the personal right to vote, by which the party sought to strengthen its political position. Although the party included social democrats, their support for equal rights for men and women was not consistent. On the one hand, they were in favor of women joining their ranks and of universal suffrage, but they were less inclined to recognize specific women's rights which reflected their battle against social, economic, and political injustice.

The efforts to secure women the vote brought together publicly active Slovenian women of different political persuasions and different ideas about how to express demands for the emancipation of women also differed. In 1901, Franja Tavčar and Josipina Vidmar established the General Women's Society (Splošno žensko društvo), which folded in 1945. It accepted all Slovenian women regardless of their class or profession, and pursued its missions within politics, education and charity. The society worked toward the equality of all women, which was intrinsically prevented by Article 30 of the 1867 state law excluding women from political societies and parties and prohibiting female membership in corporations that took decisions on education, health, culture, and other public affairs. The society assembled both active feminists and homemakers. Its fundamental tasks included the provision of general and professional education for women and the preparation of women to join the "existential struggle," or rather, to participate actively in society. The society also took part in international activities through its membership in the state-wide women's society. Its members regularly took part in mass sporting events in Prague and established close contacts with their Czech counterparts, who were their role models. During the Balkan Wars, they actively participated in charitable actions, contributing clothes, sanitary items, and other necessities to the Serbian and Bulgarian Red Cross. The outbreak of the Great War in 1914 presented new tasks and challenges for the Slovenes.

The streets of the Slovenian capital were indeed brightened by electric street lighting, but the brilliant light radiating from political, cultural, and economic progress began to be overshadowed by the approaching unknown. That is why everyone—the individual, the wider community, the family, the national community, or the state—had to be prepared for the unknown future. At the dawn of the new era that followed one of the most productive centuries in the history

---

468   Ibid., iv.

of the of the Slovenian people and others living in Carniola, (Lower) Styria, Carinthia, Trieste, Istria and Gorizia—when Slovenes were actively searching for the path to national recognition and political allies, when Slovenian culture and renowned individuals were reaching the pinnacle of contemporary European culture, and when even the Slovenian language was gaining recognition—"the years of horror" arrived.

# From the Habsburg Monarchy to the Kingdom of Yugoslavia

# Divided by the Great War

The so-called Great War stands as a landmark and a turning point in the landscape of modern Slovenian history. In Central Europe, as in many other parts of the continent, it was a noted "watershed event" that changed almost everything, from political perspectives—now focused on nationalist campaigns—to the discourse of culture. The war also exposed everyday life, art, and the economy in different ways, best illustrated in private letters and diaries. This emotionally charged material suggests, above all, that the war in Southeastern Europe was different from the war familiar from descriptions of Verdun, Ypres, or Arras. The difference becomes more apparent when the conflicts in Serbia, Macedonia, Albania, Galicia, and the Soča (It. *Isonzo*) Valley are given a human face. And the gap widens when voice is given to the soldiers in the trenches and to their families at home. This contrast also becomes obvious when one realizes that for the Austrians the war in the southeast was actually a punitive expedition against the Serbs.

Traces of this difference, and the shift from enthusiasm in 1914 to depression and frustration from 1915 onwards, are visible everywhere. Besides diaries, letters, and articles, public and private archives also contain a vast number of photographs, paintings, maps, sketches, lyrics, and amateur literature written by soldiers and their relatives. Their writings repeatedly and incessantly produce an encounter with thoughts about their inability to articulate the utter madness that surrounded them:

> No news: Death carries on its feast
> And today, as yesterday, no end to dying
> Slaughter, a thousand voices crying
> Fourth year of war—will it ever be appeased?
> No news; only death shall reap his harvest cold
> And even if for a million years
> Humanity shall heap brave deeds untold
> The world will not be saved from tears;
> Whoever is today released from mortal load,
> He is blessed with new, devoid of old.[469]

---

469  Petuškin, "Vojno poročilo." In: Janez Povše (ed.), *Oblaki so rudeči*, Trieste: Založba tržaškega tiska, 1988, 141. Translated from Slovene into English by Manca Gašperšič.

These rarely remembered battlefields created in soldiers and civilians alike the need to express their fears and horrors in the face of the extraordinary conditions of war. Almost overnight, hundreds of people who had no need to articulate their feelings before the war became poets and writers producing countless letters to dispel their anguish and worry.

In their reflections, the Habsburg Monarchy was described as a political seismic fault between Eastern and Western Europe, and any major political turbulence could undermine the existing structures. Projecting the forces of internal dissent outwards against a mutual enemy would probably not absorb them, but only lead to the ultimate undoing of the shared, established political space. From a wider sensible European perspective, Europe of 1914 was balancing on an increasingly precarious platform of old codes and beliefs that were destined to fall. On the eve of World War I, the Slovenes perceived themselves as a weak and universally vulnerable national community. Their feverish search for South Slavic political allies, whom the writer Ivan Cankar had proclaimed as brothers by blood and cousins by language, revealed deep anxieties. By the end of the belle époque, the construction of a German "bridge to the Adriatic Sea" therefore seemed only a question of time. While the German School Association (Schulverein) was beginning to establish bilingual education on purely Slovenian territory, the Austrian state discriminated against Slovenian schools. Thus, Slovenian aspirations to join forces with the ethnically related peoples of Southeastern Europe were in no way a resort to utopian ideas but reflected real and actual concerns.

Nonetheless, in the summer of 1914, no one could even remotely imagine that the war would trigger so many substantial transformations and important initiatives. For this reason, when the war broke out, the general reaction in Slovenia and Ljubljana (like in other major cities of the monarchy) was public enthusiasm, the rousing of patriotic fervor, and pledges of loyalty to the Austrian government and the monarch. After the assassination of Archduke Franz Ferdinand, the rabid political contests between Liberals and Conservatives considerably lessened, as two of the leading Slovenian newspapers of the time most clearly testified. When the declaration of war was approved as a necessary retaliatory measure, the conservative newspaper *Slovenec* (The Slovenian) even embellished its reports with stirring verses which quickly became woven into everyday use:

> Hear our cannons to salute
> You, Serbs?
> We'll lay you to the ground as food
> For herbs!

Divided by the Great War

God will send our mighty army
At your gate
To see what's all that barmy
In Belgrade.
We'll wrap your bodies under grass
In order
And seal, for many years to pass
Your border
The righteous debts from foul past
Will be squared
Our victory will then last
Uncompared.[470]

Figure 36. Austro-Hungarian position on the Isonzo Front.

The only—virtually isolated—public opponents of the war were the members of the illegal pro-Yugoslav Preporod (Rebirth) movement, a majority of the Social Democrats, and a few Liberals. They claimed that while the Conservatives were "shedding crocodile tears over the assassinated monarch," they were primarily thinking of how best to take advantage of the new situation by discrediting their political foes.

---

470   *Slovenski narod* 65, 840 (1914), 1.

The vacillation between Vienna and Belgrade was also partly reflected on the military front. Slovenian soldiers could be found on the rosters of both warring sides, and on most of the battlefields in Southeastern Europe. Those serving as regular Austrian conscripts were organized into six Austro-Hungarian regiments, while the members of the Rebirth movement, who maintained close ties with the Serbian National Defense, volunteered in the Serbian army.[471] Most of them, including their leader Avgust Jenko, died in the first two battles (at Cer and Kolubara). Those who survived the first year of war retreated to Greece and from there to Egypt. After being transferred to Thessaloniki in early 1918, they were part of the Allied forces breaking the Southern Front at Kajmakčalan, the mountain on the border between Greece and Macedonia.

**Figure 37.** In the cavern at night, Isonzo Front.

---

471  The Serbian National Defense (Narodna odbrana) was established in 1908 as a reaction to Austria's annexation of Bosnia and Herzegovina. Its main concern was to unite the representatives of all political parties in order to subdue differing party views to a higher national interest. The Serbian National Defense was also intended to provide a platform for the formation and organization of voluntary detachments in case of war between Austria-Hungary and Serbia. By 1912, it was rapidly losing support and eventually split along factional lines into the more pacifist Cultural League (Kulturna liga) and the militant Unification or Death (Ujedinjenje ili smrt).

On the Southern Front, they also met those who joined the Serbs, Croats, and Slovenes volunteer corps after deserting the Austro-Hungarian army and being taken to prison camps.

The political situation was also changing in Ljubljana. After the universal anger and regret, the Liberals and Social Democrats started to show some reservations toward Vienna. Instead of following the influential Catholic leader and the central political figure Ivan Šusteršič, who fervently accused Serbia, they sought to avoid making any public statements of support for the monarchy and opposed the declaration of war against Serbia.

Some of them left the country together with their colleagues from the Liberal Party and joined other pro-Yugoslav politicians in forming the Yugoslav Committee. Founded in Paris in April 1915, it represented Slovenes, Croats, and Serbs living in Austria-Hungary. The role of the committee, chaired by Ante Trumbić,[472] was to inform the Allied forces about the plight of the South Slavs in Austria-Hungary and to propagate their desire to unify with Serbia into a South Slavic state. Although seemingly yet another self-appointed committee, the

---

472 Trumbić, a Croatian politician and member of the Croatian Party of Rights, belonged to the party's Dalmatian faction, the least inclined to any national and political exclusivism. He would therefore later become one of those rare personalities who were able to build a consensus between Serbs and Croats in the Yugoslav Committee; in part, this was also because he had already dealt with their disputes in Dalmatia, where the Serbs were sharply opposed to the idea of unifying the Banovina of Croatia and Dalmatia. Moreover, he was one of the few supporters of the Yugoslav idea who did not fall into the traps of Austrian and Serbian politics. And even though he was mindful of the fact that "Serbia pursues its narrow egoistic interests," he still recognized the greatest threat to Croatia in Austria and "Germanism," so that he and his party leader Frano Supilo advocated a union of all South Slavs in the Habsburg Monarchy. It is therefore not surprising that as a minister in the newly formed state, he was frustrated by the drawing up of a new constitution and even more exasperated by the subsequent "unconstructive" policies of Stjepan Radić. According to Croatian historians, he was also embittered by "the king's absolutist regime of January 6, 1929," which presumably prompted him to say that "in Turkish Serbia, there were only the Turks and the underdogs, whereas the Croats were destined to become the underdogs of Yugoslavia." Despite his disappointment, even on his deathbed he refused to convert to separatism, as some insinuated, but as a realistic and sagacious politician simply came to the conclusion that "Croatia's persistence within the Yugoslav state would bear tragic consequences for the Croatian nation." All quotations from Ljubo Boban, "Ante Trumbić—Život i djelo," in: Ljubo Boban and Ivan Jelić, *Život i djelo Ante Trumbića*, Zagreb: Croatian Academy of Sciences and Arts, 1991, 9–12.

Yugoslav Committee managed to attract a small but powerful body of supporters. In London, it was approached by Robert Seton-Watson, an independent scholar and linguist, and Henry Wickham Steed, The Times correspondent in Vienna before the war. Both men looked upon Austria-Hungary with irritation as "a corrupt and incompetent anomaly and they made it their self-appointed task to put it out of its misery."[473]

When the Serbian government learned that Britain, France, and Russia had signed the secret Treaty of London with Italy to persuade it to join their alliance, Serbian Prime Minister Nikola Pašić urged the Yugoslav Committee to try and nullify the document. He was convinced that the treaty would endanger the necessary changes in the area, the most important one being the creation of one supranational state: a state that would be geographically large enough, ethnographically compact, politically strong, economically independent, and in harmony with European culture and progress. This was allegedly his position as early as December 1914, when his government delivered the first public declaration of Serbia's war aims. It became known as the Niš Declaration, named after the second largest Serbian city, where the government met following the fall of Belgrade. Later approved by the Serbian Parliament, the declaration stated that Serbia's main objective after victory was the liberation and unification of all Serbs, Croats, and Slovenes. Thus, the Corfu Declaration—signed on July 20, 1917, between Pašić on behalf of Serbia and Trumbić representing the Yugoslav Committee—may be regarded as a next step in the process of establishing a common state. However, it is also possible that Pašić signed the declaration feeling that he was left with no other choice. Since the Great Powers were still unwilling to consider a future state as one of Serbia's war aims, the declaration was partly designed to press them to accept the consequences of the dissolution of the Habsburg Monarchy. On the other hand, it was also seen as an answer to the May Declaration of the South Slavic deputies in the Austrian Parliament.[474]

The first point of the Corfu Declaration stated that the Kingdom of Serbs, Croats and Slovenes would be a "constitutional, democratic, and parliamentary monarchy headed by the House of Karadjordjević." This was a generous proposition from Serbia's perspective, since "Serbia did not demand any

---

473  Margaret Macmillan, *Paris 1919: Six Months that Changed the World*, New York: Random House, 2002, 114.
474  See Alex N. Dragnich, *Serbs and Croats: The Struggle in Yugoslavia*, New York: Harcourt Brace Jovanovich, 1992, 26.

privileged status or veto power in the new state, as had Prussia, for example, when it was the driving force in the unification of Germany. Moreover, Serbia was willing to give up its democratic constitution, convinced that the Constituent Assembly would produce a constitution that would be acceptable to all."[475] The last part of this interpretation is particularly important, since the Croatian and Slovenian committee members wanted the future constitution to be ratified by a majority of each national group or by a two-thirds majority of the Constituent Assembly. In their view, such a requirement existed in every democracy. The Serbs' standpoint was rather different. They claimed that such a question could be answered in various ways and all parties settled for the words "a numerically qualified majority." Therefore, slightly more than 50 percent of the Constituent Assembly concluded in its 1921 meeting that an absolute majority of its delegates was sufficient to meet the Corfu stipulations.

Pašić, his cabinet, and the majority of the Yugoslav Committee agreed that the future state should be centralized with certain local self-government rights. After the Niš Declaration, Trumbić paid high tribute to Serbia, saying that it "has made the greatest sacrifice for the Union [and] with that she begins the greatest of her deeds and attains the absolute right to be called the Yugoslav Piedmont."[476] A union with Serbia, whatever its drawbacks, seemed less frightening than independence, which at best would mean a country cobbled together from Slovenia, Croatia, and Bosnia, and at worst two or three small weak states. Imprudently, both sides put off discussing the constitution, so the issue of a federation or a centralized state was never settled. Quite to the contrary, the Serbs made it very clear how they viewed the process of uniting different peoples. As one Serbian government official told Trumbić, there would be, for example, "no difficulty in managing the Bosnian Muslims. The Serbian army would give them twenty-four hours—no, perhaps even forty-eight—to return to the Orthodox faith. 'Those who won't will be killed, as we have done in our time in Serbia.'"[477]

In the month following the Corfu Declaration, Pašić distanced himself from the idea of any real union. He had worked behind the scenes to make sure that the Allies did not recognize Trumbić and the Yugoslav Committee as the voice of the South Slavs from Austria-Hungary. In a London meeting with Wickham Steed, he claimed that the Corfu Declaration had been intended only for propaganda

---

475   Dragnich, *Serbs and Croats*, 26.
476   Ibid., 28.
477   Macmillan, *Paris 1919*, 115.

purposes and that Serbia was to be in control of any new state. Those Croats and Slovenes who did not like it were perfectly free to go elsewhere…[478]

Meanwhile on the "domestic political front," new divisions were arising among Slovenian parties. The main reasons were Italy's declaration of war against Austria-Hungary and Germany, and the Italian occupation of the land stretching along the Slovenian national border. The latter baffled leaders of the Pan-Slovenian People's Party to the extent that they were willing to reconsider their pre-war Croatian–Slovenian plans which had favored a tripartite reorganization of the Dual Monarchy, to accept German as the official language, and to agree to the centralist reform ideas proposed by the Austrian military authorities. In return, they at least wished to negotiate the recognition of Slovenian culture and educational autonomy. Some attribute this turn of events to an expansion of Friedrich Naumann's idea about the new *Mitteleuropean* order led by a Habsburg–German federation. Nevertheless, it is widely agreed that in the earlier half of the war, Yugoslav views were only represented by an insignificant number of Slovenian political emigrants in Western Europe. Conditions changed after the assassination of the Austrian Prime Minister Karl von Stürgkh, the death of Franz Joseph in November 1916, and the accession of Emperor Charles of Austria, who decided to reconvene the Parliament during a period of poverty, distress, and war fatigue.

It therefore comes as no surprise that an agreement on a common Yugoslav parliamentary club of South Slavic representatives from the Austrian part of the monarchy was adopted only two or three days before the Parliament resumed its sessions. And even then, a few representatives had considerable reservations about the Yugoslav question. Despite the fact that Slovenian and Croatian representatives from the western half of the monarchy and Dalmatia found themselves in the same club as their Serbian counterparts for the first time, the vast majority of the united deputies could simply not believe that their joint declaration would be met with a widespread response. After all, soon after the 1908 annexation of Bosnia and Herzegovina, they had called for something equal to the demands of the Carniolan Provincial Diet, which had also sought to unite the territories inhabited by the Slovenes, Croats, and Serbs within the Habsburg Monarchy. The May Declaration was an unequivocal demand for a

---

478 Alex N. Dragnich, "The Serbian Government, the Army and Unification of Yugoslavs," in: Dimitrije Đorđević (ed.), *The Creation of Yugoslavia, 1914–1918*, Santa Barbara, Oxford: Clio Books, 1980, 43–44.

new state "free from foreign domination and built on the basis of democratic principles." What was truly unprecedented was the reception of the declaration. By spring of 1918, the declaration had been signed by more than 200,000 people and supported by the vast majority of parties, including the Social Democrats. After that, events unfolded at a dizzying pace. At first, the government—interpreting the truce with Russian Bolsheviks at Brest-Litovsk as an important war victory—suppressed the May Declaration movement and guaranteed the monarchy's German-speaking population access to the Adriatic Sea, a valid reason for concerns whether the Slovenian provinces would be retained in the German part of the state should the monarchy be reorganized. The result was a serious decline in Austro-Hungarian patriotism among the Slovenes, who saw the May Declaration movement as a synonym for a national state that would finally fulfill their aspirations "for self-determination" and national independence. Conversely, recent research indicates that the majority of the May Declaration signatories continued to express their unreserved loyalty to the new Austrian Emperor Charles until mid-1918.[479] Archival evidence further confirms this evaluation, with fewer anti-Austrian activities detected among the Slovenes than among the Czechs.[480]

Pašić undoubtedly sensed this division. In mid-October 1918, he told *The London Times*, *The Morning Post*, and *The Manchester Guardian*: "The Serbian people cannot wish to take up a dominant position in the future Kingdom of Serbs, Croats and Slovenes." At the same time, he asserted that Serbia considered it a national duty "to liberate all Serbs, Croats, and Slovenes. […] And when they shall be free," he said, "they will be guaranteed the right of self-determination, that is, the right to declare freely whether they wish to join Serbia based on the Corfu Declaration or to create small states as in the distant past." He concluded that the Serbian government would in no way "invoke the Corfu Declaration should it go against their wishes."[481]

What was fatal for the developments after the 1918 coup d'état was the fact that the leaders of the May movement had no specific plan before the fall of the monarchy about how the desired Yugoslav state community should be organized. Indeed, they had delayed discussions until the fall of 1918. Newspapers only fully

---

479 See Vlasta Stavbar, "Izjave v podporo Majniške deklaracije," *Zgodovinski časopis*, 3 (1992): 357–381; 4 (1992): 497–507; *Zgodovinski časopis*, 1 (1993): 99–106.
480 Walter Lukan, "Slovenci in nastanek jugoslovanske državne skupnosti," *Glasnik Slovenske matice*, 1 (1989): 40–44.
481 Dragnich, *Serbs and Croats*, 29–30.

explored the controversy in October, when a conservative Fran Šuklje presented his deliberations to the Slovenian nation at Korošec's behest. Even though Šuklje asserted that the Slovenes made up one nation with the Croats and the Serbs, his conception of a state of Slovenes, Croats, and Serbs composed of the South Slavic units within the Austro-Hungarian Empire was a federal republic with boundaries based on historical, national, linguistic, and ethnic principles. According to this plan, Istria would be annexed to Slovenia, Bosnia to Croatia, and Herzegovina to Dalmatia. Each unit would be internally subdivided and granted broad autonomies.

The liberals discredited Šuklje even before he could thoroughly explain his ideas and rejected his attempt to strike a middle ground between autonomy and centralism (indeed his entire plan) as not Yugoslav enough. "If we imagine a Yugoslav state, we cannot imagine it developing separate Slovenian, Croatian, and Serbian groups… Our future state must be built out of one stone and one stone only," maintained a liberal Ivan Tavčar.[482] Such a "unitary" view was not at all uncommon in 1918, and to some degree even later on. Rather, it was shared by much of the Slovenian intelligentsia who were turning westwards, particularly toward France, firm in their belief that the unification of Serbs, Croats, and Slovenes would give birth not only to a new Slavic country but also to a new "Yugoslav nation."[483]

---

482  *Slovenski narod*, October 17, 1918.
483  Anton Loboda (Anton Melik), "Narod, ki nastaja," *Ljubljanski zvon* (Ljubljana, 1918), 476–484; ibid., "Nacionalna država proti historični," 788–797.

# The Making of the New State

During these debates in August 1918, all three Slovenian parties founded the Slovenian National Council. In early October, the National Council of Slovenes, Croats and Serbs was established in Zagreb. Although the council was to replace the Yugoslav Committee as the voice of the South Slavs within the crumbling Habsburg Monarchy, the Committee instantly welcomed its formation. Moreover, Trumbić and his colleagues sought the council's international recognition to extract concessions from Pašić. But the time for projects of this sort had already run out. Despite Emperor Charles's decision to grant the right of self-determination to all nations in the monarchy, the members of the Austro-Hungarian Parliament had already decided, and Anton Korošec's words to the emperor at their last meeting ("Ihre Majestät, es ist zu spät"[484]) were coming true. The very next day, Prague announced the founding of the Republic of Czechoslovakia, led by a provisional government that had been formed only two weeks earlier in Paris.

The State of Slovenes, Croats and Serbs was constituted on October 29. Anton Korošec and two vice presidents (the Croat Ante Pavelić and the Serb Svetozar Pribićević) formally presided over the highest representative body of the state, which was destined to exist for only a month. The administration of the Slovenian territory, the National Government for Ljubljana, was assigned by the National Council of Slovenes, Croats and Serbs. It enjoyed absolute autonomy over political affairs in much of Slovenia until the first joint Yugoslav government in Belgrade was established. However, its jurisdiction did not extend to the territories subject to the Treaty of London. These had been seized by the Italian army after it broke through the front line on the Piave River and remained separated from the rest of the Slovenian territory until 1945.

The issue of postwar borders was crucial in shaping a common state. Italians exerted pressure from the west, and in some areas even violated the Treaty of London border, while the northern border was yet to be determined. The German-speaking population resisted incorporation into the new state in southern Carinthia (which the Slovenes considered their own) and certain areas in Lower Styria, and sometimes even resorted to arms. By November, the

---

484   "Your Majesty, it is too late" (transl. note).

situation had only calmed down somewhat in Styria, where, after a few days of hesitation by the National Government for Slovenia, military control was introduced under the former Austro-Hungarian Major Rudolf Maister. It is no wonder that the Slovenes were so committed to the immediate foundation of a new state. Aided by the Serbs, whose intimidating role during the war had earned them respect in this part of Central Europe, they hoped to negotiate the most comprehensive border settlement.

As it turned out, however, most of these expectations remained unfulfilled. A relatively favorable border was negotiated only in the east of the Slovenian territory, where the Yugoslav army occupied much of the so-called Windische Mark (present-day Prekmurje). In the west, the outcome proved very different. Following the promises made in the Treaty of London, Italy occupied the former Austro-Hungarian territories (Slovenian Istria, the Trieste hinterland, the entire Gorizia region, Croatian Istria, and the islands in the Gulf of Kvarner). Since the Paris Peace Conference had left this issue to be resolved by a bilateral agreement in the Treaty of Rapallo (November 1920), all parties, not to mention the Slovenes, remained dissatisfied. Around 340,000 Slovenes and 160,000 Croats thus remained on the Italian side of the border, whereas Italy did not receive Dalmatia as promised. The loss of the entire western part of what would later be Slovenia was catastrophic for the Slovenes, who had also permanently lost Carinthia.

In October 1920, more than half (59 percent) of southern Carinthia (the so-called Zone A) voted to live in Austria. This included a considerable number of Slovenes, representing 80 percent of the population. Such a twist of events could be partly explained by the skirmishes that occurred between Carinthian Slovenes and Germans after the State of Slovenes, Croats and Serbs had been proclaimed, and partly by the poor organization of the Yugoslav army, which failed to conquer a major portion of the Klagenfurt basin until May 1919. But mostly it was the result of anti-Serbian propaganda, which predominantly relied on the local population's misery under temporary occupation; for instance, during the general shortages, Carinthia had been abandoned by the Yugoslav administration. However, some blame for the outcome may also be attributed to Slovenian politicians from Ljubljana, who were been convinced of a favorable outcome and did not make a serious effort to argue their case. Specifically, they had not bothered to advise Carinthian farmers where and how they should sell their products in the future state, since they were economically dependent on the regional center, Klagenfurt, which the plebiscite had not included in Zone A. After October 10, 1920, explanations and frustration were in vain: Carinthia had become part of Austria.

No politician could have anticipated such a drastic turn of events in the late fall of 1918, apart from the Social Democrats, who perceived state borders principally as an aggravating obstacle to the development of the international workers' movement. Without taking those events into account, it would simply be impossible to understand the haste with which politicians tried to consummate unification with the powerful Serbs, or to explain the unprincipled manner in which their previous aspirations of equal union with Serbia and Montenegro were abandoned. In November 1918, Korošec, Trumbić, and Pašić signed a new agreement in Geneva (known as the Geneva Declaration), which confirmed the unification of three separate states into "one Yugoslav nation." But, as it turned out, Pašić signed the declaration with his fingers crossed behind his back and under pressure from the French, who were threatening to terminate the unification process. A few days later, the shrewd politician resigned tactically as prime minister to render his signature (i.e., that of the Serbian government) invalid. This cunning move was translated into an annexation of the new territories by a triumphant Serbia rather than into a voluntary unification of three equal states. Moreover, the Geneva Declaration, which had left the future organization of the state to the Constituent Assembly, was also met with fierce opposition from Prince Regent Alexander Karadjordjević, who would have to relinquish his crown if the Assembly opted for a republic. When unification became annexation, the Serbian Radicals, aided by Svetozar Pribićević and his Serbo-Croatian coalition, successfully devised a new state meeting Serbian conditions. Other contributing factors were Croatian fears of a further Italian advance eastwards and, above all, Pašić's masterful policy-making.

Before the 28-member delegation from the National Council of Slovenes, Croats and Serbs arrived in Belgrade, the representatives of the Serbian Radical Party had enabled Serbia to directly annex Vojvodina at the National Assembly in Novi Sad, although a number of deputies called for annexation through Zagreb. A similar development occurred in Montenegro, where the "Zelenaši" (Greens) demanded unification based on equality and preservation of political autonomy.[485] The newly elected Great National Assembly of the Serbian people in Montenegro ensured the victory of the "Beljaši" (Whites).[486] Montenegro was subsequently annexed by Serbia on November 26, 1918, and the old Montenegrin

---

485   The name derives from the green voting cards that were used by the supporters of the Kingdom of Montenegro at the Assembly in Podgorica (transl. note).

486   As above, the name refers to the white voting cards used by the supporters of unification with Serbia at the Assembly in Podgorica (transl. note).

Petrović dynasty was forced to abdicate. Thus, Pašić and Prince Alexander had considerably strengthened their negotiating positions as they awaited the delegation of the National Council of Slovenes, Croats and Serbs. They had also known for quite a while that the delegation from Ljubljana and Zagreb would try for the last time to impose several conditions on the new state's nature and organization. These stipulated that the Constituent Assembly would decide whether the state should be a republic or a monarchy; that the future constitution should be adopted by a two-thirds vote; and that only certain specific government functions should be lodged in the central government, while the rest should be exercised by local governing units. Holding all the trumps, Pašić and the prince regent turned the negotiations into a one-way diplomatic protocol, which forced the representatives of the National Council to drop most of their demands. The situation has been compellingly described by Jože Pirjevec:

> [T]he people in Belgrade were elated with pride at the latest triumph, showing disinclination for (any) demands. The Serbs had their own army, and they knew that when the peace conference was to be convened, they would be seated at the places of honor among the victors. As for the Slovenes and Croats, they could do nothing else but to pick up the pace to pull themselves from the debris of the monarchy, for which they were fighting just a day before. After a three-day discussion, the representatives of the National Council were compelled to concede to a unification that was not in conformity with the spirit of the instructions that they had been given. "Our Austro-Hungarian reality," Krleža wrote later, "rolled drunkenly under Karadjordjević's throne like an empty beer bottle."[487]

The prince regent and Pašić did everything in their power to reduce the crucial matter to a mere protocol. The representative of the National Council read a solemn statement during an audience with Prince Alexander (December 1, 1918) regarding its decision that the State of Slovenes, Croats and Serbs should be united with the Kingdom of Serbia under the rule of Peter I Karadjordjević. Pending the convocation of the Constituent Assembly, an agreement was to be reached to establish a responsible cabinet and provisional parliament. For a transitional period, each governing unit would retain its existing authority, albeit under control of the cabinet and the Constituent Assembly, which was to be elected on the basis of direct, universal, equal, and proportional suffrage. Prince Alexander accepted their statement and proclaimed the creation of the Kingdom

---

487  Jože Pirjevec, *Jugoslavija 1918–1992. Nastanek, razvoj ter razpad Karadjordjevićeve in Titove Jugoslavije*, Koper: Lipa, 1995, 11–12. Pirjevec is quoting the prominent Croatian essayist and playwright Miroslav Krleža, who later became the leading figure of the Yugoslav literary scene.

of Serbs, Croats and Slovenes. The National Council Presidency, however, sent the following report from Belgrade:

> In compliance with the decision of the present National Council Committee of November 24, 1918, a special delegation of the National Council addressed Crown Prince Alexander with a solemn memorandum on December 1 at 8:00 p.m. proclaiming the unification of the entire nation of Slovenes, Croats and Serbs into a unitary Yugoslav state under the rule of King Peter and Crown Prince Alexander as regent. The new state shall forthwith organize a national representative body for all branches of public administration and set up a national representative body that shall serve as a temporary legislative council until the convocation of the Constituent Assembly. During this time and until individual branches are transferred under common jurisdiction, all present provincial governments shall remain in place as well. The Crown Prince has assumed regency in his address from the throne and will appoint the common government. The function of the National Council as the highest sovereign authority of the State of Slovenes, Croats and Serbs in the territory of the late Austria-Hungary is by this Act terminated.[488]

A state was therefore constituted by three nations speaking two languages and writing in two alphabets. It included 750,000 Bosnians of Slavic origin and Muslim religion, 600,000 Macedonians, 500,000 Germans, about the same number of Hungarians and Albanians each, over 200,000 Romanians, 150,000 Turks, 115,000 Czechs and Slovaks, and a few thousand Ukrainians, Poles, Italians, Jews, and Roma. Its religious composition was equally diverse. Along with 5 million Orthodox Christians, Yugoslavia was populated by about the same number of Catholics, 1,300,000 Muslims, 400,000 Uniates, 230,000 Protestants, and 36,000 Jews. Ethnic groups were also considerably mixed in certain places. National and religious diversity aside, the state was also characterized by vast discrepancies in terms of development. For instance, according to the 1921 census, illiteracy ranged from 8.8 percent in northern Slovenia to 83.8 percent in southern Serbia.

The structural problems facing the new administration were caused not only by these differences, but also by the unfathomable devastation wrought by the war. Serbia and Montenegro, which had lost more than half a million of their population between 1914 and 1918, had yet to rebuild their cities from the gutted ruins left by the Habsburg troops, who after 1915 claimed the two conquered kingdoms as their spoils of war. Cities and villages were pillaged of cattle, tools, and any item of moveable property, and large numbers of civilians were killed.[489]

---

488   Dragnich, *Serbs and Croats*, 176.
489   See John Reed, *War in Eastern Europe: Travels through the Balkans in 1915*, London: Phoenix, 1995.

# The Kingdom of Serbs, Croats and Slovenes

Against this background, the very first common Yugoslav government, constituted on December 20, resolved to implement uniform policies across the entire country, irrespective of the different legal arrangements that had existed in individual areas. Even before the constitution had been adopted, the government pursued centralized regulation, supported by the state's two most powerful political parties (the People's Radical Party and the Yugoslav Democratic Party). From a Slovenian perspective, Serbian ruling circles thereby simply "absorbed" all the other Yugoslav provinces into an expanded Serbian state and secured predominant influence.[490] After a common government in Belgrade was formed, the National Government for Slovenia resigned in Ljubljana and was replaced by a provincial government with fewer powers. The central authorities thus violated the Decree on Transitional Administration, which would normally have preserved its power to make autonomous decisions until stipulated otherwise by the Constituent Assembly.

This replacement of the National Government was also one of the reasons for the rapid and complete collapse of national unity in Slovenia, which significantly emphasized the partisan climate that had existed as the Habsburg Monarchy declined. According to the Slovenian historian Peter Vodopivec, new divides and quarrels among political parties made it increasingly clear that the Slovenes lacked any common strategy for their future in the new state. Arguments about autonomy or centralism, Yugoslav unity or a strict observance of national distinctiveness did not just reflect different views about the national question, the Yugoslav state, and the Slovenian future. Most of all, they were the result of an unequal distribution of political powers and opportunities for parties to impose their authority and influence. Under these circumstances, the Slovenes not only became "pawns of Belgrade," but—as Slovenia's foremost 20th-century historian Bogo Grafenauer suggested before World War II—were the captives of their own parties.[491]

---

490    Ervin Dolenc and Aleš Gabrič, *Zgodovina 4. Učbenik za četrti letnik gimnazije*, Ljubljana: DZS, 2002, 77.

491    Bogo Grafenauer, *Slovensko narodno vprašanje in slovenski zgodovinski položaj*, Ljubljana: Slovenska matica, 1987, 160. Here quoted from Peter Vodopivec, "Pogled zgodovinarja," in: Drago Jančar and Peter Vodopivec (eds.), *Slovenci v XX. stoletju*, Ljubljana: Slovenska matica, 2001.

However, Slovenian internal divisions had no notable influence in the new state, according to the first election results less than two years after the Kingdom of SHS had been established. The Yugoslav Democratic Party (JDS) received 20 percent of the vote, followed by the National Radical Party (NRS) with 18 percent, the Communist Party of Yugoslavia (KPJ) with 14 percent, the Croatian Peasant Party (HSS) with 12 percent, the Yugoslav Muslim Organization (JMO) with 7 percent, and the Slovenian People's Party (SLS) with 3.7 percent, among others. Out of 40 political groups, 17 entered the Parliament. These results must be seen in relation to each respective (national) constituency, yet only the absolute results counted for state policy, which gave the Slovenian parties no advantage.

Despite initial speculations as to whether the Slovenes constituted a nation or merely a tribe within the unifying Yugoslav nation, most Slovenes were inevitably disappointed by these developments. Even more disappointing was the Vidovdan Constitution,[492] which provoked general disapproval even before it was adopted in December 1921. Although this document contained some undeniably liberal policies (e.g., separating the Church and the state, and granting autonomy to the judicial branch), following the economic and social tenets of the Weimar Constitution, it nevertheless restricted the autonomy of individual provinces. Moreover, it divided the state into 33 regional administrative units (the Slovenian part now consisted of the Ljubljana and Maribor units), and created Serbian majorities in the ethnically mixed areas of Croatia and Bosnia. The non-Serbian population of the western half of the state was also troubled by the monarch's (excessively) broad powers and the establishment of a Yugoslav nation. Furthermore, the anti-centralist political public was upset by the decision to require a simple majority (over 50 percent) for the adoption of the constitution instead of the qualified (two-thirds) majority according to the Corfu Declaration, which had been reflected in the previous opinions voiced by the representatives of the National Council of Slovenes, Croats and Serbs.

This change triggered loud protests among the Slovenes and Croats, who had feared from the outset that they might simply be outvoted in crucial Constituent Assembly decisions. The representatives of the SLS, the fourth largest party in the state, did not even venture to Belgrade, since the new constitution also stipulated an obligatory pledge of allegiance to the king. The vote went according to the

---

492  The name derives from the religious holiday of Vidovdan (St. Vitus Day), observed by Orthodox Christians on June 28. In Serbia, this is also a date of major historical importance (transl. note).

Radicals' and Democrats' instructions and, in view of the power relationship in the Parliament, was merely a matter of protocol. The third largest Communist Party did not take part in the voting either, because a special government ordinance had proscribed it. In the end, only 258 of 419 representatives participated, with 223 (53 percent) voting in favor of the constitution.

Such developments made the Slovenes increasingly aware of what it meant to be one of three different "tribes" of the same (Yugoslav) nation, and of the fact that life in a culturally colorful and politically specific state would be a challenge by any measure. They gradually realized that they now lived in a country created from culturally and economically diverse units and in provinces with entirely different legal regimes. For instance, Slovenia, Dalmatia, and to some extent Bosnia were well acquainted with Austrian law; Slavonia and Vojvodina had previously applied Hungarian law; Kosovo, Sandžak, and Macedonia had remained within the framework of Muslim Turkey until 1912. This knowledge, once grasped, turned what were initially very high aspirations into disillusionment. Moreover, the recovery of the southern part of the country soon proved to be sluggish, the aftermath of the war was still strongly felt, and efforts to establish a functional and efficient administrative system were grinding to a halt, while speculators prospered amid human misery.

The Serbian Radical Party found its political orientation at once under the firm leadership of the aged Nikola Pašić, who remained at the head of his party and in the forefront of Serbian politics for more than three decades. He was an elusive politician who achieved practically everything he wanted. Under his direction, the state was unified, the Parliament adopted "his" constitutional concept, and— with the king's backing—he controlled the first seven years of Yugoslav politics. Lloyd George would later refer to Pašić in his memoirs of the peace conference as one of the "craftiest and most tenacious statesmen of Southeastern Europe. [...] The foundation of the Kingdom of Yugoslavia was largely his doing. [...] He took care that this extended realm was an accomplished fact before the peace conference even began."[493]

Indeed, together with Svetozar Pribićević (the first Interior Minister and a Serbian member of the Serbo–Croatian coalition in the former Diet in Zagreb), Pašić set the tone and style of rule for day-to-day administrative and political matters during the kingdom's first year. Although Stojan Protić was the first Yugoslav prime minister, Pašić and Pribićević issued administrative orders and appointed and dismissed local officials. They introduced a specific policy-making

---

493   Dragnich, *Serbs and Croats*, 34–35.

style and were largely responsible for its consequences, which would leave their mark on Yugoslavia until the onset of World War II. When Pašić died in 1926, little had changed. While the Serbs pushed to advance their centralist agenda on all fronts, the Slovenes, limited in their influence, resorted to a politics of compromise. The Croats, the second strongest nation in Yugoslavia, soon realized that the new state deprived them of even the limited independence they had had under Hungarian rule. In Bosnia and Herzegovina, where political affiliation was largely determined by religion, major influence was wielded by the Yugoslav Muslim Organization. The only significant political party which spanned the whole country was the Yugoslav Democratic Party—at least for a short while. Pribićević managed to establish it with members of the Independent Radical Party and the Liberal Party of Serbia, the Democrats from Slovenia, and several smaller groups from Bosnia, Montenegro, and Macedonia. But as early as 1922, the old Slovenian Liberals resigned from the party, which led to its transformation into the Independent Democratic Party a year later.

The leading role in the battle against centralism was largely taken by the Croatian Peasant Party under the leadership of Stjepan Radić. Radić was repeatedly accused of anti-state activism by the authorities and the Serbs considered him not to be a peasant but someone who had been able to identify with Croatian peasantry and galvanize its national consciousness to his political advantage. Unlike most party leaders at the time, he did not favor a federal Yugoslavia, according to his private correspondence and his public advocacy of an independent Croatia. He asked the US President Woodrow Wilson and other heads of state for help in the realization and recognition of a Croatian republic. He also wrote letters and pamphlets to enlist the support of the foreign press.

Although Radić was not taken seriously by the new government, in the spring of 1919, he was already sentenced to one year in prison. When he resumed his anti-government activities after his release, he was imprisoned again, which shows that in Croatia the postwar consensus about the Kingdom of SHS had deteriorated already before the 1920 November election. Radić's party, once small and insignificant, won nearly all the seats for Croatian delegates, decisively defeating the parties that had represented Croatia in the provisional parliament—those parties that had accepted unification under the Karadjordjević Monarchy. Radić therefore interpreted the election results as a mandate to create a separate Croatian state. Soon afterwards, at a mass rally of his supporters, he renamed his party the Croatian Republican Party and announced that it would not take part in the deliberations of the Constituent Assembly. In a letter to King Alexander in early 1921, he complained about his ministers and declared null and void the request of the National Council of Slovenes, Croats and Serbs to establish a union. By trying to rearrange the state, he successfully united the Croats, while the Serbs—divided between the Democrats and the Serbian Radicals—were unable to deal with his demands.

The last serious attempt to cooperate with Belgrade is best illustrated by the short-lived co-governance with the ruling Serbian Radical Party. Since Radić had not achieved his desired results, he returned to the opposition in 1928 and stepped up his criticism of government policy. Infuriated by the outcome, Puniša Račić, a Montenegrin representative of the Radical Party, shot at Radić and his party colleagues in the Parliament in the summer of 1928. In doing so, he not only mortally injured Stjepan Radić and killed two other members of his party but also ended the first period of the Constitutional Monarchy. The shooting in the Parliament was followed by a severe political crisis, a short-term leadership by Anton Korošec, and the dissolution of the Parliament and all political parties by King Alexander at the beginning of 1929. Two years later, the king's new constitution allowed only parties with members across the entire country, adding a new twist to the political situation.

This is also the reason why the Slovenian politics of the period was characterized by the discussion on centralism, federalism, and the so-called "Slovenian question." The latter was particularly urgent and painful after the loss of Carinthia, the Littoral, and a considerable portion of Inner Carniola. Besides, the constant adjustments to new conditions during the 1920s and several scandals in Slovenian politics changed the traditional way of doing politics. The gradual transformation of the economy was accompanied by a considerable lack of transparency, constant bickering between parties, and dissatisfaction with the excessively finicky policy-making of the party elites.

As early as 1919, the leading Slovenian geographer, Anton Melik (alias Anton Loboda), complained in the periodical *Ljubljanski zvon* (The Ljubljana bell) that Slovenia's political situation remained extremely rudimentary because an "average Slovenian individual is still a far cry from achieving political independence," drawing "too little on his own judgment" and too much "on leadership directives." According to Melik, the Slovenian political sphere was characterized by "inadequate political knowledge" and a lack of "strict political education, instruction in basic rights of human existence." He continued, "I can only support the claim that we are a politically mature nation when individual political participation has prevailed on every issue, as have the rights and obligations of man and his relations to the social organizations of a nation and state, and when political and factional governance in Slovenia have withdrawn from their present role."[494]

---

494   Anton Loboda [Anton Melik], "O našem notranjepolitičnem stanju," *Ljubljanski zvon* (1919), 19–21. See also: Vodopivec, "Pogled zgodovinarja," 7.

Those responsible for the conditions described by Melik were, of course, not to be sought in Belgrade but in Ljubljana, where the political arena was still dominated by two traditional political groups. The largest Slovenian party and the winner of most elections, the Catholic Slovenian People's Party (SLS), had been calling for autonomy ever since 1921 and—apart from Korošec's 100-day ministry in 1924—remained in opposition until 1927. Although the party never contested the legitimacy of the Vidovdan Constitution but merely wished to amend it, its resistance was based on anti-liberal and national grounds. Autonomy would guarantee Catholic political authority over Slovenian territory. SLS leaders also opposed those constitutional provisions that prohibited the clergy from active participation in political life, limited the Church's influence on education, and made religious instruction an extracurricular activity.

From the very beginning of the 1920s, the liberal camp was racked by splits and divisions, although its majority—which supported Yugoslavism and centralism—was sincerely convinced that the federalization of the Yugoslav kingdom would usher in its demise. Both young and old members of the Liberal leadership maintained that an autonomous Slovenia would turn into a papal, clerical, or Italo–German province. Pro-autonomists were dismissed as provincial "Austriacants."[495] The two feuding parties found some common ground in their choice of external allies, feeling much more at home in Belgrade than in Zagreb. While the Liberals recognized their natural allies in the Yugoslav Democrats, the Catholic leaders found Radić's views too radical and opted for winning various concessions from negotiations with the Serbian Radicals and the royal court. Between 1918 and 1940, Anton Korošec served twelve times as minister, once as vice president of the Yugoslav government, and once as its prime minister. It is nonetheless clear that neither bourgeois party was able to significantly advance a stronger awareness of democratic citizenship or harmonize the Slovenian political community through its policies.

Still, the Slovenes undoubtedly associated their interwar destiny with Yugoslavia. Under Imperial Austria, national affiliation had never been a constitutive element of Slovenian consciousness, but between 1918 and 1929 it became exactly that. The Slovenes, however, took a slightly different stance toward Belgrade once the dictatorship was introduced and the school authorities attempted to rid Slovenian textbooks of content that was extremely important for building up Slovenian national identity. They were particularly irritated by

---

495   A local term denoting a supporter of the Habsburg Monarchy.

Ivan Cankar, who notably contributed to the fact that many Slovenes thought of Yugoslavia as a dreamland before it had been founded.[496]

Nevertheless, the wave of enthusiasm for Yugoslavia that had swept over Slovenia in the latter half of World War I did not instantly turn into distrust of the new government after its initial breach of faith. Most of the population still believed that finding the right leadership would be enough to correct the situation. A prolonged political apathy ensued only when it became definitively clear that such expectations would not be fulfilled, so the election outcomes in 1923, 1925, and 1927 were predictably similar. Politics had become a circus with many grand words but little effect. And yet the Slovenes considered the new Yugoslavia their own country, irrespective of the fact that they paid by far the highest taxes. In short, despite a profound dissatisfaction with the unified state and its limited political influence, Slovenia considered unification as the least undesirable—if not best liked—solution to its national question.

**Figure 38.** Ljubljana in the 1920s.

---

496  Igor Grdina, "Samopodoba Slovencev v XX. stoletju," in: Jančar and Vodopivec (eds.), *Slovenci*, 201.

Feelings of security and faith in development were substantially strengthened by progress in education and culture. With the foundation of the University of Ljubljana, the Scientific Society for Humanities, a wide-ranging network of new schools, theaters, and galleries, numerous newspapers and publishing houses, the Slovenes enthusiastically restarted their lives. Another undeniable gain was the soaring growth of the non-agricultural economy as Slovenia was integrated into Yugoslavia. The Slovenian territory, previously part of the former empire's underdeveloped southern periphery, became part of the developed West in the new state. In the Kingdom of Serbs, Croats and Slovenes, a new market for industrial consumer products emerged for Slovenian entrepreneurs. It provided an impetus for rapid industrial growth, the development of non-agricultural activities, and expansion of the banking sector (before the 1929 recession, Slovenian banks were among the most solid financial institutions in the country). Rapid modernization amid the agricultural crisis was also one of the main reasons why Slovenian citizens accepted the proclamation of dictatorship in 1929, and the new administrative and political reorganization without significant protest. The new Drava Banovina (administrative unit), when joined with Carniola in 1931, corresponded to Slovenia and became the only one of the nine Yugoslav regions to embrace a single nation. The settlement represented a particularly generous reward for Korošec's participation in the dictatorial government.

**Map 6.** Drava Banovina (1929–1941).

Economic and social life during the first decade of the new state reflected a predominantly rural population (about 66 percent). Even between the wars, industrial development in the Slovenian provinces followed its pre-1914 pattern. Because factories and workshops mainly sprang up along railroad lines,[497] on the eve of World War II the vast majority of workers (90 percent) still lived close to railways or in their immediate hinterland. Nevertheless, creating a new state whose parts were heavily underdeveloped provided an enormous opportunity for the region with basic infrastructure, rising industry, and an elaborated trade network. It was also for this reason that Slovenian workers paid higher taxes in the first decades. When taxation was harmonized in 1928, the average industrial worker was taxed roughly as much as a small farmer. On the other hand, the same person did not earn enough to provide for a family of four, even during the period of greatest prosperity just before the economic crisis. An average family with more children therefore faced severe existential problems, as well as the lack of basic water and sewage facilities, heating, and electricity. They also suffered from chronic shortages of food, clothing, and money. Automobiles, an increasingly popular means of transport, were affordable only to a very small elite (3 percent) of wealthy businessmen, directors, and senior officials.

The Slovenian cultural landscape, on the other hand, was considerably more varied, reflecting the increasing influence of domestic intellectuals who either warned of being overwhelmed by Yugoslavism (Josip Vidmar) or urged the Slovenes to rely above all on their history and only then seek solutions in various ideologies (Edvard Kocbek). Although neither camp had a decisive impact on political thought, their considerations were an important achievement in the development of the Slovenian academic landscape. This was primarily possible because Slovene was introduced as the language of instruction in 1918 and 1919. Moreover, in 1919, the first university was established, and two national theaters and the opera house in Ljubljana became fully operational. In fourteen seasons, the Slovenian National Theater "rose from a provincial to a national theater."[498] The new Orchestral Society (1919) and the new music academy (1920) with two musical periodicals, *Cerkveni glasbenik* (Church Musician) and *Novi akordi* (New Chords), opened the space for theoretical contemplation on

---

497    One line followed the Maribor–Celje–Ljubljana–Trieste route of the Southern Railway and the other line ran toward Upper Carniola, passing through Ljubljana and Kranj to Jesenice.
498    Leon Stefanija, "Glasba in slovenska glasba XX. stoletja," in: Jančar and Vodopivec (eds.), *Slovenci*, 188; Borut Loparnik, "Poličeva doba slovenske opere: ozadja in meje," in: *Zbornik ob jubileju Jožeta Sivca*, Ljubljana: Založba ZRC, 2000, 221.

local and international production, as well as legacies concentrated on works by Satie, Debussy, Mahler, and Schönberg.[499] A decade later, the Slovenian National Gallery opened its first permanent exhibition. National radio was launched in the same year, and its broadcasts increased the availability of contemporary Slovenian instrumental music. Besides professional chamber orchestras, the interwar music scene in Slovenia was especially distinguished by the work of four orchestras: the Orchestra of the Slovenian National Theater, the Military Band of the Drava Division, the Ljubljana Radio Orchestra, and the Ljubljana Philharmonic Orchestra. The latter was established in 1935.

Yet by far the most important gain was the establishment of the University of Ljubljana, which attracted numerous Slovenian professors from Vienna, Prague, Munich, and other Central European university centers. Individual departments within various faculties promoted the formation of a wide range of disciplines, from legal to linguistic sciences, and made major contributions to the creation of Slovenian professional terminology. Still, German remained an important second language of instruction for some time and provided access to most scholarly material. Slovenian scientific practice was also crucially shaped by museums, including the National Museum and the Maribor Regional Museum, and numerous professional libraries.

Furthermore, the presence of the new state also created favorable conditions for the development of the fine arts, including literature. Soon after World War I, remarkable achievements were noted in expressionism (Veno Pilon, Ivan Čargo, France and Tone Kralj), futurism (the poet Anton Podbevšek) and constructivism (personified by the Bauhaus-inspired painter August Černigoj) as "the most characteristic and yet unique avant-garde phenomenon," and in the works of the film director Ferdo Delak and the poet Srečko Kosovel. Slovenian artists of the time began to present themselves as groups or generations of the interwar period: "The Fourth Generation—following the impressionists, *vesnani*,[500] and expressionists—was torn between the lore of the new reality and the approaching color realism."[501] Its representatives pursued their studies at the

---

499 In his summary article, Leon Stefanija explicitly emphasizes that 1,400 tickets were sold for the Orchestral Society's very first concert on December 9, 1919. Over the next 20 years, the society organized no less than 183 concerts involving more than 200 performers. See Stefanija, "Glasba in slovenska glasba XX. stoletja," 187–188.

500 A group of artists and social activists named after their central literary-scientific review *Vesna* (transl. note).

501 Milček Komelj, "Slovenska likovna umetnost v XX. stoletju," in: Jančar and Vodopivec, *Slovenci*, 162–163.

nearby Zagreb Academy, which was renovated in 1921. Its courses directed them away from the influences of the Germanic artistic milieu, toward the creative atmosphere of Paris. This transition is best symbolized by Miha Maleš, a painter who "transformed the secessionist form into Matissean linear fullness and expressive symbolization into a lyrical play."[502] During the 1930s, color realism became universally recognized as a European reaction to the avant-garde form, further strengthening the position of the related impressionists. Abandoning cosmic dramas, painting returned to traditional middle-class motifs such as landscapes, still-lifes, or nudes, introduced by the realist Matej Sternen. Sculpture similarly brushed off literary elements and drew from French culture and the ideals of antiquity (Karel Putrih, Zdenko Kalin, and Frančišek Smerdu), whereas Stane Kregar was heavily influenced by cubism and surrealism. The time immediately prior to World War II also witnessed the rise of social critique in works containing social and rural motifs, e.g., by Nikolaj Pirnat, Ivan Čargo, Tone Kralj, and France Mihelič.

In literature, which otherwise eschewed the strict use of ideal theorems, the poet Srečko Kosovel had the most profound impact on future generations. Regardless of his young age (he died at 22), he convincingly questioned the relationship between tradition and innovation. His poetry offered a precise portrayal of the tension between groundbreaking changes in the world and "inherited notions and heaped-up knowledge," which no longer have the capacity to depict a new, "de-centered" world.[503] In music, similar achievements may be attributed to Marij Kogoj and Slavko Osterc. The architect Jože Plečnik had earned a distinguished place in Central Europe by the 1920s. In the early 1920s, he returned from Vienna and took the chair at the newly established school of architecture. His work was known and respected by experts from Vienna and Prague, including Adolf Loos, Peter Altenberg, and Otto Wagner. Wagner even recommended him as his successor at the Vienna Academy. Plečnik's ultimate achievement was the design for the reconstruction of the Prague Castle in the district of Hradčany.[504]

Part of Slovenia's significant industrial development were the steel works from the Austro-Hungarian period, now enlarged and upgraded to serve the

---

502  Ibid., 162–163.
503  See also: Jože Pogačnik, "Slovenska književnost XX. stoletja," in: Jančar and Vodopivec (eds.), *Slovenci*, 171–185.
504  Quoted from Aleš Vodopivec, "Plečnik in Ravnikar," in: Jančar and Vodopivec (eds.), *Slovenci*, 152–156.

entire Yugoslav market. The same held for its railway, telegraph and telephone networks, and electrification in general.

Given its well-established capital market, its relatively good relations with Czechoslovakia and Austria, its high level of literacy (90 percent), and its satisfactory network of vocational schools, Slovenia was perceived as a highly advanced province whose postwar industrial development was further boosted by the Yugoslav market. As a result, certain industries expanded at an incredible rate, nearly doubling the number of factories. During the 1920s, at least 15 new companies a year began to operate.[505]

The most remarkable boom occurred in the textile industry, Slovenia's third most important economic sector after timber and metal. The most rapid growth was experienced by small companies with up to 250 employees, which could finance their own development. Larger companies were in a somewhat less favorable position because they depended on foreign capital markets, but they were not permitted to seek investment from countries at war with Serbia. Even after this restriction was repealed, strict control was maintained over enhanced cooperation and larger foreign investment. Thus, companies that cooperated with Austrian and Czech partners were inspected more frequently and thoroughly than those that cooperated with investors in France or Great Britain. In the 1930s, the situation took a drastic turn when Germany assumed first place among foreign trade partners, and investment by Czech and Austrian stakeholders also dramatically increased. On the eve of World War II, approximately 70 percent of Slovenian industrial production was Czech- or Austrian-owned.

Development was somewhat different in rural areas, where only large farms could offer a relatively comfortable livelihood, whereas owners of small farms, pushed into debilitating debt, could barely make a living. The crisis that peaked at the end of the decade had been brewing ever since the mid-1920s, when rural incomes on sold products dropped by half.[506]

Since most of the debtors had gone bankrupt, the state was forced to halt debt reimbursements in 1932 to save creditors from having to sell off a vast share of their farm holdings far below the actual price. Additional relief came from the gradual introduction of new and more productive seed varieties, whereas a special agency promoted agricultural production to raise the output of Slovenian farm holdings to the European level.

---

505 Dolenc and Gabrič, *Zgodovina 4*, 100.
506 In 1923, a farmer could purchase 20 pairs of shoes for a cow (500 kg); two years later, the same cow would only buy 10 pairs. Ibid., 101.

Contrary to the menacing announcements of European populist policies, there was an astonishing development of Slovenian culture, higher education, and sports during the last years before World War II. In the late 1930s, Ljubljana acquired additional assets: the National University Library and the already mentioned Scientific Society for the Humanities. The 1930s saw sound introduced to Slovenian cinemas (by the end of the decade, there were more than 60 theaters across the Drava Banovina) and various cultural societies contributed even further to cultural development. Altogether, they involved around 100,000 people who would regularly organize and attend lectures, take part in theater productions, or pursue further education. Most of them—from the Catholic Cultural Union, the Liberal Union, and up to gymnastic societies—were associated with political parties.

**Figure 39.** Men's gymnastics team at the 9th Summer Olympic Games in Amsterdam, 1928. First from the left is Leon Štukelj, the most successful Slovenian Olympic competitor.

Following the Czech example, Slovenian athletes organized themselves into two groups: the liberal Sokol society and the conservative Catholic Orel society (meaning falcon and eagle, respectively). Some of them, particularly gymnasts, achieved outstanding performances in international competitions. For over a decade, athletes surrounding the Olympic gold medalist Leon Štukelj set the pace for the international development of gymnastics and won Yugoslavia seven Olympic medals. Similar developments occurred in football, ice hockey, tennis, and ski jumping,[507] followed by mountaineering and alpine skiing, which grew into one of Slovenia's "national" sports after World War II.

## Beyond the State Borders

Slovenes outside Yugoslav borders faced enormous difficulties coping with assimilatory pressures exerted by the Italian, Austrian, or Hungarian authorities. Particularly distressing was the situation in Italy, where military and civil administrations took action against all forms and manifestations of Slovenian national affiliation. They dissolved Slovenian national councils and appointed royal commissioners to take charge of local administrations. They imprisoned former Austro-Hungarian soldiers and persecuted Slovenian intellectuals, who subsequently fled to Yugoslavia. When an affiliate of the Fascist movement was established in Trieste and Fascism spread across the borderland of Julian March, the life of Slovenes rapidly deteriorated.[508] Alongside the common attributes of anti-socialism and nationalism, the Fascist Party in this ethnically mixed area was also characterized by overt racism embellished with aggressive glorification of the Italian victims on the Isonzo Front. A typical manifestation of Fascism in border areas was promoting the feeling of cultural superiority by prohibiting all expression of Slovenian national culture. The Italians escalated their actions with deliberate attempts to undermine Slovenian economy, which—not unlike

---

507   In Planica, which later became one of the world centers of ski jumping, the first jump over 100 m was achieved in 1936. In 1938, the Drava Banovina had 40 football pitches, 12 athletic tracks, 2 velodromes, 15 swimming pools, 23 ski jumps, 17 ski centers, and 10 tennis courts. See Dolenc and Gabrič, *Zgodovina 4*, 104.

508   Six months after the second Fascist daily had appeared in Trieste, the local branch office of the Fascist Party registered 15,000 members and became the second largest in the country.

the cultural and political life of the Slovenes—was frequently more prosperous than most Italian companies and organizations.

Equally disturbing for the Slovenes was the brutality with which the Italian authorities relentlessly tried to demonstrate the superiority of Italian culture. Due to the presumed danger posed by Yugoslavia, they displaced many Slovenian officials and scholars from the border area, while many others decided to leave voluntarily. Some merely crossed the Italian–Slovenian border, while others emigrated to Argentina and the US.

One of the first public outbursts of violence against Slovenes was the burning down of the Slovenian Cultural Center in Trieste on July 13, 1920, and a series of similar actions in other locations. Next came the prohibition of all political parties except the Fascist Party in 1926 and the persecution of Slovenian clergy, which became even more forceful following Mussolini's recognition of the Vatican City State. Bishop Andrej Karlin (1911–1919) was forced to resign from his diocese in Trieste, and a similar fate awaited his successor Alojzij Fograj (1924–1938) and Frančišek Borgija Sedej, Archbishop of Gorizia (1906–1931).[509]

The Slovenes thus organized themselves into underground resistance movements, such as the Council of Priests of St. Paul, led by the Supreme Council of Christian Organizations, or the youth anti-Fascist organization TIGR (an acronym for Trieste, Istria, Gorizia, and Rijeka). The latter responded to Fascist violence with violence. TIGR members tried to instill courage in the local population and draw the world's attention to Fascist Italy's policy of de-nationalization.

Their methods included attacks on soldiers and leading Fascist figures, burning military storage facilities and Italian schools, gathering information on Italian border posts, and sabotaging railway lines to Austria. Although the Italian authorities finally crushed the movement, TIGR remained the first European anti-Fascist resistance. Between the fall of 1929 and the spring of 1930, more than 60 members of the movement were arrested, some were shot, and others sentenced to long prison terms.

By comparison, Slovenes in the post-plebiscite Austrian Carinthia seemed at first glance to encounter relatively tolerable conditions. Although there were limitations, they retained the right to establish their own associations, publish their own newspapers, and in their joint appearance in the provincial election even won two seats in the Provincial Diet. However, events changed drastically

---

509  Quoted from France M. Dolinar, "Katoliška cerkev na Slovenskem med politiko in versko prakso," in: Jančar and Vodopivec (eds.), *Slovenci*, 100–105.

when Hitler annexed Austria in 1938 and the Carinthian Slovenes were denied the status of a national minority. Their situation became identical to the conditions in the Hungarian Rába region, most notably in the municipality of Zalaegerszeg, where the Hungarian authorities never granted minority status to the Slovenes.

# Dictatorship and the Crisis

Given the pervasive, overheated political climate in the Parliament in the years leading to its dissolution, and the impotence of government cabinets, whose actions or inactions were constantly the subject of critique, it came as no surprise that the king dissolved the Parliament. The Serbs, who headed most of the cabinets, were convinced that no compromise could simultaneously meet the Croatian demands and save Yugoslavia as an integral state. Nevertheless, they respected King Alexander Karadjordjević, who became the nation's leader in the testing days of World War I, and had remained a "sincere Yugoslav."[510]

Yet Alexander would also be described as a "young and impatient" ruler, unable to grasp the seriousness of the alarm bells set off by the 1927 agreement between the perpetual rivals Radić and Pribićević, and incapable of comprehending the gravity of the Croats' threatening call for a state within the state.

His assessment after the Parliament shooting resulted in the abolition of the Vidovdan Constitution, the prohibition of all political parties, and the dissolution of the Parliament. At the same time, he consolidated the existing legislation regarding the "protection of the state" and imposed a dictatorial regime, which he believed would be short. The most important side effects of his decisions were an unchecked Serbian predominance over the limited room for maneuver in quasi-parliamentary politics, and strict police supervision that led to increasing resistance against the government and the royal court.

After a while, the regime also became increasingly unpopular in Slovenia. When the Slovenian People's Party was not part of his cabinet, King Alexander was portrayed as a powerful ruler on the verge of becoming a dictator—even though Korošec participated in several of his newly created governments, and even though the system of centralized government was replaced with banovinas (partially self-governing administrative units with elected councils), which satisfied neither the Croatian nor the Serbian opposition leaders. Equally unsatisfactory was the October Constitution and its new electoral law. While political parties were legalized,

---

510  See Dragnich, *Serbs and Croats*, 63. His devotion to the Yugoslav cause was most clearly demonstrated by the selection of names that he gave to his sons: the firstborn was named after his Serbian grandfather Peter I, his second son Tomislav after the 10th-century King of Croatia, and his third son received the Slovenian name Andrej.

the system was replete with procedural and substantive limitations. Applications to form parties required a specified number of signatures and proof of some strength in a large number of districts. No organization, political or otherwise, was permitted if it was founded on a religious or regional base, or if it opposed national unity, the integrity of the state, or the existing order.[511]

In such a situation, the cabinet's composition remained largely uncontested. Since the Croatian and Serbian oppositions could not agree either on a common program or on a statement about the new political system, they decided to boycott the election. The small Communist Party, operating underground, received instructions from its leadership abroad to call for an armed uprising, but to no avail. The most visible protest against the regime came from students of Belgrade University.

Like the majority of other Yugoslav nations, many Slovenes also perceived the new guided democracy as a continuation of the dictatorship, since not only parliamentary politics but all political gatherings provided merely a limited outlet for dissent and the expression of discontent. On the whole, the Parliament and the newly formed parties were more or less sterile institutions, unable to solve anything. The entire political process therefore moved in two separate directions: whereas the Croatian and Macedonian nationalists subsequently joined the communist underground, middle-class political parties split into two camps. The Radicals and some of the Liberals, including the Slovenian wing, supported the king, whereas most others opposed the growing Yugoslav centralism—though there was occasional fraternization. An excellent example of such double-dealing was the policy of the Slovenian People's Party, whose representatives oscillated between resistance and collaboration. Thus, Korošec could be seen participating in various Belgrade governments, while at the same time also joining the signatories of the Slovenian Declaration.[512] Consequently,

---

511 Ibid., 73.
512 The Slovenian Declaration: "Today, the Slovenian nation is divided and dismembered by four countries: Yugoslavia, Italy, Austria, and Hungary. Its fundamental demand is to be united as a single political community; only in this way will it be possible to preserve its existence and ensure its general development. The predominant part of the Slovenian nation living in Yugoslavia is presented with the task of faithfully pursuing this ideal until its final realization. For these reasons, the Slovenian nation must prevail in its struggle to obtain such an independent position within the Yugoslav state that will unceasingly serve as an attractive force for all the remaining parts of the nation living abroad. To this end, the following is required: (a) national individuality, name, flag, ethnic community, financial independence, political and cultural freedom; (b) radical social legislation that must provide for the protection

he was arrested at the beginning of 1933 along with his party's other highest representatives. His Croatian colleague Vladko Maček was sentenced to three years of imprisonment for the same offence, leaving many to feel that the SLS and other Slovenian parties made Slovenia's predicament a kind of security for their dubious purposes. Korošec, especially, was detested by the Croats as a politician who built his fortunes on the unremitting Croatian–Serbian conflict. In the opinion of Djuro Šurmin, a notable Croatian autonomist and member of Protić's cabinet in 1920, Korošec joined the same cabinet as Minister of Railways to campaign against Croatian "separatists." In Šurmin's words, he would always work to prevent a Croatian–Serbian agreement because then "the Slovenes would be nothing." But as Banac aptly puts it, Korošec "was a man for every cabinet," and as such he seemed indispensable.[513]

This was a constant in SLS politics, and was also known as the politics of the "three K's": i.e., the three unsentimental priests turned extremely successful politicians. After Janez Krek, a member of the Vienna Reichstag, Korošec became the first and only non-Serbian prime minister of royal Yugoslavia. When he died in 1940, SLS leadership was passed on to Franc Kulovec, who, according to Banac, "was not under the protection of truth-loving owls."[514] Kulovec died in the German bombardment of Belgrade in April 1941. The fourth K, Miha Krek, was the first SLS chief who was not a priest. Then again, Krek was not too far from the established political exclusivism either, as was typical of the entire Slovenian political landscape, which the SLS dominated as if it were the only Slovenian party.

Indeed, the SLS was the only party that could strike a balance between centralism and federalism without compromising Slovenian aspirations and provoking an open confrontation with the Radicals. As Banac notes, it was emblematic of the SLS to declare that important issues were not particularly urgent. Obviously, republicanism could only lead the party into the camp of hardline opposition, and the SLS's experimental republican slogans during

---

of vital interests and ensure the harmonious development of all necessary and productive vocations, especially of the agricultural and middle class. To achieve this aim, it is necessary that we, the Slovenes, the Croats and the Serbs, build, on a free agreement and democratic basis, a state composed of equal units: one being Slovenia. A state thus constructed is also called for, or at least not excluded, by the Peasant Democratic Coalition and the Radical Party."

513 Quoted from Ivo Banac, *The National Question in Yugoslavia: Origins, History, Politics*, Ithaca, London: Cornell University Press, 1984, 342.
514 Ibid., 341.

the fall of Austria-Hungary were quietly swept aside because the peoples of Yugoslavia were not yet sufficiently sophisticated for an admittedly loftier form of the state. Korošec, who described himself as a monarchist, often stressed that the Yugoslav nations "are not mature enough for a republic." The same attitude prevailed in reaction to a dispute between centralists and federalists, which was treated as singularly turgid, if not completely meaningless. But even when tactical evasion dictated periodic emphasis on one or another aspect of the SLS politics of autonomy, the principal concerns of its program remained remarkably steady. In fact, the SLS envisaged much more than cultural and economic autonomy, at least in Slovenia. For one thing, the Slovenian cultural and economic communities were meant to erase the ancient and historically established frontiers between the duchies of Carniola, Styria, and Carinthia, or at least those parts of their frontiers that were within Yugoslavia:

> The autonomy of a province or a land consists of the following: that the supreme provincial authorities have immediate, highest, and supreme power of decision and authority in political, economic, educational, and financial questions. An autonomous state also possesses the right to issue regulations on these matters, to decide on these matters according to its own discretion and reasoning, with no master to instruct it.[515]

The SLS wanted to make it clear that autonomy for the Slovenes meant that they "would not be commanded by incompetent and headless Belgrade officials."[516] Moreover, the "particularities of the [Slovenian] tribe that had their roots in history and culture" could only be defended when it was recognized that the "unity of the Serbian–Croatian–Slovenian state demands the entire Slovenian territory be united within its borders."[517] Just as Croatian parties called for a Croatian parliament, the SLS considered a Slovenian provincial parliament and government to be the highest autonomous representative body. "We demand," stated the SLS leadership in its electoral manifesto in October 1920, "the autonomy of Slovenia with an assembly and a provincial government responsible to it—a government that should also manage state administration at provincial level. The provincial assembly should have legislative power, insofar as the latter is not retained by the common parliament."[518]

On the other hand, the SLS's autonomy program was somewhat more tolerable to centralists because it insisted that economic-cultural communities should not

---

515 "Boj za avtonomijo Slovenije," *Slovenec*, September 15, 1920, 1
516 Ibid., 1.
517 "Velika manifestacija Slovenske ljudske stranke," *Slovenec*, October 25, 1920, 2–3.
518 "Slovensko ljudstvo!" *Slovenec*, October 26, 1920, 1.

Dictatorship and the Crisis 433

be confused with federal units based on nationhood, such as those advocated by the Croatian opposition. According to the SLS, the main problem with federalist proposals based on "tribal" affiliation was the impossibility of drawing a clear demarcation line between the Serbs and the Croats. The same objection, of course, did not apply to Slovenia, which would remain a distinct unit whether defined as a compact homeland of Slovenes or as a historical economic-cultural area. This advantage was certainly not lost on SLS leaders when they advocated the superiority of economic-cultural autonomies. It would be misleading, however, to leave the impression that the official SLS course always went unchallenged by the party base or among the wider public. An analysis of the SLS's characteristic strategic and tactical trends best fits Korošec's wing of the party—admittedly the dominant one. Younger SLS members reflected the more radical disposition of the Slovenian masses and favored determined steps in defense of Slovenian autonomy and national individuality. Republicanism and Christian socialism were also persistent hallmarks of the younger Slovenian Catholic intelligentsia, to the extent that these sentiments were often voiced in the SLS trade union section, much to Korošec's displeasure.[519] This does not, however, detract from the overall estimate of the SLS as a party that did not utilize its unique position to mitigate centralist excesses in order to further Slovenian interests.

The only time when the SLS might potentially have been presented with an alternative was when Korošec re-entered Belgrade politics in 1935, after the consolidation of the party's conservative core. It seemed then, at least for a moment, that the main initiative within the National Defense movement would largely be taken by smaller, left-oriented party groupings, students, intelligentsia, and the working-class Christian socialists, especially since the rifts developing throughout the 1920s in the Catholic camp had finally cracked in the 1930s. Many party members, particularly younger ones, were disturbed by Korošec's "policy of small steps" and his "opportunistic stance" according to the principle of "the most important thing is to sit in the government."[520] An increasing number of people warned of inconsistencies between the party's stance in principle— the aforementioned declaration of 1932 was already bad enough—and its actual policy-making. However, the issue did not decisively lead to an ultimate rupture.

---

519   Janko Prunk, *Pot krščanskih socialistov v osvobodilno fronto slovenskega naroda*, Ljubljana: Cankarjeva založba, 1977, 49–88.
520   Bojan Godeša, *Kdor ni z nami, je proti nam. Slovenski izobraženci med okupatorji, Osvobodilno fronto in protirevolucionarnim taborom*, Ljubljana: Cankarjeva založba, 1995, 28.

The critical turning point in the Slovenian Catholic camp came with Pope Pius XI's encyclical *Quadragesimo anno* (1931), and the idea of organizing Catholic Action for the re-Catholicization of Slovenian society. The instigators of this organization, which gradually took the central position among Catholic associations in Slovenia, imagined Slovenian society as a *Standesstaat* founded on Christian principles and Catholic social thought as presented in the papal encyclicals, most notably in *Rerum novarum* (1891), *Ubi arcano Dei consilio* (1922), and *Quadragesimo anno*. The Catholic Action received crucial support from the Bishop of Ljubljana Gregorij Rožman, who was to be renamed "Bishop of Catholic Action."[521] While in principle an apolitical, religious organization promoting a "religious and ideological movement among laity," the Catholic Action in fact supplanted the persecuted Catholic associations after King Alexander dissolved the Parliament. Moreover, communism was gradually elevated to the position of the main enemy, a role that the Catholic scenario had for decades assigned to liberalism. This was not so much a defensive stance, as could be claimed when insisting on anti-communism, but rather an extremely aggressive agenda, which over and above impeding communism aimed to promote the Catholic Action's own views.

It was therefore little wonder that the Catholic Action pointed at a wide variety of individuals as communists, communist disciples, or, at best, Masonic conspirators.[522] The militant posture of Catholicism, most pronounced after the papal encyclical *Divini redemptoris* (1937), which called for the "division of souls," was also evident in the slogan "only a good Catholic can be a true Slovene." The most important task on this front was reserved for Catholic intellectuals, who, according to the advocates of the so-called integral Catholicism, should provide for the restoration of Catholicism.

This ideological bias inevitably led to a division between "true" and "false" Catholics. Those recognized as true Catholics were students and intellectuals associated with the weekly *Straža v viharju* (Sentinel in the Storm) under the editorship of professor Lambert Ehrlich and the young ortho-Catholics who published the militant Catholic magazine *Mi mladi borci* (We Young Fighters) under professor Ernest Tomec. Both groups were established as elite organizations that would educate the sorely needed Catholic intellectual cadres in line with the principles of Catholic integration. Both organizations also made it their

---

521 Ibid., 29.
522 Ibid., 31; see also Peter Vodopivec, "Prostozidarska loža Valentin Vodnik v Ljubljani (1940)," *Kronika*, 1 (1992): 44–50.

solemn mission to "uproot communism" and "Christianize the University." An uncompromising campaign against those who did not share their views or show utmost loyalty to the Catholic Church was, however, slightly more characteristic of *Mladi borci*. *Straža* wrote already in 1935 that the battle against the Bolshevik–Communist Bloc required a "community that is united under the invisible head of Christ." A far-flung campaign against "cultural communism" was launched the next year and reinforced in 1938, when its program recognized that only Catholic totalitarianism could save humanity.

Finally, both organizations best personified the reasoning of those who considered the views of more liberal supporters of modern Christian thought, as expressed in the main Catholic periodical *Dom in svet* (Home and World), to be the most immediate danger to the Catholic camp. Some were regarded by militant Catholics as "false" Catholic intellectuals. This ultimately resulted in a clash of interests on the editorial board and the emergence of a new periodical *Dejanje* (The Action). The new crisis deepened the polarization of the Catholic intellectual sphere into conservative and Christian socialist factions. The former was characterized by the Young Fighters' authoritarian-corporative posture and the latter by members of the student society *Zarja* (Dawn), who joined the essayist and poet Edvard Kocbek and his periodical *Dejanje*. They rejected leftist and rightist totalitarianism as well as religious integralism, and promoted freedom of artistic expression.

A similar divide occurred in the liberal camp, which continued to firmly hold most of Slovenia's intelligentsia together despite its waning political influence and power. Two main streams, divided by their differing views on the national question, emerged in the 1930s. For fear of a strong SLS, a circle of liberals formed around the periodical *Jutro* (Morning) and continued to support centralist policies. Younger liberals, on the other hand, mounted a counter-campaign in the footsteps of their predecessors and co-signatories of the declaration on autonomy; the list of 43 signatories included 20 liberals. Instead of suppressing the movement, the introduction of King Alexander's dictatorship further fanned the flames of the resistance against centralism. In this respect, it is worth considering the two main reasons for a changing attitude toward the Kingdom of Yugoslavia.

The first emerged from a discussion of *Kulturni problem slovenstva* (The Cultural Problem of Slovenian Identity), written by the literary critic Josip Vidmar. The booklet caused a tremendous public stir with its sharp criticism of Yugoslav centralism and its rebuke of individual authors who, like the poet Oton Župančič, claimed that language was not necessarily a manifestation of national identity. The ensuing debate soon became a controversy that caused a crisis at the newspaper *Ljubljanski zvon* and eventually gave rise to a new periodical. After the dispute with the owners of *Ljubljanski zvon*, a few members of the editorial board launched their own periodical *Sodobnost* (Contemporaneity), whose subtitle *Neodvisna slovenska revija* (Independent Slovenian Review) signaled a shift from resisting unitarism toward defending Slovenian autonomy. Two years after its launch (1933) and under the editorship of Josip Vidmar, Stanko Leben, and Ferdo Kozak, the review began publishing articles written by communist authors, lending the movement a more socially critical tone. Later historical interpretation thus considered the pre-war *Sodobnost* to have primarily been a Slovenian leftist democratic academic review. *Sodobnost* was clearly the most faithful indicator of developments within the liberal camp, although not the only one. It was soon followed by *Ljubljanski zvon*, which "entered the leftist political stream" in the mid-1930s.[523] A somewhat different development occurred in connection with the "national democratic" periodical *Slovenija*, which vehemently opposed the pro-Yugoslav regime parties but remained equally distrustful of the communists.

The leftist influence was most strongly felt among students, and had been apparent since the mid-1930s in the Yugoslav Sokol society; the society appeared in 1931 as a state organization and was an ardent supporter of Yugoslav unitarism until five years later when the association's secessionists established an independent Slovenian Sokol. The final blow for the liberal camp came in 1935, when its parliamentary representatives found themselves among the opposition. Their political impotence, which they had previously been able to conceal as a government party, was finally laid bare; by the time that "World War II also reached our soil, the liberal camp had already ceased to exist."[524] But its legacy somehow endured when a section of the Liberal Youth published the gazette *Naša misel* (Our Thought, 1935), continuing the tradition of "constructive Yugoslavism." Others were being drawn closer toward the circle, which formed the National Communist Party in 1937. Here, the initiative of the experienced liberal Adolf Ribnikar is worth mentioning. He broke his ties with *Jutro* and launched *Mariborski Večernik* (The Maribor Evening Paper), "a liberal platform for all who promote progressive thought." The paper frequently discussed the need for Slovenian unity, and toward the end of 1938, on the recommendation of Ribnikar and his colleagues, a program for the social re-education of Slovenian society was published.

---

523 *Kronika*, 1 (1992): 44–50.
524 Vasilij Melik, "Slovenski liberalni tabor in njegovo razpadanje," *Prispevek k zgodovini delavskega gibanja*, nos. 1–2 (1982), quoted from Godeša, *Kdor ni z nami*, 39.

Another individual with similar ambitions was Dr. Dinko Puc, a politician of radically democratic leanings, who published the gazette *Slovenska beseda* (The Slovenian Word) at the beginning of 1937, calling for a united and progressive but non-Bolshevik front. One of his collaborators was the known socialist Dr. Milan Korun, who stressed that the Slovenian national question could only be solved by united Slovenes.

The course of events in the reconstruction of the Marxist camp was almost diametrically opposite. After unusually favorable election results in the early 1920s, Slovenian communists were forced to retreat underground when the Law on the Protection of Public Security and State Order was adopted. They formally remained illicit until the outbreak of World War II but actually returned to the political arena in the 1930s. A decade of utter marginalization had led to a generational shift and with it to a thorough change in the party network organization. With the Popular Front movement and an end to factional conflicts in the latter half of the decade, a numerically small party gradually strengthened its influence over political life in Slovenia. The new stream succeeded in toning down sectarianism and party dogmatism, and further enhanced its power by temporarily avoiding radical class slogans or overzealous campaigning for the democratization of society. Their primary aim was to emphasize the difference between Fascism and parliamentary democracy and, at the same time, to emphasize changes in conceptions of the Yugoslav state. Stressing the right of nations to self-determination, the Communist Party had adapted to the current situation.

Through the newspaper *Ljudska pravica* (People's Right), the communists campaigned for democratic elections and called on all "working Slovenes" to join forces in a united Slovenian front in a fight for "national equality, democracy and a better life." They believed that any group participating in this united front would have to pursue a cause derived exclusively from the Slovenian situation, drawing on solidarity both with and among the Slovenian people. The joint program "must be created especially by delivering speeches, addressing the crowds, and by maintaining direct contacts with them." At the beginning of November 1935, *Ljudska pravica* published a direct demand for establishing "the Slovenian People's Freedom Front of All Democratic Forces." The appeal welcomed the workers, groups associated with the newspapers *Slovenska zemlja* (The Slovenian Land), *Slovenija*, and *Bojevnik* (The Warrior), advocates of freedom and democracy from the SLS, as well as the social democratic and workers' left. These endeavors had *Ljudska pravica* silenced in the early 1936, but also reverberated in the liberal sphere and within the movement of workers and peasants.

In 1937, the circle associated with *Slovenija* established the Slovenian Society, whose activities asserted that a nation was an organic community of people that rested on a common culture, whereas a state was an organizational structure developed to serve the nation. The Slovenian Society soon set up affiliates in all major cities and coordinated lectures on current issues. In 1938, it adamantly called for an agreement between the Yugoslav nations allowing its signatories to pursue autonomous developments. The movement of workers and peasants pursued a similar goal, with its representative body in Ljubljana calling for a prompt unification of all democratic forces in the country and the organization of a concentric government, dissolution of the National Assembly, and free elections. The movement's leadership believed that the newly formed Slovenian progressive front, joined by the democratic forces of the brother Croatian and Serbian nations, would constitute a sufficiently strong rampart against any offensive by international Fascism. Its leaders also reached similar conclusions about the communists, namely that "the national existence" would be saved neither by particular Slovenian regions, nor by political parties, but only by "the united Slovenian nation."

The communist appeal to join the People's Front elicited a far more lukewarm response from the socialists, who had been embroiled in bitter conflict with communists ever since the end of 1935. The reason was twofold: by that time, the socialists had launched a new, large-scale action for a united socialist labor front to function as an independent labor organization, while at the same time socialist representatives thought that the People's or Slovenian Front could never become a genuine socialist association. Consequently, they called on their followers to join other democratic groups in their struggle against Fascism and for democratic freedoms, but avoided forming official alliances with any front.

The Christian socialists also followed their own path. After their 1934 dispute with the SLS leadership, one faction organized the United Workers Association in 1935, whereas the other attempted to establish the Socialistična zveza delovnega ljudstva Jugoslavije (the Socialist Alliance of the Working People of Yugoslavia). Reiterating that it did not intend to be anyone's "second locomotive," the latter faction also struggled to establish an autonomous socialist party and criticized the communists as being incapable of "proper policy-making." Accordingly, at the all-Slovenian demonstrations of the People's Front, sharp criticism was directed against the Živković government in Belgrade, resulting in a deepening rift with the communists. No less interesting, however, was the recurrent self-criticism that the discussion about an autonomous Slovenia was too often left to "writers and poets who think that nations and peoples live on gracefully refined words," constantly forgetting that "the fate of a Slovenian working man is closely tied with the destiny of the working people from all over the world."

By then, the communists had headed in the opposite direction and prioritized the national question. One result of this shift was the formation of a national party organization, the Komunistična partija Slovenije (KPS, Communist Party of Slovenia), which immediately attended to the current political situation in Slovenia. The founding statement pointed to the rise of Fascism and expressed its full commitment to the protection of human rights and freedoms. Also, the communists reiterated that a united and free Slovenia could only be accomplished in alliance with the Yugoslav brother nations in a federal state. The statement ended with a call for action to overcome political quibbling and quarreling, and for a unification of all democratic forces that "hold the fate of the Slovenian nation truly close to their hearts."

That same year witnessed the birth and death of a new, legal Slovenian communist newspaper *Delavski list* (Worker's Paper), whose primary tasks were to work for the causes of the Slovenian working class and the struggle against any form of Fascism. During its controversy with the socialists, the newspaper advocated an alliance of the workers' movement with the democratically oriented Catholic population; one month later, it provided a comprehensive explanation of the difference between the United Workers' Front and the People's Front. Before the year's end, *Delavski list* had been replaced by another newspaper *Glas delavca* (Worker's Voice).

Nevertheless, changes brought about by the generational shift were best personified by Edvard Kardelj and his book *Razvoj slovenskega narodnega vprašanja* (The Development of the Slovenian National Question). Much to the surprise of his contemporaries, Kardelj, the most important leader of the Slovenian communists besides Boris Kidrič, drew up an unprecedentedly clear and consistent program regarding the possible developments of the Slovenian national question. Moreover, both men succeeded in attracting completely new followers with a middle-class and clerical background. Their subsequent rise to notable positions in academic and cultural circles during the 1930s allowed for constant involvement in liberal periodicals (*Sodobnost, Ljubljanski zvon*), publishing houses Hram (meaning temple), and cultural institutions (Slovenska matica) from 1935 onwards. After three years, they took over the editorial boards of *Sodobnost* and *Ljubljanski zvon*.

On the verge of World War II, irrespective of its demonstrated expansion, the KPS once again proved a rather exclusivist organization. Specifically, in the summer of 1940, it dismissed a group of students calling for stronger alliances with other political forces. In a most ironic twist of fate, the KPS banned from its own ranks advocates of views it would start to defend the following year, proving that it was, after all, not so different from other contemporary parties. During this time, thoughts about autonomy were widely pushed aside, even by the reinvigorated, numerically poor (480 members in 1935, 200 in 1937, and 800 in the fall of 1940) but influential leftist elite. Also in the 1930s, the concept of the "cultural nation," which due to its numerical weakness should "activate all spiritual powers to ensure its physical survival," was "surpassed only by left- and right-wing radicals," according to the philosopher France Veber (*Nacionalizem in krščanstvo* (Nationalism and Christianity)).[525]

These two groups were comprised of less orthodox Marxists and militant rightists. The former were associated with the gazette *1551* and originated the motto "Neither Moscow nor Rome, but Ljubljana alone," and with the communists, who pointed to the transient nature of the Yugoslav frame in Kardelj's famous analysis *Razvoj slovenskega narodnega vprašanja*. Militant rightists operated at the other end of the ideological spectrum and were associated with the periodical *Straža v viharju*. The so-called "Storm Guards" aimed at integrating Slovenia into the Central European context—or more precisely, into a pan-Danubian Catholic confederation—and fervently pursued confessional exclusivism. They simply could not imagine a multi-religious Yugoslavia as the ultimate solution. They essentially pursued and improved upon the ideas developed by Ivan Šušteršič during the final months of World War I. Nevertheless, neither of the two camps could surpass the radicalism of Anton Novačan, who had already initiated a campaign for a Slovenian–Croatian–Serbian–Bulgarian republican confederation in the 1920s. Thus, even during the 1930s, the idea of creating an entirely independent nation did not receive any serious consideration. Indeed, the Slovenes had best internalized Radić's conclusion that, when giants clash, little folks should duck under the table.[526]

## The Crisis Continues

The SLS had been most conscious of this fact and had two decisive advantages over the radicals and the democrats: the sagacious Korošec, and the ability to present itself repeatedly as the sole, genuine representative of Slovenian national interests, despite periodic participation in the federal government in Belgrade. This position apparently remained unchanged even after the assassination

---

525 Grdina, "Samopodoba Slovencev v XX. stoletju," 203.
526 Ibid., 203.

of King Alexander, when his first cousin Prince Paul Karadjordjević became regent.

Having received no guidance from King Alexander, Prince Paul was a political illiterate, striving to reach an agreement with the opposition in Slovenia and Croatia that had already collapsed during the reign of King Alexander. The SLS issued a new proclamation, making it clear that its demands for the reorganization of the state should not be interpreted as an attack on the existing Yugoslav state. The party denied any affiliation with Slovenian federalist or Italian Fascist activists, and Prince Paul again found his main opposition partner in the Croats, now led by Radić's successor Dr. Vladko Maček.

King Alexander was seemingly dissatisfied with the results of guided democracy in general, particularly after two radical prime ministers had failed to find a solution to the Croatian problem. Despite the king's firm belief that federalism would lead to constant jurisdictional conflicts between the National Parliament and the Provincial Parliament of Croatia, he reconsidered a meeting with Maček upon his return from France in October 1934. According to his most confidential circles, he was deeply convinced that he would finally negotiate a solution to the Croatian question.

But the actual events took a different turn. During his visit to France, King Alexander was assassinated by an activist allegedly close to Croatian separatists. Despite popular speculations as to whether the dying Alexander still had the time to reportedly utter "safeguard Yugoslavia," his assassination triggered a positive response at home and tended to endow his memory with some sort of heroic political mystique. Yet it did not make the task of ruling any easier for his successor. As someone who had never been prepared for a leading role nor invested any interest in politics, Prince Paul found himself in a delicate position in both domestic and foreign affairs.

Nevertheless, he soon demonstrated a strong determination to bring change to the kingdom. According to Dragnich, he realized the utmost importance of working with political leaders to reach some kind of general arrangement about the pressing Croatian question. Although he was tempted to release Maček from prison, he also feared that this decision would undermine the late king's policies. Within three months of the dissolution of the cabinet, Prince Paul appointed a dissident Radical Bogoljub Jevtić the new prime minister. The initial acts of Jevtić, who declared himself a loyal defender of Alexander's program to safeguard Yugoslavia, appeared promising. In addition to granting amnesty to Maček, he succeeded in lowering interest rates and appropriating money for public works programs, and declared a moratorium on peasant debts. After he somewhat surprisingly resigned due to disagreements that erupted within the cabinet after his party won the election, his critics would call him a man of good will and sincere intentions but with no vision or political plan. Some even claimed that he behaved as if the national question had not existed, because for him there were no Serbs, Croats, or Slovenes—only Yugoslavs.

Jevtić's resignation forced Prince Paul to appoint a cabinet from groups that were neither involved with nor presumably in conflict with the opposing camps, so it consisted of the Radicals, the Slovenian People's Party, and the Yugoslav Muslim Organization. His choice for prime minister was Milan Stojadinović, a young Radical who had joined the previous cabinet without his party's consent. Before the appointment, Prince Paul consulted Maček, who was still willing to cooperate but also demanded a neutral cabinet and immediate election to the Constituent Assembly. The regent and the Serbian Radicals opposed the idea and argued that the constitution could not be changed before King Peter II was of age. In the end, the cabinet was joined by individuals from the leading Serbian, Slovenian (Korošec), and Muslim political groups, as well as three Croats, which signaled some support from Croatia.

The new prime minister decided to form one political organization, the Yugoslav Radical Union (uniting the National Radical Party, the Slovenian People's Party, and the Yugoslav Muslim Organization), to secure a majority in the Parliament. Although promises made while forming the cabinet had yet to be kept, freedom of speech was mostly respected and the press was free to provide full coverage of opposition speeches and pronouncements. Even Maček was able to deliver a belligerent speech in the center of Belgrade.

This relaxation was supposed to signal Stojadinović's intentions to create an atmosphere of confidence among Serbs, Croats, and Slovenes. Although he was in power for more than three years, he deliberately did not approach the national question. As a result, new separatist tendencies in Croatia and Slovenia started to gain momentum. Maček, who had condemned separatist activities several times, actually contributed to the growth of Croatian nationalism by permitting the establishment of Croatian paramilitary organizations. Stojadinović was in no hurry to solve the Croatian question, and neither was Maček. With Europe in turmoil amid rising Fascism, the Croatian leader believed that international tensions would lead to a much more favorable solution to the Croatian question than he could have otherwise anticipated, and he was convinced that direct negotiations with the prince regent would produce greater success.

At this time, he also exposed his dictatorial aspirations. For instance, at political rallies, many of his supporters wore green shirts, suggesting similarities to Mussolini's Black Shirts and Hitler's Brown Shirts. Although he denied authoritarian ambitions, several members of his coalition only remained with him because they feared for their own political future. Stojadinović once again strengthened his position in the next election, but this victory was short-lived. Following a pseudo-crisis in February 1939, Prince Paul replaced him with Dragiša Cvetković, a member of the outgoing cabinet. Although Stojadinović believed that his political demise had been caused by the desire of the prince and his wife Olga to become King and Queen of Yugoslavia, the real reasons were his foreign policy of appeasement with Germany and Italy and his neglect of the national question.

Cvetković, who had changed the strategy and, even more importantly, listened to Prince Paul's advice, finally reached an agreement with Maček in August 1939, only six months after his appointment as prime minister. A variety of complex political factors in Europe also helped him achieve this goal, including Hitler's dismemberment of Czechoslovakia following the Munich Agreement with Chamberlain, and Hitler's exploitation of separatist tendencies in Slovakia, which later became a puppet regime under the Catholic priest Josef Tiso.

Prince Paul could visualize a similar threat to Yugoslavia and was therefore prepared to negotiate directly with Maček, particularly after Maček had proclaimed that the Croats and the Serbs must find a solution to the national question within the boundaries of the state. Most Serbian parties were also convinced that international circumstances dictated a compromise with the Croats. But Maček remained staunch in his demands for the Constituent Assembly to write a new constitution, and the prince continued to oppose constitutional changes until King Peter was of age. Moreover, the methods prescribed for amending the basic law required time-consuming procedures and prolonged debates, which was unwise in light of the political tensions throughout Europe. At last, a legal and constitutional basis for allowing concessions to the Croats without formally altering the constitution was found in Article 116. Under exceptional circumstances such as war, mobilization, disorder or disturbance, or rebellion which endangered the security of the state or the public interest, the king could "temporarily take by decree all extraordinary and necessary measures, independent constitutional and legal provisos, in the entire kingdom or in one of its parts." The only limitation was that such measures were to be subsequently submitted to the Parliament for confirmation.

Before formal talks, Prince Paul had sought to determine the nature and extent of Croatian demands through Maček's emissary, only to find them unacceptable at first. He knew that the Muslim leaders would oppose taking territory from Bosnia as required by the Croats' territorial demands. Besides, Serbian leaders in Bosnia also suggested that any possible plebiscite would probably end up stranding a million more Serbs in a new Croatian territorial unit. On top of that, the military viewed any federal arrangement as considerably weakening its defensive capability. Under these circumstances, an agreement seemed impossible. After Prince Paul rejected the Croatian demands, Maček sent a message to the Italian Foreign Minister Count Ciano that he no longer intended to come to any agreement with Belgrade. He asked for specific assistance with separatist activities aimed at Croatian independence. Consequently, Mussolini approved a memorandum of understanding and also provided a grant of 20 million dinars,[527] but Maček refused it, saying that he was once again involved in negotiations with Belgrade.

Maček subsequently claimed that the Italians had initiated the talks and admitted that the draft agreement did propose that, in the event of war, the Croatian Peasant Party was obliged to proclaim an independent Croatian state and seek immediate assistance from the Italian army. He denied, as claimed in Ciano's report, that he had pledged in the draft to mount a revolution in Croatia. On the other hand, he again used the threat of foreign backing in the hope of exacting concessions from Cvetković, whom the prince regent had retained as prime minister mainly because he found him the most suited to continue talks with Maček. At the beginning of August 1939, in an interview with *The New York Times*, Maček declared that Croatia would secede from Yugoslavia if it did not gain autonomy, even if this led to a civil war. Twenty days after the interview was published, the Cvetković–Maček Agreement was signed. Following Prince Paul's approval, the new cabinet was formed with Maček as vice president and five other members of his party as ministers. Both Houses of Parliament were dissolved, and the crown authorized the cabinet to enact a new electoral law. Pending new elections, the cabinet would rule by decree.

For the Slovenes, the most important part of this process was a royal decree, an annex to the agreement, which declared that the provisions concerning the Banovina of Croatia (including the amalgamation of territories or other boundary alterations) could be extended to other banovinas. This would presumably guarantee the establishment of Slovenian as well as Serbian units.

The most ironic aspect of the reforms was that Maček and his party, who had succeeded in portraying themselves in the West as fighters for democracy, accepted an order calling for the persecution of opponents to the agreement. It is also ironic that Maček, who had tenaciously insisted in 1936 that Yugoslavia have a constitution, should have been decisive in its radical amendment in 1939. Finally, the Croats would use double standards after the agreement was

---

527   About US$ 465,000 at the 1938 exchange rate.

signed. Without denying that Serbs and others had justified demands for a reorganization of the state, they insisted that the reorganization should take place only after the election of a new parliament that would ratify the agreement. They supposedly believed that the establishment of Slovenian and Serbian units would preclude their realization of the anticipated additional demands. Moreover, shortly after signing the agreement, two contradictory documents represented Croatian attitudes. While the leadership expressed complete satisfaction with the implementation of the agreement by Cvetković's and the central government in a circular sent out to party offices, the other document, which circulated simultaneously, contained a completely different message. Labeled as strictly confidential, but without a party seal or Maček's signature, it asserted that by signing the agreement, the Croatian Peasant Party had not given up the idea of an independent Croatia. Rather, the agreement was the first step toward its creation and had achieved two goals: it destroyed the integrity of the state, and forced the government to move away from national unity, thereby destroying the foundations of Yugoslavia.

Since the cabinet continually postponed the election of a new national parliament, the agreement was never constitutionally sanctioned. Reactions of political parties were mixed. Those who had collaborated with Maček in the past felt betrayed, although they did not always show it immediately; others, including the SLS, were divided. The outlawed communists claimed that the agreement was a set of "minor concessions, primarily in favor of the Croatian bourgeoisie," and as such

> did little to satisfy the elementary national and social aspirations of the Croatian peasantry and other masses, let alone to settle the national question in Yugoslavia as a whole. By putting its signature to the agreement, the Croatian Peasant Party leadership finalized its decision to ally itself not with democratic forces and against the reactionary pillars of the anti-popular regime, but with the ruling reactionary establishment and against the people.[528]

The greatest problem, however, was not the communists but the militant Croatian terrorist organization, the Ustaše,[529] led by Ante Pavelić. Although Maček was inclined to seek the solution to the Croatian question within a Yugoslav

---

528 Edvard Kardelj, *Tito and Socialist Revolution of Yugoslavia*, Belgrade: Socialist Thought and Practice, 1980, 30.
529 The Ustaše (engl. insurgents) were an extreme Croatian nationalist organization that sought to create an independent Croatian state. The organization was founded in 1929 and led by Ante Pavelić until its dissolution in 1945.

framework, many of his followers in Croatia were under the increasing influence of the Ustaše, who were initially based in Italy. With its fierce attacks on Maček and the agreement, the organization gained more and more followers, primarily because the agreement package ensured that the provisions regarding the Banovina of Croatia could be extended to other units. To that end, commissions were actually set up to draw up decrees creating Serbian and Slovenian units. One of the main reasons for the failure to act was the dispute over the boundary between Croatian and Serbian regions.

On the other hand, time to settle these issues peacefully was also lacking. The agreement was signed in August, and World War II broke out in September. Poland was soon defeated, whereas Belgium, the Netherlands, and France capitulated in the spring of 1940, leaving Yugoslavia few real alternatives. The quick fall of France was a severe blow. The British were in no position to offer assistance, yet they expected the Yugoslav government and its army to rebuff Nazi pressure to join the Tripartite Pact of Germany, Italy, and Japan. Croats and Slovenes were in favor of signing the pact, while Serbs, including Cvetković, were against it. This opinion was shared by Prime Minister Winston Churchill, who tried to persuade Prince Paul that neutrality was not enough. As the negotiations with Germany dragged into March and Hitler became impatient, Prince Paul and Cvetković, knowing that entering the war meant committing national suicide, finally signed the pact—but only after asking for concessions which previous Balkan signatories could only have wished for. Cvetković received three brief notes signed by the German foreign minister, promising that Germany would respect Yugoslavia's sovereignty "for all time," that the Axis powers would "not request Yugoslavia to allow troops to march or to be transported through Yugoslav territory during the war," and that Italy and Germany would not ask Yugoslavia for any military assistance, leaving open the possibility that Yugoslavia might still find it in its interest to offer help.

The ink had hardly dried on the document when on the night of March 26–27 a military coup overthrew the Cvetković–Maček cabinet. The coup leaders declared young King Peter to be of age and ousted Prince Paul as regent. The new prime minister, General Dušan Simović, declared that the new cabinet would abide by all international agreements signed by Yugoslavia. A week later, Hitler ordered a massive attack that ended the existence of the first Yugoslav state. It is therefore impossible to know whether the Cvetković–Maček Agreement was ever a potential first step toward the establishment of a viable political system.

# "A Nation Torn Apart": World War II in Slovenia

After the coup d'état in Belgrade and mass demonstrations against Yugoslavia's accession to the Tripartite Pact, Hitler decided to destroy Yugoslavia both as a military power and as a state. The swiftness of his reaction undoubtedly indicates that plans for a possible "intervention" in the Western Balkans had already been made prior to the coup. The only change effectively made in view of the unfolding events was the sequence of operations. Before the attack on Yugoslavia, Germany assisted the Italians, whom the Greeks had pushed deep into Albania in the fall of 1940. The troops designated to attack Greece were thus placed on full alert in Bulgaria, from where they later launched their attack on Yugoslavia, with the Bulgarians attacking from the east. Hitler drew the remaining forces from Western Europe. Slovenia suffered a concerted onslaught by German forces from the north and northeast and by the Italian 2nd Army from the west.

The strategic initiative was in the hands of the German supreme command, which predicted a deeper incursion into Slovenia after the armies had advanced further toward Zagreb. The primary aim was to cut off the main body of the Yugoslav 7th Army, but this precaution proved unnecessary, as it took only four days to occupy the entire territory of Slovenia. The sudden collapse of the Yugoslav defense, however, came as a surprise to both the attackers and the attacked. The two occupying forces initially braced themselves for a prolonged and fierce resistance, but as it turned out, the entire operation was completed within eleven days. The Yugoslav army capitulated on April 17, and several hundred thousand soldiers were taken prisoner. Meanwhile, on April 10, the Croatian Ustaše had proclaimed the Nezavisna država Hrvatska or NDH (The Independent State of Croatia) and thus irreversibly sealed the fate of the state that had started disintegrating long before the invasion.

# "A Nation Torn Apart": World War II in Slovenia

**Map 7.** Occupation zones during the World War II in Slovenia (1941–1945).

In Slovenia, occupied by Hungary, Germany, and Italy, subsequent military administrations did not correspond to the areas of the April operation, as the Hungarians did not enter their occupation zone in Prekmurje and Medžimurje until one day before the capitulation of the Yugoslav army. It would therefore be rather erroneous to claim that the Slovenian territory was divided among the occupying states, since the latter held no conferences to negotiate the issue. "Which part of Yugoslavia was to be given to whom was a decision to be made by Hitler alone. And so it was. First on March 27 and ultimately on April 3 and 12, from his temporary headquarters at Mönichkirchen, Wiener Neustadt, Hitler issued the 'Current Directive for the Partition of Yugoslavia.'"[530] Following his decision, Germany obtained the entire territory of Slovenian Styria, the northern part of Upper Carniola, the Meža Valley, the Dravograd area, and the northwestern portion of Prekmurje. Italy was given Ljubljana, together with most of Inner and Lower Carniola. Hungary was allotted a substantial portion of Prekmurje. The Slovenes had lost all hope of falling within one occupation zone as expected by the National Council, founded in April under the direction

---

530 Tone Ferenc, *Okupacijski sistemi na Slovenskem 1941–1945*, Ljubljana: Modrijan, 1997, 7.

of Dr. Marko Natlačen, which until then had been the ban. Its members, representing all Slovenian parties except the communists, rushed off to the Germans in Celje, entreating them to occupy the western part of Slovenia as well.

Figure 40. Hitler in Maribor, 1941.

Slovenia remained partitioned into three provincial administrative units with their seats primarily located outside the Slovenian national territory. The Germans subdivided their zone into two administrative units governed from the headquarters in Maribor (for Lower Styria) and Bled (for the "occupied territories of Carinthia and Carniola"). The Hungarians simply annexed their territorial gains to the counties of Vas (Hun. Vasvármegye) and Zala (Hun. Zalamegye). Only the Italian-held administrative unit retained its seat in Ljubljana. In May, special inter-state agreements were concluded on the partition of Slovenia—except for the agreements between Germany and Hungary and between Hungary and the Independent State of Croatia, following their failure to reach a compromise on the border in Medžimurje. All four forces occupying Yugoslavia nonetheless attempted to formally annex the occupied territories to their respective states. In Slovenia, this plan, which was to be accomplished as early as 1941, was ultimately hampered by the resistance movement. The most successful were the Italians, who had by May 1941 incorporated the Province of Ljubljana, governed by the High Commissioner Emilio Grazioli (and later Giuseppe Lombrassa and General Riccardo Moizo). They also set up an advisory committee composed of fourteen Slovenes. Due to the lack of their own personnel, the Italians retained most Slovenian officials in the administration, placing them under the supervision of Italian "experts" and the direction of Italian Fascists.

The chiefs (Ger. *Gauleiter*) of the German civil administration for Upper Carniola and Styria were directly responsible to Hitler. Their first and foremost task was an expeditious Germanization of the people and the land (together with the expulsion of politically burdened elements and intellectuals and prompt Germanization of the Windisch). Germans gave their occupation zone the same status as Alsace, Lorraine, and Luxembourg. The two provinces were annexed at the same time on the request of the chiefs of both administrative units (Siegfried Überreiter in Maribor and Franz Kutchera in Bled, later succeeded by Friedrich Rainer). However, the question of the citizenship of the population within the occupied territories remained open until its resolution in Berlin on October 14, 1941. The annexation was delayed for at least two reasons: first, the rather slow deportation of "unwanted elements," and, second, the partisan resistance. Rainer proposed the last postponement of annexation after the crushing defeat by the partisans in Upper Carniola, until finally all plans were abandoned. Both provinces continued to be run by civil administration chiefs who transferred their seats to Klagenfurt and Graz. The district administration was thus entirely under German control, while the municipal administration (lacking German clerks) was mostly left in the hands of Slovenian mayors. Even though the annexation was not formally effected, the Germans gradually eliminated all borders between the Slovenian and Austrian provinces, and the Italians maintained a police and customs border coinciding with the former Rapallo border. Hungary annexed Prekmurje without delay.

All occupying states shared a clear determination to assimilate the Slovenes. As Ferenc puts it, this was not a classic Holocaust scenario but a combined method of deportation, assimilation, and destruction, where the Nazis proved by far the most impatient. As early as April, when he uttered his famous sentence "Machen Sie mir dieses Land wieder deutsch,"[531] Hitler presented his civil administration chiefs with a three-point program, which included the mass deportation of Slovenes, the mass settlement of Germans, and the fast and total destruction of those choosing to remain in their homes. Under directives signed that same

---

531 "Make this land German again!" (transl. note).

month in Klagenfurt and Maribor by none other than the State Commissioner for the Consolidation of German Nationhood Heinrich Himmler, "220,000 to 260,000 Slovenes—or every third Slovene inhabiting the German occupation zone—were to be deported over a five-month period between May and October 1941."[532] Even though such mass deportations were almost impossible to carry out in wartime, the Nazis proceeded to deport "elements hostile to Germans— intellectuals as well as influential and very nationally-minded persons, [...] post-1914 immigrants, [...] border populations, [and] those who either refused to register with denationalization organizations or would, for racial, political or hereditary medical reasons, be rejected by them."[533]

The majority of the deported were to be driven to Serbia (south of the Sava River) and those found suitable for Germanization were to be taken to the so-called "Old Reich." Moreover, before the end of 1941, the Germans carried out a unique census. Based on their estimates, they planned to populate their part of the Slovenian territory with 80,000 immigrants, but they ultimately succeeded in settling no more than 15,000. They proved somewhat more efficient in deportation: more than a third of the unwanted were expelled, mostly to Germany (approximately 46,000) and those remaining to Serbia (7,500). About 10,000 people were sent to Croatia, where Hitler initially refused to deport Slovenes on the grounds that they would be too close to their homeland. Areas most affected by the deportations were along the Sava and Sotla rivers, where the Slovenian population was almost completely removed. In absolute numbers, the Nazis deported more people from the occupied Slovenian territory than from any other occupation zone in Europe.[534]

Germanization was therefore marked by two simultaneous processes: the planned and brutal destruction of everything that would raise, maintain, or strengthen Slovenian national awareness and its material basis, and the rapid and systematic introduction and spread of everything that would give the land a German image and change the nationality structure of its population. Therefore, with a view to achieving the earliest possible "Umvölkung," the Germans immediately took to tearing down all signs in Slovene, ruining the Slovenian press (including prayer books), dissolving societies, organizations, associations, and funds, and confiscating Slovenian property. At the same time, German schools and kindergartens were founded, the names of persons and places changed, and

---

532  Ferenc, *Okupacijski sistemi*, 13.
533  Ibid., 13.
534  This conclusion also owes a great deal to Tone Ferenc. See Ferenc, *Okupacijski sistemi*, 14.

a string of denationalization societies and organizations emerged: the Styrian Homeland League (Steierischer Heimatbund), the Carinthian People's League (Kärntner Volksbund), which also incorporated the previously existing German association known as the Swabian-German Cultural Association (Schwäbisch-Deutscher Kulturbund), German Youth (Deutsche Jugend) in Lower Styria, Hitler Youth (Hitler Jugend) in Upper Carniola and the Meža Valley, paramilitary Wermannschaft formations, and in 1942 even conscription into the German armed forces. Not even the Church was left unaffected. The Germans first expelled the majority of Slovenian clergy and later forbade worship in both Slovene and Latin. A vast amount of ecclesiastical property was confiscated. Public institutions were covered with posters bearing slogans "You're not a Slovene! You're not a Styrian! You're a member of the great German community! You're going to become a full-blooded German!" The situation was slightly more bearable only in Upper Carniola, where instruction in Slovene was allowed in some schools, even though a local branch of the Nationalsozialistische Deutsche Arbeiterpartei (NSDAP, National Socialist German Workers Party) had been established here at a very early stage.

Italian assimilation policies seemed considerably less brutal. Since the population in the Ljubljana Province was allowed to preserve Slovenian institutions (the university, schools, theater, non-political penal judicature, and local administration), the Italians gained a rather favorable response from the intelligentsia. The situation was also made more tolerable with the high commissioner's advisory committee or consulta and the bilingual administration.[535] On the other hand, the Slovenes showed much less enthusiasm for the establishment of Fascist ancillary organizations, such as the Dopolavoro[536] (Section for Rural Housewives and Maidservants) or the Gioventù Italiana del Littorio di Lubiana or GILL (Lictorian Youth of Ljubljana), which were led by the Slovenian branch of the Fascist Party founded in 1941. These conditions swiftly changed as the resistance movement gained strength and the Italian offices drew up plans increasingly reminiscent of German conduct in Styria (massive deportation, Italian colonization, and Italianization of secular and cultural life).

---

535 On January 23, 1943, the Italian occupation authorities even negotiated the admission of the (Slovenian) Academy of Sciences and Arts to the Italian National Council of Academies (Consiglio nazionale delle Accademie). The President of the Academy of Sciences and Arts Milan Vidmar and its Secretary General Fran Ramovš thus attended the General Assembly in Rome.
536 A Fascist institution organizing workers' leisure activities (transl. note).

The situation in Prekmurje was much less perplexing. Once the province was annexed by Hungary, all immigrants from the Littoral (around 600) who had settled there in 1919 were banished from the area. The prevailing language in schools was Hungarian, sometimes complemented with the Prekmurje Slovenian dialect. The Hungarians also proved expeditious in Hungarization, degrading the Slovenes to Wenden from the outset.

Violence caused by the occupation alone was relatively soon followed by brutal retaliation for actions of the resistance movement. However, one should first take heed of a warning originally formulated by Tone Ferenc, who is considered the leading authority on the history of systems of occupation:

> Even though much has been written about the occupation of Slovenia, many are still convinced that the occupier's violence was provoked exclusively by resistance against it. The German forces committed acts of the most ferocious brutality even before the resistance movement had come into existence, when in May 1941 from Lower Styria alone [...] 583 mentally ill and physically disabled persons were transferred to the death camp in Hardheim near Linz and murdered. Systematic mass deportations of Slovenes from German-seized territory also started before the rise of the resistance. There were certainly also other, albeit less brutal, measures aimed at exterminating the Slovenian national entity in spring 1941. The Nazis would have undoubtedly executed some of their brutal measures in an even faster and more effective manner had it not been for the Slovenian national liberation struggle. In suppressing [...] the resistance, the occupiers resorted to all possible forms of violence, partly for preventative reasons and chiefly for retaliatory ones: overcrowding the existent prisons and setting up new ones ([...] with around 35,000 persons), herding people into concentration camps ([...] 20,000 in Italian and more than 10,000 in German camps), shooting hostages (the German occupier killing 194 groups of 2,860 hostages and the Italian 21 groups of 145 hostages), burning down houses and hundreds of entire villages, shooting male villagers and piling their bodies on the fire; the fact that even children were slaughtered is confirmed by the cases in the villages of Lipe in the Čičarija region and the village of Orehovice near Izlake; during deportation, children were torn away from their parents (more than 600 "stolen children"), etc. The scope and, even more so, the brutality of the occupiers' violence marked one of the fundamental characteristics of the occupation.[537]

## The Beginnings of the Resistance

In a way, the Slovenes had already shown readiness to resist during the invasion. As the Italian army advanced closer to Ljubljana, more and more volunteers of all political convictions left Ljubljana for Novo Mesto in the south, where conditions became so chaotic that the military command redirected them

---

537  Ferenc, *Okupacijski sistemi*, 21–22.

toward Zagreb. The first to find their orientation in terms of organization was the Centralni komite komunistične partije Slovenije or CK KPS (Central Committee of the Communist Party of Slovenia), which, sensing the inescapable collapse of the defense, ordered its members to start gathering weapons and equipment discarded by the retreating Yugoslav army. Interestingly, the Yugoslav or "Federal" Central Committee criticized this action, which, as so often afterwards, was due to a severe lack of insight into the specific situation in Slovenia. Nevertheless, nine days after the capitulation, the Liberation Front of the Slovenian Nation (Osvobodilna fronta slovenskega naroda), formerly known as the Anti-Imperialist Front (Protiimperialistična fronta), was organized in Ljubljana.

The founders' meeting, convened by the CK KPS, proclaimed an "all-national struggle": (1) against the dismemberment and enslavement of Slovenia, for the independence and unification of the Slovenian nation (United Slovenia); (2) for the unity and unanimity of the enslaved nations of Yugoslavia and all Balkan nations, for a brotherly community of free and equal nations; (3) against imperialistic war, for peace without annexations and indemnities based on every nation's right to self-determination, including the right to secession and unification with other nations; (4) against the economic destruction of Slovenian livelihoods; (5) against terror and persecution, against the exceptional conditions of the occupation, and against the denationalization and displacement of the Slovenian nation. The meeting was also attended by representatives of the Christian socialist camp, composed of two unrelated groups: the Yugoslav Professional Association (Jugoslovanska strokovna zveza) and a group of Catholic students. Both played a pivotal role in the resistance movement, not least by drawing a considerable portion of the Catholic population and intelligentsia to the movement and thus at least partially maintaining the continuity of the prewar policy. Other founding groups included a group of intellectual Christian socialists (who had otherwise little in common with classic Christian socialism), a democratic faction of the Sokol society, and a group of Slovenian progressive cultural workers.

The Christian socialist intellectuals, also calling themselves *križarji* (the Crusaders), should not be confused with the Crusaders from the Yugoslav Professional Association. One of the central representatives of this group was Edvard Kocbek, who sharply criticized the Ljubljana Diocese for kneeling before militarism, capitalism, and nationalism at the most crucial moment. Apart from that, Kocbek was convinced that "the only salvation for the Slovenes lies in an immediate armed resistance against the enemy." In his opinion, the nation was destined to "bleed in any event; therefore, it would be more honorable and beneficial to bleed consciously rather than passively."[538] France Vodnik advocated similar views, maintaining that the fundamental platform for cultural creation was not Catholicism but nationality, and that the Slovenian people constituted a unified national community, making it their duty to unite all Slovenian creative forces regardless of their worldviews.[539]

The pro-democratic faction of the Sokol society belonged to the old gymnastic tradition, which in Slovenia too was embodied by Masaryk's student, Dr. Miroslav Tyrš.[540] Like most other organizations in pre-war Slovenia, the Sokol society had a rich history of factional controversies. The most important role in directing this organization, as with the Christian socialists, was played by youth, in this case members of the Triglav and Young Triglav student societies. They were joined by "progressive" cultural workers that represented all the important literary authors of the time.

In addition to the aforementioned "founding" groups, the Osvobodilna fronta or OF (Liberation Front) embraced a long list of other organizations after April 26, 1941. It included the Peasant Boys' and Girls' Society, a faction formerly associated with the newspaper *Slovenska zemlja*, a faction of the Slovenian Christian Party, the group Stara Pravda (meaning "old justice"), the former Slovenian People's Party, as well as various groups of active officers, Styrian emigrants, volunteers, and numerous other societies. The OF ultimately comprised some 20 groups with their own representatives.

## The Organizational Structure of the OF

The OF was headed by the Supreme Plenum and its executive body was the Izvršni odbor Osvobodilne fronte or IOOF (Liberation Front Executive Committee), composed of the founding groups' representatives. The IOOF Secretariat, however, also included a representative of the Slovenian Communist Party and representatives of the Christian socialists and the Sokoli. Within one month of the founders' meeting, provincial committees were set up, at first for Styria and then for the Littoral, Upper Carniola, and Carinthia. Lower-ranking area and field OF Committees operated at local levels.

---

538 Quoted from Bojan Godeša, "Temeljne točke Osvobodilne fronte," in: Marjan Drnovšek, France Rozman, and Peter Vodopivec (eds.), *Slovenska kronika XX. stoletja 1941–1995*, Ljubljana: Nova revija, 1996, 22.
539 Vodnik was a poet, essayist, translator, editor of the literary review *Slovenska religiozna lirika* (Slovenian Religious Lyrical Poetry) and author of the collection of poems *Rastoči človek* (The Growing Man).
540 Miroslav Tyrš was the founder of the Czech Sokol society, who significantly influenced the formation of its Slovenian counterpart.

Especially noteworthy are the third and fourth plenary sessions. The third adopted the decision to form the Sloveniski narodnoosvobodilnii svet or SNOO, later SNOS (Slovenian National Liberation Council), authorized three fundamental tasks (gathering volunteers to join the partisans, expanding the OF's organizational network, and fighting against national traitors and undisciplined elements within the OF),[541] and decided to commemorate the establishment of the first Slovenian government on October 29, 1918. Even though all decisions were of landmark importance for future developments, the third task (to fight against national traitors) crucially contributed to the disintegration of widespread support for the OF in early 1942. If in October 1941 the population of Ljubljana had still massively embraced the OF initiative to remain in their homes after 7:00 p.m., this unity rapidly evaporated with the first liquidations of "national traitors."

The November decisions adopted at the fourth plenary session were therefore some of the resistance movement's last unanimous actions, despite the heated discussion about the continuity of the former state and whether postwar Yugoslavia should be reconstructed as it had been before the occupation. The participants finally agreed on a compromise stipulating (under Point 3) that since they understand "the community of the Yugoslav nations as a natural and historic one, the Liberation Front does not recognize the break-up of Yugoslavia. It will make every effort to fight for the good understanding and unity of all Yugoslav nations. At the same time, it will strive toward a union of all the Slavic nations under the leadership of the great Russian nation based on the right of all nations to self-determination."[542] Unfortunately, the final point concerning the

---

541   The task of fighting against national traitors further legitimized the operations of the Security Intelligence Service, which had been established at the second plenary session of the OF (July 28) under the "Decree on the Protection of the Slovenian Nation and its Movement for Liberation and Unification."

542   The remaining eight points (two were adopted on December 21, 1941) read as follows: "1. Merciless action must be carried out against the enemy. 2. This action represents the foundation for the liberation and unification of all Slovenes. 3. […] 4. With the liberation action and the activation of Slovenian masses, the Liberation Front will transform the Slovenian national character. The masses fighting for their national and human rights will create a new form of active Slovenianhood. 5. All groups participating in the Liberation Front have undertaken to be loyal to each other. 6. After the national liberation, the Liberation Front of the Slovenian nation will assume power over the entire Slovenian territory. 7. After the liberation, the Liberation Front will introduce a genuine people's democracy. All questions beyond the scope of national liberation will, after the liberation, be treated by the people and in a genuinely democratic manner. 8. According to the solemn declarations

national army signaled no such compromise. In a report on its first six months of action, the OF stepped up its threat to punish everyone who worked against it.

The Communist Party, on the other hand, had already started organizing its first fighting formations in June, immediately after its conference, which had also been attended by the members of the so-called Military Commission. At this very session, Kardelj announced Germany's attack on the Soviet Union, which also meant the beginning of armed resistance in Slovenia.

The results of the general disunity, which deepened at the turn of 1941–1942, were devastating. The Slovenes, whom Kocbek in 1943 described as the "[people who] never solved the contradiction between captivity and imagination, between poverty and ability, between provincialism and lofty insights, [and as the people who] danced all the dances of the macabre and sinned on all our pilgrimages […]," had apparently become once again "[…] tragically split apart, and unredeemed."[543]

Figure 41. Occupiers executing hostages.

of Winston Churchill, Franklin D. Roosevelt and Joseph Stalin (the Atlantic Declaration), the Slovenian nation will, after the liberation, determine the internal order of United Slovenia and its foreign relations. The Liberation Front will by all means assert and defend this basic right of the Slovenian nation. 9. The National Army in the Slovenian territory will be raised from the national liberation and national defense troops, which all nationally-minded Slovenes are invited to join."

543  Banac, *The National Question*, 340.

Just how right he was became evident from the events that followed the first liquidations, when for the second time in the 20th century the Slovenes were swept by a vehement wave of politicization. National disunity was best demonstrated by the verdict on how radical a stance to take against the occupiers. The right wing felt that the Slovenian territory had no significance whatsoever for the final outcome of the war and, despite the violent Germanization of Styria, opted for overt collaboration. The left wing, headed by the communists, organized the resistance movement. Collaboration was the result of liquidations of the OF's political opponents and, in particular, the communists' use of coercive methods to stir up militancy, understandably leaving the traditional bourgeoisie (which would not concede to any form of armed resistance) feeling seriously threatened. However, the decision to turn to the occupiers (albeit at first predominantly the Italians) for support and to stigmatize the entire movement as communist, even though it mainly embraced Catholics, soon proved regrettably shortsighted.[544]

## The Resistance Movement and Collaboration

Before the political atmosphere in Slovenia became completely polarized, the first partisan formations (the Molnik, Mokrc, Ribnica, and Borovnica companies) emerged around Ljubljana in July 1941. They mainly received support from the OF's rear in Ljubljana, whose military apparatus distributed material supplies and issued directives from the partisan army's high command. The first Styrian units (e.g., the Pohorje, Savinja, Revirje, and Celje companies) were formed soon afterwards, while in Prekmurje, after the execution of the regional resistance leaders, no significant partisan movement developed until 1944. In the Littoral, armed resistance started in the fall of 1941, when the Littoral and Pivka companies were raised with support from Ljubljana.

The first to respond to the resistance movement were the Germans, who by late July had imposed a curfew and instituted a special court, which in the next two days sentenced four captured partisans to death by firing squad. In the military sense, the Italians reacted more decisively a year later, when General Mario Robotti surrounded almost the entire Ljubljana Province with considerable military strength. The offensive lasted from July to November and had a decisive impact on the later developments. Robotti claimed in his order to cease military operations that "in the Italian portion of Slovenia, they are the only masters." He vastly contributed to the further polarization of the Slovenian

---

544  Vodopivec, "Pogled zgodovinarja," 9.

nation, but ultimately failed to destroy the partisan movement. During the fourth phase of the offensive, Kardelj (partly following Tito's advice) negotiated an IOOF proclamation against the collaborationist White Guard.[545] With this proclamation, Kardelj aimed to instill iron discipline and unity in the Slovenian nation, and threatened the White Guard leaders and organizations with severe punishment.

The offensive also led the high command to reorganize the partisan army. Detachment groups were eliminated, and the four brigades and six detachments became directly responsible to the high command. In December 1942, four operation zones were established: Lower Carniola, Inner Carniola, Alpine–Upper Carniola, and the Littoral.

## Collaboration

### The Blue Guard

The first, initially illegal, collaborationist core was established in 1941 in Urh near Dobrunje by the adherents of the Četnik leader Draža Mihailović's representative for Slovenia, Major Novak. It attracted a number of former Yugoslav commissioned and non-commissioned officers, as well as several liberals and conservatives. In the early spring of 1942, the Blue Guard (formerly known as the "Styrian Battalion" or the "Death Legion") moved via Suha Krajina to the vicinity of Novo Mesto, where it was legalized by the Italians. There were incessant disputes between Novak's representatives and the White Guard, part of which maintained an extremely anti-Yugoslav orientation. On the other hand, it is still unclear how much the Slovenian Četniks really knew about the developments in Serbia, which had sealed their fate at the very onslaught of the war—particularly, how much they knew about the meeting between Tito and Mihailović at Struganik in Šumadija, where Tito had allegedly proposed cooperation to Mihailović. There was much talking and less mutual understanding, and, according to Tito, it became clear that Mihailović was reluctant to fight against a German force that had "crushed France, Poland, Czechoslovakia and other countries overnight."[546]

---

545 The Slovenian White Guard borrowed its name from the Tsarist, i.e., anti-revolutionary, forces during the Russian Civil War.
546 From Tito's speech to prisoners and internees who had returned from Germany after the war. See Aleksandar Petković, *Političke borbe za novu Jugoslaviju. Od drugog AVNOJ-a do prvog Ustava*, Belgrade: Jugoslovenska revija, 1988. In the same speech, Tito also stated: "There were many in Serbia back in 1941 who, in their desire to

In a similar vein, Novak could not reach any agreement with the Milizia volontaria anticomunista or MVAC (Anti-Communist Voluntary Militia), since the latter was, in fact, not run from the Italian headquarters at all. After Italy surrendered, the Blue Guard was finally crushed at the village of Grčarice (September 10, 1943) by the partisan army (the Šercer Brigade). Surprisingly enough, before these developments, even Stalin insisted that the partisans and Četniks should "reach an agreement at any price [...] in order to create a joint army under the command of Draža Mihailović."[547] Kardelj recounted that despite the information they had given him about Četnik attacks on the partisans and their collaboration with the Germans, Stalin did not change his mind: "Obviously, as in many other cases, he did not believe our information."[548] In this respect, relations with the Western powers were almost identical to those with the Soviet Union until the end of 1943. As Kardelj reported, Tito decided on a tactical maneuver in which he acted as if he had accepted Stalin's demands without comment, while at the same time conducting his politics according to the requirements of the real situation in Yugoslavia.

---

win their laurel wreaths easily, set out for the woods only to await the end of the war quietly and then return as liberators. These self-proclaimed heroes were led by Draža Mihailović. Comrades, I spoke to him in 1941. I went to see him when we had just launched our territory-wide action against the occupying force in the main part of Western Serbia. At that time, he was walking freely across our territory recruiting people from village to village to join his Četnik detachments. As regards our side at the time, we were not discriminating between the Četniks and ourselves. We were saying, 'Whoever wants to join Četniks, let them, but anyone who wants to join the partisans is welcome in our midst.' Meanwhile, they were saying, 'Whoever chooses to join the Četniks will not have to fight, while all those who want to join the partisans will have to tighten their belts and go to battle [...].' Today, there are certain people who are saying that Draža Mihailović was the first to set out for the woods. Indeed, Draža Mihailović was the first to run to the woods, while we stayed in the cities. In the cities, with the majority of the population and the main part of the German forces. Here, we drew up plans for our future operations. What distinguished us from them was the fact that he was the first to go and hide in the woods, while we set out for the woods with elaborate plans and determination to launch our operations from there. I said to him, 'It is not worth waiting because the enemy will eradicate us.' He responded, 'If we put up resistance, they will exterminate the Serbian people.' To which I replied, 'The Serbian people can only save itself by fighting.' But he did not share my views." See Petković, *Političke borbe*, 162.

547 Edvard Kardelj, *Reminiscences: The Struggle for Recognition and Independence: The New Yugoslavia, 1944–1957*, London: Blond & Briggs in association with Summerfield Press, 1982, 20.
548 Ibid., 20.

## The White Guard

The first proper White Guard outposts were established in the Dolomites (Šentjošt nad Horjulom, June 17, 1942).[549] They were armed by the Italians, who formally considered both the Blue and the White Guards part of the MVAC. However, the White Guard publicly admitted in September 1942 that the anti-communist movement was permeated with ideological hatred. The Slovenian Alliance, formed after a few liberal, conservative, and social democratic leaders had retreated from the National Council and Grazioli's consulta, was caught between a rock and a hard place until Italy surrendered: it was collaborating with the anti-Fascist coalition, whereas its conservative segment was more overtly collaborating with the Italians. For example, General Rupnik, the mayor of Ljubljana at the time, had been doing so since 1942. This rapprochement also owed much to the IOOF's attacks on several priests in Lower Carniola. According to Italian reports, by June 1942 the Slovenes had been mired in a civil war, in which the Italians skillfully (and most often covertly) supported the White Guard. After August, the Slovenian Alliance in particular tried to raise an illegal anti-occupation army, but once they had succumbed to Italian pressure to publicly denounce the OF, there was no turning back. That same fall, the White Guard became actively engaged in anti-partisan actions, whereas the Slovenian Alliance leadership had to yield command of the entire MVAC to the Italians. Kocbek's diary also suggests that this was one of the most decisive moments in the war. His entry for September 11, 1941, reads: "Slovenia is thus facing a new situation. The idyll of our liberated territory is shattered. The time has come for a fight to the finish. People have started to take sides. The civil war is near. The dance macabre shall rave across Slovenia, of which our descendants will talk with a peculiar shudder."

Even though the Slovenian Alliance formally sought to maintain relations with the anti-Nazi coalition forces and accordingly opted for a "common national and political program"[550] of all Slovenes in the occupied territories, all its formations

---

549 According to Bojan Godeša, "as early as February 1942, the representatives of the prewar parties [...] negotiated how they would persuade the Italian occupation authorities not to intervene in their fight against the communist threat." Grazioli, keen on using the Slovenes as his pawns in the fight against the liberation movement, was pleased with this proposal in March but did not approve it until June. See Godeša, "Temeljne točke Osvobodilne fronte," 38.
550 "1. Restore and expand the Kingdom of Yugoslavia. 2. Free Slovenia as an autonomous and equal integral part of Yugoslavia with its corresponding economic and transport territory demarcated according to the principle of nationality.

(from village sentries to the MVAC) were already collaborating directly or indirectly with the Italians. Italy's surrender only triggered further polarization in Slovenia. One side held that the "Slovenian national army" should take to the forests and organize Slovenian guerrilla divisions, while others felt that they would be better off siding with the Germans as a voluntary gendarmerie. The German arrival in the former Italian occupation zone had above all prepared the ground for collaboration, while the competing leaderships of the former Slovenian political parties, still beleaguered by old quarrels, had come out worst. The German General Rösener in fact tricked the newly raised unit of the "Slovenian army"—the Home Guard—into subordination to his command. However, even earlier, between September 9 and September 22, the partisans had crushed most of the White Guard at Turjak Castle.

## The Home Guard

Like the Blue and White Guards, the Home Guard, too, was divided ever since its foundation. One faction attempted to minimize collaboration with the Germans, wait for their worldwide defeat, and in the end deliver the final blow to the partisans. The other faction hoped to destroy its national adversary with German support, accelerate the German retreat at the end of the war, and join the Allied forces as the resistance army. This was the greatest tragedy of the Slovenian Home Guard, which recognized a far greater adversary in the partisans than in the Germans. The first Home Guard battalions were organized no later than

---

3. Internal regulation of the renewed Yugoslavia only in agreement with all its integral parts holding equal rights and obligations at the federal level. 4. The tasks of the common state shall be as follows: the regulation of international relations, the defense of national sovereignty and territorial inviolability, and the determination of general guidelines for the harmonious coexistence of all its integral parts. All other matters shall be subject to the jurisdiction of the respective federal parts. 5. Yugoslavia shall constitute a uniform economic territory, and its economic and social systems must subordinate all self-interested tendencies of individuals to its nations and public welfare. 6. Yugoslavia should unite with Bulgaria in the spirit of consummate equality and consent; in its northern territories, it should establish common borders and ally as closely as possible with Poland and the Czechoslovakian Federation, should a common federation of North and South Slavic countries fail to emerge. 7. The Slovenian members of the Yugoslav government in London and other members of the Slovenian National Committee (from 1941) in emigration should apply themselves to realizing the above points within the framework of the Allied Peace Program, whereas the Slovenes in the Americas should universally support their efforts."

the fall of 1943. In the Littoral, they were raised at the turn of 1943-1944 as the Slovenian Security Assembly. Apart from deepening rifts, however, the period between the fall of 1942 and mid-1944 also yielded some incentives for reconciliation between the opposing camps. The two most important incentives appeared in the first half of 1943 after the OF leadership asked the Bishop of Ljubljana Gregorij Rožman to dissuade his priests from acting against the OF. The bishop initially received letters from the partisans with an invitation to visit the liberated territory, yet the contacts came to an abrupt end once the Diocesan Curate France Glavač demanded that the bishop terminate all agreements with the OF and publicly condemn the resistance movement. The curate's demand was met quickly: in an interview with *Slovenec*, Rožman again condemned the OF as a godless communist organization. Partly at the initiative of the Slovenian representatives of the government in exile in London, the second mediatory effort was coordinated by Lojze Ude, at the time one of the most ardent advocates of Slovenian national unity. The correspondence lasted for several months, but, as in the former case, brought no tangible results.

The intentions of the Slovenian Home Guard leadership and its dubious role were formally clarified almost a year later, in 1944, when the Home Guards pledged to fight "in the common struggle with the German armed forces under the command of the Führer of Great Germany, the SS troops, and the police against the bandits as well as communism and its allies [...]."[551] Until the end of the war, they formed six battalions within the German armed forces.

The last fateful action stemming from this tragic decision was the Home Guard's withdrawal to Carinthia at the rear of the retreating German forces toward the end of the war (May 8, 1945). It continued with their repatriation and concluded with mass killings by the partisan army at Kočevski Rog on the outskirts of the village of Vetrinje, on the southwestern slopes of Pohorje, in the Zasavje region, and in several other places across Slovenia.

However, one of the main reasons for collaborating with the Italian occupier was the OF's loss of its democratic platform with the signing of the so-called Dolomite Declaration. This document forced the Christian socialists and the Sokoli to denounce the establishment of their own political structures and cede the leading role to the Communist Party. The OF was steadily becoming a mere transmission belt for the policies of the Communist Party, thus losing the character of a "pan-Slovenian organization allowing equal access to all Slovenes and truly patriotic groups, irrespective of their political, ideological, traditional,

---

551  From the Home Guard's pledge to Hitler. See Dolenc and Gabrič, *Zgodovina 4*, 153.

and social differences, who had not been marred by treason and collaboration with the occupying forces."[552] The declaration was publicly proclaimed at Kočevski Rog on the OF's second anniversary, and was followed in mid-June 1943 by the restructuring of the partisan army leadership, which, after the appointment of Franc Rozman-Stane and Commissar Boris Kraigher, fell completely under communist influence.

**Figure 42.** The Home Guard and its leader Leon Rupnik, the spring of 1944.

In the meantime, Italy had surrendered. The supreme command of the German military forces, which had long been expecting it, deployed its troops in Italy to occupy the most important strategic points. In the Slovenian territory, the German army took control over the westward sections of the rail lines. Hitler incorporated the Province of Ljubljana into two German operation zones, the Alpine Foreland (Ger. Alpenvorland) and the Adriatic Littoral (Ger. Adriatisches Küstenland), with their respective centers in Bolzano and Trieste. At the end of September 1943, the Germans divided the liberated territory in the central part

---

552   Taken from the first page of the declaration.

of the Littoral, somewhat relieving the cities of Trieste and Gorizia from partisan pressure. For the administration of these newly established occupied provinces, Friedrich Reiner appointed prefects of Italian nationality, while the governance over the Province of Ljubljana was entrusted to General Leon Rupnik. In October, the Germans further occupied strategic areas in Lower and Inner Carniola, where the provincial government was attempting to establish its own municipal administrations. The provincial administration had limited room for maneuver, as the Italian "counselors" were now replaced by their German counterparts and the chief of administration was "assisted" by the police and SS General Erwin Rösener, administrative adviser Dr. Hermann Doujak, and economic adviser Dr. Friedrich Jacklin. A similar scenario unfolded in the Adriatic Littoral, where the Germans imposed mandatory military service, yet another way of trying to pacify this strategically extremely important area as much as possible.

## The Kočevje Assembly

At about the same time, the election to the assembly of representatives of the Slovenian nation and the National Liberation Movement took place within liberated territories and among partisan and OF district organizational units. The Kočevje Assembly was held on October 1–3, 1943, in the town of Kočevje. The two chief spokespeople were again the central figures of the resistance movement and leading Slovenian communists, Boris Kidrič and Edvard Kardelj. Kardelj's report in particular may serve as a political document concerning the OF's assumption of the leading role within the resistance movement as well as the further fate of the Slovenian nation. The Kočevje Assembly concluded by resolving that the OF was the sole legal national authority in the Slovenian territory and that the Narodnoosvobodilna vojska or NVO (National Liberation Army) constituted an integral part of the NOV of Yugoslavia; by publicly denouncing the White and Blue Guards; and by condemning the exiled Yugoslav government in London as the fifth column of the occupiers and General Rupnik as a national traitor. It also reiterated the need for voluntary incorporation of United Slovenia into a federal Yugoslavia and its belief in the victory of the democratic authority under the leadership of the Slovenian Communist Party, combined with a "group of Slovenian patriots, democrats and freedom-loving people." The Assembly elected a 120-member Slovenski narodnoosvobodilni odbor or SNOO (Supreme Plenum of the Liberation Front) with a 40-member delegation to the Anti-Fascist Council of National Liberation, and passed a vote of confidence in the IOOF.

## The First Session of the Slovenian National Liberation Council (SNOS)

The establishment of the "national authority" in Slovenia was further encouraged by the Second Session of the Antifašističko vijeće narodnog oslobođenja Jugoslavije or AVNOJ (Anti-Fascist Council of National Liberation of Yugoslavia), which rejected the royal government in exile at the end of November 1943. The AVNOJ Assembly, after duly appointing a provisional government and banning the king's return to the country at least until the end of the war, promoted Tito to the (previously nonexistent) rank of marshal.

In turn, the British government in 1944 initiated the establishment of a coalition government consisting of Tito and his ministers and a few ministers from the royal government in London, headed by Ivan Šubašić. This initiative was invented to bestow international recognition on Tito's government, which, following the Soviet-aided partisan liberation of Belgrade in October 1944, in any case secured control over the main part of the Yugoslav territory.[553]

These fairly dynamic developments imposed two concrete tasks on the Slovenian resistance movement: to step up political activities in areas where the movement had so far been non-existent or extremely weak (Styria, Upper Carniola, and Prekmurje), and to establish a republican government, the Narodni komite osvoboditve Slovenije or NKOS (National Committee of the Liberation of Slovenia). This was largely owing to the SNOO's or IOOF's conclusion that the AVNOJ decisions provided a platform for the establishment of Slovenia's own national government that would be set up as soon as the circumstances permitted it. In the preparatory period, its function was assumed by the Slovenski narodnoosvobodilni svet (SNOS), which integrated the political and authoritative functions as per the Kočevje Assembly. At the beginning of 1944, the SNOS elected its legislative committee in the town of Črnomelj and passed the Decree on Election to the National Liberation Committee. The same presidency stressed the need to form appropriate administrative departments that would later become commissariats. The SNOS also passed the Declaration on the Rights and Duties of the Slovenian Nation, and the Decree on the Increase in the Number of Members and on Step-by-Step Election to the Slovenian National Liberation Council. Finally, the presidency set up the "Commission for the Investigation of Crimes of the Occupiers and their Collaborators."

---

553   See also: Aleksandar Pavković, *The Fragmentation of Yugoslavia: Nationalism and War in the Balkans*, London: Macmillan, 2000, 42.

The popularization of the SNOS's session held in Črnomelj paved the way for accelerating preparations for the election to Narodnoosvobodilni odbori or NOO (National Liberation Committees), announced at the beginning of March and held between March 25 and April 30, 1944. The Decree on the Election involved passing decrees to establish the following special sections of the SNOS Presidency: for the construction of national authority, internal affairs, secular affairs, economy, finances, reconstruction, public health, information, and propaganda.

Given the circumstances, the turnout within the liberated territory was relatively high, especially in Carniola, where the political influence of the SNOS was strongest. The election also appeared successful in the Littoral, but it did raise some doubts: on the day when the writs were issued, messages started pouring from the province that the election ought to have been postponed due to German incursion into the liberated territory. The situation was much grimmer in Upper Carniola and Styria, where the partisan movement could not make any significant progress until it was clear that the Germans were going to lose the war or, in other words, until one of the most experienced partisan divisions from Inner Carniola finally advanced into Styria.

In a political sense, the election was marked by two central questions regarding the postwar Yugoslavia's foundations and the status of the Slovenes in it. Kardelj's letters and speeches always contained numerous statements about the Slovenes' right to establish a state that would not compromise their national interests. Moreover, from Kocbek's records, later published in the book *Pot v Jajce* (Journey to Jajce, 1964), it can be gathered that at the Second Session of the AVNOJ, Tito even promised the Slovenian delegation a national army. In the second half of 1944, these deliberations were for some time overshadowed by the Tito–Šubašić agreement, which also had tangible consequences for the Slovenian resistance movement. British planes, which were landing more and more frequently in Inner and White Carniola, supplied Slovenian troops with substantial military assistance and created an air bridge for transporting the wounded to Italy.

With the war drawing to an end, however, numerous questions concerning postwar life came to the forefront. As important as the accelerated operations of special commissions were the efforts of the partisan-run Scientific Institute of the SNOS Presidency, which had been intensely discussing the future Slovenian borders since its foundation (September 9, 1944). According to the proposals of its members, notably the historian Fran Zwitter, the future united Slovenia was to comprise the Littoral with Trieste, Gorizia, and Udine in the west, whereas the northern border was to run north of Klagenfurt, the Zollfeld, and Völkermarkt.

From a military point of view, the end of 1944 was marked by the growth of the partisan army and German and Home Guard incursions into the liberated pockets (by December even into the whole Slovenian territory). The beginning of 1945 brought about the reorganization of the NOV and partisan detachments of Slovenia, as well as the preparations for the final liberation, which began its concluding phase with the battles for Trieste at the end of April. At about the same time, numerous representatives of those who had collaborated with the occupiers in one way or another only a few weeks before, initially established the so-called National Committee (October 29, 1944).[554] During the final military events (May 3, 1945), they proclaimed in Ljubljana "the first Slovenian Parliament," "the Slovenian government," and the "nation state of Slovenia as an integral part of the democratic and federal Kingdom of Yugoslavia." At the same time, the Home Guard was simply renamed "the Slovenian national army," whereas the partisan detachments were called upon to cease hostilities and the Slovenian nation to reach "general reconciliation."

## The Fate of the Home Guard

The Home Guard did not believe the utopian plan of the National Committee and, despite its flattering new name, continued its retreat at the rear of the German troops toward Austrian Carinthia. Just before (or straight after) the end of the war, they engaged in a fierce clash with the partisans, who were determined to thwart their joint attempt at withdrawal from Slovenia. The main body of the retreating army somehow managed to break through the blockade— only to be sent back to Slovenia a little more than a month later by the British authorities, which were progressively taking control over Austrian Carinthia. While we may never really know the role of the only just elected first Slovenian

---

554 Even though it was formally established on the anniversary of the foundation of the State of Slovenes, Croats and Serbs, the National Committee did not commence operations before mid-December, when the Catholic and liberal parties' representatives reached an agreement on the number of their members in it and drew up the so-called "National Declaration." What particularly distinguished this document from previous national-political programs (e.g., the London Points or the Program of the Slovenian Alliance) was that it left supreme authority in the hands of the National Committee instead of conferring it on the royal government. The founders of the National Committee addressed Petkovič's postwar assumption of power in an extremely systematic, if not over-methodical manner, since besides their meticulous organization of gathering signatures for support, they also deliberated on a new national coat of arms, anthem, passports, and so on.

partisan government in their subsequent tragic fate, the truth remains that the extra-judicial killings of about 12,000 Home Guards caused a long-lasting rift in the Slovenian nation. The general assertion that the spring and summer of 1944 witnessed one of the most tragic episodes in modern national history is further supported by the fact that even 75 years after World War II, the issue continues to polarize the Slovenian political and intellectual public. This became all the more evident after the 2004 elections, when the Slovenes found themselves caught between two different worlds, cultivating two different ways of understanding their past and promoting two different perceptions of national interests and the correct course of future development.

The decision of the Yugoslav and Slovenian communist leaderships to resort to all means possible, including murder, and to destroy any potential opposition proved fatal. Not least because full vengeance for the Home Guards' war crimes had not yet been exacted; the killings of the Home Guards were followed by brutal retaliatory acts committed against their family members, the expulsion of Germans, and the exile of a vast number of Italians. Even simple statistical calculations clearly show that the decline in population during the last year of the war and the first year afterwards was all but equal to that observed in all previous wars. By the first half of the 1950s (with the emigration of the *optants* from the Littoral and the Yugoslav zone of the Free Territory of Trieste), it had probably already exceeded 100,000 people. The killings of the Home Guards returned by the British authorities from Austrian Carinthia may be compared to the killings of collaborators in Western Europe. There, especially in France, extrajudicial settling of scores with the enemy took place before a new government was established, whereas in Slovenia and Yugoslavia summary mass killings and convictions were unleashed by order of the new communist government.[555]

The liberation thus had a bitter aftertaste, even though the annexation of the Littoral had finally unified the vast majority of the Slovenes in their common country. In light of the postwar killings, the end of the war meant liberation primarily for the adherents of the resistance movement, whereas political developments demonstrated that the communists were clearly unwilling to share their exclusive authority (built up since the forcibly imposed Dolomite Declaration) with any other political option.

---

555   Vodopivec, "Pogled zgodovinarja," 12. See also: Boris Mlakar, *Slovensko domobranstvo (1943–1945)*, Ljubljana: Slovenska matica, 2003.

## The Arts and the War

Although several Slovenian artists had initially negotiated some benefits from the Italian occupation authorities, the vast majority renounced their subsidies after Ljubljana became a nest of intellectuals from the German occupation zone and after the first hostages were killed.[556] This decision was also based on the OF's call for cultural silence, which demanded boycotting anything reminiscent of cooperation with the occupying forces and refraining from organizing or participating in cultural events. The response to the imposed boycott was striking. The inhabitants of Ljubljana systematically ignored concerts, theater and cinema performances, Italian newspapers and magazines, and by clearing the streets and markets at agreed hours clearly demonstrated their opposition to the occupation.

After Italy's capitulation, the criteria for cultural silence became even stricter. The OF banned any form of cultural activity and started inviting cultural workers to move to the liberated territory. Some responded while others remained in Ljubljana. Intellectuals who left Ljubljana for the liberated territory "were mostly politically prominent individuals [who were in] one way or another compromised or even threatened by the arrival of the German occupier." In fear of the Germans, many of those who until then had no dealings with the OF joined the partisan troops. On the other hand, there were also many who had been previously associated with the OF but were now reluctant to take "'to the hills,' as the popular expression went […] for joining the partisans."[557] The majority were held back by the fear of reprisals against their families and possessions, being well aware that a priority of General Rupnik's provincial administration was to confiscate the property of the OF collaborators.

A most unique example was the greatest Slovenian poet of the time, Oton Župančič, who responded to the OF's call soon after the beginning of the war, even though he never left Ljubljana. His poem, "Veš, poet, svoj dolg?" (Poet, do you know your due?), signed Incognito,[558] is particularly noteworthy. Its final stanza ("Spring will come again,/Another dawn will break/Then the wolves

---

556 The Italians tried to win the Slovenian intelligentsia's favor by providing them with financial and other resources. For example, the University of Ljubljana, which had opened only three weeks after the occupation, received substantial material support. The Italian government allotted Slovenian students 100 scholarships to study at Italian universities. See Godeša, *Kdor ni z nami*, 81–88.
557 Godeša, *Kdor ni z nami*, 252–253.
558 The poem was first published in September 1941 as "Pojte za menoj" (Sing after me).

will come and have their hunters slain/[...] a tooth for a tooth and a head for a head!") rather unintentionally foretells what ultimately became reality at the end of the war.

Artists and scientists who joined the partisans in the liberated territory eventually set up a theater and an orchestra, formed folklore dance groups, established a scientific institute, edited two collections of literary compositions (1942 and 1945), organized exhibitions, and even convened a congress of Slovenian cultural workers at the end of January 1944. One of the most prominent art exhibitions was organized in Črnomelj, which a year later hosted prominent painters Nikolaj Pirnat, France Mihelič, and Božidar Jakac. Predominantly on display were drawings and graphics in the style of social realism, which later became better known as such in Slovenia as well.

**Figure 43.** Partisan army arriving to Ljubljana, spring of 1945.

In the winter of 1944, the opposite side responded with an anthology printed by the Zimska Pomoč publishing house. The organizers rallied more than a hundred writers under the pretense of a major charity campaign, which they later presented as an effective opposition to the imposed cultural silence. However, one major difference between the literary productions of the resistance and collaboration movements is best seen in two poems by the youngest generation of poets. On the one hand, there was the social realist, almost propagandistic call to resist by the partisan poet Karel Destovnik Kajuh ("Oh, mother, there are countless reasons to live for/but for what I died, I would die/once more!"), and on the other hand, the expressionist melancholy of one of the best Slovenian lyric poets France Balantič, pleading: "Don't veil the light with my eyelids/Let me see tender death rippling far and wide/Let me feel the night falling 'tween the graves as their bride!"

# Slovenia after the Liberation: The "People's Republic" and the Time of Socialism

# The Establishment of the "New Order"

Elections for the Local National Liberation Committees took place as early as May and August 1945, first in the Littoral and finally in fourteen districts around Ljubljana. Their principal aim was to legitimize the Osvobodilna fronta slovenskega naroda or OF (Liberation Front) determinations, the first phase of which ended in mid-July with the OF's First Congress in Ljubljana. As so often before, Kardelj and Kidrič were the most prominent speakers. Kardelj talked about the OF's historical role and its new tasks, as well as about the opponents to the new order. To some extent, he even foreshadowed the Home Guards' fate. Kidrič focused on the OF's attitude toward individual parties and the clergy, and the development that "transformed the OF from a coalition to a united movement." Further political developments were profoundly influenced by the political climate across the entire "Democratic Federative Yugoslavia," which acquired its first "provisional people's assembly" at the third Antifašističko vijeće narodnog oslobođenja Jugoslavije or AVNOJ (Anti-Fascist Council for the National Liberation of Yugoslavia) on August 10, 1944, in Belgrade.

One of the first major laws passed by the new authorities was the Law on Agrarian Reform and Colonization. The document, issued by the Slovenski narodnoosvobodilni svet or SNOS (Slovene National Liberation Committee) Presidency in December, took away most Church land holdings and those of nonfarmers (everything over 3 hectares), liquidated large feudal estates, and expropriated all landed property from German citizens and "enemies of the state." A considerable share of confiscated land was distributed among 10,000-odd "persons with agrarian interests" and colonists (around 10 percent of the confiscated land). The rest of the land was distributed to the state, cooperatives, and private owners. One important chapter of this law was the partial cancellation of peasant debts incurred before 1941. After that, the Law on the Election to the Constituent Assembly of "Democratic Federative Yugoslavia" followed, which unleashed a wave of pre-election demonstrations across Slovenia.

The election to the Constituent Assembly ushered in the so-called "period of administrative socialism," which lasted until 1952 and undoubtedly represented a pivotal period in national history, not least because of the Agrarian Reform, with which the communists tried to win farmers' sympathies. The central slogan of land nationalization was that "the land belonged to those who worked it," whereas the reform, first and foremost, targeted agricultural estates worked by hired farmers. These were principally the large estates owned by banks, the

Church, and proprietors to whom agricultural activities were not the most important source of income. The Church was thus left with up to 10 hectares and up to 30 hectares of landed property in exceptional cases. Expropriated land, principally taken without compensation, and land confiscated on the basis of nationality (Germans) or collaboration with the occupying forces, went into the Agrarian Reform's Land Fund and was distributed according to social and political criteria to peasants who owned a small amount of land or none. This method of distribution "exhibits the political and economic purpose of the Agrarian Reform." After the first distribution of land to small farmers, the major part remained in the hands of the state, which thus became the owner of 54.6 percent of the total land and as much as 86 percent of the Land Fund. Land was also given to Slovenian agricultural cooperatives, mainly composed of former vintners who had cultivated vineyards before World War I, which were often owned by Germans (in southern Styria) and Italians (in the Goriška Brda region). The Agrarian Reform also played a significant role in terms of national liberation, "since a large portion of landed proprietors, especially big landowners, were non-Slovenes, with their ownership rights stemming from feudal relations."[559]

Even before the Law on Agrarian Reform was passed, the federal election to the Constituent Assembly of the Democratic Federative Yugoslavia had been called for late August. It triggered a flurry of intensive pre-election campaigning in Slovenia organized by the national liberation committees; at pre-election gatherings assemblies of voters listened to the "statements about the work done by the National Liberation Committee and drew up programs of future tasks." Although the normal functioning of the opposition was prevented, according to a pre-1989 historical interpretation, it seems that the opposition soon realized that it could not join the elections with its own list of candidates. It therefore chose to follow a route of "abstention, intrigue, and false propaganda at home and abroad, on the grounds that the elections were unfree and irregular, that there was only one list of candidates and thus one ballot box." The state leadership appeared to have "successfully refuted" this statement by setting up the so-called "blind ballot box." Today, there historians agree that the November election was undemocratic, and they no longer ascribe the negative outcome in Prekmurje to "the activities of individual reactionaries."[560]

---

559 Zdenko Čepič, "Zemljo tistemu, ki jo obdeluje," in: Drnovšek and Bajt (eds.), *Slovenska kronika*, 125.
560 Bogo Grafenauer, *Zgodovina Slovencev*, Ljubljana: Cankarjeva založba, 1979, 889–890. In Prekmurje, the majority of voters cast their ballots in the blind ballot box,

The elected deputies met their colleagues from other Yugoslav constituencies for the first time on the symbolic date of November 29 to proclaim Yugoslavia as "a federal national state of republican structure, a community of peoples enjoying equal rights, who, on the basis of the right to self-determination, including the right to secede, have expressed their will to live together in a federated state." On January 30, 1946, the Constituent Assembly adopted the Constitution of the Federativna narodna republika Jugoslavija (FNRJ, Federative People's Republic of Yugoslavia), after which the SNOS Presidency adopted the following new names in February: the Ljudska republika Slovenija (LRS, People's Republic of Slovenia), the Government of the LRS, Regional, District, and the Local People's Committee. The government comprised fourteen members. The presidency, previously occupied by Boris Kidrič (succeeding Josip Vidmar), was handed over to Miha Marinko.

The last SNOS session, held in September 1946, concluded with the presidency dissolving the SNOS and calling an election to the LRS Constituent Assembly for October 27, 1946. The election was won by the OF, with 95 percent of the votes, whose deputies adopted the first constitution of the People's Republic of Slovenia in January 1947. One of its most important articles was certainly Article 2, which emphasized not only the freely expressed people's will to unite in Yugoslavia but also the right to secede.[561]

Previously, on May 21, 1946, a general law on people's committees had been passed, meaning that even before the adoption of the Constitution, the People's Committee was the supreme state administrative body responsible for issues of local importance and also entrusted with settling matters of general interest. The first all-Slovenian election (which included the Littoral) of people's committees was held in November 1947.

---

which prompted the Slovenian Central Committee to replace almost the entire provincial leadership. This was one reason why, even after the war, Prekmurje was considered a volatile backwater where the new authorities often sent experts (teachers, engineers, etc.) who had somehow fallen out of their favor.

561 "Based on the liberation struggle and the joint struggle of all Yugoslav nations, the Slovenian nation has created the People's Republic of Slovenia. Expressing its free will to live together with its brotherly nations in a common federative state, the Slovenian nation has, on the basis of the right to self-determination, including the right to secession from and unification with other nations, and on the basis of the principle of equality, united with other nations of Yugoslavia and their people's republics [...] into a common state called the Federative People's Republic of Yugoslavia."

## Borders

On the whole, the developments in the Littoral (and Carinthia) regions had a profound impact on the political climate in Slovenia. There were numerous protests against new borders held in the form of meetings, rallies, or congresses. These years saw the rise in catch phrases such as "Udine, Trieste, and Gorizia are rightfully ours," which characterized the period until the mid-1950s, particularly 1948, when mass demonstrations were organized in Koper demanding that Yugoslavia annex the Free Territory of Trieste. This concluded the first phase of the process initiated in early May 1945, when the partisan army reached ethnic border areas in western Slovenia that had a considerable Italian population (especially in urban centers). Regardless of their political orientations, all Yugoslav army units were welcomed by the Slovenes as liberators. Leftist Italian laborers who saw the new Yugoslavia as a realization of their political ideals reacted similarly. The population of Trieste, however, was much more reserved, especially when the occupying units began implementing the Slovenian Central Committee's decisions about the need for "immediate purges" of "Fascist" collaborators. In line with these orders, in early May 1945, the so-called National Guard units arrested, executed, or deported several thousand people to Yugoslavia, based on previously drawn up lists. The bodies of the executed (around 1,500), including Fascist collaborators and even several Italian anti-Fascists, were thrown into foibas.[562]

---

562 "Foiba" is a term denoting a deep natural sinkhole characteristic of the Karst region shared by Italy and Slovenia. Its name is also commonly associated with killings of Italians attributed to Yugoslav partisans shortly after World War II (transl. note).

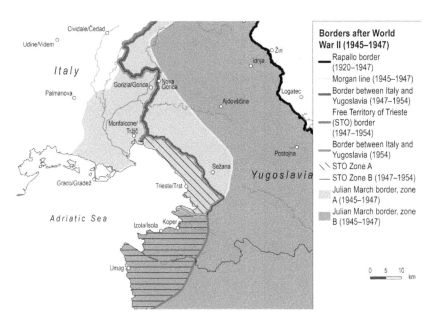

Map 8. Borders after World War II, Free Territory of Trieste, Zone A and Zone B (1945–1947).

After June 1947, the border territory claimed by both Italy and Yugoslavia was divided into zones A and B by the so-called Morgan Line. By the end of that month, the Yugoslav troops were required by the Belgrade and Duino agreements to withdraw from Zone A, comprising the counties of Trieste, Gorizia, and Pula. The provisional administration was taken over by the Allied Military Government. Zone B, encompassing the western part of Julian March, was assigned to the Military Government of the Yugoslav Army. The border was ultimately defined by a peace treaty with Italy (September 1947) which granted Yugoslavia Zone B and part of Zone A but not the Free Territory of Trieste, which was later annexed to Italy. The final demarcation of the western border was only made possible by the memorandum of understanding or the so-called London Memorandum, which gave Yugoslavia the whole of Zone B and part of Zone A. Slovenia thus obtained Koper and its surroundings, its outlet to the Adriatic Sea. Having lost Trieste to Italy, Slovenia immediately undertook to build its own port around the Slovenian largest city, which by the mid-1980s had become the largest commercial port in the northern Adriatic. The political side of the story did not conclude until 1975, when both states signed an agreement in the

Italian town of Osimo (hence the Osimo Agreement) officially terminating the Free Territory of Trieste. The part of the Slovenian population remaining on the Italian side of the border organized its own societies and primary and secondary schools soon after the war, but the problem regarding the use of the Slovenian language was yet to be solved. Since the Italians systematically avoided this issue, it is little wonder that the Law on the Protection of the Slovenian Minority was not passed by the Parliament until 2001. The Italian authorities proved somewhat more expedient in rebuilding the Slovenian cultural center, which burnt down in 1920 and reopened at the end of 1964.

**Map 9.** Socialist Federative Republic of Yugoslavia (1947–1991).

The northern border had not moved an inch. During the war, the Allies had already decided that Austria was to be restored within its pre-war boundaries. Their conclusions were confirmed in 1947 and enacted in 1955 with the Austrian State Treaty Article 7 (in particular) related to the minority rights of Slovenes in Carinthia and Croats in Burgenland. And what was the situation of Carinthian Slovenes after seven years of Nazi rule and five years of war? First and foremost, they were still there despite Nazi attempts to eliminate them as a national group. After Austria's borders were confirmed, the treatment of the minority local authorities (generally backed up by the British occupation authorities) was initially based on a pragmatic accommodation. It increasingly became influenced by policies that prioritized loyalty to the province and assumed assimilation to German culture to be essential to this. The Cold War lent this provincial polarization an international dimension and influenced the prolongation of the border dispute between 1947 and 1949 in two ways. It aggravated tensions between Catholic and communist parts of the minority leadership until they created an unbridgeable fissure, and allowed some of the most extreme forms of loyalism, barely distinguishable from that promoted by the Nazi regime, to regain a political role and to set the agenda for ethnic politics. Yet influence was not dominance; the Cold War did not realign politics around ideology or submerge existing ethnic cleavages; Catholic Slovenes and German nationalists would not unite in their shared dislike of communism, whether Stalinist or Titoist. Meanwhile, the long-term assimilation of the Carinthian Slovenian minority, having reached a low point of brutality under National Socialism, continued under the changed conditions of postwar Austria.[563]

Bilingual signs in villages and towns remain the central subject of disputes between the minority and the provincial administration. The Slovenian grammar school in the provincial capital Klagenfurt was regularly attacked by anti-Slovenian extremists during the 1960s and 1970s. Austria seriously infringed human rights with a minority census in 1976, which the entire Carinthian Slovenian population boycotted. In this event, the Austrian democratic tradition was only preserved by a group of Viennese intellectuals who made sure that greater numbers of people declared themselves as Slovenes in Vienna than in Carinthia. As for local authorities, not even Slovenia's accession to the EU has led to any changes in the conditions of the Slovenian minority.

The Slovenes in Hungary obtained their minority rights soon after the war, including the right to education in their mother tongue, the use of Slovene in courts, and bilingual signs. The main hindrance in preserving their national identity was therefore their restricted contact with their home country. After the dispute with the Cominform, minority representatives had no direct contacts with their relatives in Slovenia for an entire decade.

---

563 Robert Knight, "The Carinthian Slovenes: Ethnic Actors in Bit Part Roles?" (paper presented at a project meeting in Ljubljana in March 2006). The three-year project sponsored by the British Academy focused on Central European minorities policies during the Cold War.

## Settling Scores with the Home Guards

By far the largest spate of killings of captured Home Guards (2,000) or Home Guards returned from Austria (8,000–12,000) took place immediately after the war, in the summer of 1945. On May 27, the British authorities dispatched the first transport carrying 600 persons from the refugee camp at Vetrinje under the pretense of transferring them to northern Italy. The returned Home Guards were allocated to reception camps and prisons, particularly in the suburbs of Ljubljana, and to the former German recruit training center near Celje. Minor prisoners were soon released. After a brief interrogation, the rest were divided into three groups: group A was designated to be released; group B was to be turned over to the military tribunal; and group C was designated for execution. A major part of group C was executed at Kočevski Rog and in a few abandoned coalmines in the Zasavje region, with over 200 killing sites subsequently registered. Something similar might be said of the number of the executed. If, in the mid-1990s, historians stated figures between 7,000 and 8,000, later research by the Institute of Contemporary History showed that the number of executed Home Guards was around 12,000. The question of responsibility remains unanswered; even more than 30 years after Slovenia's independence, it has not been possible to determine accurately who ordered these mass killings. It is only clear that the Allied headquarters at Caserta decided on May 14 that the quisling sections, under their guard after surrender, were to be turned over to the Yugoslav army.[564] Irrespective of who was actually responsible for the greatest single crime in modern Slovenian history, it became evident soon after the war that this event would continue to fatally divide the Slovenian people. Together with the Home Guards, approximately 15,000 civilians emigrated from Slovenia to North and South America and Australia.[565] The majority (more than 5,000) decided to settle in Argentina, where they soon formed a considerable Slovenian community whose representatives preserved the memory of these postwar developments and in the 1980s, with several colleagues in Slovenia, called for an open debate. Moreover, after Slovenia's independence, they initiated and participated in a special parliamentary commission that conducted a careful investigation of the circumstances surrounding the postwar killings and thus contributed to a

---

564    Božo Repe, "Vračanje domobrancev in obračun z njimi," in: Drnovšek, Rozman, and Vodopivec (eds.), *Slovenska kronika XX. stoletja*, 100.
565    According to Marjan Drnovšek, this emigration wave started to subside after 1951, only to be succeeded by another wave of emigrating family members and those who refused to serve in the Yugoslav army.

so-called "national reconciliation." Regrettably, three decades later, there is still no accurate answer about who decided to execute the returned Home Guards. Some see a harbinger of this "solution" in Tito's speech in Ljubljana, in which he stated, among others, that "the avenging hand" had reached "the greater part of the traitors," while the few who nevertheless managed to escape would no longer behold "our beautiful mountains" and "blooming fields."[566]

## The Trials

The second wave of score-settling events came with a series of trials that began at the end of 1945 (the so-called Christmas trials) and ended with the 1948 Cominform show trials. Officially, 2,000 people were convicted altogether. The long-term tragic consequences marred the entire 1950s, since most of those convicted were not released until 1953–1955, and some not until after 1959. While the primary aim of the trials before the "Slovenian National Court of Honor" was to prosecute war criminals, a series of trials was also initiated against "class enemies" and "Church opposition." Most of these trials received considerable publicity, as the proceedings and related developments were often broadcast and regularly commented on in newspapers. A series of explicitly political trials also took place against collaborationist organizers and military leaders. Those of Leon Rupnik, Erwin Rösener, Dr. Miha Krek, and Bishop Gregorij Rožman captured the most attention, despite the absence of some of the accused.

## The Church under Socialism

Some of the most crucial trials for later developments in Slovenia were those of priests and nuns. A considerable number of them (266) were convicted—despite the Slovenian government's principled commitment to the freedom of opinion, conscience and religion, and irrespective of the imposed Declaration of Loyalty, which the representatives of the Diocese of Ljubljana had read before the Prime Minister Boris Kidrič and the Minister of Internal Affairs Zoran Polič.[567]

---

566 Quoted from: Repe, "Vračanje domobrancev in obračun z njimi," 101.
567 The authors of this declaration stated, among other things, that they wished to "extinguish the fire of hatred, revenge, and injustice, which threatens to further divide the unity of our family, village and nation" and expressed their "belief that the government will gladly guarantee the Catholic community the right to religious instruction, church marriage, necessary religious press, education of future priests and property required for the purposes of the Church." Božo Repe, "Škofovska

Nevertheless, the pressure on the Church intensified gradually until it became critical in 1952, when contacts with the Vatican were completely terminated. There were no groundbreaking developments immediately after the war. Voluntary religious instruction was more or less permitted; the Theological Seminary and the Theological Faculty, then still part of the University of Ljubljana, continued educating future priests without interruption, so worshippers at first barely detected the formal separation of the Church and the state. The change came after the publication of an epistle written by Yugoslav bishops under the direction of the Archbishop of Zagreb (later Cardinal) Alojzije Stepinac, who refused to consent to Tito's decision to settle the relationship between the Yugoslav government and the Church without the Vatican's intervention. Consequently, Yugoslavia prohibited the assembly of worshippers, isolated them from foreign contacts, repressed the religious press, and banned religious instruction from schools (February 1, 1952), while the Vatican intensified anti-Yugoslav propaganda. The atmosphere did not change until the end of 1952, when Slovenia witnessed the appearance of the religious periodical *Družina* (Family), the establishment of a theological seminary in Vipava, and a governmental amnesty for more than forty priests. One of the most tragic victims of this period was Bishop Anton Vovk, who was drenched with gas and burnt alive by an assassin in Novo Mesto. Conditions did not significantly improve until the Vatican and Yugoslavia eventually signed a special agreement in 1966. According to Klaus Buchenau,[568] Yugoslavia adopted a rather liberal stance in the 1960s toward its religious communities. After the removal of the Yugoslav Minister of the Interior Aleksandar Ranković, the state interfered less and less with the churches' internal affairs. Federal and republican commissions for religious affairs, closely connected to the Secret Service, limited themselves to the "diplomatic" aspect of religious policy, although the police, Secret Service, and Interior Ministry archives generally remained closed. The documents contained in these archives might show why neither Protestant nor Catholic clergy could feel safe on the streets between 1945 and 1953, particularly if they displayed their dislike of radical secularist policies. As shown in the case of Bishop Vovk, attacks on the clergy were staged as "spontaneous demonstrations" of "popular" wrath. This practice, which was in most instances arranged by the political police, came to an end in 1953, after

---

izjava o lojalnosti in pastirsko pismo," in: Drnovšek, Rozman, and Vodopivec (eds.), *Slovenska kronika XX. stoletja*, 121.
568 Klaus Buchenau, "What Went Wrong? Church–State Relations in Socialist Yugoslavia," *Nationalities Papers* 33, 4 (2005): 547–568.

Tito publicly declared that physical attacks against priests and worshippers must be stopped. Persecution subsequently continued in a different, usually repressive and arbitrary way, especially when it came to the "separation of the Church and the state or the abuse of religion for political purposes."[569] The liberal tendencies of the Communist Party program adopted in 1958 received worldwide attention, but religious communities could still not rejoice. The Party's gradual retreat from administrative compulsion was to be compensated by more intense ideological work in society, which meant another form of pressure. Over and over again, religion was explained as a result of intellectual and material backwardness.

After the Belgrade Protocol between the Yugoslav government and the Vatican and Ranković's dismissal in 1966, the commissions for religious affairs, once closely associated with the UDBA or Uprava državne bezbednosti (State Security Administration), were separated from their main source of information and had to use more civilian forms of gathering knowledge. The "commissioners" had to more carefully read the religious press, which developed rapidly in the liberal climate of the 1960s. After this reform, the churches enjoyed a rather generous freedom of action in the countryside and were not blocked systematically from communicating with their traditional clientele. Under the protocol, the Catholic Church had to ensure that the clergy was not active in political life and Yugoslavia recognized the Vatican's jurisdiction over the Church in spiritual matters. The state granted the Holy See the freedom to appoint bishops and bishops the freedom to establish contacts with the capital of the Roman Catholic Church. In addition, both states vowed that they would sustain their dialogue and announced the establishment of diplomatic relations, which became effective six years later in 1970. The gradual thaw of relations also had concrete consequences for the situation in Slovenia. In 1961, Pope John XXIII had raised the Ljubljana Diocese to an archdiocese; seven years later, Pope Paul VI constituted the Slovenian ecclesiastical province with its metropolitan in Ljubljana and suffragan bishop in Maribor. In the spirit of communication, the state allowed a department of the Ljubljana Theological Faculty to be founded in Maribor (in the 1970s) and even agreed to provide social insurance for priests and members of Church orders.[570] Priests, however, still did not have the opportunity to work as they would have liked in urban areas, among elites, in intellectual discourse, or in politics.[571]

Nonetheless, the interaction between the Vatican and the Yugoslav "third way" did produce dialogue between Christians and Marxists. Although the

---

569   Ibid., 549.
570   Vodopivec, *Od Pohlinove slovnice do samostojne države*, 395–396.
571   Buchenau, "What Went Wrong?," 550–551.

interaction was limited to a small elite of liberal theologians and university professors of a reformist Marxist leaning, its progress was still significant. This became especially clear after the "neutralization" of the so-called Croatian Spring, when the liberal-minded intellectuals, who led the dialogue for the Marxist side, lost their support in the Party. The breakdown had serious consequences for the internal development of the Catholic Church. The Church hierarchy, which had been critical of the "experiments" of the Second Vatican Council, nevertheless enabled the traditionalists to withdraw their support for the modernist theologians engaged in dialogue. The situation in Slovenia was somewhat different. The dialogue remained alive, although the Yugoslav authorities had been able to observe how the growing activity of the Catholic Church, particularly in Croatia, was setting a chain reaction in motion ever since the beginning of the 1960s. Especially in multi-confessional areas such as Bosnia and Herzegovina, Orthodox priests and Muslim imams followed the example of their active Catholic colleagues and intensified their concern about the religious practice of worshippers. Wherever the Catholic Church asserted its presence by organizing mass events, other religious communities tried to promote similar activities. This practice later led to a greater problem because various churches understood the relations between the atheist administration and the Church as a national issue, not just a religious one. If there were no Croats or Slovenes without Catholicism and no Serbs without Orthodoxy, then atheization could not be anything but a danger to the nation.[572] Communication channels established in the 1960s and the late 1970s gained substantial autonomy on both republican and local levels. Moreover, during the 1980s, these channels provided an important base for the alliance between churches and communists-turned-nationalists. Therefore, in the late 1980s and early1990s, "the greatest damage was done by supposed 'anti-communists' and 'dissidents' who were, in fact, nationalists, pragmatic enough to conclude alliances with (ex-)communists when necessary."[573]

## The "Dachau Trials"

The most noteworthy of all trials were undoubtedly the eleven Dachau show trials involving over 30 falsely convicted "war criminals," ten of whom were also sentenced to death in the Diehl–Oswald trial (officially on May 12, 1948). The mock trials took place from April 26, 1948 (the first and the major Dachau trial

---

572   Ibid., 560.
573   Ibid., 561.

known as the "Diehl and Codefendants" trial) to October 11, 1949 (the "Fakin's Group" trial at the District Court of Ljubljana). The only reopened case whose proceedings were conducted with consistency and the indictment was reversed and dismissed, was the trial of Jože Marčan, who was convicted on June 17, 1949.

There was still no call for a final and conclusive reversal of the Dachau trials' convictions until the Tenth Congress of the League of Communists of Slovenia (April 17–19, 1986). At the same time, an initiative was launched for a scientific investigation of the trials' circumstances, course, and consequences. This was also the first decision to politically rehabilitate all the arrested, accused, and convicted. By the end of October 1989, the Zveza komunistov Slovenije (ZKS, League of Communists of Slovenia) and the City of Ljubljana had unveiled a cenotaph in memory of the victims at the central Ljubljana cemetery of Žale. The actual dates of executions, the numbers of executed persons, and the locations of their burial sites remain unknown to this day.

The Dachau trials were classical Stalinist trials distinguished by three important characteristics: the struggle to gain absolute control within the established authority, the automatically ensuing persecution of the members of one's own party or its adherents, and the resultant need for unfounded and false charges. Hardly any individual among the convicted could have (even remotely) posed a threat to the then authority structure. The first characteristic could therefore not hold for this example, despite the trials' embodiment of the "euphoric, paranoid and schizophrenic political climate and the Yugoslav Stalinist milieu."[574] The inexorability toward the indicted most certainly also arose from international tensions (the conflict with the Cominform) and from the zenith of Yugoslav Stalinism and etatism.

Typical of all the Dachau trials was that all the defendants, without exception, had been inmates of the Dachau and Buchenwald concentration camps. They were charged and sentenced for alleged involvement in Gestapo-run or related operations at the aforementioned camps and for "deliberate injurious conduct" after the war as agents of foreign intelligence services.

The entire "project" resulted from the so-called "Pufler case" (an alleged sabotage plot at the glassworks in Hrastnik), that is, the trial of Janko Pufler, Jože Benegalija, Karol Savrič, and Jože Percl, which took place in Celje on May 24, 1947. Several of the accused and indicted in the subsequent major Dachau trial (Karel Barle, Stane Oswald, Boris Kranjc, Branko Diehl) had already

---

574  Branko Ziherl, "Promemoria," in: Martin Ivanič (ed.), *Dahauski procesi. Raziskovalno poročilo z dokumenti*, Ljubljana: Komunist, 1990, 25–32.

been indirectly involved in this first trial. But the Dachau trials began upon the arrest of Barle, a pre-war communist, volunteer in the Spanish Civil War, an inmate of the Graz prisons, and a Dachau internee, who had appeared in Pufler's trial as a witness for the defense. Barle was reproached, based on mere speculation, for collaborating with the Gestapo. Later, the investigators of the Urad državne varnosti (UDV, State Security Office) focused on intensified information-gathering about the operations of experimental stations in German concentration camps and the possible criminal responsibility of all surviving internees who had been employed in them. Thus, before the end of October 1947, almost every single Slovenian chemical engineer and several doctors were arrested on charges of participating in human experiments in concentration camp medical centers or malaria stations. Only then would the investigation turn toward finding evidence of postwar espionage, organized action against the national government, and sabotage, the most severe crimes against the nation.[575]

## The Year of the Cominform

The Dachau trials concurred with a series of "Cominform trials" orchestrated by the UDV. The Cominform trials led to 731 arrests and 334 individuals were sentenced to a maximum two years' imprisonment with possible extension. The majority of Slovenian Cominform agents did not have a pro-Soviet attitude but were "one way or another critical of the government and the situation in both Slovenia and Yugoslavia,"[576] while in certain instances, they may have even taken revenge for old grudges and conflicts. On the other hand, the anti-Cominform campaign in Slovenia could never compare with the ruthlessness of intra-Party score-settling in other parts of Yugoslavia where Stalin's idea of a Belgrade-based Consultative Bureau of the European Communist Parties met with a much wider response. It is interesting to note how the most important Slovenian figure in Yugoslavia at the time, the Minister of Foreign Affairs Edvard Kardelj, perceived the events. He remembered the Yugoslav "definitive no" to Stalin in his *Reminiscences* published in 1982:

> When we review relations between the CPY [Communist Party of Yugoslavia] and CPSU [Communist Party of the Soviet Union] from the First Session of AVNOJ [in 1942] through Zhdanov's intrigues at the meeting in Jelenia Gora to the final break with Stalin, it becomes apparent that Stalin was the whole time mentally planning the

---

575 Ibid., 25–32.
576 Božo Repe, "Informbiro v Sloveniji," in: Drnovšek, Rozman, and Vodopivec (eds.), *Slovenska kronika XX. stoletja*, 174.

end of Tito's independent Yugoslavia. He forgave nothing, not even forgetting those issues over which he himself had backed down. We had clashed a number of times: over Draža Mihailović in 1941; over the Second Session of AVNOJ held without Moscow's knowledge and of which Moscow had been very critical; over the problems with the Soviet experts and joint companies, which were quickly resolved by Stalin's acceptance of the Yugoslav request—though only because he considered it too early to launch his campaign against Yugoslavia. These differences remained a thorn in Stalin's side, and he spent four years preparing for his final onslaught. About the end of March 1948, a special envoy, an officer of the NKVD,[577] arrived from Moscow with a sealed letter for Tito. It was addressed to Tito and myself with the words "To Comrades Tito and Kardelj." The letter, although written in a fairly polite tone, contained a whole catalogue of errors, which they claimed we had made in our domestic and foreign policies, and which were a cause for real concern to the Soviet leadership. At the meeting of the Politburo of the Central Committee of the CPY, we judged the accusations slanderous, and decided to reject them. The only dissenter at the stage in our quarrel with Stalin was Sreten Žujović, who pressed that we admit the errors of which Stalin was accusing us and then went on to demonstrate that we really had been guilty. However, he stood alone, while Hebrang, who was to join him later, remained silent, either because he was waiting to see the outcome or perhaps even because he had been directed to do so by Moscow. All the other members of the Politburo strongly supported Tito and his proposal to reject the letter.[578]

A brief correspondence (three letters from each side) followed in which Stalin extended his former accusations "to include all possible policy areas," called on Tito and Kardelj "to admit our errors and correct them, and added that the CPY would have to elect a new leadership." The Cominform Resolution in Bucharest followed at the end of June, in which all communist parties present joined Stalin's accusation. According to Kardelj, mindful of the Politburo's decision to publish all the letters, "the peoples of Yugoslavia instinctively felt that [Stalin's] letters represented a mortal threat to their independence, to which they remained loyal. Anti-Stalin and anti-Soviet feelings were whipped up, and the CPY received the widest possible level of popular support."[579] Kardelj had completely different expectations of the relations with "brotherly communist parties," although the members of the Yugoslav Politburo hoped that some of them would support them.

> We had counted, for example, on Albania, but she was amongst the first to turn against us. We were counting on Hungary, with whom we had very good relations, but Rakosi

---

577   People's Commissariat for Internal Affairs (Rus. *Narodny Komissariat Vnutrennikh Del*). The NKVD was the Soviet Union's leading secret police organization (transl. note).
578   Edvard Kardelj, *Reminiscences*, 115–116.
579   Ibid., 117.

wrote to us, "Not every cockerel can be king of its own dung heap." Before long, the communist parties of Italy and France, and then of the whole world, united against us. Only a few individual groups, who were aware of the significance of Stalin's campaign and who had broken away from their own official parties, tried to help us. But they were just voices crying in the wilderness, which no one heard [...] The battle shifted rapidly from the political to the economic arena. Frontiers were closed, one after the other all economic agreements with Yugoslavia were cancelled, rail and postal links were cut; on the other side of the border high wire fences and observation posts were built, and the earth around the fences was ploughed up so that the footprints of those who crossed secretly into or out of Yugoslavia might be seen. In short, from the western edge of Hungary to the southern tip of Bulgaria and along the border with Albania, all forms of communication had been cut, as if neither they nor we existed. All that remained were the continual frontier incidents, which left behind the dead and the wounded.[580]

The situation in the West, as seen by Kardelj, was quite different:

> Some thought that this was the end of Tito's Yugoslavia, and that it was absolutely impossible for her to withstand Stalin's pressure, while others—I would say the minority—believed that Yugoslavia could resist and defeat Stalin. It was these countries, which began to give us our first economic and material aid in hastily agreed treaties, etc. There was also a third group who were so anti-communist that they thought the communists were staging the whole affair in order to spread their influence in the West more easily.[581]

Therefore, establishing the first economic contacts "with the West was a slow business":

> At first we pursued a temporary policy, which actually prevented us from speeding up the process. The point was that we did not want to come out immediately 'with all our guns trained' on the Soviet Union, lest we gave them an excuse for military intervention in Yugoslavia. Indeed I remember, at a meeting of the General Assembly of the United Nations in 1948, that I spoke as though nothing had happened in our relations with the Soviet Union, although our quarrel had by then reached its climax.[582]

Here Kardelj refers to the Third Session of the General Assembly of the United Nations held from September to December 1948, from where he wrote an interesting letter to Tito saying that, from his contacts with various left-wing intellectuals,

> [i]t seems that [Soviet] arguments are considered more unconvincing than ever before. Even from those who support the Cominform position, one usually only gets the reply that we were wrong in so far as we allowed a quarrel to break out. But only a few people

---

580   Ibid., 118–119.
581   Ibid., 119.
582   Ibid., 120.

believe in the charges themselves, and it seems certain that, with every day that passes, these will be increasingly deemed unconvincing and false [...].

It is clear to all honest men that the day-to-day facts disprove the claims of the Cominform's Resolution, which are supported only by empty slander. Vilfan has spoken with some American communist intellectuals, who say that they support the Cominform out of a sense of international discipline, and out of fear for the fate of the international anti-imperialist movement should the authority of the Central Committee of the CPSU be undermined—and they think this would happen if we were proved to be right. At the same time, they criticize the line taken by the Cominform against us. Our case, say these people, is crucial in the International Worker's Movement because of the similar situation regarding the Chinese Communist Party and Mao Tse-Tung. They say that the Chinese Communist Party's views, which they call "Maoism," have long been considered a "worrying" development.[583]

**Figure 44.** Josip Broz Tito and Edvard Kardelj at a meeting in Okroglica, Slovenia.

---

583   Ibid. In his letter, Kardelj refers to the Slovenian diplomat Joža Vilfan, who was, among other functions, Head of the Yugoslav Mission to the UN, Yugoslav Ambassador to India, Secretary General to the President of the Republic, and a member of the Permanent Arbitrary Court in The Hague.

# The First Five-Year Plan and Self-Management

Subsequent claims that the 40 years of Yugoslav and Slovenian communism—until its end in the second half of the 1980s—were a period of "sheer terror," when public conduct and everyday behavior were above all determined and directed by the "fear of (omnipotent) authority," turned out to be exaggerated, biased, and historically erroneous. As in all periods and countries, different people in communist Yugoslavia feared the authorities and political moguls in different ways. It is also true that the reasons for this fear were sometimes more obvious and sometimes considerably less so, primarily because of the significant changes in political conditions and climate during particular postwar periods. The communists, however, were not all cowards, despots, and careerists; many were genuinely self-sacrificing enthusiasts who truly believed that they were building a better and fairer world. Following Tito's break with Stalin and Yugoslavia's withdrawal from the Eastern Bloc—which was also largely provoked by the power stand-off between the two communist leaders and parties—the Yugoslav communist leadership looked for an alternative model of socialism. At the turn of the 1940s and the 1950s, they gave the "green light" to Kardelj's "self-management" ideas. The Yugoslav system of self-management was a peculiar, if not utopian, form of "conceptional syncretism" leaning toward a fusion of Marxist, Proudhonist, Blanquist, and other socialist ideas that were often mutually antagonistic and as such continuously created disparities and conflicts.

In Kardelj's opinion, the turning point for Yugoslavia came after the split with the Soviet Union, when the key actors of the Komunistična partija Jugoslavije (KPJ, Communist Party of Yugoslavia) encountered the dilemma of whether the Party was to identify itself on the one hand with the state and:

> [w]hat is more, with the financial resource and through that with the state bureaucracy and administration, or whether it remained on the other hand the advance guard of the working class which, as the dominant force in the country, was itself the dominant force in the state authority. More than ever before it became clear that the Party must guide the working class and not rule it. The self-managing working class must have decisive influence on all questions of state and economic power, and the Party must be the one to inject knowledge, theory and experience into their deliberations, and show how to fight the enemies of socialism and the working class itself.[584]

---

584  Ibid., 123–124.

In Kardelj's words, the Politburo supposedly knew "that this was a long road […], but it was the only possible one" to secure themselves "against the deformations, which" they had met in their "quarrel with Stalin."

According to Kardelj's account, the first decision on how to shape self-management into a system of social organization was made while several members of the Politburo of the Yugoslav Communist Party were visiting Tito during his spring holiday in Split. In Kardelj's words, Tito not only agreed with the proposal "but was of the opinion that self-management offered the only possible route to socialism. On the other hand, he had considered the problem at length and suggested various organizational measures to help put the proposal into practice." On that occasion, the inner core of the Politburo also decided to "draft a law on workers' councils as a basis for the whole system, which would have to develop gradually, since it was obviously impossible to implement such a far-reaching social reform overnight."

The meeting's second important conclusion was the proposal to change the KPJ's name to the Savez komunista Jugoslavije (SKJ, League of Communists of Yugoslavia). After initial reservations, Tito "accepted this proposal too," and the name was changed after the Sixth Congress on the KPJ's approval in November 1952.

In 1950, Tito made his "historic speech" introducing the law on workers' councils, which provided a formal platform for worker self-management until the end of the 1980s. This constituted the introduction of one of the most radical changes in the entire period of the Federativna narodna republika Jugoslavija (FNRJ, Federal People's Republic of Yugoslavia) and Socialistička federativna republika Jugoslavija (SFRJ, Socialist Federal Republic of Yugoslavia) respectively, necessitating the adoption of a new constitution "which, albeit not worked out in detail and still containing parts of the old constitution, did introduce self-management into Yugoslav society and changed the role of the state, of the technocratic administration, and of the working man, not only in factories but in […] all areas of life."[585]

Kardelj's ideas undoubtedly had a profound and lasting impact on both Yugoslav and Slovenian reality after 1948. He was not only "the father of self-management" but also the key author of all postwar constitutions and constitutional laws, including the last ones adopted in 1974 and 1976. In view of the Party's increasing political power and extreme political and economic centralization, these constitutional documents ultimately revealed the unrealistic

---

585   Kardelj, *Reminiscences*, 124–125.

nature of Kardelj's socio-political projects. All the same, many old communists still clung fast to them in the first half of the 1980s, rejecting any prospect of thoroughly reforming the complicated constitutional and political system. They argued that the problems were not in "the constitution and the system" but in the backwardness of "self-management and self-managerial relations." This was the final act of the processes which began in the early 1950s and would have a major impact on the political, economic, and cultural life in the "New" Yugoslavia.

## The Reconstruction of the State and Cultural Institutions

The 1950s were also the period of overall reconstruction. Transport and other public infrastructure had been destroyed or become utterly obsolete. With the State Treasury completely depleted, the state was only able to provide a basic standard of welfare (paid leave and basic health insurance), despite its ideologically driven endeavors to achieve social equilibrium. Maternity leave was short and childcare was provided by public kindergartens. Women, now officially equal to men, paid for their emancipation with a double working week; housework was now coupled with regular employment (in three shifts). Other than that, there was also a shortage of clothes, housing, equipment, food. Only work was sufficiently available. Besides regular work, there was a high demand for a workforce in the so-called work actions, in which young people (high school and university students) in particular had to take part. In the beginning, these actions were organized to "eliminate the consequences of [the] occupation."[586] After 1946, the so-called federal work brigades started to participate in actions all over Slovenia and Yugoslavia. The first postwar local actions were directed at rebuilding and renewing the existing buildings and later nationwide actions were mainly organized to build a brand-new infrastructure, which included roads, railways, electricity, and water. Subsequent estimates show that in the first postwar years such projects were economically viable because the country was not adequately mechanized. On the other hand, later analyses clearly indicate that the "working brigades" were foremost a political and ideological enterprise designed to construct "brotherhood and unity." One of the first great projects of this kind was the building of the Brčko–Banovići railway in northern Bosnia, in which some foreign students also participated. One of their number, a certain E.P. Thompson, was to become a famous British historian and the author of *The Making of the English Working Class*.

---

586 "Študentje in obnova—dve osnovni nalogi ljudske študentske mladine," *Ljudski študent*, March 30, 1946.

**Figure 45.** Members of the youth work brigades.

In this period, private initiative and entrepreneurship were rendered completely impossible. The economy was wholly submitted to the state economic policy and the Law on the Five-Year Economic Plan passed in 1947. Under this law, most investment funds were appropriated for the construction of electricity and transport infrastructure, and for basic industries (iron, construction, and chemicals). Manufacturing, intended to provide consumer goods, was much neglected, as became especially apparent in everyday life. The situation was so dire that there was even a shortage of brooms, so it was little wonder that once the border with Italy was opened in 1950 most of the first shoppers to cross the border organized the so-called "March of the Brooms."[587]

---

[587] In the opinion of Branko Marušič, a historian from Nova Gorica, this march represented a form of protest with which the first cross-border shoppers wanted to draw attention to general shortages. See Vida Zei and Breda Luthar, "Shopping across the Border" (paper presented at the conference Everyday Socialism: States and Social Transformation in Eastern Europe 1945–1965, The Open University Conference Centre, London, April 24–26, 2003). According to Zei and Luthar, the March of the Brooms took place on August 13, 1950, when "the border authorities allowed relatives and friends from both sides of the Slovenian–Italian border to meet and be

In the early 1950s, it became more or less clear that the first five-year plan, with its rigid provisions and disregard for the market situation, would not yield a serious economic project but, at best, an unfulfilled wish list. Given the centralized financial policy, Slovenia faced another problem arising from the decision that the most backward regions should expedite their progress. As a result of the extremely poor infrastructure, lack of skill, and difference in lifestyles, this decision only led to a series of misguided projects and a dramatic loss of funds. The situation only started to change after 1953, when the Agrarian Reform was amended and the unilateral funding was redirected to heavy industry. Considerable attention was also given to transport and commerce, and to the manufacture of consumer goods. The positive effects were almost immediate. The gross national product soared to 10 percent for several consecutive years, so that in the late 1950s Yugoslavia was considered the fastest growing economy in the world.

## The First "Cultural Revolution" and the Emancipation of Women

Similar developments could also be observed in the field of culture, especially after the abolition of strict censorship that had been embodied in agitprop,[588] disseminating the ideas of social realism. But even immediately after the war, the Slovenes in particular had hardly any reason to complain. As pointed out by Ervin Dolenc and Aleš Gabrič, post-1945 Slovenian culture found itself in a completely different situation, especially because individual republics had the highest level of authority over culture, education, and science. Slovenia, which had already had a well-developed network of cultural institutions and societies before the war, could also benefit from the new situation by elevating certain establishments only of regional importance in the old Yugoslavia (radio, the archives, the national and university library) into central Slovenian state institutions.

---

reunited for a few hours as they had also done a week ago. The word got around, and some 5,000 inhabitants of border villages and towns waited on the Yugoslav side of the border from early morning on; they finally tore down the wooden barrier in front of the helpless border police and continued walking to the Italian part of Gorizia, where they spent a few hours and then returned with all kinds of consumer goods. Most of them carried brooms in their hands and combs in their pockets—the two items that were badly needed yet unavailable in Yugoslavia at the time."

588  Agitprop is a Russian contraction derived from two terms, agitation and propaganda. It denotes an activity with which the Soviet authorities tried to elevate their people's revolutionary awareness under socialism and win their active support for the communist movement.

Thus, in 1945, the national theaters in Maribor and Ljubljana were renamed Slovenian national theaters and the Academy of Sciences and Arts became the Slovenian Academy of Sciences and Arts. University enrollment soared, especially among female applicants. The Party's official declarations that female participation was essential for real democracy appeared unconvincing, or rather, contributed to the formation of the so-called second or shadow society. It was characterized by the image of a woman as a "socially sensitive, loyal to the socialist regime, educated, employed, politically active, a responsible mother, and an equal partner to her husband."[589]

In these images propagated by Party officials, education played an important role, as confirmed by the sharp increase in female students during the first five postwar years. For example, in the school year 1950/51, the number of female students in Slovenia was proportionally higher than in the US. Employment figures show a similar trend, although in Slovenia the percentage of working women was higher than in other parts of Yugoslavia. This may have contributed to the fact that women in Slovenia recognized slightly earlier that they shouldered a threefold burden (as workers, housewives, and mothers). Consequently, the number of women who participated in political or special-interest organizations halved. Nevertheless, domesticity was increasingly becoming central to the definition of socialist femininity and feminine identity in the official discourse, which was constructed in relation to working as "caring-for-others." The transformation of gender politics after 1945 showed dramatic ruptures as well as subtle continuities.

Similarly, in terms of their public role, for instance, women experienced an increasing under-representation in different political organizations. While the number of female deputies in the National Assembly increased, their representation in local bodies halved. The leading members of "women's organizations" soon noticed this and drew women's attention to the fact that, if they failed to participate, "politics [...] will ignore the particularity of women's needs"[590] and additionally enhance the discrepancy between official (public or front stage) and private (backstage) behavior and opinion. Therefore, the failure of the socialist institutions to shape "socialist preferences" resulted in widespread cynicism and withdrawal into private life. This low level of social trust and support for the formal system resulted in strong differences between the public and the private spheres and a lack of coherence between the official version of

---

589   Boris Kidrič in Mateja Jeraj, "Položaj in vloga žensk v Sloveniji (1945–53)," PhD dissertation, Filozofska fakulteta, Univerza v Ljubljani, 2003, 118.
590   Jeraj, "Položaj in vloga žensk v Sloveniji," 290.

reality and people's own experiences of the gender regime. Consequently, the "social contract" once again became a gendered "fraternal social contract."

In the first postwar years, mass amateur cultural activities were organized under the Ljudska prosveta Slovenije (People's Cultural Organization of Slovenia), the predecessor of the future Zveza kulturnih društev (Association of Cultural Organizations), while part-time education was organized by Delavske univerze (Workers' Universities), which above all provided foreign language courses and programs for advanced professional training. As early as the end of the 1940s, traditional forms of mass culture were gradually replaced by popular culture. This progress could be noted most clearly in music, where the popular trends of the 1950s were largely determined by radio broadcasting and the (initially very modest) record production. In such circumstances, until the beginning of the 1960s people mainly learned about the latest music productions from ballroom dances and folklore concerts. After 1962, they also learned about them from the Slovenian Song Festival, which was first held in Bled.

**Figure 46.** One of the most popular Slovenian singers Majda Sepe performing at the first Slovenian Song Festival, Bled, 1962.

The number of periodicals appearing in Slovenia increased greatly, especially before the mid-1950s. They stirred heated debates and encouraged dealing with contemporary topics and intimate problems. The thematization of the partisan struggle and the glorification of collective projects gradually gave way to a discussion of current problems and concepts. The most tangible progress was visible with the Slovenska matica literary association.[591] Its work was suspended by a German decree in 1944 and resumed in 1950 after appointing the geographer and historian Anton Melik as its new president. Three years later, Slovenska matica returned with a fresh impetus to compiling its regular collection of fundamental philosophical and historical texts. Thus, as many books were published in one year during the mid-1950s as in the first five years after the war, with the production even doubling in the mid-1960s.

The early 1950s also witnessed the first protest by the younger generation of Slovenian writers who had gathered at an extraordinary general meeting to criticize the policy pursued by the president of their society, who was said to act on behalf of the Party. One year later, Edvard Kocbek published a book entitled *Strah in pogum* (*Fear and Courage*), which broke with the prevailing social realist traditions. A similar transformation was signaled by the newly founded literary journal *Naša sodobnost*, which many saw as a result of the changing cultural and political climate ushered in by the 6th Congress of the League of Communists in 1952. More precisely, it was part of the standpoints presented by Miroslav Krleža, who had lobbied for a break with the Yugoslav version of social realism at the Third Congress of Yugoslav Writers in 1952. His decision notwithstanding, at the end of that decade, publishing houses still tenaciously rejected manuscripts on account of their "ideal inappropriateness," while the state even extended censorship to projects of self-published authors.[592]

---

591   Established in Maribor (1864) as a cultural and educational society, *Slovenska matica* aimed to support research, higher education, and literary production in the Slovenian provinces of the Habsburg Empire.

592   In mid-1958, a central publishing house (Cankarjeva založba) rejected three poetry collections by the younger generation of Slovenian poets. Two (*Požgana trava* by Dane Zajc and *Jalova setev* by Veno Taufer) were later self-published, whereas Jože Snoj was even prevented from self-publishing his collection *Mlin stooki*. This signaled the definitive enforcement of the idea promoted by the first Slovenian President Boris Kidrič that the state must exercise full control and supervision over the "material that is to be put into print." See also: Aleš Gabrič, "Samozaložba v enostrankarskem sistemu," in: Drnovšek, Rozman, and Vodopivec, *Slovenska kronika XX. stoletja*, 239.

Viewed from a distance, 1953 was a ground-breaking year in many respects: March saw the first postwar census, which revealed the ongoing intensive deagrarization of the population, and a month before that, the former Yugoslav Liberation Front was transformed into the Socialistična zveza delovnega ljudstva (SZDL, or Socialist Union of the Working People of Yugoslavia), an independent democratic alliance that would serve as a broad-based political tribune devoted to exchanging opinions. Since the SZDL's leading positions were still occupied by the most notable and deserving communists (Tito as president and Kardelj as secretary general; Miha Marinko and Boris Kraigher in Slovenia), it cannot be called an alternative political organization. Last, but not least, the official explanation also included that the new organization would endeavor to "build socialist relations." The only one to recognize the SZDL as a potential successor to the League of Communists was Milovan Djilas, a member of the Executive Committee of the Central Committee of the SKJ, ranked fourth in the Party hierarchy. In his groundbreaking articles in the Belgrade-based newspaper *Borba* (The Struggle), he first dealt with the so-called Marxist aesthetics and then characterized the new socialist political elite as "the new class." The man who only a year before had presided over agitprop and demanded a tight rein on artistic expression now took a completely different stance, for which he paid a high price. He was excluded from the League of Communists and later received long prison sentences. His ideas met with no major response in Slovenia, even though he had duly presented them in Maribor three months before his demise. Some Slovenes surrounding the periodicals *Naši razgledi* (Our Views) and *Revija 57* (Magazine 57) were discredited as "Djilasists" in the second half of the 1950s.

The departure from social realism in the fine arts took a slightly different track. The younger generation—best personified by Marij Pregelj with his expressionist illustrations of Homer's epics, Miloš Požar with his abstract art, and the surrealist Stane Kregar—opened the door to an abstract conception of the world. The polemics from which the so-called *Grupa 53* (Group 53) later emerged, "converged in particular on [...] the notion of abstract art"[593] and on ambitions to express interpretations of the subjective or intimate perceptions of the world. The foundation of what is now the world's oldest graphic biennial showed that their efforts had not been in vain. In 1955, Ljubljana saw the opening of the first international graphic exhibition organized by Božidar Jakac and Zoran Kržišnik, following the example of the international graphic exhibition Bianco e Nero (Black and White) in Lugano. The beginning of

---

593 Milček Komelj, "Kregarjev poseg v abstrakcijo," in: Drnovšek, Rozman, and Vodopivec (eds.), *Slovenska kronika XX. stoletja*, 202.

the future Biennial of Graphic Art was extremely important to Slovenian artists because it enabled them to keep abreast of contemporary international productions. Later, the biennial significantly contributed to the popularization of the Slovenian graphic arts, which became "the most flourishing discipline in Slovenian art" in the 1960s.[594] Another testament to its distinguished reputation was the new term "Ljubljana School of Graphics," which was celebrated above all for its characteristic colorfulness, experimentation with color, and aestheticism. Finally, the biennial paved the way for the foundation of the International Center of Graphic Arts with its printmaking workshops, where, since 1987, overview graphics exhibitions are regularly housed.

The development of Slovenian theater in this period took a completely different course. Its beginnings were institutionally ambitious: by 1955, Slovenia had no less than twelve professional theaters and its own regular festival in Celje. However, ten years later, only four in Ljubljana, Maribor, and Celje remained operational. Despite the efforts of artists—such as Slavko Jan, Fedor Gradišnik, Lojze Filipič, and Herbert Grün, joined later by Bruno Hartman, Franci Križaj, Bojan Štih, and Andrej Hieng—Slovenian theater, which had relatively quickly transcended Austro–Hungarian formalism, found itself gridlocked in pronounced realism for years to come. This is probably one reason why Jan, the director of the Slovenian National Theater, wrote dispiritedly in the early 1950s that the Slovenian National Theater in Ljubljana,[595] under his direction (as well as all other Slovenian theaters), suffered from too much realism and not enough imagination and style. The situation finally started to improve toward the end of the 1950s, when Albina Brankovič and Draga Ahačič established their respective experimental theaters. The development that followed created many opportunities to stage new Slovenian and foreign plays, including the premiere of Eugene Ionesco's *The Lesson* and premieres of other avant-garde dramatists. Yet this development was far less consequential than the further trends in Slovenian theater: a series of emerging and disappearing small experimental or alternative theaters, which ensured that until the end of the 1980s, even major theaters kept pace with new trends. On the other hand, it was chiefly these theaters and theater groups which seriously politicized drama.

---

594 Milček Komelj, "Mednarodni grafični bienale in ljubljanska grafična šola," in: Drnovšek, Rozman, and Vodopivec (eds.), *Slovenska kronika XX. stoletja*, 222–223.

595 Jan's criticism of the situation is quoted from Aleš Gabrič, "Prvi festival sodobne slovenske drame," in: Drnovšek, Rozman, and Vodopivec (eds.), *Slovenska kronika XX. stoletja*, 221.

Before the end of the decade, this politicization was matched by politically contaminated debates over the nature of the first Slovenian television station, which had first broadcast into Slovenian homes in late 1958. Politicians energetically rejected the proposed name RTV Slovenia and the Slovenian television station started broadcasting as RTV Ljubljana. The program was initially received by 800 TV sets, rising only three years later to reach more than 10,000 families. By the end of the 1960s, television had become an integral part of people's everyday routines. Between 1953 and 1964, Yugoslavia had an extremely high rate of economic growth as a result of a low starting point and of many structural changes influenced by industrialization, urbanization, and modernization in general. Moreover, the increased productivity brought a certain degree of prosperity which could not be overlooked, especially in comparison to other Eastern European countries. In 1965, for instance, Yugoslavia had more motor vehicles per capita than some of the people's democracies where national income and per capita consumption were substantially higher. Savings deposits increased 25-fold between 1955 and 1965 and helped keep demand for consumer goods at a high level, thus making it less dependent on current incomes. Moreover, by 1965, the possession of durable goods represented a much more significant element of personal wealth than any time since World War II. In Yugoslavia, the index of consumption per capita rose from 103.6 in 1954 to 130.1 in 1957.

These political and economic processes were accompanied by social and cultural transformations, including a rearrangement of social groups: differentiation and industrialization necessarily brought with them new modes of community, new forms of social etiquette in the cities, and a distinctive new sociality or structure of feeling. Moreover, new forms of self-understanding and self-cultivation—in short, new forms of individuality with distinctive ways of life—were emerging. The increased differentiation in earnings and occupational reclassification were only one aspect of the changes in the "social opportunity structure." The latter—as a socio-structural process that opens up social space for class differentiation—is composed of educational, income, lifestyle, and occupational elements. A class structure based on "quantity of competence" began to emerge, while education and lifestyle differentiation rather than income marked barriers between social classes. The changes in the "social opportunity structure" produced an emerging middle class with a specific internal differentiation and enough available economic and cultural capital (qualifications, taste, and morals) to be spent on "marking services."[596]

---

596 See also Breda Luthar, "Remembering Socialism: On Desire, Consumption and Surveillance," *Journal of Consumer Culture*, 2 (2006): 229–259.

## Contested Loyalties

The fact that Yugoslavia was officially a federation could hardly be noticed in the first years after the war. The federal government in Belgrade had very broad powers and was soon embroiled in its first serious clash with the Slovenian government. The Slovenian communists were especially irritated by its explicitly centralized economic policy and the unequal status of the Slovenian language, which the federal authorities had excluded from the military and all other spheres of public life. Basic administrative information in public places in Slovenia, as in all other republics, was thus often provided in Serbo-Croatian. Slovenian exasperation, however, peaked at the end of the 1950s, when the federal administration proposed the unification of regulations, most conspicuously on the question of a common cultural milieu. Some officials even contemplated the elimination of republics as unnecessary intermediate administrative units between the federal government and municipal authorities, whereas the 1961 population census first listed the special "Yugoslav" category as a nationality. In opposition, the Slovenes most often allied with the Croats, who would adopt a declaration on the Croatian language six years later, following the Slovenian example in protesting against the general socialist criteria underlying Yugoslav culture, or joining the Slovenes in their claims "that Yugoslavia is inhabited by different nations that speak different languages and have different cultural traditions stemming from separate historical developments.

The Slovenian politicians made every effort to convince the centralists about the senselessness of narrowing down inter-republican relations within the state to those between various administrative units, inasmuch as the republics are constituted in agreement with national borders and that for this reason it would be more appropriate to talk about international relations."[597] The Slovenes enjoyed not only Croatian but also Macedonian support in these discussions. In the early 1960s, the growing resistance to centralism allowed for the rapid and wide-ranging development of several Slovenian cultural institutions. Television, which broadcast its first evening news in Slovene in 1968, only ten years after its establishment, now aired more and more programming in Slovene. The number of films with Slovenian subtitles similarly increased, while the most crucial advance was the establishment of the Cultural Fund of the Republic of Slovenia in 1962. The major share of cultural allocations had previously gone to Belgrade, which had also enraged the Croats. Therefore, it is clear that the discussion of language and culture foregrounded everything that had been

---

597   Dolenc and Gabrič, *Zgodovina 4*, 223.

excluded from consideration during economic planning. Whatever economists had been precluded from addressing while discussing the (un)reasonableness of the assistance funds for the developing republics was now reasoned by writers—in association with liberal politicians. In the increasingly fiercer dispute between the so-called federalists and centralists, it was largely due to liberals that the new Yugoslav constitution (again composed by Kardelj) effectively bridled the centralist ambitions.

# "Liberals" vs. "Conservatives"

The 1963 Constitution precluded the complete predominance of Belgrade and at the same time diverted the controversies over national relations into a surprisingly well-articulated discussion on the cautious liberalization of the economy. In spite of the huge gap between classical Western European liberalism and the Party liberalism pursued by Yugoslav communists, which veered toward an increasing market influence and greater independence of national economies, endeavors of the Yugoslav liberals were far from risk-free. This risk further increased after the leadership of the Slovenian government was taken over by Stane Kavčič, who nourished ideas of concluding loose and free inter-republican agreements, enhancing independence of the republics in establishing foreign contacts, financing federal services and institutions based on the principle of participation fees, reorganizing the army,[598] and developing a more extensive pluralism within the socialist federation. The old generation of communists, alarmed by the thought that such efforts might soon lead to the restoration of a multi-party system, backed the conservatives and with their help simply rid themselves of the unwanted reformers. Somewhat surprisingly, Kardelj was one of the leading personalities (besides Tito) who fervently aimed to undermine Slovenian and Croatian efforts, although it did provide them with an opportunity to eliminate some of their unpleasant adversaries. The impact of this principally political controversy on the economy was even more striking, since it began to stagnate after top executives were purged from successful companies. The subsequent retreat of Party liberals from Slovenian, Croatian, and Serbian politics was therefore also a final defeat of the adherents of a more comprehensive economic reform within a one-party system, and all further reformist attempts came from proponents of the rule of law and the multi-party system, as well as economic reform. Yet, even before their demands were voiced, conservatism had prevailed, extending the dominance of those Party leaders who still clung

---

598   Kavčič's government was committed to ensuring that conscripts would do military service in their home republics instead of being sent to other parts of the country. It also strove to give the republics more say in defense issues. After the Soviet military intervention in Czechoslovakia in 1968, the federal government in Belgrade partly adopted this idea and assigned some of its defense tasks to the republics. In addition to the regular Yugoslav army units, Slovenia thus also established its own Territorial Defense force, commanded by Slovenian officers and manned by Slovenian troops.

tenaciously to the planned economy for the next 25 years. In practice, this meant that companies were required to allocate 50 percent of their income to various funds (15 percent to the General Contribution Fund, 10 percent to the Spatial Unit Fund, 20 percent to the Republican Fund, and the remaining share to the Non-Recurring Income Contribution Fund).[599] In other words, events took an entirely different course, which was contrary to the reformers' hopes.

Nevertheless, a chronological outline of the first two decades after the war still allows the conclusion that there were quite a few initiators and advocates among Slovenian communists who promoted modernization and economic reforms which would increase competitiveness, particularly in industrial branches, to a level close to that of Western economies. The question yet to be solved is whether the potential and actual reforms could have prolonged the life of communist Yugoslavia or at least ensured its more peaceful disintegration. Namely, by 1970, it had become clear that the Yugoslav economy had already reached the extreme limits of modernization within the bounds of the "socialist self-management" framework and that every further step toward modernization would inevitably require more radical changes in the political system.

On the other hand, the progress in Yugoslavia since the late 1940s had taken a completely different course from any other country of the Socialist Bloc. Especially from the 1950s onwards, it was by no means merely a period of narrow focus and single-mindedness, but very often extremely rich and fruitful in its cultural and creative aspirations. Yet despite recurrent political pressures, it was also a time of numerous attempts by small and generally misunderstood groups of cultural creators and intellectuals to broaden and expand the space of liberty and democracy.[600] It is also true that, regardless of political pressures and their accompanying crises, Slovenia evolved into a modern industrial society in the 1960s and 1970s. At the end of the 1960s, and especially during the 1970s, this development brought about the emergence of a new middle class, whose only remaining connection with communism was the Party membership card and whose efforts were now entirely dedicated to the acquisition and expansion of property. In the early 1980s, there were over 120,000 more or less inactive members of the League of Communists (more than 20 times the number in 1945), so self-managing socialism was very slow to lose its adherents. The results of public opinion polls showed that even six years before Slovenia's independence,

---

599  Jurij Perovšek, "Spremembe v gospodarskem sistemu," in: Drnovšek, Rozman, and Vodopivec (eds.), *Slovenska kronika XX. stoletja*, 251.
600  Vodopivec, "Pogled zgodovinarja."

nearly 60 percent of respondents trusted a "self-managing democracy," despite the growing influence of critical opinion toward the communist government.

In light of this fact, it is not so surprising that there was practically no serious opposition until the 1980s. Edvard Kocbek never succeeded in attracting a wider circle of supporters, whereas the ideas of Milovan Djilas had no consequential influence on the developments in Slovenia. Seemingly promoting the critical intelligentsia, Slovenian political leadership attempted to create a "modus vivendi" by permitting and financing the printing of various newspapers and periodicals, but was quick to withdraw from its flexible cultural policy by abolishing periodicals that had gone out of its control. Critical and opposition movements only gained more momentum and more followers at the beginning of the 1980s. The first strides were taken by young people, who organized demonstrations against outdated communist symbolism and strove for conceptual and ideological freedom, civil control over military forces and society, and respect for human rights. Likewise, they opposed any manifestations of the pledge of allegiance to Yugoslavism and called for a critical confrontation with the perplexing and contradictory Yugoslav reality.

## From Liberals to Democrats

The socialist period was a kaleidoscope of various, often conflicting cultural, economic, and political undertakings. Thus, the diversification of the efforts to achieve democracy can by no means be interpreted as a linear transition from communism to democracy. How else can one otherwise explain why the "liberal-leaning" Stane Kavčič decided to terminate the liberal periodical *Perspektive* (Perspectives) and why the fervent democrats could simultaneously advocate "original Marxism"?

Until the mid-1980s, the regime's major critics pursued the following three causes above all: more room for maneuver in the economy, cultural autonomy, and more personal freedom. The latter often coincided with unlimited freedom of religious belief and freedom of movement, although freedom of movement was not really called into question until unreasonable austerity measures were introduced in the early 1980s.[601] In everyday life, these three causes were closely intertwined and covered an extremely wide sphere of activities. Autonomy, for instance, could involve Slovenia's own territorial defense system, which the republic had already obtained in 1968/69, or further independence of the

---

601   See "The economic crisis," this chapter.

republican administration, which the 1974 Constitution eventually provided for. The most important novelty of the Constitution, the third one in only three decades, was that in one of the Federal Assembly's two chambers, the Chamber of Republics and Provinces, decisions could only be passed upon the agreement of all republics and provinces. The two autonomous provinces within Serbia, Kosovo and Vojvodina, thus obtained the right to veto; the state became explicitly confederal; and the status of Kosovo and Vojvodina, as well as the republics, was strengthened. Serbian politicians, unsurprisingly, soon became the major opponents of the Constitution and demanded radical amendments. The Slovenian communists, who had achieved a high degree of independence through the Constitution, began to spearhead the defense of a confederalist concept of the state.

However, their efforts saw several activists lose their positions or be pushed into political isolation. The 1965 reform, which took place in several stages, was only half-complete, and left many advocates of enhanced economic freedom utterly dissatisfied. While, as acknowledged, the devaluation of the dinar, the tax reform, and the alignment of prices did somewhat constrain inflation, companies were still not guaranteed enough freedom to make a substantial breakthrough. The reform was finally abandoned in the late 1960s and early 1970s, when the general economic trend was accompanied by the conservatives' reaction. Socialist economics was reluctant to introduce a full market economy, hence reluctant to give capital, the market, labor, increased productivity, and free corporate governance a decisive role. Whereas it admittedly ensured full employment and a high level of social security, it also became a pawn in negotiations or, even worse, the subject of political decisions.

Nevertheless, Slovenia was much better placed to take advantage of the reform than other Yugoslav republics. The government headed by Stane Kavčič inherited Sergej Kraigher's reform policy, which paid more attention to consumer products, service activities, commerce, transport, and tourism. It also made heavy investments in education and science with a view to being able to introduce comprehensive economic programs. The republican government started to promote new energy sources and in the 1970s built the first (and so far the only) nuclear power plant in Slovenia. Significant funds were invested in modernizing the road system on the east–west line. But like so many other projects, this one too was stopped in its tracks.

Kavčič, the then prime minister, was the driving force behind Slovenia's development. However, before taking the reins of the government, he had presided over the so-called Ideological Commission. In this role, just three years before becoming prime minister, he had suppressed *Perspektive*, one of the most critical reviews "for cultural and social issues" of the period. This was incidentally the sixth such suppressed periodical: The Press Commission under the SZDL, and from time to time even the Centralni komite komunistične partije Slovenije (CK KPS, Central Commitee of the Communist Party of Slovenia), had already suppressed *Mladinska revija* (Youth magazine) in 1951, *Svit* (Dawn) in 1954, *Bori* (Pines) in 1956, *Beseda* (Word) in 1957, and *Revija 57* (Magazine 57) in 1958. The latter two were considered especially menacing and Edvard Kardelj took it upon himself to intervene personally with an article in *Naša sodobnost* (Our Contemporaneity). The members of both editorial boards were reproached for being contaminated with Djilas' ideas and for plotting to take over power.

Kavčič had to stave off similar reproaches eight years later, when he was criticized for fostering liberalism in politics and the economy. The reproach regarding the economy was made after a dispute about the allocation of an international loan for road-building, when Kavčič's group struggled to obtain a greater share of the loan in the late 1960s on the grounds that Slovenia had found itself at a disadvantage in Yugoslavia. This time the intervention came not only from Kardelj but also from Tito, who ultimately threatened the insubordinate Slovenes with the "most severe measures."[602]

Two years later, Kavčič was partially ascribed the so-called objective responsibility for the disobedience of 25 members of the Slovenian Assembly. At the election to the membership of the presidency, the Slovenian Assembly refused to choose among the Party's predetermined favorites and, abiding by the rules of procedure, elected their own candidate. The Party criticized their disobedience, as well and considered it primarily an attempt to "manipulate the deputies."[603]

## New Trends in Art

Even artists, after their initially universal engagement, accompanied every possible group performance with assurances that their cooperation and performance served strictly artistic purposes. Such assurances were most expressly reiterated by the painters of Grupa 69,[604] who prepared a very striking survey

---

602 Božo Repe, "Cestna afera," in: Drnovšek, Rozman, and Vodopivec (eds.), *Slovenska kronika XX. Stoletja*, 315.
603 Božo Repe, "Akcija 25 poslancev," in: Drnovšek, Rozman, and Vodopivec (eds.), *Slovenska kronika XX. stoletja*, 329.
604 The members of Grupa 69 were Janez Bernik, Jože Ciuha, Riko Debenjak, Andrej Jemec, Kiar Meško, Adrian Maraž, France Rotar, Gabrijel Stupica, Marko Šuštaršič, Slavko Tihec, and Drago Tršar (later joined by Zdenko Kalin, France Mihelič,

exhibition in Bled. Its special guests were Dušan Džamonja and Vladimir Veličković, then two of the most prominent Yugoslav artists.

Although the painters of Grupa 69 were critical of the conditions surrounding their artistic production, their performances could not be compared to the all-pervasive engagement and criticism of the OHO Group.[605] The latter infused the mid-1960s Slovenian fine arts, sculpture, photography, and film with elements of conceptualism, which promoted the ideas of land art, arte povera, and body art, as well as introduced the principles of reism to local literature. Their exhibitions were frequently forbidden; its ten members performed all across Yugoslavia and even exhibited at the Museum of Modern Art (MoMA) in New York, in 1970. Immediately afterwards, they took the collective decision to refrain from further appearances. Their work had a major impact on future generations, especially because it very often came in for extreme denunciation by the regime. *Poker*, the collection of poems by Tomaž Šalamun, which was published in 1966 and signaled the beginning of avant-garde modernism in Slovenian poetry, was even criticized by none other than Josip Vidmar, Slovenia's most important literary critic and long-standing President (1952–1976) of the Slovenian Academy of Sciences and Arts.

## The Student Movement

Despite—or perhaps because of—such reactions, the OHO Groups' influence was particularly strong during the student movement of the late 1960s and early 1970s. The movement has been described by some as a poorly reflected repetition of the events of 1964 and by others as a questionable response to the developments in Paris and Berlin.[606] Like all other forms of activism at that time, it voiced the desire for change and, above all, for modernization and reform.

---

and Štefan Planinc). The main reason for their collective performances lay in dissatisfaction with the current position of Slovenian fine arts in general and with its critical evaluation, which the said painters found to be "not learned enough and too idyllically off-track." See Milček Komelj, "Grupa 69," in: Drnovšek, Rozman, and Vodopivec (eds.), *Slovenska kronika XX. stoletja*, 313.

605 OHO was an avant-garde art group (1966–1971) that had a distinctive influence on the Slovenian art landscape. The members of the group were Marko Pogačnik, Marjan Ciglič, Iztok Geister, Milenko Matanović, Franci Zagoričnik, Tomaž Brejc, David Nez, Tomaž Šalamun, Aleš Kermauner, and Naško Križnar.

606 Questionable because student demonstrations in 1968 across Yugoslavia differed from similar demonstrations all over Europe. Students elsewhere were struggling against the system and its institutions, whereas Yugoslav and Slovenian students were calling for the introduction of the authentic values of socialist self-management. See

Its similarity with other emancipatory efforts by the Slovenes in that period is clear. Its initially radical posture (e.g., demanding the university reform and criticizing social inequality) soon developed into a kind of syndicalism, which the government and the Party were able to contain by increasing the Scholarship Fund, and promising new student halls of residence. Nevertheless, students successfully negotiated the founding of their own radio station (named Radio Študent), which later influenced the promotion of ideas voiced by alternative movements and was crucial in transforming the system. Along with scarce articles appearing in various literary journals, Radio Študent also furthered the tradition of quality criticism of all sorts of artistic production, which began in the 1960s. In this respect, Radio Študent was on a par with the film magazine *Ekran* (Screen), which first appeared when Slovenian cinema was finally reaching beyond the short-lived phase of so-called social realism.

Figure 47. Student protests, May 1971.

also: Božo Repe, "Spremembe da, cirkus ne," in: Drnovšek, Rozman, and Vodopivec (eds.), *Slovenska kronika XX. stoletja*, 305.

## Cinema and Theater

Nowhere was this shift in the Slovenian cultural landscape more clearly marked than in cinema,[607] and many see the 1960s as by far the most important period in Slovenian film—not least because, unlike other socialist countries, Yugoslavia had allowed both Eastern and Western productions to be screened ever since the first half of the 1950s. Thus, Slovenian viewers could follow not only Andrei Tarkovsky and American Westerns but also Michelangelo Antonioni, Claude Chabrol, or Ingmar Bergman. Boštjan Hladnik, the director of one of the most prominent Slovenian films of all time, even worked on two Chabrol films as an assistant director. Although Hladnik's film *Ples v dežju* (Dance in the Rain, 1961) polarized the Slovenian audience, particularly with its alternation between reality and dream sequences, its use of expressionistic lighting, and the lucidity with which it presented its protagonists' traumatic world, marked the birth of modern Slovenian cinema. Indeed, Hladnik's colleague Jane Kavčič had already celebrated it with his film *Akcija* (The Action, 1960). Other pioneers of Slovenian film included Jože Babič, Jože Pogačnik, Matjaž Klopčič, and France Štiglic. Slovenian cinema also presented a major stimulus to Slovenian writers, who often took up writing screenplays. Moreover, film became the principal source of income for many, including the former boxer, partisan, and journalist Vitomil Zupan, one of the greatest Slovenian writers. Zupan's literary circle, which also included Marjan Rožanc, Dominik Smole, Beno Zupančič, and several others, endowed Slovenia with a generation of writers and wide-ranging artists who were also willing to take risks for the sake of their artistic integrity. This was also the generation that paved the way for the Slovenian version of musical, fashion, and sexual revolution. Jeans, rock'n'roll, and miniskirts could only become a viable fashion statements once jazz, Jacques Brel, Ingmar Bergman's treatment of free love, and "existential black" turtlenecks were already part of the everyday lifestyle. The artistic community in Ljubljana, and to some degree in Maribor, was quick to embrace this lifestyle with Vespa scooters and Beatlemania. It kept up with the latest art trends through reference books and professional literature bought on study visits to Paris.

It is difficult to ascertain what kind of sources inspired the always opulent and progressively more radical artistic productions in Europe or in parts of Yugoslavia. However, this intensive development grew increasingly more indeterminate and even less restrainable by potential censors. While this does not mean that the

---

607  The first postwar Slovenian film was titled *Na svoji zemlji* (On Our Own Land, 1948), set during World War II and the partisan resistance.

regime relinquished control over artistic production, it nevertheless signifies that the new generations already appeared to have mastered the art of ambiguity by skillfully masking their ideas so that even the least knowledgeable members of their audience could read between the lines and infer sharp political critiques from seemingly the most innocent expression. Newspaper editors, institute directors, and theaters learnt to translate potentially controversial content into the language of each respective supervisor and censor.

Such survival techniques could only evolve with the support of a sufficient volume of artistic or professional production. During the early 1970s, this certainly occurred, at least in the theater. At the beginning of the decade, the experimental theater Pupilija Ferkeverk emerged and was shortly followed by the theaters. Unlike the first wave of avant-garde theaters, which were based on dramatic texts, the new theaters undertook to experiment with the text or sometimes even omitted it completely. Contemporary theater directors still vividly recall the unforgettable performances staged by Lado Kralj, when he interpreted Dane Zajc's *Potohodec* (The Wanderer), Dušan Jovanovič with his adaptation of Bojan Štih's *Spomenik G* (Statue G), or Rudi Šeligo directing Milan Jesih's *Grenki sadeži pravice* (The Bitter Fruits of Justice). Now and then, professional theaters briefly came to life, and their numerous outstanding actors (e.g., Duša Počkaj, Stane Sever, Arnold Tovornik, Zlatko Šugman, and Branko Miklavc) were sure to delight even the most demanding audience.

It is hard to imagine that a time of such vigor and engagement was succeeded by a new period of censorship and restriction. In other words, perhaps it is finally time to reconsider whether the 1970s were really the so-called "leaden years" or whether, even in the 1970s (Party policies aside), different trends were converging as they had done in the 1960s. It is becoming increasingly clear that the 1970s were above all a logical extension of the 1960s and that they were a time of unsuspected changes which were not definitively articulated before the 1980s. However, the 1970s were also a time when faith in socialism gradually gave way to the realization that, as Pierre Bourdieu would put it, the ethos of necessity and morality of self-sacrifice and duty began to be replaced by the ethos of desire and the morality of fun.[608]

---

608   Pierre Bourdieu, *Distinction: A Social Critique of the Judgement of Taste*, London: Routledge, 2000, 367.

## A Period of Double Standards

It is probably by no coincidence that in the 1970s jokes started to be spread about the decaying capitalism, which, although obsolete, was very reluctant to crumble. A major catalyst in the ensuing debates was the freedom to travel, offering citizens abundant opportunities to compare both systems. These comparisons made many realize that the Slovenian version of socialist self-management was not such a bad solution after all. Full employment was followed by an apparent growth in the standard of living, which was a result not only of Yugoslavia's excessive borrowing from abroad but also of loans granted under extremely favorable terms.

This was the only time when working families could build their own homes and the middle class could afford holiday houses and regular shopping trips abroad. The peculiar and somewhat schizophrenic climate was partly also created by the political elite, which decided to uphold many liberal policies, even though it had eliminated the undesirable liberals. Therefore, the state apparatus, determined to maintain the tradition of grandiose yet senseless manifestations, flatly ignored the fact that the projects commemorated by ceremonies and festivals had long been void of their original meaning.

While there was much talk of equality, brotherhood, and the legacy of the revolution, this period was also characterized by a radical advancement in class differentiation, widening disagreements among the republics, and a growing weariness of lessons from contemporary history. Another testament to the system losing much of its credibility in the eyes of the Slovenes was cinema, especially comedy, whose main protagonists were usually portrayed as typical socialist characters. But if the federal constitution and the Slovenian constitution a month later in 1974 were perceived by the Slovenes as indicators of things finally moving in the right direction, a completely opposite signal was sent by the ultimate invention of socialist self-management, the Associated Labor Act. The concepts involved therein heralded a peculiar reinvention of planned economy, whereas the proposed method of integrating economic and social activities, designed to "transcend the incessant frictions between 'manufacturing' and 'consumption' activities" in a specific corporativist manner,[609] bordered on sheer science fiction. The project's utopian nature was made manifest in the basic idea of a so-called mutual harmonization or pluralism of self-management interests. The act, passed by the Federal Assembly in late 1976 and termed by some the "little constitution," in fact overthrew the liberal legacy of the 1960s. The former reformist orientation

---

609 Božo Repe, "Zakon o združenem delu: ozdi, tozdi in sozdi," in: Drnovšek, Rozman, and Vodopivec (eds.), *Slovenska kronika XX. stoletja*, 360.

toward a market economy was largely abandoned and individual companies were transformed or fragmented into the so-called Temeljna organizacija združenega dela (TOZD, Basic Organization of Associated Labor), Organizacija združenega dela or (OZD, Organization of Associated Labor), and Sestavljena organizacija združenega dela (SOZD, Composite Organization of Associated Labor). OZDs were founded for all forms of economic and non-economic activities with a view to becoming basic cells of associated labor where workers, who now assumed the roles of both employers and employees, were granted direct or indirect access (through delegates) to decision-making on all important matters. The market mechanism was to be substituted by dialogue between self-managing companies.

The new system, which remained formally in place until the end of 1988, was ineffective, isolationist, non-competitive, and as such devoid of any practical value. More importantly, it left many with the impression of living in two parallel worlds. One was the real world of everyday needs propelled by competition and private interests, the other the world of the impractical normative system, which even transformed banks into associated labor services. The system's futility was demonstrated less than two years later when the unrealistic interest rates applicable to the internal crediting of (usually) unfeasible economic projects could no longer cover expensive international loans. In the 1980s, the unstable equilibrium consequently pushed the banking system into a serious crisis which finally disillusioned even the regime's most ardent supporters. The last in the series of ground-breaking documents, quite unusual for socialism, was certainly Edvard Kardelj's book *Smeri razvoja političnega sistema socialističnega samoupravljanja* (The Courses of Development of the Socialist Self-management System), published in 1977 and adopted as the ideological platform for the Eleventh Congress of the League of Communists of Yugoslavia, when it was still in its study phase. As envisioned by Kardelj, the Yugoslav socialist system, based on the said pluralism of self-management interests, provided the maximum possible number of citizens with the possibility of direct participation in social life. This possibility came from the so-called delegation system, under which individual representatives (from local to federal levels) were replaced by delegations or, as Kardelj called them, "authentic representations of citizens […] expressing the community of authentic self-management interests." In Kardelj's opinion, "self-management can only exist if workers are represented and administered by a community of interests to which they objectively belong rather than a political power operating outside the sphere of their direct influence."[610]

---

610 Quoted from: Božo Repe, "Socializem po meri človeka in demokracije," in: Drnovšek, Rozman, and Vodopivec (eds.), *Slovenska kronika XX. stoletja*, 362.

However, despite these conspicuous and rather utopian views, the schizophrenia was not so much the result of the ideas promulgated by the chief Party ideologist of socialist Yugoslavia. Rather, it stemmed from the fact that real power rested in the hands of the SKJ, despite theoretical assurances of the broadest possible democracy. Because the delegation system was so complex, individual responsibility was completely dispersed. This did not go unnoticed by the public, which was also aware of, and irritated by, the government's continuous practice of allocating the largest share of resources from international loans to wasteful projects in the most disadvantaged parts of the common state.

How, then, could the system persevere right until the end of the 1980s? There are three reasons: inertia, international support (based on the West's interest in preserving Yugoslavia as a convenient buffer state between the two blocs), and the fact that the Kardeljian self-management model, despite its utopian objective, delivered more democracy than any of the Eastern European systems of state socialism—precisely the trait that won it much attention from Western theoreticians. In any event, its author died in 1979, soon after the publication of his first ground-breaking study. A year later, in May 1980, came the death of his fellow fighter and comrade Josip Broz, with whom, interestingly enough, Kardelj had never developed a genuinely close friendship. The majority of the population did not yet know, however, that the world was spiraling into a deep energy crisis—just when Yugoslavia had accumulated debts equal to the total amount incurred over all previous decades, therefore finding itself among the most indebted countries in the world.

On top of that came the death of an icon. Tito's charisma had made it possible until then to subdue national differences and alleviate the economic crisis with international loans. After Tito's death all the long-concealed mistakes and weaknesses burst into the open in all their complexity, awakening many to the realization that the self-management agreements had not been a proper solution. For the aged rulers and their conformist, career-driven co-workers of the middle generation, the 1980s could not have had a worse start.

# From Crisis to Conflict and Beyond

The final turning point came as late as 1987–1988; not just because of the economic crisis, the political pressures of Serbian nationalists, the bureaucratic rigidity of state officials, and the arrogant posture of the Jugoslovenska narodna armija (JNA, Yugoslav People's Army) but also because of the complete disunity within the so-called Yugoslav opposition intelligentsia of "democratic" leanings. Like the Slovenian "independence movement," the opposition intelligentsia had older origins. This, of course, does not mean that Slovenia's condemnation of the JNA, police brutality against the Albanians in Kosovo in 1981, and its opposition to a uniform curriculum[611] was automatically its first move toward secession from Yugoslavia. However, the indecision and incapacity of the state leadership, which after Tito's death constituted a collective presidency of representatives from all the republics and autonomous provinces, allowed enough room to reflect on proper democratization. It should be stressed, however, that both the continuity camp and the circles, which now demanded the rule of law and true republican autonomy, were once again mainly composed of communists or people who had been socialized within the framework of megalomaniac delegation and self-management structure.[612]

## The Economic Crisis

After Tito's death, the state of the Yugoslav economy escalated from a long-dormant crisis into full-blown and uncontrollable agony. The suspension of the economic reform in 1971 ushered in a 15-year era during which efforts to introduce a market economy were abandoned. The so-called "consensual

---

611 The proposal to formulate so-called "common programming nuclei" in education was the last serious attempt at some form of unification of Yugoslavia. In Slovenia, it met with considerable indignation because, for example, the selection of proposed set texts was clearly biased in favor of the literature produced by numerically stronger nations. The opponents to the proposal were further alarmed by the authorities' systematic concealment of the program and its new arrangements from the public.
612 In the early 1980s, when party membership was at its record high, the Party had 2,111,731 members (200,000 had only joined in the year of Tito's death), amounting to 9 percent of the Yugoslav population and almost 25 percent of all employed. In Slovenia, this percentage was lower by 50 percent. In 1982, the number of members in Slovenia soared to 126,432, while in 1988 it dropped to 110,000.

economy," as economists called the system of "basic" and "composite" organizations of associated labor, collapsed as early as the beginning of the 1970s. The annual inflation rate exceeded 20 percent, but until Tito's death the Yugoslav government successfully concealed its effects by borrowing abroad. Tangible signs of the crisis first became apparent in the early 1980s, with the 30-percent devaluation of the dinar in June of 1980, followed by growing inflation and shortages of basic necessities such as oil, sugar, coffee, or detergent, for which the authorities issued coupons. Automobile traffic was initially restricted by vehicle license plate number under the so-called "odd-even system," then by petrol coupons. Besides the energy crisis (at the beginning of the 1980s oil imports barely sufficed to fuel 290 days a year), the population was also adversely affected by limitations on the import of "luxury" goods (tropical fruits, foreign magazines and newspapers, cosmetics, etc.) and the imposition of special border-crossing taxes. The Slovenes, accustomed to an open border regime, understood this measure as a severe encroachment on their freedom of movement.

The crisis led to a decline in living standards, which in the late 1980s regressed to the level of the 1960s. By 1980, Yugoslavia's external debt accounted for more than 40 percent of foreign currency inflows. Top state officials, refusing to acknowledge the gravity of the situation, talked of "nascent problems in the economy" and "stabilization." A special "Kraigher Commission" (headed by Sergej Kraigher, the former President of the SFRJ Presidency) was set up, consisting of 300 politicians and economists from the entire country. Its task was to find a way out of the crisis, although the word "crisis" was almost never used until the mid-1980s. The results of the commission's work were rather dismal: foreign debt exceeded 20 billion US dollars and was followed by the socialization of debts (imposed by the Federal Assembly in July 1983). Hopes that the crisis would be solved through political measures (e.g., exchange-rate and interest-rate policies) and fiscal tightness proved delusory. This was, after all, one reason why the Slovenian Executive Council concentrated its efforts on reorienting the Slovenian economy toward the West by ensuring the supply of basic necessities, preventing foreign currency outflows, providing funds to service the debt, and by pursuing desperate administrative efforts.

## Between Ljubljana and Belgrade

The victory of the Greater Serbian orientation in 1987 brought the domestic political dispute in Serbia to an end and simultaneously fueled the first and last major Slovenian–Serbian conflict between two different development concepts. The Slovenes, who had enjoyed the support of many reputable intellectuals in Belgrade in the early 1980s, now gained fresh impetus through civil and alternative movements which initiated a serious discussion on political pluralization and raised new demands for the rule of law, whereas the Serbian political elite insisted on maintaining the patriarchal and egalitarian model. The antagonism between Slovenia and Serbia was therefore not so much a clash of two nationalisms—as was usually claimed by Western politicians and media at the time—but a disagreement between two models of development.

Ever since the beginning of the 1960s, Slovenia had generally been recognized as the most problematic republic in the federation: firstly, because it had called for the abolition of the centralized economy and for an amendment to the Constitution granting more rights to the republics, secondly, because of its protests in the late 1960s and early 1970s against raising loans for road construction, and thirdly, because of the "liberal" views of the president of the Slovenian government.

In the 1980s, Slovenia further consolidated its protest with efforts of decentralization. The dilemma was whether it should continue addressing its problems via Belgrade through maximum commitment to the federation or whether it should reduce inter-republican cooperation to the necessary minimum and use what little room for maneuver the system allowed to deal first and foremost with its own problems. The first option especially appealed to a few politicians of the old generation, whereas the second enjoyed the support of the younger generation headed by Milan Kučan. What Slovenian politics actually did was combine the two strategies, supervising whom it sent to Belgrade while keeping the most influential officials at home, just as Serbia did. For example, in 1989, Kučan refused to run for a position in the federal presidency.

The Slovenian politics of the time may thus be regarded as defensive maintenance of the status quo. In the economic sphere, it unequivocally articulated the stance of the economically most advanced republic in the federation, refusing to acquiesce to the socialization of debts, investments in the so-called Reciprocity and Solidarity Fund (which was mainly used by other republics to cover their losses), higher contributions to the federal budget, or the growing percentage of direct financing of the federation.

In foreign policy, Slovenia aspired to win more opportunities for direct contacts with other countries (which especially served its economic interests) and to achieve enhanced equality in terms of diplomatic representation and language. One of its (unsuccessful) efforts in this regard was for international conventions relating to Slovenia to also be written in Slovene. For instance, the convention with Italy about maintaining the land border was only written in Serbo-Croatian, even though it solely concerned Slovenia.

Notwithstanding all efforts along these lines, the Slovenian political establishment's credibility was rapidly eroded by its inability to solve the most crucial economic and political problems, and especially by the fact that even the so-called liberal stream had succumbed to reformism, ultimately forfeiting its political incentive. At the same time, the liberal supremacy unintentionally opened the door to the emerging democratic opposition in the shadow of youth subculture, which was later most often interpreted as a set of alternative movements.

The democratization process started to take on clearer political connotations after 1982, when a new journal for the promotion of critical thought called *Nova revija* (New Review) was launched. Cultural groups were the first to benefit from the new situation, as often before, followed by academics and Catholic intellectuals. Given the type of individuals involved, it was clear to the entire intellectual public from the outset that *Nova revija* had made it its mission to uphold the traditions of the suppressed periodicals *Beseda*, *Revija 57*, and *Perspektive*. Its symbolic name was the ultimate confirmation that it was first and foremost a political project, masked by its founders with the traditional Eastern European dissident title *A Cultural Newspaper*.

Several other important events showed that these developments were more than another experiment in "liberalization."[613] But the fundamental shift in criticism of the authorities was brought about by alternative movements that, in general, united young people who had not experienced authoritarian socialism and therefore did not stand in such awe of the regime as the generation surrounding *Nova revija*.

## Punk Rock, the Alternative, and Political Appropriations

The emergence and flourishing of youth subcultures and alternative social movements in the late 1970s and early 1980s was also a consequence of the regime's declared tolerance toward rock music and its relative disinterest in cultural expression in general. This "social deafness," ignorance, and dismissal

---

613   1984 saw the publication of France Balantič's poetry collection *Muževne steblike* (The Sappy Stream), which was first printed as early as 1966 but destroyed at the behest of the Communist Party. In 1983, Igor Torkar wrote an extraordinary novel about the postwar Dachau trials, *Umiranje na obroke* (Dying in Installments), following a novel about the Goli otok island in the Adriatic Sea, the site of a concentration camp for political prisoners. The novel is called *Noč do jutra* (Night Till Morning) by Branko Hofman.

of critical art production sparked off increasingly vocal forms of resistance in music, film, literature, and theater, which were sometimes well-orchestrated but mostly quite random events. Paradoxically, these subcultures and movements initiated the opening-up of cultural space and enabled the expression of different ideas and views.

These movements lacked any explicit political agenda, let alone dissidentism, but rather "understood their action as directed toward the production of alternative social spaces of otherness." A wide array of movements was offered shelter by the Zveza socialistične mladine Slovenije (ZSMS, League of the Socialist Youth of Slovenia), which provided a communication channel between the "alternatives" and the authorities. This significantly contributed to "the promulgation of different views about a broader spectrum of social problems than those recognized by the authorities."[614] The weekly *Mladina* (Youth) and Radio Student were the first to seize and consequently offer space to promote freedom of the press and information. They primarily targeted younger generations, but *Mladina* in particular soon attracted wider circles of readers, thus considerably influencing the "production" of public opinion and the Slovenian political agenda. The Slovenian authorities, too, "resisting the realist-socialist temptation and not declaring alternative movements as 'counterrevolution' [...] opened space for legitimate public action of powers that were questioning the legitimacy of the regime itself."[615]

The banalization and trivialization of rock music reflected a certain impasse in the broader Yugoslav socio-political situation. Because of the "liberalization of the Party, marked by attempts at economic reform, greater political tolerance, and artistic autonomy with the creative impact of The Beatles and Rolling Stones," the question of rock'n'roll had not been on the main political agenda since the 1960s.[616] The potentially rebellious charge of rock music was thus immobilized by making it the mainstream. Space for social action was reduced, only leaving room for politically harmless and often explicitly pro-regime incantations that praised the endless opportunities in the realm of the exquisite Yugoslav socialist project. This, however, was hardly a reflection of reality. The historical moment was ripe for the "social deafness" to be answered.

---

614   Tomaž Mastnak, "From Social Movements to National Sovereignty," in: Jill Benderly and Evan Kraft (eds.), *Independent Slovenia: Origins, Movements, Prospects*, Basingstoke: Macmillan, 1994, 95.
615   Slavoj Žižek, *Druga smrt Josipa Broza Tita*, Ljubljana: DZS, 1989, 68.
616   Gregor Tomc, "The Politics of Punk," in: Benderly and Kraft (eds.), *Independent Slovenia*, 117.

A new and even more influential stimulus emerged with punk rock and the new wave scene. In Slovenia, the members of the band Buldožer, formed by Marko Brecelj in 1975, are considered the founding fathers of new wave. Although firmly anchored in the 1970s' legacy of rock music, Buldožer introduced a stylistically distinct and textually provocative approach that was later to become the seminal new sound of Yugoslav rock music. The approach and attitude of Buldožer later even spread to other artistic endeavors, of which perhaps the most prominent ones were Sarajevo's New Primitivism and *Top Lista Nadrealista* (Surrealists), a television comedy program. Throughout the 1980s, such attitudes gained momentum as a form of resistance by ridicule. The "scene" was developed most eloquently, but not at all exclusively, in the republics' capitals. Throughout the country, numerous bands were formed whose sound echoed British punk rock. Ljubljana heralded punk with Pankrti, Otroci socializma, and Lublanski psi, Zagreb nurtured the socially engaged and sometimes overtly anti-regime Azra, Sarajevo was home to Zabranjeno pušenje, Belgrade produced prominent new wave musicians Ekatarina Velika, as well as many others. For quite some time, Yu-rock in general became a transnational forum for expressing discontent with the contemporaneous state of affairs that managed to group people in categories other than national ones.[617] Indeed, Yu-rock attracted large numbers of Yugoslavians and became an all-Yugoslav, anti-totalitarian urban network.

An additional stimulus for the rise of alternative movements was the aggressive attitude of the Slovenian political authorities and mainstream media toward punk rock and local punk bands, whose image and performances led pro-regime journalists to condemn them for flirting with Nazism. These utterly absurd accusations culminated in the so-called Nazi Punk Affair, which was characterized by the police maltreatment of punks and several months long prison sentences for two punks, who were charged with attempting to establish the Fourth Reich. Nevertheless, the Secretariat for Internal Affairs, which was responsible for monitoring these developments, determined in 1982 that these movements were not unconstitutional and that the first "proper" oppositional phenomenon was the appearance of *Nova revija*, whose concept and content were considered to advocate political pluralism. On the other hand, both the Služba državne varnosti (SDV, State Security Service) and the top political leadership were correct to observe that social movements had a much more profound

---

617 Mirjana Laušević, "The Ilahiya and Bosnian Muslim Identity," in: Mark Slobin (ed.), *Retuning Culture: Musical Changes in Central and Eastern Europe*, Durham, London: Duke University Press, 1996, 118–120.

impact on public life than the circle around *Nova revija*. In the mid-1980s, the peace, ecological, and feminist movements and many other social movements combined with an extremely engaged musical, fine art, and performance scene to establish a very diverse, wide-ranging network of people and actions whose far-reaching effect caught even the most vigilant pro-regime critics off guard.

In Slovenia, as everywhere else, punk rock was refreshingly loud and fast, the lyrics provocative, daring, dirty, and obscene. The sound was demolishing the apparent idyll of the numbing status quo. And yet, it was a fierce rejection of any relation to the past, musical or otherwise. Its lyrics expressed utter boredom, disappointment, and disillusionment with the present situation. In these songs the only concern was with the individual in the here and now, not with any grand story of a better, collective future. Bands were singing about having fun, getting wasted, girls, poseurs, being streetwise, the spoils of developing consumerism, and the ever-increasing presence of materialism. "Pesem za Mandič Dušana," a song by a Ljubjana band, Otroci socializma, is worth quoting in full:

> The wind is crawling in the street,
> It's a national holiday.
> Fog is blowing under the sleet.
> People are walking the streets,
> A speaker on every corner,
> The sound of marches bouncing off the walls,
> severing their heads.
> It's a national holiday.
> The fog is thick,
> The streets are grey with mud,
> Look at these cold people,
> Flags are oozing blood.
> A youth group parading from house to house:
> "Long live!"
> "We decide!"
> "Hurray!"
> "Which lawn will the road cut through?"
> "Hurray! We decide!"
> Fog dragging over sidewalks. Morning.
> Tiny rain drizzling down people's necks,
> chilling them to the bone.
> There are red stains all over the ground. Silence.
> It's a national holiday.[618]

---

618   Otroci socializma, "Pesem za Mandič Dušana," in: *Otroci socializma*, Ljubljana: Dallas 1998 [1981].

It should be noted that the bands had little or no political ambition and did not consider themselves part of a grander political agenda. According to many, they just wanted to have fun. In an interview, sociologist and member of the punk band Pankrti, Gregor Tomc,[619] states:

> We didn't fight for political freedom in Slovenia or Yugoslavia. We lived our personal freedom and consequently extended the space of autonomous socializing to others as well [...] we didn't fight against the system; we played rock music with subversive connotations—mostly because we enjoyed provoking the ruling political paranoiacs.[620]

Nevertheless, there were political implications and these should not be neglected. Indirectly, by singing about and exposing taboo topics, such as oppression, corruption, or homosexuality, public space was gradually being extended, enabling more social action and making room to express otherness outside official institutions.[621] Punk rock and new wave as musical genres and social phenomena became political as soon as the music was heard, recorded, disseminated, and then publicly and politically problematized. In turn, every subsequent action, attitude, song, or text was rendered political, as it presented a reaction against oppression.

Moreover, punk developed almost as an intellectual endeavor at the very beginning of the 1980s and thereafter, with the publication of three special editions of the Slovenian academic journal *Problemi* (Problems), which dealt with the emergence and wider social implications of the new artistic expressions in music and art in general. In 1984, an edited volume *Punk pod Slovenci* (Punk under the Slovenes)[622] was published compiling theoretical analysis and a vast amount of documents and testimonies. In light of the growing instability of the state, punk (and other movements) eventually became a tool for cultural elites to articulate their discontent politically and offered an opportunity to define and promote the nationalist agenda. Now largely politicized and publicized, the debate drew in some up-and-coming Slovenian intellectuals and critics who were able to use the emergence of punk and new wave in general as a way to voice their own issues with the state. By means of this wider socio-political engagement, punk rock entered a broader social discourse.

---

619 He was also a founding member of the punk group Pankrti and the heavy metal group Bombe, an activist, and a university professor.
620 Branko Kostelnik, *Moj život je novi val. Razgovori s prvoborcima i dragovoljcima novog vala*, Zagreb: Fraktura, 2004, 29, 34.
621 Božo Repe, *Jutri je nov dan. Slovenci in razpad Jugoslavije*, Ljubljana: Modrijan, 2002, 59.
622 Tomaž Mastnak and Nela Malečkar (eds.), *Punk pod Slovenci*, Ljubljana: KRT, 1985.

**Figure 48.** A popular punks' meeting place, appropriately named Johnny Rotten Square, Ljubljana, 1981.

The politicization of punk and other movements allowed for an institutionalization of the alternative, which was able to develop and extend its discourse to articulate views and demands within the field of a new political culture. Importantly, it managed to attract and mobilize a significant number of activists and thinkers who had access to the media and were therefore able to articulate and present their views to the increasingly disoriented public. The attempts to repress the subculture failed for three reasons: the protagonists were intelligent enough to initiate public discussion and suspended their ideological divisions to oppose the use of violence as a means to solve social problems, while the ZSMS decided to listen to its social base and not to participate in the repression, since no evidence could be brought against the punks. Eventually, in 1986, the ZSMS "ceased to be the umbrella organization for the alternative and transformed itself to become the counterpart of the alternative scene in the political system."[623]

---

623   Mastnak, "From Social Movements to National Sovereignty," 100.

The sentiments spread to other domains of society and into everyday life, resulting in feelings of belonging to a global setting beyond the confines of the crumbling Yugoslav reality, particularly in light of the Chernobyl and Challenger disasters, which had ended the postwar dream of endless progress. Moreover, media coverage of these events resulted in an increasing awareness of the fragility of the world, natural and social. Yet idealism had not vanished completely: "The dream is mine, don't want to stand in line/I want at least a dash of the American Dream."[624] It was because of the ambiguity and intrusive presence of such topics in everyday life that punk and new wave played such a significant role in shifting the boundaries of freedom in society. At a time of growing interethnic tensions between the Yugoslav nations, the results of the movement were soon appropriated by Slovenian nationalist political figures (who often held high positions in institutions such as social organizations or editorial boards) who safely followed the fresh path of the newly expanded socio-political environment.

## The "Poster Affair"

Given this situation, the editorial board of *Nova revija* fittingly concluded that the time was ripe to launch an open discussion of democratization and (at least partial) autonomy. In early 1987, it published contributions to the so-called Slovenian national program. Although the selection of philosophical and broader social scientific discussions was neither the first nor the only attempt at dealing with the Slovenian national question, these contributions raised by far the greatest political attention and were ultimately accepted as the central Slovenian national program because several authors were unequivocal in their demands that the Slovenian nation achieve its statehood through original sovereignty, free from that of Yugoslavia, and introduce a new rule of law that would deprive the League of Communists of control over the Slovenian nation.

On the other hand, the thematic issue 57 of *Nova revija* gave prominence to the fact that the intellectual opposition had taken the initiative in developing the Slovenian national program. Its contents evoked condemnation from government circles and elicited changes in the editorial board, but at least for the time being, the regime's censors appeared hesitant to impose stricter administrative measures. One reason might have been the sudden outbreak of another scandal, this time involving the alternative scene. What began as a minor artistic diversion turned into a highly inflammable political controversy known

---

624 Videosex, "Jesen." *Arhiv*, Dallas Records, 1997 [Lacrimae Christi, Ljubljana: ZKP RTL, 1985].

as the "poster affair," which called into question the very notion of the annual celebration of Tito's birthday. The traditional festivity had begun in 1957, when Tito renamed his birthday the Day of Youth and dedicated it to young people throughout Yugoslavia. Thereafter the so-called Relay of Youth, a symbolic relay race, was held across the whole of Yugoslavia every year and became a major media event, bringing the images of Tito with thousands of young people into the homes of Yugoslav citizens. Photographs, news reports, books, and TV shows covered the event for those who did not have the opportunity to see it in person. In its 40-year history, about 20,000 relay batons had been carried across Yugoslavia. Each year, every republic and province had its own representatives carrying the batons, which came in countless different shapes and designs. The batons were always carried by young people who were recognized for their outstanding achievements in their academic, cultural, or sports activities. The annual Day of Youth celebration continued even after Tito's death in 1980 and was preceded each year by a nation-wide open competition for the Youth Day Poster.

In 1987, the Slovenian avant-garde collective Neue Slowenische Kunst (NSK, New Slovenian Art) submitted its proposal, which was pronounced the winner out of thousands of other designs. However, the story has an interesting background: the poster was initially created to publicize a theatrical Day of Youth Art Event by the Gledališče sester Scipion Nasice (Sisters Scipion Nasice Theater, 1983–1987), a part of NSK. According to its director Dragan Živadinov, "The Day of Youth Art Event was conceived as an act of self-destruction of the Scipion Nasice Sisters Theater, already announced in the [collective's] 1983 manifesto. At the same time, this was meant to be an act of abolition of the state."[625] Immediately afterwards, the public found out that the poster had actually been based on the painting The Third Reich by a German painter Richard Klein. The irony was that the government officials had accepted a work that drew on the Nazi past and its ideologies, and the scandal brought the problematic and complex relationship between the representation of power and socialist ideologies to the surface. The year 1987 was also the last year of the Day of Youth: exactly 40 years after the first relay baton had set off on its journey across Yugoslavia from Tito's birth town of Kumrovec in Croatia, the event ceased to take place.

The long-lasting Slovenian controversies regarding the appropriateness of artistic actions caricaturing totalitarianism were now followed by criticism

---

625  Jela Krečič, "'Plakat je kovinsko črne barve, ker je bil tudi tovariš Tito kovinar!,'" *Sobotna priloga—Delo*, May 19, 2005, 24.

from Yugoslavia. Other republics were given a perfect opportunity to become acquainted with the NSK art collective, composed of the group that designed the poster, *Novi kolektivizem* (New Collectivism), the new wave music group Laibach and the Scipion Nasice Sisters Theater. Its members, who had already earned a great deal of infamy for choosing German names, were now battered by a wave of disqualifications. They were criticized by Slovenian nationalists and received threats from Yugoslav (mainly Serbian) internationalists, both of whom identified their simulations of the spectacular glorification of collectivism, discipline, a new world, and a new mankind as neo-Nazi tendencies. The collective's very few supporters were a handful of emancipated university professors, whereas the only sympathizers from everyday politics were the members of the ZSMS, which had opened the door to ecological groups, anti-war activists and groups fighting for gay and lesbian rights as early as 1983. There was also some sympathy and moral support from the Slovene Writers Association, which was generally known for its commitment to promoting free discussion on forbidden issues and opposition to severe measures against critically-minded and engaged individuals and groups.[626]

Nevertheless, the political landscape in Slovenia appeared extremely liberal in comparison with other parts of Yugoslavia. This was not so much the result of a conscious decision but of the divisions within the communist political elite, which agreed with a number of the ideas of the opposition. This was especially evident in their defense against the reproaches from Belgrade. Moreover, the evaluations of the political situation in Slovenia by the Slovenian politicians also displayed more and more criticism of federal politics. Slovenia entered the crucial year of 1988 with widespread anticipation of change, but the general tension was partly also due to the pressures from Belgrade, which triggered a flurry of irreversible and decisive events.

## New Social Movements and the *Mladina* Affair

On May 31, 1988, the SDV arrested three journalists from the critical weekly *Mladina* magazins and one JNA non-commissioned officer on "suspicion of revealing state secrets." Janez Janša (the future defense minister and later prime minister), the journalist David Tasić, and *Mladina*'s editor Franci Zavrl obtained

---

[626] A similar spirit of solidarity was also expressed for two members of the Slovenian Writers' Association who were prosecuted. Firstly, for Igor Torkar and for his book on the Dachau trials, and the secondly, for Drago Jančar, because he was also condemned and imprisoned solely for possessing forbidden emigrant literature.

a confidential document from Staff Sergeant Ivan Borštner which provided for a possible declaration of a state of emergency in a politically unstable Slovenia. Their attempt to disclose its content created much fear that this move would only inflame the already volatile situation in the republic, but the ensuing turmoil was much more the result of their arrests and subsequent trial. During the next six months, both events became the central subject covered by all Slovenian media, since the public response across the whole republic was no less striking. The entrance to the military prison, where the arrested were kept pending trial, was surrounded by masses of demonstrators who peacefully expressed their opposition to the conduct of the military authorities.[627]

The Yugoslav Army leadership exhibited similar but diametrically opposed sentiments toward Slovenia and the Slovenes. The generals were irritated by the Slovenian political leadership and its support for the offenders, as well as by its insistence upon preserving the territorial defense system (established in 1968) and its continued struggle for linguistic rights in the army, as well as other rights. Several demands which became known to the public in the 1980s had been raised with the Yugoslav authorities by individual Slovenian politicians since the late 1960s—albeit extremely rarely and behind closed doors until some members of the Slovenian media, most notably *Mladina*, the Maribor-based student periodical *Katedra* (Rostrum), and *Nova revija*, started to raise these issues in public.

In 1988, the Slovenian political authorities were barraged by analyses of the destructive and hostile activities of new social movements against the JNA and evaluations of their own ineffectiveness in neutralizing them. However, Slovenia rejected such presumptions, while occasionally noticing that the peace movements belied a wider critique of the regime and the JNA, and even promoted the so-called "alternative" or civil service. In the eyes of military authorities, this was yet another indicator of the unreliability of Slovenian leadership, which prompted the supreme command to seriously consider the possibility of declaring a state of emergency.

---

627 Notwithstanding the massive and long-lasting protests, Ivan Borštner was sentenced to 4 years, Franci Zavrl (who had been bailed) and Janez Janša to 18 months, and David Tasić to 5 months in prison. They were granted provisional release pending their final ruling, while the Slovenian authorities continued to suspend their sentences until the very last moment and filed amnesty applications with the SFRJ Presidency. The final result of these joint efforts was that the convicted served only one third of their sentences.

The tensions dramatically escalated in the first half of 1988, when the media attacked the State Secretary of Defense Branko Mamula. *Delo* (Labor) and *Mladina* journalists criticized him for concluding an agreement with the Ethiopian regime, which preferred to spend far more money on weapons than food, and revealed that conscripts were building him a villa free of charge. The highest military authorities interpreted such writing as an overt attack on the army and the ongoing developments in Slovenia as a counter-revolution and called for strong condemnation from the Central Committee of the League of Communists. A series of official and unofficial meetings was initiated between the representatives of the Military Council and the Slovenian Party leadership, which included France Popit, Milan Kučan, and Stane Dolanc. Not only did the meetings fail to produce tangible results, but—to the contrary—led the military leaders to regard Slovenia as the second Kosovo.

A further furor was caused by the leak of confidential discussions, starting with a segment of the transcript from the late-March Central Committee session that contained Kučan's exposé on the subject. As so often before, the material came into the hands of *Mladina* journalists, who were prevented from immediately publishing their commentary by the SDV.[628] Shortly afterwards, the SDV arrested the four men in the state secrets case. Three of them—two journalists and an editor—worked for *Mladina*.

Events began to gather pace from then on. Less than a month after the arrests, the Committee for the Defense of the Rights of Janez Janša was established at the initiative of Tomaž Mastnak[629] and later renamed the Committee for the Defense of Human Rights. Since this marked the beginning of the mass movement for democratization, it should perhaps be pointed out that this section of Slovenian civil society consisted of more than 3,000 different institutions, associations, and organizations that represented more than 100,000 individual members through

---

628 Under the pseudonym Majda Vrhovnik, the later *Delo* and *Dnevnik* (The Daily) columnist Vlado Miheljak wrote a critical article entitled "Noč dolgih nožev" (The night of the long knives), which pointed to the possible declaration of emergency rule mentioned in the shorthand report. The record was supplied to *Mladina* through Igor Bavčar, a then specialized political consultant of the SZDL, who had acquired the documents from his superior Jože Knez, Vice-President of the SZDL. For a detailed treatment of the subject, see Ali H. Žerdin, *Generali brez kape. Čas odbora za varstvo človekovih pravic*, Ljubljana: Krtina, 1997.

629 Sociologist, researcher at ZRC SAZU's Institute of Philosophy and the Institute of Advanced Study in Princeton. Visiting researcher at Harvard University, New York University, Johns Hopkins University, University of California, Irvine, and other universities and institutes in the UK and Egypt. Author of the book *Crusading Peace: Christendom, the Muslim World, and Western Political Order*.

300 commissioners. After its foundation in early June 1988, the Committee organized several demonstrations which culminated in the mass meeting of several thousand people on June 21, in Trg osvoboditve, which was so known from 1947 to 1991, in Ljubljana.[630] The committee was in operation until April 1990 and participated in negotiations on the future electoral legislation launched by the Socialist Union of the Working People (SZDL) in the fall of 1989.

In less than a year, the committee became one of the central political partners in the dialogue with the Slovenian Party authorities and subsequently an indispensable agent in discussions about Slovenia's future organization and its position within Yugoslavia. In October, when it entered the final stage of the discussion on amending the federal and Slovenian constitutions, it also became involved in another pressing issue which had been under consideration since January 1987, when the Yugoslav presidency had submitted the amendments proposal for approval by the republics.

In Slovenia, the first opposition to the proposed constitutional changes, which would make the federation more centralized, came from the Društvo slovenskih pisateljev (Slovenian Writers' Association). In February 1987, it held an expert discussion and in mid-March it opened a public debate that attracted 800 participants. The main speaker and initiator of the public discussion was France Bučar.[631] Legal experts from the opposition felt that the changes would lead to more unitarization and appealed several times to the National Assembly not to support the amendment proposals. Their most resounding demand *Za demokracijo* (For Democracy) was signed by the Coordination of New Civil Movements.[632] The document drew on various drafts prepared by Bučar, Janša, Matevž Krivic,[633] and Grega Tomc, and was signed both by (over 10,000) individuals and by representatives of a broad spectrum of groups from various social movements to student and cultural institutions. On March 20, the Slovenian Assembly nevertheless voted in favor of the proposals to

---

630  The square changed its name often: Capuchin Square (1606–1821); Congress Square (1821–1945); Revolution Square (1945–1974); Liberation Square (1974–1991); Congress square (1991–).
631  "O ustavnih spremembah," a public tribune of the Slovenian Writers's Association, Cankarjev Dom, Ljubljana, March 16, 1987.
632  The statement was first read on March 8 in front of the Military Court in Ljubljana (during the trial of Franci Zavrl regarding the article "Mamula go home!" published in *Mladina*).
633  Lawyer and judge, member of the Constitutional Court of the Republic of Slovenia and member of the no longer existent Slovenian Committee for the Defense of Human Rights.

amend the federal constitution, whereas the Slovenian Writers' Association and the Slovenian Sociological Association established their own respective constitutional commissions.

In the opinion of the opposition, any constitutional changes should aspire toward greater independence for the republics (confederation), liberalization of private economic activities, abolition of the League of Communists of Yugoslavia's (SKJ) monopoly position, and the introduction of political pluralism (direct election with multiple candidates). Despite the critical views voiced by Slovenia and some other republics, the Federal Assembly passed the draft amendments to the SFRJ Constitution on December 27, 1987, followed by a public discussion and the harmonization process. The Slovenian representatives in federal agencies, reluctant to take a single collective vote on the amendments, suggested voting on each amendment separately. However, their proposals were outvoted, barring a few exceptions where they managed to push through a negligible number of minor modifications. The discussion of the amendments definitively undermined unity in Slovenian leadership, which was now divided into those who defended Slovenian interests and those who insisted that Slovenia should remain part of Yugoslavia.

**Figure 49.** A protest rally in Congress Square, June 21, 1988. A poster with the crisscrossed Article 133 of the Yugoslav Penal Code demanding the abolition of restrictions on freedom of speech (verbal delict).

The tensions were further aggravated by the concurrent debate about changes to the Slovenian Constitution. The opposition organized a special Convention for the Constitution (composed of the representatives of the Slovenian Sociological Association and the Slovenian Writers' Association, newly emerging societies, various organizations, and the League of Socialist Youth of Slovenia), and undertook to prepare its own proposal for the Slovenian Constitution. The basic platform was first presented in the journal *Časopis za kritiko znanosti* (ČKZ, Journal for the Critique of Science) and in a series of professional articles in *Nova revija*, and other publications soon followed. The opposition's list of basic constitutional demands included a confederation of entities bound by contractual obligations, the introduction of traditional democratic freedoms and the separation of powers into legislative, executive and judicial branches of government. A few opposition writers and legal experts allowed for non-partisan democracy and viewed Yugoslavia with fewer reservations. The authorities rejected the opposition proposals for constitutional amendments, but both sides continued the discussions at different levels. The government's legal experts aimed to simultaneously adopt the Yugoslav and Slovenian constitutions in order to maintain a moderate form of self-management and, to a limited extent, the assembly system. The opposition, on the other hand, preferred to see the Slovenian Constitution adopted first and only then make arrangements with the federation. The government did allow for party pluralism in the form of "alliances," which, however, should be forged on a supra-party basis, i.e., under the SZDL umbrella. They also agreed to the introduction of market conditions, equality of property, and the elimination of "socialist" from the republic's name. Their ideal was an "asymmetric federation" permitting the republics to acquire different positions or different levels of conferring jurisdictions on the federal agencies.[634]

Official politics were under intense pressure when responding with counter-proposals, which largely rejected the opposition proposals. The situation was further exacerbated by the overheated discussion on the constitutional amendments and by the foundation of the Council for the Protection of Human Rights, whose primary task was to protect the rights of all individuals living in Slovenia, regardless of their nationality and beliefs, and irrespective of who was responsible for the violation of these rights.

---

634 The leading constitutional expert among the government officials was Dr. Ciril Ribičič, who organized a number of discussions on constitutional changes. These discussions were chiefly confined to legal experts supporting the authorities' views, and rarely extended to civil society and representatives of the opposition.

The council was also expected to contribute to establishing a rule of law whose constitutional basis would promote human rights in compliance with international standards. The second turning point was the forging of political "alliances" which entered the political arena with diversified programs.[635] Some of them placed particular emphasis on democracy, while others prioritized the national question. The most prominent was the Slovenian Democratic Alliance, which mainly recruited rather conservative and well-established writers and intellectuals, although a considerable number of its members had made their careers as members of the Communist Party.

From Belgrade's perspective, the gravest single offense was Slovenia's unanimous and overt support of the miners' strike in Kosovo. In February 1989, more than 1,300 Albanian miners took over the Trepča mine and threatened to blow themselves up in protest against the new Serbian constitution, which would abolish Kosovo's autonomy. The strike ended with the declaration of a state of emergency throughout the entire province. The conflict, which turned into the Serbian occupation of Kosovo and was not even really ended by NATO's military intervention during the 1990s, continued to ravage the province for the next 20 years.

As often during this crucial period, the initiative to extend support to the Albanians came from the Committee for the Defense of Human Rights. At a demonstration later organized by the SZDL at Cankarjev Dom, the Cultural and Congress Center in Ljubljana. Speeches were delivered by both the government (Milan Kučan), and the opposition leaders (representatives of associations and groups). The declaration "Against the Introduction of a State of Emergency— toward Peace and Coexistence in Kosovo" was signed by all social and political

---

635 The Slovenian Farmers' Alliance (led by Ivan Oman) and the Alliance of Young Slovenian Farmers were established on May 12, 1988. The Slovenian Democratic Alliance (Dimitrij Rupel) was founded on January 11, 1989; the Social Democratic Alliance of Slovenia (France Tomšič) on February 16, 1989; the Yugoslav Alliance (Matjaž Anžurjev) on June 5, 1989; the Slovenian Christian Social Movement (Peter Kovačič) on March 10, 1989; the Civic Green Party (Marek Lenarčič) on March 31, 1989; the Greens Movement (Dušan Plut) on June 11, 1989; and the Association for the Yugoslav Democratic Initiative (Rastko Močnik) on September 21, 1989. The same year also saw the creation of a number of individual groups such as Grupa 88 (Franco Juri) and the Debate Club 89, which subsequently joined the ZSMS (the League of Socialist Youth of Slovenia, later renamed the Liberal Democracy of Slovenia), as well as the Academic Anarchist Anti-Alliance (Iztok Saksida), the Workers' Alliance, the Anti-Communist Alliance, the ŠKUC Association (Student Cultural Center), and the Slovenian Students' Alliance.

organizations (The Red Cross, the SZDL, the Trades Union Congress, the ZSMS, the League of Communists of Slovenia, and the League of Combatants) and all the existing associations, unions, societies, and groups. The demonstration drew fierce reactions from Belgrade, where there was a general sentiment that all forces in Slovenia (including the League of Communists) had united against Serbia and Yugoslavia. The masses attending a major rally in Belgrade on June 28 demanded a reckoning with the Albanian leaders, amid a growing fear that "Slovenia and Kosovo are tearing Yugoslavia apart." Consequently, the Slovenian representatives on federal bodies found themselves facing the worst pressure they have ever faced, while shortly afterwards Belgrade unleashed an uncompromising economic war against Slovenia and unilaterally terminated all political contacts.

However, these threats and repressions had the completely opposite effect. The Slovenian communist leadership, so far extremely wary and cautious, now undertook to forge closer relations with the opposition, which suggested the formulation of a common Slovenian political program during initial talks. After intensive consultations, the proposal was ultimately rejected. The alternative part of the newly established Coordination Committee, which had organized the demonstration at Cankarjev Dom cultural center, adopted the so-called May Declaration, first read at a protest rally on May 8, when Janez Janša was imprisoned. At this crucial moment, the authors of the declaration unequivocally called for: (1) the establishment of a sovereign state of the Slovenian nation, whose citizens would have (2) the freedom to decide whether or not they wish to unite with South Slavic or other nations within the framework of a new Europe. Finally, the newly created Slovenian state (3) would be founded on respect for human rights and freedoms, on democracy and political pluralism, and on a social system that would ensure spiritual and material welfare in line with Slovenian citizens' natural conditions and human capacities. The declaration was not signed by the Committee for the Defense of Human Rights and the ZSMS. The ZSMS based its rationale on universal human rights and citizens' rights (the sovereignty of the people) while the "nationalist" faction of the Committee for the Defense of Human Rights prioritized the sovereignty of the nation.

Although the state authorities declared the document as inadmissible because of Item 2's "negation of Yugoslavia as a national community," the Coordination Committee working group continued its activities and in 1989 produced a considerably less radical document, *Temeljna listina Slovenije* (The Fundamental Charter of Slovenia). Its moderate tone was reflected in its principled commitment to seeking a solution within the framework of the Yugoslav federation, which is why it was met with much less enthusiasm than the May Declaration, despite its substantial formal support.[636] One of the most critical points deemed unacceptable by the opposition was its consent to the federal framework, which stood as a major hindrance to Slovenia's political independence.[637]

## Amendments to the Slovenian Constitution

The divisions between the opposition and the government that had become apparent in discussions of the May Declaration also arose when it came to the Constitution. The government's legal experts considered that the constitution proposed by the Slovenian sociological and writers' associations required the elimination of social property and self-management, the substitution of the principle of unity of government for a separation of powers, the substitution of the delegation system for that of representation, the replacement of the assembly system by a presidential—parliamentary regime, and, in federal relations, the construction of a confederation as a contractual community and the establishment of the Republic of Slovenia as an entity in international law.

Nevertheless, the opposition's efforts had already borne fruit, as seen in the final versions of the constitutional amendments. Most of them were passed by the Constitutional Commission of the Slovenian Assembly in early September. The Constitutional Commission's work was directly or indirectly influenced by the Convention for the Constitution, which regularly communicated its views and played a crucial role in generating public pressure. This was one reason why the proposed constitutional amendments encountered an extremely hostile response from all authorities within the federal government and the SKJ leadership. It was also why the expert service of the Central Committee concluded that the amendments signified the introduction of an asymmetric federation and confederation and as such contravened the SFRJ Constitution. In their opinion, the solution was to be sought in political pressure.

The amendments particularly criticized by the Central Committee involved the right to self-determination, cessation and unification, as well as those relating to economic sovereignty and federal jurisdiction in the territory of Slovenia. Slovenia's substantive arguments and its claims that Serbia had itself changed the Yugoslav constitutional system in February, forbidding other republics to

---

636  The SZDL initiated a well-organized action that brought the document over 420,000 signatures, whereas the May Declaration was signed by no more than 10,000.

637  Tine Hribar, "Odločitev za samostojnost," *Mladina* (Ljubljana), December 29, 1989 (also published in Tine Hribar, *Slovenci kot nacija*, Ljubljana: Enotnost, 1994).

interfere with its "internal" affairs, were in vain. All federal bodies declared themselves against the amendments.

The entire Slovenian political sphere, including all prominent politicians, reacted very differently, and the Croatian members of the Central Committee also supported the Slovenian views. The heartened Slovenian Assembly adopted the amendments on September 27, defying threats of emergency rule from Belgrade and demonstrations held in other parts of Yugoslavia. The solemn proclamation was also attended by Janez Drnovšek, President of the SFRJ Presidency, who arrived directly from an official visit to the United States and thus, at least for the time being, alleviated the pressure from Belgrade.

The constitutional amendments were even more important. In principle they announced the end of communism, while in practice they ensured a formally regulated transition from a socialist economy to a market economy and from a one-party system to multi-party democracy. By passing the amendments, Slovenia also secured a wider jurisdiction over the economy within the federation. During the process of adopting the amendments, which resumed as early as March 1990, the adjective "socialist" was removed from all documents; arrangements were already underway to create new state symbols. France Prešeren's *Zdravljica* (A Toast) was proclaimed the Slovenian national anthem, but a heated debate surrounded the national flag and coat of arms, which were eventually, and reluctantly, adopted on the eve of the proclamation of independence.

Notwithstanding the triumphant adoption of the amendments to the Slovenian Constitution, the Slovenian leadership (partly based on information from the SDV) was well aware that the final showdown was yet to take place around the issue of existence, political and otherwise. They also knew that a most crucial role in this finale would be played by the JNA, whose top officials dared not declare emergency rule, although they had agreed to do so with the Serbian representatives in the federal leadership. Even though the Federal Secretary of Defense, Veljko Kadijević, tried to deflect criticism by claiming that there was no legal basis for such a measure, the true reason for inaction was most probably that the top generals were torn between Serbia and Yugoslavia. But the idea of settling the situation by force was not abandoned; the Slovenian constitutional amendments were only one pretext considered by the top military and Serbian leaders during the three years between the Trial of the Four in the summer of 1988 and the proclamation of Slovenia's independence in June 1991.

But now not even the overt threats from Belgrade could stop the process, which had started to gain momentum even before the "amendment crisis." On the contrary, after the amendments, dialogue continued between the representatives of the departing communist authority and the opposition in order to negotiate an agreement about multi-party elections. Given that several groups which had previously been participating in the coordination arrangements did not take part in the Round Table, it disintegrated within a month, while the opposition compiled its views in the manifesto *What Kind of Election Do We Want*.[638]

## The Formation of Demos and the Transformation of the Slovenian Political Landscape

Fourteen days after the disintegration of the Round Table, the Slovenian Farmers' Alliance, the Social Democratic Alliance, the Democratic Alliance of Slovenia, and the newly established Slovenian Christian Democrats united into the pre-election coalition DEMOS, which stands for the Democratic Opposition of Slovenia, which was subsequently joined by the Zeleni Slovenije (Greens of Slovenia), the Slovenska kmečka zveza (Farmers' Party), Liberalna stranka (Liberal Party), and the pensioners' Sivi panterji (Gray Panthers Party). Jože Pučnik was elected president, and the central points in the DEMOS election program were a sovereign Slovenia and a parliamentary democracy.

One month later, on December 27, 1989, the association and electoral acts laid the groundwork for alliances to transform into parties, allowing the Slovenian political party landscape to finally take shape. Former alliances turned into regular political parties and a similar trend was observed among former "socio-political organizations." The ZSMS was renamed the Liberalna demokracija Slovenije (LDS, Liberal Democracy of Slovenia), the League of Communists changed into the Stranka demokratične prenove (SDP, Party of Democratic Renewal), and the Socialist Alliance became Socialistična stranka (Socialist Party). The League of Combatants retained the status of a non-party organization but offered silent support to the communists in particular. In the face of increasing competition from related socio-political organizations, the former socialist trade unions were compelled to strengthen their commitment to tackling trade union issues.

The final important episode unfolded shortly before the parliamentary elections (late January 1990), when the Slovenian representatives walked out of the last Congress of the Yugoslav Communist Party in Belgrade. The Slovenian

---

638   The manifesto was issued by the Slovenian Writers' Association, Klub 89, the Committee for the Defense of Human Rights, the Center for the Culture of Peace, the Slovenian Democratic Alliance, the Slovenian Farmers' Alliance and the Alliance of Young Slovenian Farmers, the Slovenian Christian-Social Movement, the Social Democratic Party of Slovenia, the Greens of Slovenia, and the Alliance of Socialist Youth of Slovenia. The undated manifesto was made public in early September 1989.

communists had drawn up a proposal for democratic reforms for the entirety of Yugoslavia (guarantees of human rights, a multi-party system, the abolition of the crime of "verbal delict" and termination of political trials, a resolution of the situation in Kosovo in accordance with the Yugoslav Constitution, direct elections, the reform of the federation, and the transformation of the Yugoslav League of Communists into an association of autonomous entities). After all their proposals were outvoted, they left the congress and focused on the pre-election situation in Slovenia. After changing its name and winning the first democratic election, the Slovenian Communist Party (under the alias: the Party of Democratic Renewal) could not form a government and was, moreover, financially bankrupt. A vast majority of Party senior officials thus lost their previous positions and pursued new opportunities in publishing, universities, and management. This was largely made possible by the almost complete absence of the promised, sometimes well-elaborated retributions. The election campaign itself proceeded without any major incidents, despite the considerable inexperience of all the parties involved.

Based on the assembly system, which still drew on the 1974 Constitution (with revisions), the parties proposed candidates to three chambers: the Socio-Political Chamber, the Chamber of Municipalities, and the Chamber of Associated Labor. Altogether this comprised 240 deputies. Parliamentary seats were won by ten parties, two minority representatives, and a few independent candidates. The presidency of the republic was elected by direct vote (Milan Kučan as president and Matjaž Kmecl, Ivan Oman, Dušan Plut, and Ciril Zlobec as members). The already mentioned France Bučar, a lawyer and one of the few Slovenian dissidents, was elected President of the Parliament at the new constitutive session of the Parliament held on May 17, 1990. As a former partisan, a sacked professor of law, and an intransigent critic of the totalitarian and anti-democratic regime in Yugoslavia, he was most certainly considered the perfect man for this position. Even though his speech announced the end of the 50-year-old civil war, it (unintentionally) contributed to later historical revision of the World War II period.

Amid this turbulent atmosphere, the mandate to form a new government was entrusted to Lojze Peterle, President of the Slovenski krščanski demokrati (SKD, Slovenian Christian Democrats). The new 27-member government, which also included a few ministers from the opposition parties, was elected without major opposition. The change of government caused no major disturbance, but it still set off a few sparks when the presidency asked Peterle to determine the new government's program and composition. Peterle refused to submit the candidates' names and argued that the presidency had no right to know because their role in appointing the mandate-holder was only a formality. The incident was finally settled with an unofficial compromise list of future government members, which Peterle revealed after an extraordinary meeting with the members of the presidency of the republic.

The initial period of the new government was marked by severe mutual distrust between the coalition and the opposition. Several government parties believed that it was up to them alone to bring the "independence project" to completion and thought that the opposition, made up of former communists, merely impeded the independence struggle. Facing the disastrous economic situation, Peterle was never enthusiastic about this option and focused on the preparations for independence, starting with the Declaration of Slovenia's Sovereignty, which also received support from the opposition. However, discontent among opposition deputies was caused by the government drawing up the sovereignty declaration without consulting either them or the president of the republic. The heated political climate was cooled to some degree by a joint motion for a referendum on independence. The motion won a majority vote in the Parliament after a protracted discussion once it was clear that the new constitution would not be adopted before the end of December 1990, which could ultimately bring the independence process to a halt.

Despite the opposition's initial apprehensions, 88.2 percent of a 93.2 percent turnout voted for independence in the referendum. Both the United States and the European Community opposed Slovenia's independence aspirations and refused to send official representatives and observers to the republic at the end of 1990. However, the new government's main concern was not the referendum. Like their predecessors under the leadership of the last socialist president Dušan Šinigoj, who had been confronting the consequences of the economic war with Serbia since 1988,[639] the new ministers were too primarily facing what Jože Mencinger (the vice president of the government in charge of economic affairs) described as "catastrophic" conditions.

Peterle's cabinet was also compelled to adjust to changes in Belgrade. After Mamula's government had collectively resigned at the end of 1988, the helm of the last federal government of the SFRJ was taken by the Croat Ante Marković, who in March 1989 made the first serious attempt to advance the economic reform.[640] With the aid of Western European countries and the US, he reconstructed the reform program in 1990. It took Marković a few months

---

[639] According to a questionnaire prepared by the Chamber of Commerce and Industry of Slovenia, 229 Serbian companies severed business relations with their Slovenian partners.

[640] "Program ekonomske reforme i mere za njegovu realizaciju u 1990 godini," in: *Ekonomska reforma i njeni zakoni*, Belgrade: Federal Executive Committee—Federal Secretariat for Information, 1990, 5–44.

to improve Yugoslavia's foreign liquidity, increase foreign reserve assets (at 5.8 billion US dollars), and decrease total debt by nearly 4 billion US dollars. Much less effective, however, were his attempts to curb hyperinflation, which stood at approximately 2,700 percent from mid-December 1988 to December 1989.

This development partially explained the Slovenes' serious reservations about the Marković reforms, especially given that power was concentrated in the Savezno izvršno vijeće or SIV (Federal Executive Council) and the Narodna banka Jugoslavije (NBJ, National Bank of Yugoslavia). In their eyes, the program was also an opportunity for an organized, conscious, and planned redistribution of all internal debt and an even greater amount of external debt. It was clear that there was no way to recover the debt from actual debtors and that the main burden would be carried by the most developed parts of the country which were more capable of payment. Since the basic principle of financing the federation in proportion to the republic's share in national product or national income had been transferred to financing debt, it also became patently evident that the heaviest burdens of the reform would be shouldered by the most developed republics.

Slovenian economists had reservations about the spectacular introduction of the so-called convertible dinar, finding the exchange rate of 7 dinars for 1 German mark too low. However, their skepticism also arose from their experiences over the last few decades, which had taught them that reform could not instill financial discipline in all republics. This was one reason why the Slovenian Assembly concluded that the proposed program might offer the final opportunity to develop common objectives for solving the Yugoslav crisis within the framework of federal bodies and applicable legislation. Should the reforms fail, the competent authorities of the republic of Slovenia were to prepare concrete proposals to guarantee political and economic sovereignty.

After the exchange market ceased to exist in the fall of 1990, the reforms did collapse, prompting the Slovenian government to consider monetary independence. Even though the first provisional banknotes were ready for use as early as December 1990, the new currency, called the Slovenian tolar, went into circulation only after independence—in October 1991.

The collapse of the Yugoslav financial system to a large extent also resulted from the "incursion" of the Serbian authorities into the Yugoslav banking system at the end of 1990. Aided by Serbian banks, Serbia "borrowed" 1.4 billion US dollars from the NBJ, a sum which, in 1991, amounted to half the planned total amount of all borrowing by Yugoslav banks from the NBJ. Serbia had entered its own pre-election period, and Milošević needed the resources to pay pensions and salaries. This drastically increased the amount of money in circulation. Marković demanded that

Serbia return the money to the NBJ, but took no specific measures to ensure that his demand was actually met.

In Slovenia, these developments led to another heated discussion, this time on privatization and denationalization. Two models of privatization were developed: one by Jože Mencinger and one by Harvard professor Jeffrey Sachs. Sachs started from the consideration that all main economic measures were to be taken simultaneously and insisted on a fixed exchange rate for the new currency. The state would become the chief titular owner of public property and as such receive the mandate to appoint directors and administrative boards for the transition period. Internal buy-outs would be restricted, and workers and management would be granted the right to buy up to 15 percent of shares, but no voting rights. The remaining shares were to be transferred to the Development Fund and, through it, the Pension Fund (15–20 percent), to Ljubljanska Banka, the largest Slovenian bank, with its affiliates (5–10 percent), and various investment companies or funds (4–5 percent, set up by either domestic or foreign investment groups) in which all Slovenes would acquire shares. According to Sachs, such a system would prevent the old management structures from obtaining legal control over all major companies and simultaneously boost the Slovenian authorities' popularity through the free distribution of shares (around 30 percent in major companies). The Development Fund would sell approximately 25 percent of its shares to domestic and foreign investors. Ultimately, major companies would not be controlled by internal shareholders but by the representatives of the state, banks, various funds, and the Pension Fund.[641]

The government's keen interest in Sachs's model led Mencinger to resign, although the cabinet ultimately combined both models. The approach to denationalization was a little different, and a law supported by the conservative faction of the DEMOS coalition easily passed through the Parliament.

The independence process was thus well under way in thirteen different fields (dissolution, the economic system, foreign economic relations, foreign financial relations, foreign affairs, cooperation with other Yugoslav republics, supply, infrastructure, industry, borders, defense, legislation, and the media), steered by the Project Council, first headed by Lojze Peterle and later by the Minister of the Interior Igor Bavčar.

---

641 *A program for Economic Sovereignty and Restructuring of Slovenia. A proposed Policy Framework*, March 21, 1991, Archives of the Government of the Republic of Slovenia. The Slovenian version of the document is available in Neven Borak, Žarko Lazarević and Jože Prinčič (eds.), *Od kapitalizma do kapitalizma*, Ljubljana: Cankarjeva založba, 1997, 628–640.

## From the Idea of an Asymmetric Confederation to the Struggle for Independence and Beyond

It should be emphasized that the Slovenes espoused the idea of an independent Slovenian state only when life in Yugoslavia became unbearable. This was first and foremost due to the disastrous economic conditions,[642] but also because, apart from the unrealistic plans of Yugoslavia's last prime minister and the energetic Serbian opposition to the concept of asymmetric federation, there was no other alternative in sight. These were perhaps two key reasons why the disintegration of the common state was so dramatic and its consequences utterly devastating. At the beginning of 1991, no one could have imagined that the price was to be so high. The number of victims, which ultimately amounted to more than 200,000 dead and at least a million displaced persons, could not have been foreseen even by the darkest pessimists. The course of action taken by the Slovenian government and the Parliament provided hope for a much more peaceful outcome.

On March 8, 1991, the Slovenian Parliament passed a constitutional act making military service in the JNA no longer compulsory for Slovenian citizens. It also enabled Slovenian citizens to do their military service in Slovenia. The act also allowed for alternative service. A similar but exclusively defensive purpose was signaled by the subsequent establishment of the defense headquarters and the coordination group, which in the middle of 1991 effectively organized the actions of the Territorial Defense and police forces in all their engagements with the JNA. The most volatile moments (e.g., the capture of the Chief of the Regional Territorial Defense Headquarters or the blockade of a Slovenian Territorial Defense recruiting center) culminated in the cutting off of electricity and water supplies for the army barracks and inevitable offers for negotiations.

Yet, amid constant uncertainties, the first completely independent Slovenian budget was adopted on April 1. It was soon succeeded by several acts of independence, including the Citizenship of the Republic of Slovenia Act, the Passports of the Citizens of the Republic of Slovenia Act, the Aliens Act, the National Border Control Act, the Customs Service Act, the Bank of Slovenia Act, the Banks and Savings Banks Act, the Foreign Exchange Act, and the Defense and Protection Act. Two days before the official declaration of independence, the emerging state obtained its new coat of arms and flag. After a protracted

---

642  In 1985, the disparity in living standards between Yugoslav republics grew wider. Serbia's gross domestic product per capita was equal to Turkey's, Kosovo's to Pakistan's, Croatia's to Greece's, and Slovenia's to Spain's and New Zealand's.

and fierce debate, a decision was made in great haste that the former Slovenian coat of arms on the old tricolor should be replaced by a new one designed in the form of a shield by the sculptor Marko Pogačnik. The author remained largely faithful to the symbolic design of the previous coat of arms. A blue shield bears a white stylized image of Triglav, the highest mountain in Slovenia and in former Yugoslavia, beneath it run two wavy blue lines symbolizing rivers and the sea, and above it there are three six-pointed stars arranged in an inverted triangle, resembling the stars on the coat of arms of the Counts of Celje. On the eve of the proclamation of independence, the Parliament adopted the Temeljna ustavna listina samostojnosti in neodvisnosti Republike Slovenije (Basic Constitutional Charter on the Independence and Sovereignty of the Republic of Slovenia) and the Deklaracija ob neodvisnosti (Declaration of Independence).[643] Based on these and previously adopted acts, Slovenia assumed the former federal jurisdiction over its territory. On June 25, Prime Minister Lojze Peterle informed customs posts that from that day forward they would operate under the jurisdiction of the Republic of Slovenia. Officials from the federal administration were transferred to the Slovenian customs service and several provisional crossing points on the Slovenian–Croatian border were set up, with 3,000 police and Territorial Defense officers deployed there a few days earlier. Slovenia and Croatia arrived at an agreement on the land border, while the maritime border was left to be decided by future negotiations. On that day, Slovenia also took over air traffic control over Slovenian airspace.

These developments added yet another tone to the fierce reactions from Belgrade. The Federal Executive Council responded to Slovenia's last steps by enforcing the Decree on Implementing Regulations Relating to the Payment of Customs Duty and ordered the deployment of customs officers from other parts of Yugoslavia to the Slovenian border. Until then, the main part of the dialogue between Belgrade and Ljubljana took place on two seperate tracks: on the one hand, the War Council demanded that Slovenia continue to comply with enlistment procedures soon after it had passed the Military Service Act, while, on the other hand, the Federal Executive Council headed by Marković cautiously delayed retaliation measures for fear of a possible armed conflict. Two weeks before independence, Marković had even come to Ljubljana at the invitation

---

643 The Charter declared Slovenia an independent and sovereign state, whereas the act on its implementation imposed on the Slovenian authorities to exercise the powers that had until that moment been vested in the federation by the constitutions of the Socialist Republic of Slovenia and the Socialist Federative Republic of Yugoslavia.

of the Slovenian prime minister and tried to reason with the government and the Parliament to shelve its declaration of independence and continue participating in negotiations to determine the new organization of Yugoslavia. On this occasion, Marković suggested that the JNA should act as a kind of "blue helmet" peacekeeping force, but warned that the federal government would use all legal means at its disposal to prevent any unilateral changes to both internal and external borders, should Slovenia further its independence aspirations. Marković unequivocally repeated this threat soon afterwards at the session of the Federal Chamber of the Yugoslav Assembly. As if by agreement, the members of the European Community supplemented his words with a warning two days later that they would not recognize the independence of any republic declaring a unilateral secession from Yugoslavia.

The Croats found themselves in a similar situation and the two republics, which had completed their final transformation into independent states, decided to synchronize their efforts of concluding the independence processes. Therefore, it was no coincidence that the two republics adopted their formal decisions almost concurrently. Slovenia justified its declaration of independence with the result of the plebiscite on December 23, 1990.

Even though this was a historic decision and a singular moment, there was no time for celebration. Almost immediately after Slovenia declared its independence, the JNA was ordered to secure the borders of the common state. A JNA anti-aircraft armored battery from Karlovac crossed the Slovenian-Croatian border and at about 4:00 a.m. engaged in the first skirmish with Territorial Defense units.[644] Shortly before that an armored battalion had set off for Brnik international airport from Vrhnika near Ljubljana, and the longest border crossing with Austria (the Karawanks mountain range) was seized by air assault units. Although the JNA at first did not encounter any major resistance and Admiral Stane Brovet announced that same afternoon that the Yugoslav armed forces had achieved their objective and secured the state borders, the situation on the ground was far more complex. That same evening, the Territorial Defense forces shot down one of the helicopters that were to transfer the members of the Yugoslav Brigade for Internal Affairs to the border crossings. Almost simultaneously, the Slovenian Territorial Defense and police forces surrounded most army barracks and cut off electricity, water, and telephone connections. The JNA's plan to "secure" all border crossings started to fail. By June 28, Slovenian

---

644 Božo Repe, "Začetek vojne v Sloveniji," in: Drnovšek, Rozman and Vodopivec (eds.), *Slovenska kronika XX. stoletja*, 468.

units had recaptured some border crossings, army barracks, and warehouses. The JNA launched a series of air-raids targeting barricades, TV transmitters, and the platform of the Karavanke border crossing.

Another immense and very effective contribution to the war was made for Slovenia by the Slovenian Press Agency and the media, coordinated by the then Minister of Information Jelko Kacin. Largely owing to Kacin's team, the European and world community was quick to learn that the smooth restoration of law and order (which initially received EC and US approval) had rapidly grown into a serious military conflict that stirred the international public to put pressure on their governments.

Another reason for the failure of the Yugoslav People's Army and the special federal units may have been because the military campaign against Slovenia was mostly waged by inexperienced conscripts. The major surprise was the high tactical and organizational readiness of the Slovenian Territorial Defense, which caused the Yugoslav army much unexpected embarrassment. In addition, the JNA's role and mission in the early 1990s were riddled with profound dilemmas. The JNA was a crucial institution for preserving the federation and upholding Yugoslav interests. If the country were to dissolve, the JNA would be left without a state and its very identity would be thrown into question. On the other hand, it is important to know that in the early 1990s the JNA's composition was predominantly Serbian. But the ten-day war for Slovenia was, most importantly, brought to a rapid conclusion because by then the Serbs had already shifted their primary attention to the situations in Croatia and Bosnia–Herzegovina. This could already be discerned during the talks between the presidents of all the republics held in Belgrade in January 1991,[645] when Slobodan Milošević made it palpably clear that if any attempts were made to replace the federal structure of Yugoslavia with some looser arrangement, he would seek to annex the Serbian-populated areas of Croatia and Bosnia. The favorable outcome for the Slovenes certainly also resulted from the efforts of the international political community, even though the subsequent wars in Croatia and Bosnia–Herzegovina showed that without Serbia's consent to the JNA withdrawal from Slovenia four months

---

645 On January 10, the presidents of the Yugoslav republics met for the first time and opened a discussion on the future of the federation. Their meetings then rotated throughout the republics' capitals. At a meeting in April in Slovenia, they formed two possible solutions about the future organization of the state. One envisaged a union of sovereign states, i.e., a confederation, and the other sought to preserve the single federal state.

later, the ten-day war and the whole independence enterprise could never have come to such a fortunate closure.

The EC sent the foreign ministers of Italy, Luxembourg, and the Netherlands (Gianni de Michelis, Jacques Poos, and Hans van den Broek) to Slovenia and Yugoslavia as observers. The ministers held meetings with Milan Kučan, Ante Marković, Franjo Tudjman, and Dimitrij Rupel in Belgrade, Zagreb and Ljubljana, and adopted the rather vague formula of a "moratorium" on the implementation of the Constitutional Charter on Independence. The first truce was reached on the same day, but unfortunately it did not last. Ten days into the war, the number of dead had risen to 65, while the Slovenian armed forces had captured an additional 3,000 soldiers, military, police, and customs officers. The cessation of hostilities was finally confirmed at the Brioni meeting, which was attended by the representatives from Serbia, Croatia and Slovenia, and the EC mediators. Slovenia was represented by Drnovšek, Kučan, Bučar, Peterle, and Rupel. The EC representatives held talks with each delegation separately and all the disputing parties held bilateral meetings with each other. The protracted negotiations finally yielded the Brioni Declaration, which also provided that the Slovenian police would retain control over border crossings in compliance with federal regulations, that customs duty would form a component of federal revenue, that air traffic control would remain a federal government function, and that the JNA units would unconditionally withdraw to the barracks. The Brioni Declaration was then ratified by the Slovenian Parliament. One week later, on July 18, the Presidency of the SFRJ resolved that the JNA should withdraw all of its units and equipment from the territory of Slovenia.

Both parties had signed a three-month moratorium of secession at Brioni through European Community mediation. After the moratorium expired, independent Slovenia was first recognized by Croatia and the Baltic states, and a short while later, on January 15, 1992, by the European Community. The state, which in the first decade of its independent existence was often mistaken for Slovakia and the Croatian province of Slavonia, became a full member of the United Nations on May 22, 1992.

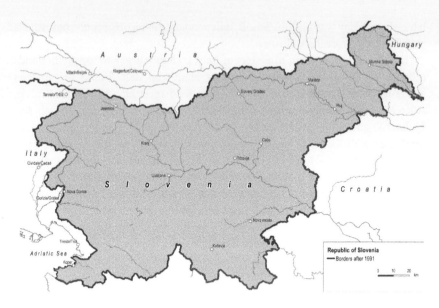

**Map 10.** Map of the Republic of Slovenia after independence.

Nevertheless, according to its Constitution, Slovenia had come into existence as a democratic republic as early as December 23, 1991. From this date onwards, legislative power has been vested in the 90-member National Assembly, which elects the government and supervises its work. The president of the republic, elected by direct popular vote, represents Slovenia in international relations. The first election, which took place in December 1992 in accordance with the new constitution, was won by the Liberal Democracy of Slovenia, the most powerful party throughout the entire first decade of independent Slovenia. For much of that time, its leader Janez Drnovšek, who later became president of the republic, served as prime minister. When Slovenia was granted UN membership, Milan Kučan, the first president of the republic elected for two consecutive terms, delivered a speech offering a concise summary of the reasons for Slovenia's recent independence and defined its future objectives. He said:

> "Slovenia did not declare independence in order to become an island in the middle of the world [...], but to ensure an appropriate role and just treatment in the processes of integration which we are joining." He concluded his address with the promise that Slovenes as citizens of the former Yugoslavia were willing to contribute "to the assertion and respect of the national diversities of, and coexistence among, all nations."[646]

---

646 Quoted from Dolenc and Gabrič, *Zgodovina 4*, 247.

From the Idea of an Asymmetric Confederation ... 551

The road toward the European Union was open, and the steps were ever faster and more confident. However, the period from December 1999 to November 2000 brought about a turbulent time in a new democracy: in less than a year, Slovenia switched three governments, two voting systems and one constitutional article. In the first half of the year, the united SLS+SKD (also known as the Slovenian People's Party) dominated the political horizon as a "genuine conservative party." It seemed that the merger between the SLS and the SKD had paved the way for the emergence of a firm center-right coalition in the form of Coalition Slovenia, which, apart from the united party, also included the SDS. But as soon as the coalition was built anew, yet a caretaker government, it fell apart after a modified system of proportional representation was set out in the Constitution. It only consisted of the SDS and the newly formed non-parliamentary party NSi, headed by Andrej Bajuk, a long-time emigrant to Argentina, who became the prime minister for a period from May to October. After center-left parties posted success in the October parliamentary elections, a broad coalition was set by the LDS, ZLSD, SLS+SKD and DeSUS, which then formed a government, Janez Drnovšek as prime minister once again.

**Figure 50.** The Slovenian Territorial Defense takes over the Ljubelj border crossing between Slovenia and Austria, June 1991.

The first year of Drnovšek's fourth government, and the country's sixth, was surprisingly calm—perhaps because of a relatively strong coalition government (the LDS alone gained 34 seats within the 90-seat Parliament) and a very determined goal to enter Euro–Atlantic integrations at the top of the group of first new members. Constitutional changes were worked on more intensively and there were several steps taken toward marking mass graves and adopting a different attitude toward the postwar killings. June was marked by a referendum on the in-vitro fertilization of single women, which was accompanied by a heated debate accross all levels of society. Although the referendum initiative was filed by lawmakers of both the opposition and the coalition, the act failed to raise public support. What also caused quite a stir on both sides was the (non-)return of nationalized Church property. Both the Ministry of Agriculture's decision to return some 8,000 hectares of forests in the Triglav National Park and the decision of the Ministry of Culture to return in kind the church on the Lake Bled island, and not the entire island, resulted in several lawsuits and severe public furor. Slovenia also had to face the first major wave of refugee migrations, which were illegal and a regular threat to the country's stability, according to Slovenian authorities. Due to police brutality and inappropriate conditions in refugee centers and in the asylum-seekers' homes, Slovenia also found itself on the Amnesty International's annual report on human rights violations, with the increasing police brutality also mentioned in November's European Commission report on EU candidate countries' progress toward accession.

In 2002, the simultaneous holding of local and presidential elections—an occurrence of every 20 years—provided for a very diverse year in domestic affairs. Because Milan Kučan, who had been the only president since the country's independence eleven years earlier, could not run again, Prime Minister Janez Drnovšek was elected president. In turn, Drnovšek's Minister of Finance Anton Rop became the new prime minister, and a total of 193 Slovenian municipalities elected new mayors and city councils as the country held its third post-independence local elections—somewhat overshadowed by the presidential ballot.

The results of the first census after independence in 1991 showed that Slovenia had 1,964,036 inhabitants, which was 2.6 percent more than in the previous census carried out 11 years earlier. With negative natural population growth, the main reason for the increase was immigration and the regulation of the legal status of people who moved to Slovenia before the 1991 census. The 2002 census also showed interesting changes in the structure of religious groups, with the number of Roman Catholics falling from 71.6 percent in 1991 to 57.8 percent in 2002. Nevertheless, Roman Catholics remained the largest single religious group in the country, followed by the Islamic community; with as much as 2.4 percent of inhabitants citing Islam as their religion as compared to 1.5 percent in 1991, the Muslims had outnumbered the previously second largest religious group in Slovenia, that is, the Protestants. As for ethnic origin, as many as 83.06 percent of Slovenian inhabitants declared themselves Slovenes, (88.31 percent in 1991, 96.52 percent in 1953), followed by Serbs (1.98 percent) and Croats (1.81 percent).

For the third time since its formation 11 years earlier, the Constitution had been changed. The Parliament assumed its rarely seen constitution-drafting role as it amended the supreme legislative act, the Constitution. The constitutional act, which was passed in order to enable Slovenia to become a member of supranational institutions and cede part of its sovereignty to them, was required to allow Slovenia to join the EU and NATO. On March 23, 1993, the membership of the EU and NATO was confirmed in a referendum in a traditional plebiscite manner: the turnout was 60.4 percent, and over 89.6 percent of the voters supported joining the EU. NATO membership was backed by 66.08 percent of the vote.

Twelve years after becoming a full member of the United Nations, Slovenia also joined the European Union, on May 1, 2004. Unfortunately, its role in pacifying the crises in the southwestern Balkans did not bear fruit. Quite the contrary: Slovenia was never really able to take strategic advantage of its familiarity with the situation and the language or its long-standing intimate experience of coexistence, despite incessant references to doing precisely that.

Efficient Slovenian mediation in this field was also thwarted to some extent by the domestic situation, which was far from simple, not least because the two most important post-independence legislative acts which would partially rectify postwar injustices and boost economic growth were both slow to be implemented. The government needed 15 years to ensure the proper implementation of the Denationalization Act, while, even at the beginning of 2008, it was still impossible to assess the effectiveness of the Privatization Act with any certainty. In spite of these drawbacks, Slovenia was able to maintain its image of the most successful new EU member. It also became the first among the former Eastern European states to adopt the common European currency in January 2007 and to take over the EU Presidency one year later. It was Janez Janša's government that, following the celebration of the victory in the 2004 parliamentary election, took the EU Presidency as its pre-election campaign—and then lost the new election in 2008. That same year, in February 2008 and just after the end of his term, came the death of Janez Drnovšek, the country's second prime minister and the second president of the republic. Not only symbolically, this marked the end of an important period of Slovenian history. Danilo Türk, a long-time diplomat and the first Slovenian permanent representative to the United Nations, became the third president of the republic.

Five years later, he lost the presidential election to the career politician Borut Pahor, whose term as Slovenian prime minister ended at the early parliamentary election in the fall of 2011. Regardless of the fact that the latter was won by the newly established center-left party Positive Slovenia, the tenth government of the independent Republic of Slovenia was once again formed by Janez Janša and his SDS. One of the reasons for this unexpected turn of events may undoubtedly be attributed to the consequences of the global crisis. Specifically, Janša succeeded in making the voters believe that the economic situation was largely the result of the wavering policies of the Social Democrats and the Liberals, who had been slow to provide the reduction of public spending and the pension system reform. The longer the crisis persisted, the clearer it became that the great economic downturn 20 years after the independence was the doing of all political parties. Rather than providing structural changes (controlled privatization of some economic sectors, the pension system and labor market reforms, etc.) and a serious development strategy, they were interested in economic benefits, striving for control over state-owned companies and looking for a reinterpretation of World War II history. For this reason, the twentieth anniversary of independence was indeed dedicated more to the memory of the secession from Yugoslavia than to the introduction of parliamentary democracy. What is more, the rule of law, the market economy, and the role of alternative movements for democracy, which were so prominent in the critical 1980s, were almost completely ignored.

Post-communist nationalism and archaic neoliberalism crushed the endeavors to preserve the welfare state. And when, at the end of 2012, protesters took to the streets again—including many of those who some twenty-five years ago expressed their support for the imprisoned Janez Janša, who was now in the role of the prime minister, and was calling them left fascists in his last public address—it suddenly became clear that a man who was considered one of the main architects of Slovenia's democratization and independence suddenly turned into Big Brother, whose sole objective was to hold political power.

Therefore, the major difference between the demonstrations of the late 1980s and early 2010s was that in 2012, people demonstrated against their own political elite. Some protesters, however, also drew attention to corruption in the Slovenian Catholic Church which was a consequence of the Church's political Catholicism expressed in "direct interventions into political matters by the Church leaders."[647]

Two decades after the creation of the independent state and formation of its own political elites, politics seeped its way into every pore of social life. This even led to political bias and polarization among the veterans of the war for Slovenia's independence, otherwise grouped into three rival veteran organizations. Based on a conflict that erupted between the armed and police forces soon after the independence, the Association for the Values of Slovenian Independence was formed on the eve of its 20th anniversary. The reason for this division can already be seen in its name, which places ideological or political orientation above people or veterans. Also testifying to the fact that this was more than a coincidence was the festival of patriotic songs, which the above-mentioned association organized for the first time in 2012 and which Janez Slapar, the first chief of general staff of the Slovenian armed forces, described as the wrong path toward patriotism.[648]

And finally, culture, too, became laced with political implications as a self-evident reflection of social developments on the one hand and as one of the stages of the above-mentioned cultural struggle on the other. This gave rise to more or less direct and more or less witty comments about several politicians as well as to synchronized actions within the framework of the so-called "all-Slovenian uprising" against corruption in politics and against the dismantling of the welfare state.

The most attention-grabbing among individual projects was the work of three Slovenian artists who legally changed their names to "Janez Janša" and shot a documentary entitled *Jaz sem Janez Janša* (My name is Janez Janša, 2012). What is more, with their last project, the Janez Janša trio drew on the post-World War II tradition of socially engaged art, a tradition started by the art collective OHO in the 1960s and continued by the Neue Slowenische Kunst movement. Just as the late socialist cultural commissars attacked the works of the NSK, the

---

647 Peter Kovačič Peršin, "Cerkev in družba. Kam je (s)krenil slovenski katolicizem?" *Sobotna priloga—Delo*, February 16, 2013, 10.

648 "The best path toward patriotism is a friendly state, objective presentation of independence, and not singing patriotic songs." In: "Intervju Janez Slapar," *Objektiv—Dnevnik*, February 16, 2013, 9.

minister of culture in Janša's second government, too, fell for the provocation of Janez Janša, Janez Janša, and Janez Janša by describing the political satire as a project "that incites political ridicule and hatred and discredits an individual."[649]

**Figure 51.** Demonstrations in Ljubljana, February 7, 2013.

The minister's reaction can partly be understood as the ranting of a frustrated representative of the government that enjoyed the lowest level of public support in the brief history of the independent Slovenia. The reason for this can undoubtedly be found in "the staunch neoliberal austerity policies"[650] that

---

649 Janez Janša, Janez Janša and Janez Janša, "Politična samovolja ne sme vplivati na izvajanje kulturnih programov," *Delo*, February 16, 2013, 19.
650 At the beginning of 2013, one of the most famous Slovenian economists, the long-standing Director of the Institute for Economic Research Lojze Sočan concluded, based on a comparison of anti-crisis programs in 27 EU member states and nine OECD members, "that the program of the [Slovenian, author's addition] government pursues the staunchest neoliberal austerity policy." In his opinion, it was "the most primitive version of austerity measures." Miha Jenko, "Bojim se, da je izgubljeno celo več kot desetletje," an interview with Lojze Sočan, *Sobotna priloga—Delo*, February 16, 2013, 4–7.

continued to drain the entire economy rather than develop economic policies with added value. Instead of undertaking a serious development strategy, economists say, "(all) governments in general made it a priority to place their people in high positions and implement their own obscure economic projects in complete disregard of expertise."[651] That was also the trademark of Janša's last term in government and the same goes for the majority of political elites in power at the beginning of the country's third independent decade. This led many to look for opportunities abroad,[652] while it was a herald of the third Slovenian spring for others.

---

651   Ibid.
652   According to the data from overseas recruitment agencies, about three times as many mostly young and well-educated people left Slovenia in 2012 than in the previous decade.

# History Does (Not) Repeat Itself

Despite our firm belief that history does not repeat itself, the turbulent year of 2020 seems to have proven us wrong at least once. Janez Janša, and his third term in office (2020–2022), reminded us why some people, including some historians, tend to use the *history repeats itself* slogan. Much like his second term, Mr. Janša's third term as prime minister showed us how he and his supporters are continuing to fail to see life in independent Slovenia as a collective national endeavor, which came about as a result of long-term and sustained social, cultural and political actions by a plethora of democratic movements and independent or alternative cultural initiatives. Instead of stressing the rule of law and solidarity, he and his political partners have been striving for further monopolization of political power and mythologization of his public and political persona. On the top of that, his government was marked with the extremely poor, nontransparent and autocratic management of the Covid-19 pandemic.

All of this forced many Slovenes to once again ask themselves what went wrong in a society that was praised as the first one in Eastern Europe to develop a "functioning democracy" and used to be presented as a model for other countries in transition to aspire to?

## What Went Wrong?

The first part of the answer can be sought in the change in the political vocabulary and practice that embedded Slovenia in EU and US political and economic institutions, as well as cultural milieu. The rise of neoliberal lingo, accompanied by heavy-weight populism, subjected the Slovenians, as well as other other post-socialist countries, to a paternalistic approach of EU institutions. Not only was the Bank of Slovenia forced to spend prohibitive amounts of money to bail out the public bank system, the government was then also compelled to sell most of their banks. This whole process almost ended up in bringing in the infamous "troika" (the EU Commission, the European Central Bank, and the International Monetary Fund), which was followed by the final stage of a decidedly opaque process of privatization. This erratic process rid the country of its last successful state-owned companies, such as Mercator, the biggest grocery chain in the country, which was also an important regional player, as well as Elan, the internationally renowned ski and yacht factory, along with Gorenje, which represented the entire home appliance industry in Slovenia. After Elan

Ski & Yacht was dismantled and Gorenje was sold to a Chinese appliance chain, while the pharmaceutical plant Lek was sold to Novartis, the story of industrious Slovenians was seriously damaged.

Yet another part of the answer, however, lies with the political elites (as does elsewhere in the region). Although the writers of this book share the belief that political figures in a modern society no longer are, or should no longer be, the driving force of structural changes, we nevertheless have to admit that in 2020s Slovenia, politics has been all about individual politicians, much like in other parts of the former socialist bloc, nowadays often referred to as New Europe.

## Soldier, Model, Actor, Scholar

Heading deep into 21st century, all the leading figures were men once again. Men, who not only were incapable of producing a consistent strategy or implement coherent policies, but also tried to sideline women, who saved the country from more embarrassment, while envisaging a political counterweight to Janšaism (the illiberal democracy spearheaded by Janša), and saved it from "troika" and a destiny similar to that of Greece.

Katarina Kresal, a lawyer and politician, was the Minster of the Interior from 2008 to 2011, who ended the agony of the "erased,"[653] while other female politicians, a Christian-democrat Ljudmila Novak, who is the president of the Nova Slovenija party, and a liberal Alenka Bratušek, who runs the Alenka Bratušek Party, openly opposed Janšaism: Novak, with her modern understanding of Christian-democracy, and Bratušek, with her opposition to Janša's hysterical calling for economic (and political) intervention from Brussels.

In addition to Janša, there are at least three other politicians who were the ones to be held accountable at the turn of the 21st century, for the fustiness pervading Slovenian politics, hindering the artistic, entrepreneurial, and social energies.

The first was Borut Pahor, Prime Minister (2008–2011) and President of the Republic for two consecutive terms (2012–2022). According to his critics, he is the culprit for emptying the political discourse and shifting the relevance of political action. Starting as a model and an ambitious student politician, he was also the youngest member of the Slovenian Central Committee of the Communist Party of Slovenia, which makes him one of the oldest career politicians in independent Slovenia. Pahor was also among the first to realize that political accountability

---

653  Barbara Beznec et al., *I Feel Slovenia, Časopis za kritiko znanosti, domišljijo in novo antropologijo*, 228.

was increasingly being relegated to external, supranational factors outside one's control, represented by international organizations such as the European Union and NATO. He not only managed to evade the traps of unpredictable party politics during the transition but also succeeded in executing a thorough ideological transformation.

His method importantly relied on the use of social media that helped him survive the dynamic and occasionally rough period following the financial crisis of 2008. As the president of the Social Democrats or SD (the Communist Party's successor party), he took over the government in 2008–2012, and transitioned to the President of the republic that same year. In this period, Pahor attempted to sculpt his public figure as the "President of all Slovenes." On the other hand, the transformation of his mediated and political persona is not only a unique case study for understanding the transformation of politics in post-socialism, but can also serve as a good example of the reconstruction of changes in the European political landscape. Therefore, the Slovenian model/president is not an example of a particularly Slovenian situation, but rather serves as a broader illustration of the mediatization, eventization and personalization of the political field that effectively leads to the depoliticization of politics.

A slightly different strategy was adopted by Miro Cerar, a university professor and parliamentary adviser on constitutional law, who acted as Prime Minister from 2014 to 2018. His expertise and his image as a thoughtful and reserved academic, however, could not dent the militancy of Janšaism. Under constant attacks by Janša and his followers, his restrained academic appearance was replaced with an image of a determined politician who does not fear even such radical moves as the erecting of a barbed wire fence along the Croatian border.

Therefore, instead of building on the figure of a moderate and compassionate leader, and following the example of his mother, Zdenka Cerar, who served as the minister of justice in 2004, he, too, started courting popularity in order to gain votes. In an environment defined and dominated by the masters of the game described above, however, this was not an easy task. Much like his predecessors, he was unable to coordinate and promote the ambitions of center politics, characterized by central indecision and the endless attempt to harmonize the particular interests of political partners.

In the end, it was the political pragmatism of these partners that enabled the third Janša government come to power, following a similar mistake made by Marjan Šarec, a university educated actor-turned-politician (mayor of the mid-size town of Kamnik from 2010 to 2018 and Prime Minister from 2018 to 2020), who confused his political audience with a theater stage. Much like Cerar, who did not understand that the parliament, the government, and the party are not his usual student audience, Šarec failed to grasp the vast differences between running a town and running a government. This became particularly apparent when he was confronted by more experienced actors such as Pahor or Janša, whose third term as Prime Minister coincided with the beginning of Covid-19 pandemic. In less than 14 days after the pandemic was declared, Slovenians were faced with two dangerous developments: a pandemic that threatened physical lives of citizens as well as the political appropriation of a socio-medical situation that undermined democratic processes.

# Epilogue

This book, as stated in the introduction, aspires to present a concise, structured, and intelligible history of the peoples living at the crossroads between the Eastern Alps and the Northern Adriatic. In addition, we wanted to embed it in the history of the territory that connects the Balkans with Central Europe, to thus give the reader an insight into the interplay of Romanic, Germanic, Slavic, and Hungarian influences, which shaped and created a language and culture with more than 250 dialects. In a small area between the Friulian and the Pannonian plains, the neighbors not only communicate in two, or even three languages, but also share a plethora of their dialects. All this and more is reflected and archived in popular songs, dances, rituals and religious practices.

Starting with the period that began long before the first Slavic settlements, we paid special attention to the Roman Empire and the time just before it. In doing so, we wished to emphasize that our ancestors did not settle an empty territory, but rather coexisted ever since their arrival in the Eastern Alps with other peoples and cultures. Over the past 1,300 years, this has allowed them to build a community shaped by countless influences. At the same time, the Slavs who ventured farthest westward exerted from the outset a dynamic impact on the original populations, as did they on their new neighbors. For this and a multitude of other reasons stemming from such a long period of living in a melting pot of languages, cultures, and landscapes, they were well known for their chameleonic abilities from the very beginning of the Central European nation-building processes. Thus, since the end of the 19th century, if not before, foreigners described them as people who shared plenty with their German, Romance, and South Slav neighbors.

The same reasons have most probably also contributed to the perception of their land as a "land in between," which has persisted for more than a century. Such feelings became even clearer in the late 1980s and yet more distinct after independence. Moreover, judging by its visitors' perceptions, post-1991 Slovenia has increasingly been becoming a middle ground between different worlds, or rather a point where these worlds meet, mix, and converge. Thus, in addition to constituting a link between Eastern and Western Europe, Slovenia has increasingly been perceived as the meeting point of Central Europe and the Balkans.

According to this perception, it started to become a country that shared the "picturesque scenery of Switzerland and Austria with the Mediterranean climate of Italy." The fascinated gaze, typical of Western Europeans when looking at its "peripheral neighborhoods," started to picture it as a pleasant land on the edge of the Julian Alps, as a "jewel in the wilderness," as "Europe in miniature, a pleasant surprise," which is "small, scenic and flexible."

Slovenia does have a relatively small population, and in the last decade it has been accorded a series of attributes. Despite the fact that Carantania, the first state-like entity of the "Slovenes," occupied a territory three times the size of the present-day Republic of Slovenia, and even though the Counts of Celje played a crucial role in Central Europe and the Balkans, today Slovenia is considered a small state. However, the interaction and overlapping of western, eastern, and southern histories, events, as well as arts and cultures make it an interesting space to explore further.

This typically Central European, Balkan, and (perhaps not so much) ex-communist state was, as already mentioned, the first of the post-socialist states to join the EU in the major enlargement of 2004, to adopt the Euro in 2007, as well as to assume twice the EU Presidency. It prides itself on its Lipizzaner horses, ardent athletes, inventive artists, and one of the best philosophers of (post-)modern Europe, Slavoj Žižek. It is increasingly portrayed as "a place so pastoral that it seems an entire century has gone missing," while the descriptions by Western culinary adventurers note that "this tiny Alpine country has a reputation as the quiet sister" of Italy, Austria, and Croatia. Slovenia thus "remains as puzzling as ever," with its "matchbox-size airport," and "its history as a messy and overturned board game," which is "still lightheaded from its [...] young sovereignty and independent identity, with people from the outside still trying to figure out what it means to be a Slovene."[654]

The Slovenes like to see themselves as "industrious, pliant [and] little inclined to resist or complain," and therefore sometimes treatable "with moderation showed to no other subject Slavs."[655] Moreover, in the first three decades of living in an independent country, they have more or less systematically created an image of themselves as entrepreneurial, ingenious, kind, and thrifty people who do not differ significantly from their northern neighbors. Since their independence, they have also assumed the role attributed to them in the early 1990s as the best "students in market democracy," who like to think of themselves as wise enough to take advantage of the crucial moments in history.

---

654 Quotes in this paragraph are from Alexander Lobrano and Andrea Fazzari, "A Land in Between," *Gourmet* (February 2006), 112–119.
655 *The National Geographic Magazine* 34, 6 (December 1918), 489.

On the other hand, Slovenes have had to experience for the very first time what it means to be responsible for the long-term challenges that come with a state of one's own. They now have to look inward to find appropriate ways to act in the world, whereas in the past, cultural preservation and explanations for their frustrations could always be found in reference to others, to empires or to collective states.[656] Within the home environment, however, they have to face an important structural problem, the solution to which will crucially determine the country's future development. In this respect, Slovenes will be bracing themselves for a no less turbulent period of challenging the unsurpassed essentialist notions of identity in creating a thoroughly open society.

---

656   James Gow and Cathie Carmichael, *Slovenia and the Slovenes: A Small State and the New Europe*, London: Hurst & Company, 2001, 211.

# A Note on Pronunciation and Regional Toponyms

The following guidelines may help the reader pronounce the names of people and places in the text.

C, c – ts, as in its
č, č – tsh, as in patch
Ć, ć – ty, as in nature
Đ, đ – g, as in schedule
Dj, dj – dy, as in bad year
Dž, dž – dg, as in edge
J, j – y, as in yes
Lj, lj – ly, as in halyard
Nj, nj – ny, as in canyon
Š, š – sh, as in nation
Ž, ž – zh, as in pleasure

The following Slovenian regions are referred to in English as:

Gorenjska – Upper Carniola
Dolenjska – Lower Carniola
Štajerska – Styria
Koroška – Carinthia
Kras – Karst
Kranjska – Carniola

# List of Figures

| | | |
|---|---|---|
| Figure 1. | Neanderthal musical instrument from Divje babe I. Courtesy of the National Museum of Slovenia. Photo: T. Lauko. | 17 |
| Figure 2. | Pile dwellings on the Ljubljana Marshes. Dragan Božič et al. (eds.). From: *Zakladi tisočletij. Zgodovina Slovenije od neandertalcev do Slovanov*. Ljubljana: Modrijan, 1999, 65. | 21 |
| Figure 3. | Anthropomorphic clay statuette from Ig on the Ljubljana Marshes. Courtesy of the National Museum of Slovenia. Photo: T. Lauko. | 22 |
| Figure 4. | Iron Age house at Kučar near Podzemelj in the Lower Carniola (Dolenjska) region. From: *Zakladi tisočletij*, 108. | 30 |
| Figure 5. | Golden female attire found in grave at Stična from the beginning of 6th century BCE. Courtesy of the National Museum of Slovenia. Photo: T. Lauko. | 34 |
| Figure 6. | The Vače situla. Courtesy of the National Museum of Slovenia. Photo: T. Lauko. | 35 |
| Figure 7. | Celtic necklace from Podzemelj made of glass beads. Courtesy of the National Museum of Slovenia. Photo: T. Lauko. | 39 |
| Figure 8. | Ocra Pass below Ocra Mt., present day Mt. Nanos. Courtesy of the Archives of the Institute of Archaeology, ZRC SAZU. | 43 |
| Figure 9. | Celtic silver coins from Celeia, 1st century BCE. Courtesy of the National Museum of Slovenia. Photo: T. Lauko. | 45 |
| Figure 10. | Tombstone of Plaetor and Moiota, carved into natural rock at Staje near Ig. From: *Zakladi tisočletij*, 237. | 57 |
| Figure 11. | Partly preserved baptismal chapel at Emona. Archelaus and Honorata donated 1.8 m² of the mosaic floor. From: *Zakladi tisočletij*, 207. | 59 |
| Figure 12. | Reconstructed third Mithraeum at Zgornji Breg in Ptuj, the Roman Poetovio. From: *Zakladi tisočletij*, 220. | 64 |
| Figure 13. | Aerial view of the late-Roman Ad Pirum fort in the Claustra Alpium Iuliarum defensive system. Courtesy of the Archives of the National Museum of Slovenia. Photo: J. Hanc. | 73 |

Figure 14. Aerial view of the late-Roman settlement at Ajdovski gradec near Vranje. Courtesy of the Archives of the National Museum of Slovenia. Photo: J. Hanc. ............................ 79

Figure 15. Ostrogothic and Lombard money, and an Ostrogothic fibula, 5th and 6th century CE. Courtesy of the National Museum of Slovenia. Photo: M. Pavlovec. ............................... 83

Figure 16. Lombard jewelry from middle and late 6th century found in the Lajh cemetery in Kranj. Courtesy of the National Museum of Slovenia. Photo: T. Lauko. ..................................... 84

Figure 17. Glagolite ponaz redka zloueza ("Say after us [these] few words"). Beginning of first Freising Manuscript. From: France Bernik (ed.), *Brižinski spomeniki*, Ljubljana: SAZU, ZRC SAZU, 1993, first illustrated supplement. ........................................................................ 97

Figure 18. The oldest known illustration of the Installation of a Carantanian Prince in Leopold Stainreuter's Österreichische Chronik von den 95 Herrschaften. The original is kept in City Library, Bern, Switzerland, Cod. A. 45, from c. 1480. From: Peter Štih, *Ustoličevanje koroških vojvod: najstarejša upodobitev 1480*, Ljubljana: Slovenska knjiga, 1999, 9. ............................................................. 98

Figure 19. Ptuj, panorama with castle. From: Marjeta Ciglenečki, *Ptuj, starodavno mesto ob Dravi*, Maribor: Umetniški kabinet Primož Premzl, 2008, 7. ............................................ 112

Figure 20. "Mantled Virgin Mary—the Protector," relief from c. 1410 in the Church of the Mantled Virgin Mary—the Protector on Ptujska Gora (near Ptuj): under Virgin Mary's mantle, the artist portrayed the church's founder Bernard of Ptuj with his wife, the Patriarch of Aquileia, Hungarian King Sigismund with his wife Barbara of Celje, Bosnian King Tvrdko, Count Herman II of Celje, and other dignitaries from the broader region. From: Marijan Zadnikar, *Ptujska gora*, Ljubljana: Družina, 1992, 62–63. ......................... 140

Figure 21. "Dance of Death," fresco by Janez of Kastav from 1490 in the Church of the Holy Trinity at Hrastovlje (Istria). From: Marijan Zadnikar, *Hrastovlje. Romanska arhitektura in gotske freske*, Ljubljana: Družina, 1988, 128. ..................... 144

Figure 22. Remains of Carthusian Monastery in Žiče. From: Marijan Zadnikar, *Žička kartuzija*, Maribor: Obzorja, 1973, cover. ........ 152

List of Figures 571

Figure 23. "Holy Sunday," fresco from c. 1460 (workshop of Janez of Ljubljana) on façade of the Church of the Annunciation at Crngrob near Škofja Loka; scenes of various occupations that may not be performed on Sundays. From: *Ars Sloveniae*, Ljubljana: Mladinska knjiga, 1972, 48–49. ............... 156
Figure 24. Remains of Šalek castle near Velenje, c. 12th century. Photo: Ivan Jakič. ........................................................... 160
Figure 25. Cod. 685 in Österreichische Nationalbibliothek, Wien (S. Hieronimus, Commentarius in prophetas), produced in the Stična Monastery at the end of 12th century. From: Nataša Golob, *Stiški rokopis*, Slovenska knjiga, Ljubljana, 1994, 98. ..... 170
Figure 26. Celje. From: Primož Premzl (ed.), *Topographia ducatus Stiriae*, Maribor: Umetniški kabinet Primož Premzl, 2006, fig. 11. ........................................................................ 172
Figure 27. Peasants' uprising. From: Janez Vajkard Valvasor, *Slava vojvodine Kranjske*, Ljubljana: Mladinska knjiga, 1978, 309. .... 198
Figure 28. Cover of Dalmatin's *Biblia, tu je, vse Svetu pismu, Stariga inu Noviga testamenta*. From: Jurij Dalmatin, *Biblia, tu je, vse Svetu pismu, Stariga inu Noviga testamenta* (fascimile) Ljubljana: Mladinska knjiga, 1998, 1. ........................... 220
Figure 29. Composer Jacobus Gallus Carniolus. From: Dragotin Cvetko, *Jacobus Gallus Carniolus*, Ljubljana: Slovenska matica, 1965, 1. ..................................................... 225
Figure 30. Dramatist, poet, and historian Anton Tomaž Linhart. From: Ivo Svetina, Francka Slivnik, and Verena Štekar Vidic (eds.), *Anton Tomaž Linhart. Jubilejna monografija ob 250. letnici rojstva*, Ljubljana, Radovljica: Slovenski gledališki muzej, Muzeji radovljiške občine, 2005, cover. Image: Jurij Kocbek. ........................................................................ 263
Figure 31. Mentor and reformer baron Žiga Zois. From: Marija Kacin, *Žiga Zois in italijanska kultura*, Ljubljana: Založba ZRC, 2001, 6. ......................................................... 269
Figure 32. Napoleon's Illyria Memorial, French Revolution Square, Ljubljana: National University Library, courtesy of Cartographic and Image Collections. .......................... 277
Figure 33. Poet France Prešeren. From: Janez Cvirn (ed.), *Slovenska kronika XIX. stoletja*, 1800–1860, Ljubljana: Nova revija, 2001, 27. ................................................. 294

## List of Figures

Figure 34. Peter Kozler's Map of Slovenia, first printed in 1853. From: *Slovenska kronika XIX. stoletja*, 422. ................ 311
Figure 35. Ljubljana, latter half of the 18th century. From: *Slovenska kronika XIX. stoletja*, 109. ................ 347
Figure 36. Austro–Hungarian position on the Isonzo Front. From: Jože Dežman, *The Making of Slovenia*, Ljubljana: Muzej novejše zgodovine, 2006, 12. ................ 399
Figure 37. In the cavern at night, Isonzo Front. Courtesy of the Ciril Prestor Collection. ................ 400
Figure 38. Ljubljana in the 1920s. From: *The Making of Slovenia*, 18. ........ 419
Figure 39. Men's gymnastics team at the 9th Summer Olympic Games in Amsterdam, 1928. First from the left is Leon Štukelj, the most successful Slovenian Olympic competitor. From: *The Making of Slovenia*, 58. ................ 425
Figure 40. Hitler in Maribor, 1941. From: *The Making of Slovenia*, 65. ...... 449
Figure 41. Occupiers executing hostages. From: *The Making of Slovenia*, 67. ................ 457
Figure 42. The Home Guard and its leader Leon Rupnik, the spring of 1944. From: *The Making of Slovenia*, 2006, 71. ................ 464
Figure 43. Partisan army in Ljubljana, the spring of 1945. From: *The Making of Slovenia*, 76. Photo: Čoro Škodlar. ................ 471
Figure 44. Josip Broz Tito and Edvard Kardelj at a meeting in Okroglica, Slovenia. Courtesy of the Museum of Contemporary History. Photo: Vlastja Simončič. ................ 491
Figure 45. Members of the youth work brigades. From: *The Making of Slovenia*, 85. Photo: Marjan Pfeifer. ................ 496
Figure 46. One of the most popular Slovenian singers Majda Sepe performing at the first Slovenian Song Festival, Bled, 1962. Courtesy of RTV Slovenia. Photo: Milan Kumar. ................ 499
Figure 47. Student protests, May 1971. From: *The Making of Slovenia*, 92. Photo: Edi Šelhaus. ................ 513
Figure 48. A popular punks' meeting place, appropriately named Johnny Rotten Square, Ljubljana, 1981. Courtesy of the Photon Gallery. Photo: Vojko Flegar. ................ 527
Figure 49. A protest rally in Congress Square, June 21, 1988. A poster with the crisscrossed Article 133 of the Yugoslav Penal Code demanding the abolition of restrictions on freedom of speech (verbal delict). From: *The Making of Slovenia*, 93. Photo: Marjan Ciglič. ................ 534

List of Figures

Figure 50. The Slovenian Territorial Defense takes over the Ljubelj border crossing between Slovenia and Austria, June 1991. Photo: Mirko Kunšič. .................................................................... 551

Figure 51. Demonstrations in Ljubljana, February 7, 2013. Photo: Sunčan Stone. .................................................................... 556

# List of Maps

Map 1. Division of present-day Slovenia under Roman Empire. Courtesy of the Institute of Archaeology ZRC SAZU. Map design: Mateja Belak. ............................................................ 49
Map 2. Map of "Great Carantania" and its Marches, about 1000. Adapted after: Bogo Grafenauer et al. (eds.), Zgodovina Slovencev, Ljubljana: Cankarjeva založba, 1979, 157 and Peter Štih, Vasko Simoniti, Slovenska zgodovina do razsvetljenstva, Ljubljana, Celovec: Mohorjeva družba, Korotan, 1995, 256–257. ......................................................................................... 133
Map 3. Ecclesiastical organization of Slovenian regions before 1777. Adapted after: Bogo Grafenauer et al. (eds.), Zgodovina Slovencev, Ljubljana: Cankarjeva založba, 1979, 373. ................... 149
Map 4. Map of the Ilyrian Provinces (1809–1813). Map design: Manca Volk Bahun. ZRC SAZU, Anton Melik Geographical Institute. ........................................................................ 276
Map 5. Austro-Hungarian Empire (1867–1918). Map design: Manca Volk Bahun. ZRC SAZU, Anton Melik Geographical Institute. ........................................................................ 316
Map 6. Drava Banovina (1929–1941). Map design: Manca Volk Bahun. 420
Map 7. World War II Occupation Zones (1941–1945). Map design: Manca Volk Bahun. ZRC SAZU, Anton Melik Geographical Institute. ........................................................... 448
Map 8. Borders after World War II, Free Territory of Trieste, Zone A and Zone B (1945–1947). Map design: Manca Volk Bahun. .... 479
Map 9. Socialist Federative Republic of Yugoslavia (1947–1991). Map design: Manca Volk Bahun. ZRC SAZU, Anton Melik Geographical Institute. ........................................................................ 480
Map 10. Map of the Republic of Slovenia after independence. Map design: Manca Volk Bahun. ZRC SAZU, Anton Melik Geographical Institute. ........................................................................ 550

# Selected Bibliography

## From Prehistory to the End of the Ancient World

Božič, Dragan. "Die Erforschung der Latènezeit in Slowenien seit dem Jahr 1964 (Raziskovanje latenske dobe na Slovenskem po letu 1964)." *Arheološki vestnik*, 50 (1999): 189–213.

———, et al. *Zakladi tisočletij. Zgodovina Slovenije od neandertalcev do Slovanov.* Ljubljana: Modrijan, 1999.

Bratož, Rajko. *Il cristianesimo aquileiese prima di Costantino: fra Aquileia e Poetovio.* Udine, Gorizia: Istituto Pio Paschini; Istituto di Storia Sociale e Religiosa, 1999.

———. *Severinus von Noricum und seine Zeit. Geschichtliche Anmerkungen.* Vienna: Österreichishe Akademie der Wissenschaften, 1983.

———. (ed.). *Slowenien und die Nachbarländer zwischen Antike und karolingischer Epoche. Anfänge der slowenischen Ethnogenese.* 2 vols. Ljubljana: Narodni muzej Slovenije and SAZU, 2000.

Ciglenečki, Slavko. *Höhenbefestigungen aus der Zeit vom 3. bis 6. Jh. im Ostalpenraum.* Ljubljana: SAZU, 1987.

———. "Romani e Longobardi in Slovenia nel VI secolo." In: *Paolo Diacono e il Friuli altomedievale (secc. VI–X)*, vol. 1, 179–199. Spoleto: Centro italiano di studi sull'alto medioevo, 2001.

Cuscito, Giuseppe (ed.). *Aquileia dalle origini alla costituzione del ducato longobardo. Storia—amministrazione—società.* Trieste: Editreg, 2003.

Dular, Janez. "Ältere, mittlere und jüngere Bronzezeit in Slowenien—Forschungsstand und Probleme (Starejša, srednja in mlajša bronasta doba v Sloveniji—stanje raziskav in problemi)." *Arheološki vestnik*, 50 (1999): 81–96.

——— and Sneža Tecco Hvala. *South-Eastern Slovenia in the Early Iron Age. Settlement—Economy—Society/Jugovzhodna Slovenija v starejši železni dobi. Poselitev—gospodarstvo—družba.* Ljubljana: Založba ZRC, 2007.

Gabrovec, Stane. *Stična I. Naselbinska izkopavanja/Siedlungsausgrabungen.* Ljubljana: Narodni muzej, 1994.

———. *Stična II/1. Grabhügel aus der älteren Eisenzeit—Katalog.* Ljubljana: Narodni muzej Slovenije, 2006.

Gassner, Verena, Sonja Jilek and Sabine Ladstätter. *Am Rande des Reiches. Die Römer in Österreich*, vol. 2 of *Österreichische Geschichte 15 v. Chr.–378 n. Chr.*, ed. Herwig Wolfram. Vienna: Ueberreuter, 2002.

Guštin, Mitja (ed.). *Zgodnji Slovani. Zgodnjesrednjeveška lončenina na obrobju vzhodnih Alp/Die frühen Slawen. Frühmittelalterliche Keramik am Rand der Ostalpen.* Ljubljana: Narodni muzej Slovenije, 2002.

Horvat, Jana. *Nauportus (Vrhnika).* Ljubljana: SAZU, 1990.

———. *Sermin. Prazgodovinska in zgodnjerimska naselbina v severozahodni Istri/A Prehistoric and Early Roman Settlement in Northwestern Istria.* Ljubljana: ZRC SAZU, Inštitut za arheologijo, 1997.

———, et al. "Poetovio. Development and Topography." In: *The Autonomous Towns of Noricum and Pannonia/Die autonomen Städte in Noricum und Pannonien—Pannonia II*, Marjeta Šašel Kos and Peter Scherrer (eds.), 153–189. Ljubljana: Narodni muzej Slovenije, 2004.

Kos, Peter. *The Monetary Circulation in the Southeastern Alpine Region ca. 300 B.C.—A.D. 1000.* Ljubljana: Narodni muzej Slovenije, 1986.

Lazar, Irena. "Celeia." In: *The Autonomous Towns of Noricum and Pannonia/Die autonomen Städte in Noricum und Pannonien—Noricum*, Marjeta Šašel Kos and Peter Scherrer (eds.), 71–101. Ljubljana: Narodni muzej Slovenije, 2002.

Lotter, Friedrich, Rajko Bratož, and Helmut Castritius. *Völkerverschiebungen im Ostalpen-Mitteldonau-Raum zwischen Antike und Mittelalter (375–600).* Berlin, New York: De Gruyter, 2003.

Lovenjak, Milan. *Inscriptiones Latinae Sloveniae 1. Neviodunum.* Ljubljana: Narodni muzej Slovenije, 1998.

Parzinger, Hermann. *Studien zur Chronologie und Kulturgeschichte der Jungstein-, Kupfer- und Frühbronzezeit zwischen Karpaten und Mittleren Taurus.* Mainz am Rhein: von Zabern, 1993.

Petru, Peter and Jaroslav Šašel (eds.). *Claustra Alpium Iuliarum 1. Fontes.* Ljubljana: Narodni muzej, 1971.

Pleterski, Andrej. *Župa Bled. Nastanek, razvoj in prežitki (Die Župa Bled. Entstehung, Entwicklung und Relikte).* Ljubljana: SAZU, 1986.

Šašel, Jaroslav. *Opera selecta.* Ljubljana: Narodni muzej, 1992.

Šašel Kos, Marjeta. *Appian and Illyricum.* Ljubljana: Narodni muzej Slovenije, 2005.

———. "The End of the Norican Kingdom and the Formation of the Provinces of Noricum and Pannonia." In: *Akten des IV. intern. Kolloquiums über Probleme des provinzialrömischen Kunstschaffens/Akti IV. mednarodnega kolokvija o problemih rimske provincialne umetnosti. Celje 8.–12. Mai/maj 1995*, Bojan Djurić and Irena Lazar (eds.), 21–36. Ljubljana: Narodni muzej Slovenije, 1997.

———. *A Historical Outline of the Region between Aquileia, the Adriatic, and Sirmium in Cassius Dio and Herodian.* Ljubljana: SAZU, 1986.

———. *Pre-Roman Divinities of the Eastern Alps and Adriatic.* Ljubljana: Narodni muzej Slovenije, 1999.

Teržan, Biba (ed.). *Hoards and Individual Metal Finds from the Eneolithic and Bronze Ages in Slovenia.* Ljubljana: Narodni muzej Slovenije, 1995/96.

———, Fulvia Lo Schiavo, and Neva Trampuž-Orel. *Most na Soči (S. Lucia) II. Szombathyjeva izkopavanja/Die Ausgrabungen von J. Szombathy.* Ljubljana: Narodni muzej Slovenije, 1984/85.

Turk, Ivan (ed.). *Divje Babe I: Upper Pleistocene Palaeolithic Site in Slovenia.* Ljubljana: ZRC SAZU, inštitute za arheologijo, 2007.

———. *Images of Life and Myth.* Ljubljana: Narodni muzej Slovenije, 2005.

——— (ed.). *Moustérienska "koščena piščal" in druge najdbe iz Divjih bab I v Sloveniji.* Ljubljana: ZRC SAZU, 1997.

Velušček, Anton. "Neolithic and Eneolithic Investigations in Slovenia (Neolitske in eneolitske raziskave v Sloveniji)." *Arheološki vestnik*, 50 (1999): 59–79.

——— (ed.). *Hočevarica: An Eneolithic Pile Dwelling in the Ljubljansko Barje.* Ljubljana: Založba ZRC, 2004.

Vičič, Boris. "Colonia Iulia Emona: 30 Jahre später." In: *The Autonomous Towns of Noricum and Pannonia. Pannonia I*, Marjeta Šašel Kos and Peter Scherrer (eds.), 21–45. Ljubljana: Narodni muzej Slovenije, 2003.

## The Early Middle Ages

Benedik, Metod, et al. *Zgodovina Cerkve na Slovenskem.* Celje: Mohorjeva družba, 1991.

Benussi, Bernardo. *L'Istria nei suoi due millenni di storia.* Trieste: G. Caprin 1924.

Bernik, France (ed.). *Brižinski spomeniki. Znanstvenokritična izdaja.* Ljubljana: SAZU, 1993.

Bodin, Jean. *Les six livres de la Republique.* 1576. Reprint, Paris: Fayard, 1986.

Bratož, Rajko. "La chiesa istriana nel VII e nell'VIII secolo (dalla morte di Gregorio Magno al placito di Risano)." *Acta Histriae*, 2 (1994): 65–77.

———. "La cristianizzazione degli Slavi negli atti del convegno 'ad ripas Danubii' e del concilio di Cividale." In: *XII centenario del concilio di Cividale (796– 1996)*, 145–190. Udine: Convegno storico-teologico—Atti, 1998.

——— (ed.). *Slovenija in sosednje dežele med antiko in karolinško dobo. Začetki slovenske etnogeneze/Slowenien und die Nachbarländer zwischen Antike und karolingischer Epoche. Anfänge der slowenischen Ethnogenese*, Vols. 1–2. Ljubljana: SAZU, Narodni muzej Slovenije, 2000.

―――. *Vpliv oglejske cerkve na vzhodnoalpski in predalpski prostor od 4. do 8. stoletja*. Ljubljana: Zveza zgodovinskih društev Slovenije, 1990.

―――. "Začetki oglejskega misijona med Slovani in Avari: sestanek škofov 'ad ripas Danubii' in sinoda v čedadu 796." In: *Vilfanov zbornik*, Vincenc Rajšp and Ernst Bruckmüller (eds.), 79–111. Ljubljana: Založba ZRC, 1999.

"Conversio Bagoariorum et Carantanorum." In: *Conversio Bagoariorum et Carantanorum und der Brief des Erzbischofs Theothmar von Salzburg*, Fritz Lošek (ed.). Hannover: Hahnsche Buchhandlung 1997.

Dolinar, France Martin (ed.). *Sveta brata Ciril in Metod v zgodovinskih virih. Ob 1100 letnici Metodove smrti*. Ljubljana: Teološka fakulteta, Inštitut za zgodovino Cerkve, 1985.

Dopsch, Heinz. "Arnolf und der Südosten—Karantanien, Mähren, Ungarn." In: *Kaiser Arnulf. Das ostfränkische Reicham Ende des 9. Jahrhunderts*, Franc Fuchs and Peter Schmid (eds.), 143–185. Munich: C. H. Beck, 2002.

Dvornik, Francis. *The Slavs: Their Early History and Civilization*. Boston: American Academy of Arts and Sciences, 1956.

Ferluga, Jadran. "L'Istria tra Giustiniano e Carolo Magno." *Arheološki vestnik*, 43 (1992): 175–190.

Frankl, Karl Heinz and Peter G. Tropper (eds.). *Heilige Nonosus von Molzbichl*. Klagenfurt: Verlag des Kärntner Landesarchivs, 2001.

Fräss-Ehrfeld, Claudia. *Geschichte Kärntens 1. Das Mittelalter*. Klagenfurt: J. Heyn, 1984.

Giesler, Jochen. *Der Ostalpenraum vom 8. bis 11. Jahrhundert. Studien zur archäologischen und Schriftlichen Zeugnissen*. Vol. 2. Rahden/Westf.: Verlag Marie Liedorf, 1997.

Grafenauer, Bogo. "Oblikovanje severne slovenske narodnostne meje." In: *Zbirka Zgodovinskega časopisa 10*. Ljubljana: Zveza zgodovinskih društev Slovenije, 1994.

―――. *Od naselitve do uveljavljenja frankovskega reda*, vol. 1 of *Zgodovina slovenskega naroda*. 3rd ed. Ljubljana: DZS, 1978.

―――. "Razmerje med Slovani in Obri do obleganja Carigrada (626) in njegove gospodarsko-družbene podlage." *Zgodovinski časopis*, 9 (1955): 145–153.

―――. *Ustoličevanje koroških vojvod in država karantanskih Slovencev*. Ljubljana: Inštitut za zgodovino SAZU, 1952.

―――. "Vprašanje konca Kocljeve vlade v Spodnji Panoniji." *Zgodovinski časopis*, 6–7 (1952/1953): 171–190.

Grivec, Franz. *Konstantin und Method. Lehrer der Slawen*. Wiesbaden: Otto Harrassowitz, 1960.

Kahl, Hans-Dietrich. "Der Staat der Karantanen. Fakten, Thesen und Fragen zu einer frühen slawischen Machtbildung in Ostalpenraum (7.–9. Jh.)/Država Karantancev. Dejstva, teze in vprašanja o zgodnji slovanski državni tvorbi v vzhodnoalpskem prostoru (7.–9. stol.)." In: *Slovenija in sosednje dežele med antiko in karolinško dobo. Začetki slovenske etnogeneze. Dopolnilni zvezek/ Slowenien und die Nachbarländer zwischen Antike und karolingischer Epoche. Anfänge der slowenischen Ethnogenese.* Ljubljana: Narodni muzej Slovenije, SAZU, 2002.

Kos, Janko, Franc Jakopin and Jože Faganel (eds.). *Zbornik Brižinski spomeniki.* Ljubljana: SAZU, 1996.

Kos, Milko. *Zgodovina Slovencev od naselitve do petnajstega stoletja.* Ljubljana: Slovenska matica, 1955.

Krahwinkler, Harald. "Ausgewählte Slawen-Ethnonyme und ihre historische Deutung." In: *Slovenija in sosednje dežele med antiko in karolinško dobo. Začetki slovenske etnogeneze/Slowenien und die Nachbarländer zwischen Antike und karolingischer Epoche. Anfänge der slowenischen Ethnogense I*, Rajko Bratož (ed.), 413–418. Ljubljana: Narodni muzej and SAZU, 2000.

———. *Friaul im Frühmittelalter. Geschichte einer Region vom Ende des fünften bis zum Ende des zehnten Jahrhunderts.* Vienna, Cologne, Weimar: Böhlau, 1992.

———. *...In loco qui dicitur Riziano... Zbor v Rižani pri Kopru 804. Die Versammlung in Rižana/Risano bei Koper/Capodistria im Jahre 804.* Koper: Univerza na Primorskem, Znanstveno-raziskovalno središče, Zgodovinsko društvo za južno Primorsko, 2004.

———. "'In territorio caprense loco qui dicitur Riziano,' il 'Placito' di Risano nell'anno 804." *Quaderni Giuliani di Storia*, 27 (2006): 255–330.

Lotter, Friedrich, Rajko Bratož, and Helmut Castritius. *Völkerverschiebungen im Ostalpen.Mitteldonauraum zwischen Antike und Mittelalter (375–600)*, 149–155. Berlin, New York: Walter de Gruyter 2003.

Ludwig, Uwe. *Transalpine Beziehungen der Karolingerzeit im Spiegel der Memorialüberlieferung. Prosopographische und sozialgeschichtliche Studien unter besonderer Berücksichtigung des Liber vitae von San Salvatore in Brescia und des Evangeliars von Cividale*, 175–236. Hannover: Hahnsche Buchhandlung, 1999.

Mitterauer, Michael. "Slawischer und bayrischer Adel am Ausgang der Karolingerzeit." *Carinthia I*, 150 (1960): 693–726.

Pertz, Georg Heinrich et al. (eds.). *Monumenta Germaniae Historica (MGH).*

Pohl, Walter. *The Avars: A Steppe Empire in Central Europe, 567–822.* Ithaca, London: Cornell University Press, 2018.

Šašel, Jaroslav. "Der Ostalpenbereich zwischen 550 und 650 n. Chr." In: *Opera selecta*, Jaroslav Šašel (ed.), 821–830. Ljubljana: Narodni muzej, 1992.

Škrubej, Katja. *"Ritus gentis" Slovanov v vzhodnih Alpah. Model rekonstrukcije pravnih razmerij na podlagi najstarejšega jezikovnega gradiva*. Ljubljana: Založba ZRC, Pravna fakulteta, 2002.

Šmitek, Zmago. *Mitološko izročilo Slovencev*. Ljubljana: Študentska založba, 2004.

Štih, Peter. "Carniola, patria Sclavorum." *Österreichische Osthefte*, 37 (1995): 845–861.

———. "Istra na začetku frankovske oblasti in v kontekstu razmer na širšem prostoru med severnim Jadranom in srednjo Donavo." *Acta Histriae*, 13/1 (2005): 1–20.

———. "Karantanci—zgodnjesrednjeveško ljudstvo med Vzhodom in Zahodom." *Zgodovinski časopis*, 61 (2007): 47–58.

———. "Kranjska (Carniola) v zgodnjem srednjem veku." In: *Zbornik Brižinski spomeniki*, Janko Kos et al. (eds.), 13–26. Ljubljana, Trieste: Mladika, 1996.

———. "Madžari in slovenska zgodovina v zadnji četrtini 9. in prvi polovici 10. stoletja." *Zgodovinski časopis*, 37 (1983): 171–201.

———. "Plemenske in državne tvorbe zgodnjega srednjega veka na slovanskem naselitvenem prostoru v vzhodnih Alpah (Die frühmittelalterlichen Stammes- und Staatsbildungen im slawischen Siedlungsraum in den Ostalpen)." In: *Slovenci in država*, 21–45. Ljubljana: SAZU, 1995.

———. "Priwina: slawischer Fürst oder fränkischer Graf?" In: *Ethnogenese und Überlieferung*, 213–215. Vienna, Munich: Oldenbourg, 1994.

Štih, Peter and Vasko Simoniti. *Slovenska zgodovina do razsvetljenstva*. Graz, Ljubljana: Mohorjeva družba, Korotan Ljubljana d.o.o., 1995.

Toporišič, Jože (ed.). *Obdobje srednjega veka v slovenskem jeziku, književnosti in kulturi*. Ljubljana: Filozofska fakulteta, 1989.

Van Heck, Adrianus (ed.). *Enee Silvii Piccolominei postea Pii PP. II De Europa*. Vatican City: Biblioteca apostolica vaticana, 2001.

Vilfan, Sergij. "Koseščina v Logu in vprašanje kosezov v vzhodni okolici Ljubljane." In: *Hauptmannov zbornik*, 179–216. Ljubljana: SAZU, 1966.

———. *Pravna zgodovina Slovencev od naselitve do zloma stare Jugoslavije*. Ljubljana: Slovenska matica, 1961.

———. *Rechtsgeschichte der Slowenen bis zum Jahre 1941*. Graz: Leykam, 1968.

Waldmüller, Lothar. *Die ersten Begegnungen der Slawen mit dem Christentum und den christlichen Völker vom VI. bis VIII. Jahrhundert. Die Slawen zwischen Byzanz und Abendland*. Amsterdam: Adolf M. Hakkert, 1976.

Wolfram, Herwig. *Conversio Bagoariorum et Carantanorum. Das Weißbuch der Salzburger Kirche über die erfolgreiche Mission in Karantanien und Pannonien.* Graz: Hermann Böhlau, 1979.

———. *Grenzen und Räume. Geschichte Österreichs vor seiner Entstehung. Österreichische Geschichte 378–907.* Vienna: Ueberreuter, 1995.

———. *Salzburg, Bayern, Österreich. Die Conversio Bagoariorum et Carantanorum und die Quellen ihrer Zeit.* Vienna, Munich: Oldenbourg, 1995.

## Feudalism

Baš, Franjo. "Celjski grofi in njihova doba." In: *Prispevki k zgodovini severovzhodne Slovenije. Izbrani zgodovinski spisi*, Franjo Baš (ed.). Maribor: Založba Obzorja, 1989.

Benedik, Metod, et al. *Zgodovina Cerkve na Slovenskem.* Celje: Mohorjeva družba, 1991.

Blaznik, Pavle. *Škofja Loka in Loško gospostvo (973–1803).* Škofja Loka: Muzejsko društvo, 1973.

———, et al. (eds.). *Agrarno gospodarstvo*, vol. 1 of *Gospodarska in družbena zgodovina Slovencev. Zgodovina agrarnih panog.* Ljubljana: DZS, 1970.

———, et al. (eds.). *Družbena razmerja in gibanja*, vol. 2 of *Gospodarska in družbena zgodovina Slovencev. Zgodovina agrarnih panog.* Ljubljana: DZS, 1980.

Darovec, Darko. *A Brief History of Istra.* Yanchep: ALA Publications, 1998.

Dolinar, France Martin. *Slovenska cerkvena pokrajina.* Ljubljana: Teološka fakulteta, Inštitut za zgodovino Cerkve, 1989.

Dopsch, Heinz. "Die Stifterfamilie des Klosters Gurk und ihre Verwandtschaft." *Carinthia I*, 161 (1971): 95–123.

Galántai, Elisabeth and Julius Kristó (eds.). *Johannes de Thurocz Chronica Hungarorum. I Textus.* Budapest: Akadémiai Kiadó, 1985.

Gestrin, Ferdo. *Slovenske dežele in zgodnji kapitalizem.* Ljubljana: Slovenska matica, 1991.

———. *Trgovina slovenskega zaledja s primorskimi mesti od konca 13. do konca 16. stoletja.* Ljubljana: SAZU, 1965.

Grafenauer, Bogo. *Doba zrele fevdalne družbe od uveljavljanja frankovskega fevdalnega reda do začetka kmečkih uporov*, vol. 2 of *Zgodovina slovenskega naroda.* 2nd ed. Ljubljana: DZS, 1965.

Grdina, Igor and Peter Štih (eds.). *Spomini Helene Kottanner.* Ljubljana: Nova revija, 1999.

Grmek, Mirko Dražen. *Santorio Santorio i njegovi aparati i instrumenti*. Zagreb: Institut za medicinska istraživanja Jugoslavenske akademije, 1952.

Gruden, Josip. *Cerkvene razmere med Slovenci v XV. stoletju in ustanovitev ljubljanske škofije*. Ljubljana: Leonova družba, 1908.

Hauptmann, Ljudmil. "Entstehung und Entwicklung Krains." In: *Erläuterungen zum historischen Atlas der österreichischen Alpenländer 1*, 4 (1929): 309–484.

———. "Grofovi Višnjegorski." *Rad JAZU 250, Razreda historičko-filologičkoga i filozofičko-juridičkoga*, 112 (1935): 215–239.

———. "Mariborske studije." *Rad JAZU 260, Razreda historičko-filologičkoga i filozofičko-juridičkoga*, 117 (1938): 57–118.

Hausmann, Friedrich. "Die steirischen Otakare, Kärnten und Friaul. Besitz, Dienstmannschaft, Ämter." In: *Das Werden der Steiermark*, Gerhard Pferschy (ed.), 225–275. Graz, Vienna, Cologne: Veröffentlichungen des Steiermärkischen Landesarchives 10, 1980.

Hoensch, Jörg K. *Kaiser Sigismund. Herrscher an der Scwhelle zur Neuzeit 1368–1437*. Munich: Beck, 1996.

Janko, Anton and Nikolaus Henkel. *Nemški viteški liriki s slovenskih tal. Žovneški, Gornjegrajski, Ostrovrški/Deutscher Minnesang in Slowenien. Der von Suonegge, Der von Obernburg, Der von Scharpfenberg*. Ljubljana: Znanstveni inštitut Filozofske fakultete, 1997.

Komac, Andrej. *Od mejne grofije do dežele. Ulrik III. Spanheim in Kranjska v 13. stoletju*. Ljubljana: Zgodovinski inštitut Milka Kosa, ZRC SAZU, 2006.

Kos, Dušan. *He Who Does Not Suffer with the Town, Shall Not Reap the Benefits Thereof*. Ljubljana: Ministry of Culture of the Republic of Slovenia, Cultural Heritage Office, 1998.

———. *In Burg und Stadt. Spätmittelalterlicher Adel in Krain und Untersteiermark*. Vienna, Munich: R. Oldenbourg, 2006.

———. *The Tournament Book of Gašper Lamberger/Das Turnierbuch des Caspar von Lamberg*. Ljubljana: Viharnik, 1997.

———. *Zgodovina morale: 1. Ljubezen in zakonska zveza na Slovenskem med srednjim vekom in meščansko dobo*. Ljubljana: ZRC SAZU, 2015.

Kos, Milko. *Srednjeveška kulturna, družbena in politična zgodovina Slovencev*. Ljubljana: Slovenska matica, 1985.

———. *Srednjeveški rokopisi v Sloveniji*. Ljubljana: Umetnostno zgodovinsko društvo, 1931.

Krones, Franz. *Kronika grofov Celjskih*. Translated by Ludovik Modest Golia. Maribor: Založba Obzorja, 1972 [1883].

Lechner, Karl. *Die Babenberger. Markgrafen und Herzoge von Österreich 976–1246*. 3rd ed. Vienna, Cologne, Graz: Böhlau, 1985.

Mitterauer, Michael. "Burg und Adel in den österreichischen Ländern." In: *Die Burgen im deutschen Sprachraum. Ihre rechts- und verfassungsgeschichtliche Bedeutung II*, Hans Patze (ed.), 353–386. Sigmaringen: Jan Thorbecke, 1976.

Mlinarič, Jože. *Kartuziji Žiče in Jurklošter*. Maribor: Pokrajinski arhiv, 1991.

Niederstätter, Alois. *Die Herrschaft Österreich, Fürst und Land im Spätmittelalter*. Vienna: Ueberreuter, 2001.

Ogris, Alfred. *Die Bürgerschaft in den mittelalterlichen Städten Kärntens bis zum Jahre 1335*. Klagenfurt: Verlag des Kärtner, 1974.

Orožen, Ignacij. *Celska kronika*. Celje: J. Jeretin, 1854.

Orožen, Janko. *Zgodovina Celja in okolice. I. del. Od začetka do leta 1848*. Celje: Council of Culture and Science of the Celje City Municipality Assembly, 1971.

Peršič, Janez. *Židje in kreditno poslovanje v srednjeveškem Piranu*. Ljubljana: Oddelek za zgodovino Filozofske fakultete, 1999.

Pirchegger, Hans. *Geschichte der Steiermark 1282–1740*. 2 vols. Graz: Lüschner and Lubensky, 1931, 1936.

Santonino, Paolo. *Popotni dnevnik 1485–1487*. Klagenfurt, Vienna, Ljubljana: Mohorjeva založba, 1991.

Schmidinger, Heinrich. *Patriarch und Landesherr*. Graz, Cologne: Hermann Böhlau Nachf., 1954.

Simoniti, Primož. *Humanizem na Slovenskem in slovenski humanisti do srede 16. stoletja*. Ljubljana: Slovenska matica, 1979.

Simoniti, Vasko. *Turki so v deželi že. Turški vpadi na slovensko ozemlje v 15. in 16. stoletju*. Celje: Mohorjeva družba, 1990.

Spreitzhofer, Karl. *Georgenberger Handfeste*. Graz, Vienna, Cologne: Styria, 1986.

Stanonik, Janez. *Ostanki srednjeveškega nemškega slovstva na Kranjskem*. Ljubljana: Filozofska fakulteta, 1957.

Štih, Peter. "Dežela Grofija v Marki in Metliki." In: *Vilfanov zbornik. Pravo, zgodovina, narod = Recht, Geschichte, Nation*, Vincenc Rajšp and Ernst Bruskmüller (eds.), 123–145. Ljubljana: Založba ZRC, 1999.

———. "Die Grafen von Cilli, die Frage ihrer landesfürstlichen Hoheit und des Landes Cilli." *Mitteilungen des Instituts für Österreichische Geschichtsforschung*, 110 (2002): 67–98.

———. *Studien zur Geschichte der Grafen von Görz. Die Ministerialen und Milites der Grafen von Görz in Istrien und Krain*. Vienna, Munich: Oldenbourg Verlag, 1996.

———. "Ulrik II. Celjski in Ladislav Posmrtni ali Celjski grofje v ringu velike politike." In: *Spomini Helene Kottanner*, Igor Grdina and Peter Štih (eds.), 31–40. Ljubljana: Nova revija, 1999.

———. "Ursprung und Anfänge der bischöflichen Besitzungen im Gebiet des heutigen Sloweniens." In: *Blaznikov zbornik/Festschrift für Pavle Blaznik; Loški razgledi*, Matjaž Bizjak (ed.), 37–54. Ljubljana, Škofja Loka: Založba ZRC, 2005.

———. "*Villa quae Sclavorum lingua vocatur Goriza.*" Studie über zwei Urkunden Kaiser Ottos III. aus dem Jahre 1001 für den Patriarchen Johannes von Aquileia, und den Grafen Werihen von Friaul (DD. O. III. 402 und 412)*. Nova Gorica: Goriški muzej, Grad Kromberk, 1999.

Valenčič, Vlado (ed.). *Ljubljanska obrt od srednjega veka do začetka 18. stoletja*. Ljubljana: Mestni arhiv, 1972.

Vilfan, Sergij. "Die deutsche Kolonisation nordöstlich der oberen Adria und ihre sozialgeschichten Grundlagen." In: *Die deutsche Ostsiedlung des Mittelalters als Problem der europäischen Geschichte*, Walter Schlesinger (ed.), 567–604. Sigmaringen: Jan Thorbecke, 1975.

———. "Stadt und Adel—Ein Vergleich zwischen Küsten- und Binnenstädten zwischen der oberen Adria und Pannonien." In: *Die Stadt am Ausgang des Mittelalters*, Wilhelm Rausch (ed.), 63–74. Linz: J. Wimer, 1974.

———. *Zgodovinska pravotvornost in Slovenci*. Ljubljana: Cankarjeva založba, 1996.

Voje, Ignacij. "Romanje Ulrika II. Celjskega v Kompostelo k Sv. Jakobu." *Zgodovinski časopis*, 38 (1984): 225–230.

———. *Slovenci pod pritiskom turškega nasilja*. Ljubljana: Znanstveni inštitut Filozofske fakultete, 1996.

Von Liechtenstein, Ulrich. *Frauendienst*. Translated by Viktor Spechtler. Klagenfurt: Wieser Verlag, 2000.

Weiss, Norbert. *Das Städtewesen der ehemaligen Untersteiermark im Mittelalter. Vergleichende Analyse von Quellen zur Rechts-, Wirtschafts- und Sozialgeschichte*. Graz: Historische Landkommission für Steiermark, 2002.

Wiesflecker, Hermann. *Jugend, burgundisches Erbe und Römisches Königtum bis zur Alleinherrschaft 1459–1493*, vol. 1 of *Kaiser Maximilian I. Das Reich, Österreich und Europa and der Wende zur Neuzeit*. Munich: Oldenbourg, 1971.

———. *Maximilian I. Die Fundamente des habsburgischen Weltreiches*. Vienna, Munich: Oldenbourg Verlag, 1991.

Žontar, Josip. "Banke in bankirji v mestih srednjeveške Slovenije." *Glasnik Muzejskega društva za Slovenijo*, 13 (1932): 21–35.

Zwitter, Fran. "K predzgodovini mest in meščanstva na starokarantanskih tleh." *Zgodovinski časopis*, 6-7 (1952/1953): 218-245.

———. *Starejša kranjska mesta in meščanstvo*. Ljubljana: Leonova družba, 1929.

## The Early Modern Period

Cvetko, Dragotin. *Slovenska glasba v evropskem prostoru*. Ljubljana: Slovenska matica, 1991.

Drnovšek, Marjan. *Nakljanec Gregor Voglar (1651-1717), zdravnik v Rusiji*. Naklo: Občina, 2002.

Golia, Ludovik Modest (transl.). "Herbersteinovo življenje." In: *Moskovski zapiski*, Žiga Herberstein (ed.). Ljubljana: DZS, 1951.

Grafenauer, Bogo. *Boj za staro pravdo v 15. in 16. stoletju na Slovenskem. Slovenski kmečki upor 1515 in hrvaško-slovenski kmečki upor 1572/73 s posebnim ozirom na razvoj programa slovenskih puntarjev med 1473 in 1573*. Ljubljana: DZS, 1974.

———. *Doba prve krize fevdalne družbe na Slovenskem od začetka kmečkih uporov do viška protestantskega gibanja*, vol. 3 of *Zgodovina slovenskega naroda*. Ljubljana: Kmečka knjiga, 1956.

———. *Doba začasne obnovitve fevdalnega reda pod okriljem absolutne vlade vladarja ter nastajanja velikih premožen od protireformacije do srede XVIII. stoletja*, vol. 4 of *Zgodovina slovenskega naroda*. Ljubljana: Kmečka knjiga, 1961.

———. *Kmečki upori na Slovenskem*. Ljubljana: DZS, 1962.

———. *Začetki slovenskega narodnega prebujenja v obdobju manufakture in začetkov industrijske proizvodnje ter razkroja fevdalnih organizacijskih oblik med sredo XVIII. in sredo XIX. stoletja*, vol. 5 of *Zgodovina Slovenskega naroda*. Ljubljana: Kmečka knjiga, 1974.

———. *Zgodovina slovenskega naroda vol. 2. Doba zrele fevdalne družbe od uveljavljanja fevdalnega reda do začetka kmečkih uporov*. Ljubljana: DZS, 1965.

Grdina, Igor. *Od Brižinskih spomenikov do razsvetljenstva*. Maribor: Založba Obzorja, 1999.

Gspan, Alfonz. *Cvetnik slovenskega umetnega pesništva do srede XIX. stoletja*. Vol. 1. Ljubljana: Slovenska matica, 1978.

——— (ed.). *Anton Tomaž Linhart. Zbrano delo*. Ljubljana: DZS, 1950.

——— et al. *Zgodovina slovenskega slovstva I*. Ljubljana: Slovenska matica, 1956.

Jerman, Frane. *Slovenska modroslovna pamet*. Ljubljana: Prešernova družba, 1987.

Koruza, Jože. *Slovstvene študije*. Ljubljana: Filozofska fakulteta, 1991.

Kos, Janko. *Primerjalna zgodovina slovenske literature*. Ljubljana: Znanstveni inštitut Filozofske fakultete, Partizanska knjiga, 1987.

Kovačič, Franc. *Slovenska Štajerska in Prekmurje. Zgodovinski opis*. Ljubljana: Slovenska matica, 1926.

Kranjc-Vrečko, Fanika, Jonatan Vinkler and Igor Grdina. *Zbrana dela Primoža Trubarja II*. Ljubljana: Rokus, 2003.

Magenschab, Hans. *Jožef II. Revolucionar po Božji milosti*. Maribor: Založba Obzorja, 1984.

Melik, Vasilij. "Slovenci v času Marije Terezije." In: *Marija Terezija. Od baroka do razsvetljenstva*, Victor Lucien Tapié, (ed.), 363–377. Translated by Vital Klabus. Maribor: Založba Obzorja, 1991.

Rajhman, Jože. *Pisma Primoža Trubarja*. Ljubljana: Slovenian Academy of Sciences and Arts, 1986.

Reisp, Branko. *Kranjski polihistor Janez Vajkard Valvasor*. Ljubljana: Mladinska knjiga, 1983.

Rupel, Mirko. *Primož Trubar: Življenje in delo*. Ljubljana: Mladinska knjiga, 1962.

———. *Slovenski protestantski pisci*. 2nd ed. Ljubljana: DZS, 1966.

Simoniti, Primož (ed. and transl.). "Spremna beseda." In: *Akademske čebele ljubljanskih operozov*. Ljubljana: SAZU, 1988.

Sivec, Jože. *Opera skozi stoletja*. Ljubljana: DZS, 1976.

Spektorskij, Evgenij Vasilevič. *Zgodovina socialne filozofije*. Vol. 1. Ljubljana: Slovenska matica, 1932.

Struna, Albert. *Naši znameniti tehniki*. Ljubljana: Zveza inženirjev in tehnikov Slovenije, 1966.

Tapié, Victor Lucien. *Marija Terezija. Od baroka do razsvetljenstva*. Maribor: Založba Obzorja, 1991.

Trubar, Primož. *Cerkovna ordninga. Slowenische Kirchenordnung*. Munich: R. Trofenik, 1973.

———. *Zbrana dela Primoža Trubarja*. Vol. 2. Ljubljana: Rokus, 2003.

Valvasor, Janez Vajkard. *Slava vojvodine Kranjske*. Ljubljana: Mladinska knjiga, 1978.

Vodopivec, Peter. *Od Pohlinove slovnice do samostojne države. Slovenska zgodovina od konca 18. stoletja do konca 20. stoletja*. Ljubljana: Modrijan, 2006.

## Modernization and National Emancipation

"Slava bogu v višavah in na zemlji mir ljudem dobrega serca." *Kmetijske in rokodelske novice*, March 29, 1848. Quoted from Granda, "Revolucionarno leto 1848," 303–312.

Bezenšek, Anton. *Svečanost o priliki sedemdesetletnice Dr. Janeza Bleiweisa.* Zagreb: Uredništvo "Jugoslavenskog stenografa," 1879. Quoted from Grdina, *Slovenci med tradicijo in perspektivo.*

Cvirn, Janez (ed.). *Slovenska kronika XIX. stoletja.* Ljubljana: Nova revija, 2001.

———. *Trdnjavski trikotnik. Politična orientacija Nemcev na Spodnjem Štajerskem (1861–1914).* Maribor: Založba Obzorja, 1997.

——— and Andrej Studen. *Zgodovina 3.* Ljubljana: DZS, 2007.

Delavec, Mira. "Fani Hausmann." In: *Pozabljena polovica. Portreti žensk 19. in 20. stoletja na Slovenskem*, Alenka Šelih et al. (eds.), 31–34. Ljubljana: Založba Tuma, 2007.

Fischer, Jasna. "Slovensko narodno ozemlje in razvoj prebivalstva." In: *Slovenska novejša zgodovina*, Zdenko Čepič et al. (eds.), 17–21. Ljubljana: Mladinska knjiga—Inštitut za novejšo zgodovino, 2005.

Granda, Stane. "Od razcveta v streznitev." In: *Slovenska kronika XIX. stoletja*, Janez Cvirn (ed.), 121–134. Ljubljana: Nova revija, 2001.

———. "Predmarčno obdobje." In: *Slovenska kronika XIX. stoletja*, Janez Cvirn (ed.), 113–121. Ljubljana: Nova revija.

———. *Prva odločitev Slovencev za Slovenijo.* Ljubljana: Nova revija, 1999.

———. "Revolucionarno leto 1848 in Slovenci." In: *Slovenska kronika XIX. stoletja*, Janez Cvirn (ed.), 303–312. Ljubljana: Nova revija.

———. "Ženske in revolucija 1848 na Slovenskem." In: *Splošno žensko društvo 1901–1945*, Nataša Budna Kodrič and Aleksandra Serše (eds.), 6–15. Ljubljana: Arhiv Republike Slovenije, 2003.

Grdina, Igor. "Ilirci na pohodu." In: *Slovenska kronika XIX. stoletja*, Janez Cvirn (ed.), 175–176. Ljubljana: Nova revija, 2001.

———. *Ipavci. Zgodovina slovenske meščanske dinastije.* Ljubljana: Založba ZRC, 2002.

———. "Mlad umre, kdor je bogovom drag." In: *Slovenska kronika XIX. stoletja*, Janez Cvirn (ed.), 202–203. Ljubljana: Nova revija, 2001.

———. *Slovenci med tradicijo in perspektivo. Politični mozaik 1861–1918.* Ljubljana: Študentska založba, 2003.

———. "Življenje ječa, čas v nji rabelj hudi." In: *Slovenska kronika XIX. stoletja*, Janez Cvirn (ed.). Ljubljana: Nova revija, 2001.

———. "Življenje svetnikov—ena prvih uspešnic." In: *Slovenska kronika XIX. stoletja*, Janez Cvirn (ed.). Ljubljana: Nova revija, 2001.

Jurčič, Josip. *Zbrano delo*. Vol. 11. Ljubljana: DZS, 1984.

Kalc, Aleksej, Mirjam Milharčič Hladnik, and Janja Žitnik Serafin. *Doba velikih migracij na Slovenskem*. Ljubljana: Založba ZRC, 2020.

Keber, Katarina. *Čas kolere. Epidemije kolere na Kranjskem v 19. stoletja*. Ljubljana: Založba ZRC, 2007.

Kodrič, Nataša Budna and Aleksandra Serše. "Žensko gibanje na Slovenskem do druge svetovne vojne." In: *Splošno žensko društvo 1901–1945*, Nataša Budna Kodrič and Aleksandra Serše, (eds.), 27–34. Ljubljana: Arhiv Republike Slovenije, 2003.

Kresal, France. "Struktura slovenskega od 1851–1914," *Časopis za zgodovino in narodopisje*, 2 (2002): 101–124.

Mal, Josip. *Zgodovina slovenskega naroda*. Celje: Mohorjeva družba, 1993.

Melik, Vasilij. "Načrti za reformo Avstro-Ogrske in Slovenci." In: *Slovenci 1848–1918. Razprave in članki*, Viktor Vrbnjak (eds.), 643–646. Maribor: Litera, 2002.

———. "Problemi slovenske družbe 1897–1914." In: *Slovenci 1848–1918. Razprave in članki*, Viktor Vrbnjak (ed.), 359–368. Maribor: Litera, 2002.

———. "Slovenci v času Cankarjevega predavanja o jugoslovanstvu." In: *Slovenci 1848–1918. Razprave in članki* Viktor Vrbnjak (ed.), 687–695. Maribor: Litera, 2002.

———. "Ustavna doba in Slovenci." In: *Slovenska kronika XIX. stoletja*, Janez Cvirn (ed.), 13–15. Ljubljana: Nova revija.

Paternu, Boris. "Prešeren France." In: *Enciklopedija Slovenije*, vol. 9. Ljubljana: Mladinska knjiga, 1995.

Prešeren, France. *Poezije*. Introduction and explanatory notes by Anton Slodnjak (ed.). Ljubljana: Slovenski knjižni zavod, 1952.

Šorn, Jože. *Začetki industrije na Slovenskem*. Maribor: Založba Obzorja, 1984.

Štih, Peter, Luka Vidmar, Jernej Kosi, and Aleš Gabrič. *Temelji slovenstva*. Ljubljana: Cankarjeva založba, 2019.

Šumrada, Janez. "Poglavitne poteze napoleonske politike v Ilirskih provincah." *Zgodovinski časopis* 61, 1–2 (2007): 75–84.

Verginella, Marta. "Mesto žensk pod steklenim stropom." In: *Splošno žensko društvo 1901–1945*, Nataša Budna Kodrič and Aleksandra Serše (eds.), i–viii. Ljubljana: Arhiv Republike Slovenije, 2003.

Vilfan, Sergij. *Pravna zgodovina Slovencev*. Ljubljana: Slovenska matica, 1961.

Zwitter, Fran. *Nacionalni problemi v Habsburški monarhiji*. Ljubljana: Slovenska matica, 1962.

## "From the Habsburg Monarchy to the Kingdom of Yugoslavia" and "Slovenia after the Liberation: The 'People's Republic' and the Time of Socialism"

"Program ekonomske reforme i mere za njegovu realizaciju u 1990 godini." In: *Ekonomska reforma i njeni zakoni*. Belgrade: Federal Executive Committee—Federal Secretariat for Information, 1990.

Banac, Ivo. *The National Question in Yugoslavia: Origins, History, Politics*. Ithaca, London: Cornell University Press, 1984.

Boban, Ljubo, and Ivan Jelić. *Život i djelo Ante Trumbića*. Zagreb: Hrvatska akademija znanosti i umjetnosti, 1991.

Borak, Neven, Žarko Lazarević, and Jože Prinčič. (eds.). *Od kapitalizma do kapitalizma. Izbrane zamisli o razvoju slovenskega gospodarstva v XX.stoletju*. Ljubljana: Cankarjeva založba, 1997.

Bourdieu, Pierre. *Distinction: A Social Critique of the Judgement of Taste*. London: Routledge, 2000.

Buchenau, Klaus. "What Went Wrong? Church-State Relations in Socialist Yugoslavia." *Nationalities Papers* 33, 4 (2005): 547–568.

Čepič, Zdenko and Dušan Nećak. *Zgodovina Slovencev*. Ljubljana: Cankarjeva založba, 1979.

Dolenc, Ervin and Aleš Gabrič. *Zgodovina 4. Učbenik za 4. letnik gimanzije*. Ljubljana: DZS, 2002.

Dragnich, Alex N. "The Serbian Government, the Army and Unification of Yugoslavs." In: *The Creation of Yugoslavia, 1914–1918*, Dimitrije Đorđević (ed.), 37–50. Santa Barbara, Oxford: Clio Books, 1980.

——. *Serbs and Croats: The Struggle in Yugoslavia*. New York: Harcourt Brace Jovanovich, 1992.

Drnovšek, Marjan and Drago Bajt. (eds.). *Slovenska kronika XX. stoletja 1900–1941*. Ljubljana: Nova revija, 1997.

Ferenc, Tone. *Okupacijski sistemi na slovenskem 1941–1945*. Ljubljana: Modrijan, 1997.

Godeša, Bojan. *Kdor ni z nami, je proti nam. Slovenski izobraženci med okupatorji, Osvobodilno fronto in protirevolucionarnim taborom*. Ljubljana: Cankarjeva založba, 1995.

Gow, James and Cathie Carmichael. *Slovenia and the Slovenes: A Small State and the New Europe*. London: Hurst & Company, 2001.

Grafenauer, Bogo. *Slovensko narodno vprašanje in slovenski zgodovinski položaj*. Ljubljana: Slovenska matica, 1987.

Hribar, Tine. *Slovenci kot nacija*. Ljubljana: Enotnost, 1994.

Ivanič, Martin (ed.). *Dahauski procesi. Raziskovalno poročilo z dokumenti*. Ljubljana: Komunist, 1990.

Jančar, Drago and Peter Vodopivec (eds.). *Slovenci v XX. stoletju*. Ljubljana: Slovenska matica, 2001.

Jeraj, Mateja. "Položaj in vloga žensk v Sloveniji (1945–53)." PhD diss., Filozofska fakulteta, Univerza v Ljubljani, 2003.

Kardelj, Edvard. *Reminiscences: The Struggle for Recognition and Independence. The New Yugoslavia, 1944–1957*. London: Blond & Briggs in association with Summerfield Press, 1982.

———. *Tito and Socialist Revolution of Yugoslavia*. Belgrade: Socialist Thought and Practice, 1980.

Knight, Robert. "The Carinthian Slovenes: Ethnic Actors in Bit Part Roles?" Paper presented at the *Central European Minorities Policies in the Cold War* meeting in Ljubljana, March 2006.

Kostelnik, Branko. *Moj život je novi val, razgovori s prvoborcima i dragovoljcima novog vala*. Zagreb: Fraktura, 2004.

Krečič, Jela. "'Plakat je kovinsko črne barve, ker je bil tudi tovariš Tito kovinar!'" *Sobotna priloga –Delo*, May 19, 2005.

Laušević, Mirjana. "The Ilahiya and Bosnian Muslim Identity." In: *Retuning Culture: Musical Changes in Cental and Eastern Europe*, Mark Slobin (ed.), 117–135. Durham, London: Duke University Press, 1996.

Lienhard, Thomas. "Slavs, Bulgarians, and Hungarians: The Arrival of New Peoples." In: *Rome and the Barbarians: The Birth of a New World*, Jean-Jacques Aillagon (ed.), 578–579. Milan: Skira, 2008.

Lobrano, Alexander and Andrea Fazzari. "A Land in Between." *Gourmet* (February 2006): 112–119.

Loparnik, Borut. "Poličeva doba slovenske opere: ozadja in meje." In: *Zbornik ob jubileju Jožeta Sivca*, 193–204. Ljubljana: Založba ZRC, 2000.

Lukan, Walter. "Slovenci in nastanek jugoslovanske državne skupnosti." *Glasnik Slovenske matice* 53, 1 (1989): 40–44.

Luthar, Breda. "Remembering socialism: On desire, consumption and surveillance." *Journal of Consumer Culture* 6, 2 (2006): 229–259.

Luthar, Oto. *O žalosti niti besede. Uvod v kulturno zgodovino velike vojne*. Ljubljana: Založba ZRC, 2000.

Macmillan, Margaret. *Paris 1919: Six Months that Changed the World*. New York: Random House, 2002.

Mastnak, Tomaž. "From Social Movements to National Sovereignty." In: *Independent Slovenia: Origins, Movements, Prospects*, Jill Benderly and Evan Kraft (eds.), 93-111. Basingstoke: Macmillan, 1994.

Mlakar, Boris. "Slovensko domobranstvo od ustanovitve do umika iz domovine." PhD diss., Filozofska fakulteta, Univerza v Ljubljani, 1999.

———. *Slovensko domobranstvo (1943-1945)*. Ljubljana: Slovenska matica, 2003.

*National Geographic Magazine* 34, 6 (December 1918).

Nećak, Dušan. *Avstrijska legija II. Maribor 1995. Die österreichische Legion II.* Vienna, Cologne, Maribor: Založba Obzorja, 1995.

Pavković, Aleksandar. *The Fragmentation of Yugoslavia: Nationalism and War in the Balkans*. New York: Macmillan, 2000.

Petković, Aleksandar. *Političke borbe za novu Jugoslaviju. Od drugog AVNOJ-a do prvog Ustava*. Belgrade: Jugoslavenska revija, 1988.

Pirjevec, Jože. *Jugoslavija, 1918-1992. Nastanek, razvoj ter razpad Kardjordjevićeve in Titove Jugoslavije*. Koper: Lipa, 1995.

Povše, Janez (ed.). *Oblaki so rudeči*. Trieste: Založba tržaškga tiska, 1988.

Prunk, Janko. *Pot krščanskih socialistov v osvobodilno fronto slovenskega naroda*. Ljubljana: Cankarjeva založba, 1977.

Mastnak, Tomaž and Nela Malečkar (eds.). *Punk pod Slovenci*. Ljubljana: KRT, 1985.

Reed, John. *War in Eastern Europe: Travels through the Balkans in 1915*. London: Phoenix, 1995.

Repe, Božo. *Jutri je nov dan. Slovenci in razpad Jugoslavije*. Ljubljana: Modrijan, 2002.

Stavbar, Vlasta. "Izjave v podporo Majniške deklaracije." *Zgodovinski časopis*, 3 (1992): 357-381; 4 (1992): 497-507; 1 (1993): 99-106.

Tomc, Gregor. "The Politics of Punk." In: *Independent Slovenia: Origins, Movements, Prospects*, Jill Benderly and Evan Kraft (eds.), 113-134. Basingstoke: Macmillan, 1994.

Vodopivec, Peter. "Pogled zgodovinarja." In: *Slovenci v XX. stoletju*, Drago Jančar and Peter Vodopivec (eds.), 5-15. Ljubljana: Slovenska matica, 2001.

———. "Prostozidarska loža Valentin Vodnik v Ljubljani (1940)." *Kronika. časopis za slovensko krajevno zgodovino*, 1 (1992): 44-50.

Zei, Vida and Breda Luthar. "Shopping across the Border." Paper presented at the conference *Everyday Socialism: States and Social Transformation in Eastern Europe 1945-1965*, The Open University Conference Centre, London, April 24-26, 2003.

Žižek, Slavoj. *Druga smrt Josipa Broza Tita*. Ljubljana: DZS, 1989.

# Notes on Contributors

**Oto Luthar** is a historian and director of the Scientific Research Center of the Slovenian Academy of Sciences and Arts (ZRC SAZU), and a member of its Institute of Culture and Memory Studies. He is the author of the chapters "Divided by the Great War," "The Making of the New State," "The Kingdom of Serbs, Croats and Slovenes," "Dictatorship and the Turmoil of the 1930s," "A Nation Torn Apart: World War II in Slovenia," "The Establishment of the 'New Order,'" "The First Five-year Period and Self-management," and "'Liberals' vs. 'Conservatives.'"

**Marjeta Šašel Kos** is a Senior Research Associate (for epigraphy and ancient history, retired since 2021) at the Institute of Archaeology, ZRC SAZU. She is the author of the chapters "Prehistory: History Created by Archaeology" and "The Roman Empire: Conquest and Pax Romana."

**Petra Svoljšak** is a historian, research advisor and head of the Milko Kos Historical Institute, ZRC SAZU and a lecturer at the Postgraduate School ZRC SAZU. She is the author of "French Rule," "The Pre-March Era, the Time of Non-freedom," "The Year of Freedom, the 1848 Revolution, and United Slovenia," "The Slovenes in the Constitutional Era," "Unity and National Existence," "In the Shackles of Political Parties," and "The Other Side of History."

**Martin Pogačar** is a Research Fellow at the ZRC SAZU's Institute of Culture and Memory Studies. He is the author of the subchapter "Punk Rock, the Alternative, and Political Appropriations."

**Peter Štih** is a Professor of Medieval History and Auxiliary Sciences of History at the Faculty of Arts, University of Ljubljana, President of the Slovenian Academy of Sciences and Arts, Member of the European Academy of Sciences and Arts, and Corresponding Member of the Austrian Academy of Sciences. He is the author of "The Early Slavs" and "The Carolingian Period in the 9th Century."

**Dušan Kos** is a historian and Senior Scientific Associate at the Milko Kos Historical Institute, ZRC SAZU. He is the author of "From Autonomy to the Unification of the Alpine and Danube Basin Regions" and "'Tres Ordines Slovenorum': Society, Economy, and Culture."

**Peter Kos** is a Senior Research Associate and Professor of numismatics at the Department of Archaeology, University of Ljubljana (since 2019 retired). He is author of the chapter "From the Marcomannic Wars to the Settlement of the Slavic Tribes."

**Igor Grdina** is a historian and Slovenist, and a member of the Institute of Cultural History at the ZRC SAZU. He is the author of "The Stars of Celje," "The bloody Fall of the Middle Ages," "From Humanism to Reformation," "From Counter-Reformation Rigor to Baroque Exuberance," and "Scholars, Officials, and Patriots Changing the World."

**Alja Brglez** is a historian and historical anthropologist, one of the founders of the Institute for Civilization and Culture in Ljubljana. Currently, she serves as a Head of the Cabinet of the President od the Republic. She is the author of "Reorganization of the Marches and a Shift of Ethnic and Language Borders" and "From Crisis to Conflict and Beyond."

# Index

*1551*, periodical  440

Aachen, Peace of  108
Absolutism  234, 253, 283, 289, 300, 308–310, 314
– critics of  235
– monarchial  232
Ad Pirum  72
Adalvin, Archbishop of Salzburg  113
Adriatic Sea  41, 132, 137, 143–144, 175, 186, 197, 213, 398, 405, 479
Adsalluta  60
Aecorna  58
Aegean world  27
Aemilius Aemilianus, Marcus  73
Aemilius Scaurus  50
Agnes of Andechs  136
Agrarian Reform  475–476, 497
Agricultural societies  257, 286, 313
Aguntum  74, 89
Ahacel, Matija  290
Ahačič, Draga  502
Aio  109
Ajdovska jama  20
Alemanni  74, 82
Alexander  37
Alexander I  288
Alexander I, tsar  283
Alliance of Socialist Youth of Slovenia, ZSMS  523, 527, 530, 535, 537, 540
Alpine defensive system  77–78
Alps  18, 33, 35, 37, 41–42, 47–48, 50, 56, 60, 65, 70, 72, 77, 89, 121–122, 124, 127, 130, 136, 138–139, 141, 161, 166–167, 172, 176, 187, 196–197, 227, 230, 236–237, 240, 246, 248, 261, 268, 284, 310, 564

– Eastern  89, 91, 100, 124, 127
– Julian  75
Alzeco, Bulgar leader  92
Amber Route  37
Ambrož  321
Ambrož, Miha  321
Amendment crisis  539
America, United States of  427, 542
Anabaptists  215–216
Ancient urban centers, collapse of  90
Andechs, counts  123, 135
Andechs-Meran  172
Andreae, Jacobus  223
Andrej of Turjak  223
Andrioli, Franc Ksaver  295
Annexation, to the Kingdom of Italy  330
Anti-Communist Voluntary Militia, MVAC  460–462
Anti-Fascist Council of National Liberation of Yugoslavia, AVNOJ  466–467, 475, 488
Antonius Primus, Marcus  65
Appian  45
Appianus of Alexandria  36
Apulian ceramics  33
Aquileia  37, 41, 43, 67, 72, 89, 95, 109
– Patriarchate of  122, 135, 137, 141, 148, 152
Archaeology  15
Argonauts  56
Aribo, Margrave  115
Ariovistus  37
Arnefrit, son of Lupus  109
Arno, Bishop of Salzburg  95
Arnulf, King and Emperor  110, 115, 121

# Index

Aškerc, Anton 382
Assembly DFJ, Constituent 475
Assembly FNRJ, Constituent 477
Assembly SHS, Constituent 403, 409, 411, 413–414, 442–443
Associated Labor, Act 516
Atrans (Trojane) 61
Attila the Hun 81
Attire 32
Auersperg, Anton Aleksander 287
Auerspergs, Lords 123
Augsburg 167, 201–202
– Confession 222
– Religious Peace of 213
Augustan Tenth Region 48
Aurelius Cotta, Lucius 49
Austria
– Inner Austria 139, 211–213, 216–217, 219, 222–225, 227–228, 231–235, 284, 295, 329–330
Austria–Hungary 349, 367–371, 373, 376, 378, 400, 403–404, 406, 432
Austrian Monarchy 210, 249–254, 256–259, 262–263, 265–266
Austro-Slavism 268
Avant-garde 422–423, 502, 512, 515, 529
Avars 84, 89–92, 94, 109
– Avar khaganate 91–92
Ažbe Anton 384

Babenbergs, Margraves and Dukes of Austria 131, 136, 173, 186
Babič, Jože 514
Bach, Alexander 309, 315
Balantič, France 472
Balanus 48
Balaton, Lake 111
Balbinus 72
Baldric, Duke of Friuli 104

Balkan, Mt. (Mt. Haemus) 47
Balkans 19, 27, 36, 45–47, 52, 56, 68, 72–74, 82, 85, 154, 171, 176–177, 192, 212, 266, 369, 447, 553, 563–564
Bamberg, Diocese of 134
Barle, Karel 487
Basar, Jernej 240
Bato 54
Bauer, Martin 241
Baumkircher, Andrej 194, 199
Bavaria 94–95, 99, 103–105, 110–111, 115–116, 121, 123, 266
Bavarian Eastern March 104
Bavarians 89, 92, 94, 99, 103, 114–115, 127
Bavčar, Igor 532
Beatlemania 514
Beaumarchaias 267
Beck, Max Wladimir von 365
Bedriacum 65
Bela IV, Hungarian King 131
Belcredi, Richard 329
Belgrade 141, 400, 417, 444, 536
– Protocol 485
Beljaši 409
*Beneficiarii* 61
Benegalija, Jože 487
Bergant, Fortunat 247
Berlin, Congress of 349–350
Bernadotte, Jean-Baptiste 273
Bernhard, Duke of Carinthia 136
Bernik, Janez 511
Bertold, Patriarch 148
*Beseda*, periodical 511, 522
Betal Cavern (Betalov spodmol) 15
Beust, Friederich Ferdinand von 331
Bismarck, Otto von 329
Bistra, Carthusian monastery 264
Black Sea 36
Blaznik, Jožef 295

Bled 125
Bleiweis, Janez 295–297, 300–303, 310, 320–322, 334, 338, 340, 344, 355
Blue Guard 459–462, 465
Bogen, Counts of 123
Bogenšperk 242–243
Bogenšperk graphic collection 242
Bohemia 132, 136, 179–181, 184, 209, 233, 239, 254, 328, 360, 368, 376
Boii 51
Bonaparte, Napoleon 274–275, 278–281
– Napoleonic Wars 237
– Napoleon's Illyria 255
Bonomi, Guiseppe Clemente 247
Bonomo, Petrus 210, 215–218
Borgija Sedej, Frančišek 427
*Bori*, periodical 511
Borna, Prince of Guduscans 105
Borštner, Ivan 531
Bosnia 25, 28, 54, 279, 349, 403, 414–416, 443
Bosnia and Herzegovina 9, 349, 368–369, 404, 416, 486
Boundary, stone 51
Branković Smederevac, George 177
Brankovič, Albina 502
Braslav, Slavic Prince 110
Brecelj, Marko 524
*Brencelj*, periodical 328
Brenner, Martin 231
Brežice 163
Brioni, Declaration 549
British Royal Society 241
Brixen
– Diocese of 122
Broek, Hans van den 549
Bronze Age 23
Brotherhood and unity 495
Bučar, France 533, 541, 549

Budna, Nataša 590
Bulgars 92, 111
Bullinger, Heinrich 222
Byzantium 86, 171
– Byzantine Empire 83

Cadaloh, Prefect of the March of Friuli 104
Caesar 50
Calvinist, communities 233
Cankar, Ivan 370, 383–384, 398, 419
Capistranus, Johannes 184–185
Carantania, Duchy of 121, 125, 129
Carantanian March 121, 130
Carantanians, Carantania 91–97, 99–100, 103–105, 109–110, 113–115, 121, 147, 172, 564
– Baaz 99
– Borut 94
– Cacatius (Gorazd) 94
– Carantanum 93
– Etgar 105
– Ethonogenesis 94
– Gorazd 96
– Hotimir 94–96
– Nobility 99
– Prince's Stone 96
– Principality of 94, 100
– Samo 91
– Samo's political union 92
– tribal king, knjaz, knez 96
– Valtunc 95
Carbonarius de Biseneg, Gregorius 248
Carinthia 91, 116, 122–124, 127, 129–130, 132–134, 136–137, 139, 141–142, 147, 156, 163, 167, 175, 186, 198–199, 202, 219, 224, 241, 251, 253, 256, 258, 264, 284, 290, 300, 312, 317–318, 340, 343,

351–352, 358, 361, 363, 372, 374,
381, 408, 427, 449, 481
– Duchy of 99, 130, 133
– Duke of 122, 134
Carinthian Plebiscite
– Zone A 408
Carloman 110, 112
Carni 37
Carniola 96
– March 122, 134
Carniola Historical Society 310
Carniolans 100, 105
– Vojnomir the Slav 103
Carolingian, state 104
Carthusian order 151, 189, 264
Casimir III the Great 174
Cassiodorus 107
Cassisu Longinus, Gaius 44
Cassius Dio 53
Catholic
– Action 434
– Church 148, 223, 234, 264, 328, 335, 350, 355, 435, 486
– Conservatives 332
Catholic Cultural Union 425
Catholic National Party 356, 372
Catholic Political Association 356
Catholic Society of Women 392
Catholicism 217, 231, 264, 364, 368, 434
Catmelus 44
Cebej, Anton 247
Celestin, Fran 390
Celje 125, 162
– Celeia 37, 58, 89
– Counts of 134, 139, 142, 150, 153, 169
Anna 175
Barbara 140, 171, 179, 570
Catherine 183, 189–190
Elizabeth (Frangepán) 176

Frederick II 178–179
Herman I 174–175
Herman II 140, 171–172, 175–176, 178, 189–190, 570
Herman III 178
Louis 178
Prince Ulrich 180, 182–187, 190
Ulrich II 141, 169
Ulrich II of Žovnek 173
Veronika of Desenice 178
William 174–175
Celts 35–36
censorship 248, 273, 289, 295, 299, 308, 497, 500, 515
Censorship 294
Cerknica 137
– Lake 241
Chamber of Republics and Provinces 510
Charlemagne, Emperor 95, 103–104, 106
Charles III (the Fat) 110
Chernobyl 528
Chest, military 76
Christalnick, Michael Gotthard 224
Christian churches, early 79
Christian Women's Society 392
Christian, cult 90
Christianization 95, 150
Christians 58, 75
Churchill, Winston 446
Ciganska cave 18
Cigler, Janez 295
Cimbri 48
Cincibilus 44
Ciuha, Jože 511
Cividale del Friuli 92, 207, 281
Cividale, Gospel of 110
Civil Code, Austrian 281, 307
Claudius 54
Claustra Alpium Iuliarum 73, 569

Clergy  153, 159, 199, 230, 233,
    265, 280, 291, 314, 317, 332, 338,
    427, 485
Clodius Albinus  71
Coins  52
- hoards of  71–72
Colapiani  40
Cold War  481
Collaboration  430, 458, 460,
    462–463, 476
Colonatus  155, 257
Coloni  80
Colonia  57, 63
Colonization  55, 99, 123, 126–127,
    151, 153, 159, 475
Color Realism  422
Combatants, League of  537, 540
Cominform  481, 487–491
Commodus  63
Communism, Cultural  435
Communist Party of Slovenia,
    KPS  439, 455, 465, 541
- Central Comittee, CK KPS  454
- Central Committee, CK
    KPS  454
Communist Party of Yugoslavia,
    KPJ  414
Communists of Slovenia,
    League of  487
Concordat  314, 328, 332–333, 336
Confederation, assymetric  545
Conservatives  398, 459, 507
Constans  76
Constantine II  76
Constantine the Great  75
Constantine, missionary in Moravia
    and Pannonia  113
Constantinus Porphyrogenetus,
    Byzantine Emperor  99
Constantius II  76
Constitution, December  324

Constitution, FNRJ  477
Constitution, Kingodm of
    Yugoslavia  443
Constitution, New
    Habsburg  300, 305
Constitution, October  429
Constitution, SFRJ  494, 534
- 1963  507
- 1974  510, 516, 541
- amendments  534–535
- Little  516
Constitution, Slovenian
- amendments  538–539
Constitution, Vidovdan  414,
    418, 429
Constitution, Weimar  414
Constitutional Association  337
Constructivism  422
Contemporary History,
    Institute of  482
*Conversio Bagoariorum et
    Carantanorum*  92
Copper Age  18
Corfu, Declaration  402–403
Cornelius  48
Costa, Etbin  334
Counter-Reformation  219, 223–224,
    227, 232, 234
Coup, military  446
Croatia
- Croatian question  441–442, 445
- Independent State of,
    NDH  447
Croatian Party of Rights  367, 371
Croatian Peasant Party  414,
    416, 444
Croatian Republican Party  416
Croats and the Slovenes in Istria,
    Political Association of the  361
Cro-Magnons  15
Cultural Fund  504

Cultural struggle (Kulturkampf) 337
Culture
- Literary 169
- Oral 170
Currency 520, 544, 553
Currency 37
Cvetković, Dragiša 443
Cvetković-Maček Agreement 444
Cyril. See Constantine, missionary in Moravia and Pannonia
Časopis za kritiko znanosti, periodical 535
Čargo, Ivan 422-423
Černe Franc 286
Černigoj, Avgust 422
Četniks 460
Čobal, Melhior 366
Čop, Matija 292-294, 297

Dachau, trials 486-488
- Diehl-Oswald trial 486
Daesitiates 54
Dagobert I, Frankish king 91-92
Dalmatia 23, 52, 147
Dalmatin, Jurij 219-220, 224-225, 571
Danube, River 36, 47, 67, 89
Danubian region 23, 28, 35
Day of Youth 529
Debenjak, Riko 511
Declaration of Loyalty 483
Defense of Human Rights, Committee for the 532, 536
Defense regiments (Landswehr, Honvéd) 331
Defensive
- System 75
- Zone (praetentura Italiae et Alpium) 67
Dejanja, periodical 435
Delak, Ferdo 422
Delavski list, periodical 439

Delmatae 52
Delo, periodical 532
DEMOS 540-541, 544
- Democratic Alliance of Slovenia 540
- Gray Panthers Party 540
- Greens of Slovenia 540
- Liberal Party 540
- Slovenian Christian Democrats 540
- Slovenian Farmers' Alliance 540
- Social Democratic Alliance 536, 540
Denationalization, Act 553
Dermota, Anton 364
Deschmann, Karl 20
Destovnik Kajuh, Karel 472
Detela, Oton 318
Dev, Janez Damascen 261
Development demographic 373
Dictatorship 418, 420, 429-430, 435
Diehl, Branko 487
Dinar, convertible 543
Diocletian 64, 75
Divini redemptoris 434
Djilas, Milovan 501, 509
Dolanc, Stane 532
Dolar, Janez Krstnik 239
Dolenc, Matija 301
Dolničar, Janez Gregor 246
Dolomite Declaration 469
Dom in svet, periodical 356, 435
Domenkuš, Ferdinand 323
Dopolavoro 452
Doujak, Hermann 465
Dragolič, Jurij 216
Drava Banovina 420, 425
Drava March 131
Drava, River 104
Drnovo near Krško
- Neviodunum 48
Drnovšek, Janez 539, 549-550, 553
Duino, Lords of 137

Eagle society (Orel) 326, 361, 386, 426
Eastern March 113, 115, 121
Eastern Prefecture, Bavarian 106, 110
Eberhard II, Archbishop 147
Economic Plan 496
Economy, consensual 520
Edinost
– Periodical 326, 328, 382, 391
– Political society 341, 361
Edling, Janez Nepomuk Jakob 260
Eggers 260
Elagabalus 71
Eneolithic (Copper Age) 19
Engilschalk 110
Enlightenment 197, 248, 254, 262, 266, 268, 270, 273, 278, 291, 296
Epidemics 55, 154, 166, 193
– cholera 378
Episcopal see 61, 64
Eppensteins, Dukes 123, 132
Erasmus the Knight 194
Ernest, Duke of Inner Austria 139
Etruscans 33
Eugene of Savoy 248
Eugenius 78
Eugippius 81
European Community, EC
– Conflict mediation 549
European Union, EU 553
– Presidency 553
Even-odd system 520
Expressionism 422

Fabiani, Maks 385
Falcon society (Sokol)
– Southern Falcon (Južni sokol) 326
Falcon Society (Sokol) 426, 436, 455, 463
Farms 125, 153–154, 193, 287, 343, 359, 424

Fascism 426, 437–438, 441–442, 452
February Patent 317–318, 320, 328–329
Federal Executive Council, SIV 543, 546
Feminist, movements 525
Ferdinand IV, King of Naples 288
Ferenc II Rákóczy 237
Feudal
– Order 122
– System 125–126, 197–199, 243, 256, 282
Filipič, Lojze 502
Finžgar, Fran Saleški 383
Flacius, Matthias 222, 232
Flavia Solva 74
Flavian, Emperors 66
Flavius Aëtius 80
Flis, Janez 385
Fluvius Frigidus 78
Foederati. See Foedus
Foedus 80
Foerster, Anton 342, 384
Forchheim, Peace of 115
Fortunatus, Patriarch of Grado 105
Four Emperors, Year of 65
France 275, 288, 315, 402
Francis IV, Duke of Modena 288
Francisci, Erasmus 242
Frangepán, dynasty 176
– Ferenc Kristóf 236
Frankfurt Parliament
– Boycott of elections 303
Franks 82, 96, 103, 105
– Avar, wars 103, 107
– Byzantine, treaty 83, 105
Franz Ferdinand 398
Fredegar, Chronicle 91
Frederick I Barbarossa 195
Frederick the Great 252
Free Territory of Trieste, STO 469, 478

- Zone A  479
- Zone B  479
Freising  95
- Church of  99
- Diocese of  122, 127
- *Manuscripts*  95
French
- Army  273–274
- Empire  273, 275
- Occupation  274
- Revolution  266, 273, 306
Frischlin, Philipp Nikodemus  224
Friuli, Friulian  39, 92, 101, 103–104, 106, 109–110, 117, 124, 135, 143–144, 150, 241, 311
- March  137, 141
- Plain  100
Funtek, Anton  384
Futurism  422

Gabrščcek, Andrej  361, 390
Gaiger, Janez Adam  240
Gaj, Ljudevit  296
Gaj's alphabet  297
Galatians  36
Galba  65
Galenus  68
Gallienus  63, 74
Gallus, Jacobus  225
Gauleiter  450
Gauls, Gaul
- Cisalpine  45
- Transalpine  42
Gauls, Gauls. *See* Celts
Gebhard, Archbishop of Salzburg  95
General Women's Society  392
Geneva Declaration  409
Genthius  45
George of Poděbrad  183
Georgenberg Pact  131
German operation zone
- Adriatisches Küstenland  464

- Alpenvorland  464
German School Association  398
German Supreme Command  447
German Youth (Deutsche Jugend)  452
Germanic, tribes  67
German–Slovene strife  359
Gestapo  487
Glagolitic, Slavic alphabet  113
*Glas delavca*, periodical  439
Glaser, Karel  386
Gold  44
Gołuchowski, Agenor  315
Gorizia  122, 141–142, 164, 173, 198–200, 211, 213, 219, 224–225, 227–228, 230, 234–236, 241, 246, 251, 264, 273, 275, 277, 281–282, 284–285, 288, 290–291, 298, 302, 311–312, 314–315, 321, 328, 331, 336–338, 340–341, 346, 352, 355–356, 360, 366, 373–374, 381, 386, 391, 465, 467, 478–479
- Counts of  132–134, 136–139, 141–144, 173, 187, 200, 207
- Counts, County  124
- County of  132, 134, 141–142, 200
Gorizia–Tyrol, Counts of  136, 164–165, 174
- Mainhard IV  136
Gornji Grad
- Monastery  148
Goths  64, 73
Govekar, Fran  383–384
Governor of the province (Landeshauptmann)  319
Gradišnik, Fedor  502
Graphic Art, Biennial of  502
Graphic Arts, International Center of  502
Graz  127, 139, 169, 203, 213, 223, 225, 228, 233, 262, 281, 290–291, 298, 302, 364, 376, 389, 488

– Pacification of 223
Grazioli, Emilio 450, 461
Grbec, Marko 246
Grčarice 460
Greece 28
Gregorčič, Anton 356, 360
Gregorčič, Simon 342, 355, 382
Gregorčič, Vinko 360
Gregorić, Ilija 229
Grimoald, Lombardian king 92
Grohar, Ivan 384
Grossman, Karol 384
Gruber, Gabriel 257, 263
Grün, Herbert 502
*Grupa* 53 501
Grupa 69 511
Gubec, Ambrož (Matija Gubec) 229
guild system 252
Gurk
– Convent 124
– Diocese of 95, 147
Gutsman, Ožbalt 260
Györ, Diocese of 145

Habsburg dynasty 132, 134, 139, 142, 158–159
– Albert II 158
– Charles of Austria 404–405
– Franz Joseph 404
– Frederick II 143, 158
– Frederick III/IV/V 138, 153
– Joseph II 153
– Maximilian 169
– Maximilian I 142
– Rudolf 158
– Rudolf IV 137
– Rudolf of Habsburg 136
Habsburg–German federation 404
Hacquet, Balthasar 268
Hajdrih, Anton 342
Hallstatt 28
– Groups 28

– Period 28
Hartman, Bruno 502
Haugwitz, Friederich Wilhelm 252–253
Hausmann, Fani 390
Hein, Viktor 379
Hellenic civilization 37
Helmwin, Bavarian Count 105
Hemma
– Countess of Friesach 124
– Dynasty 173
Henry I, Emperor 121
Henry II 121
Henry IV, Emperor 122–123, 135
Henry of Gurk, Bishop 151
Herbard VIII 223
Herberstein, Jurij 202–203
Herberstein, Karel Janez 265
Herberstein, Žiga 209–210
Herbert, Franz Paul 270
Hermann, Mihael 336
Herodian 72
Heunburg, Counts of 134, 158
Hieng, Andrej 502
Histri 28, 44
Hitler Youth (Hitler Jugend) 452
Hitler, Adolf 428, 443, 446, 448, 450, 464
Hladnik, Boštjan 514
Hoards, Bronze Age 27
Hočevar, Janez Jurij 248
Höffer, Johann Berthold von 247
Hohenwart, Karl 339
Holy Alliance 283, 288–289
Holy Roman Empire 177, 182, 190, 194, 203, 207, 213, 218, 221, 223, 232
Holy See 485
Holzapfel, Ignacij 295
Home Guard 462, 468, 475, 482
*Hospitium publicum* 47
Hren, Tomaž 223, 231–232, 239

Hribar, Ivan  352, 357, 365, 379
Hungarian crown  176, 181–182,
 184, 190–191, 212, 235
Hungarian-Croatian
 Kingdom  127
Hungary  51, 53, 163, 182–184, 202,
 209, 211, 213, 235, 263, 302, 306–
 307, 310, 331, 341, 380, 448–450,
 453, 481, 490
– Hungarion Kingdom  165
Huns  79
Hunyadi
– János  182, 184, 186
– Ladislaus  185–186
– Matthias Corvinus  189, 194, 196

Iapodes  28
Iazyges  74
Ice Age  15
Idrija  167, 212, 238–239, 268, 273,
 284, 290, 381
Ig  22, 569
Illyria
– French Illyria  237
– Illyrian Kingdom  281–282, 307
– Illyrian movement  296–297
– Illyrian Provinces  275, 277–
 278, 281
– Illyriann provinces  275
Illyrians  36
– Illyrian Wars  52
– peoples  46
Illyricum  45–46, 48, 52, 55, 64, 72,
 89, 114
Impressionism  383, 385, 422
Incursions  121
Independence  403, 454, 539
– Basic Charter on the
 Independence and Sovereignity of
 the Republic of Slovenia  546
– Declaration of  546
– Movement  519

– process  544
– project  542
– referendum  542
Industrialization  503
Industry
– Chemical  345
– Iron  251, 257, 284, 286, 345, 376
– Mining  178, 344
– Textile  284, 345, 424
– Timber  345, 376
Innichen, monastery  95
Ipavec, Benjamin  327, 342, 384
Ipavec, Josip  384
Iron Age  25
Istria  24, 44, 89, 106
– Byzantine  103
– Province, March  122, 135, 141,
 144, 147
Italo-Corinthian objects  33
Iulius Vepo, Gaius  53
Iunius Brutus Callaicus, Decimus  50
Iuthungi  74, 80
Izola  161
*Izvestje*, periodical  385

Jacobins  273
Jager, Ivan  385
Jakac, Božidar  471, 501
Jakopič, Rihard  385
Jama, Matija  385
Jan, Slavko  502
Janežič, Anton  309
Janša, Anton  259
Janša, Janez  530, 532, 537
Japelj, Jurij  261
Jarnik, Urban  297–298
Jason  57
Jeans  514
Jeglič, Anton Bonaventura  379
Jelačić, Josip  306
Jelovšek, Fran  247
Jemec, Andrej  511

Jenko, Avgust 400
Jenko, Davorin 342
Jenko, Simon 309, 323, 327
Jeran, Luka 309, 321
Jesih, Milan 515
Jesuits 227–228
Jevtić, Bogoljub 441
Jews 165, 175, 190, 264, 280, 311, 411
John Paul XXIII, Pope 485
John, Duke of Istria 106
Josephinism 264, 266
– Joseph's reforms 264
Jovanovič, Dušan 515
Julian March 426, 479
Julianus Apostata 76
Jupiter Dolichenus, Temple of 71
Jurčič, Josip 322–323, 327, 342, 356
Juri, Franco 536
Jurklošter, Carthusian monastery 151
Justinian I 83
*Jutro*, periodical 328, 435

Kacijanar, Ivan 211
Kacijaner, Franc 217–218
Kacin, Jelko 548
Kadijević, Veljko 539
Kalin, Zdenko 423
Kalister, Janez 286
Kamnik 135, 163, 168, 201, 216, 239, 261, 378
Kant, Immanuel 270
Kapodistrias, John 288
Karadjordjević
– Dynasty 368, 371, 402
– King Alexander 409, 429, 441
– King Peter 410–411
– King Peter II 442
– Paul 441
– Prince Alexander 410–411

Kardelj, Edvard 439–440, 457, 459–460, 465, 467, 475, 488–489, 493–494, 501, 505, 507, 511, 517
Karlin, Andrej 427
Karnburg 96, 115
Karpe, Franc Samuel 270
Karst 19, 24–25, 27, 30, 48, 51, 74, 90, 103, 122, 141–142, 234, 252, 310, 344, 478
Kastelec, Matija 240
*Katedra*, periodical 531
*Katoliški glas*, periodical 338
Kavčič, Jane 514
Kavčič, Matija 306–307
Kavčič, Stane 507, 509–511
*Kazina*, society 389
Kempf, Nicholas 214
Kempf, Nicolas 168
Kepler, Johanes 231
Kersnik, Janko 342, 356
Kette, Dragotin 383
Kidrič, Boris 439, 465, 475, 477, 483
King Matthias (Corvinus) 194–195
– Good 195
Klagenfurt 97, 132, 151, 168, 216, 228, 260, 266, 285, 290–291, 302–303, 308, 312, 345, 374, 408, 450–451, 467
Klombner, Matija 216
Klopčič, Matjaž 514
Kmecl, Matjaž 541
*Kmetijske in rokodelske novice*, periodical 295
Kocbek, Edvard 435, 454, 457, 500, 509
Kocel, Count and Prince of Pannonia 111–114
Kočevje Assembly 465–466
Kočevski Rog 463, 482
Kogoj, Marij 423
Kolpa, River 40, 99, 127, 135, 212, 310

Kopač, Josip 366
Koper 75, 106, 108, 137, 143–144, 148, 161–162, 168, 227, 277, 282, 290, 478–479
Koper, convent 150
Kopitar, Jernej 268–269, 278–279
Korošec, Anton 372, 407, 409, 417–418, 430–431, 433, 440
Kos, Franc 386
Koseski, Jovan Vesel 297
Kosezi, Erdlinger, arimanni 99, 127, 157
Kosovel, Srečko 422–423
Kosovo 519, 532, 536
Kostanjevica 123
Kottanner, Helen 181
Kozak, Ferdo 436
Kozler, Peter 301, 310
Kraigher Commission 520
Kraigher, Boris 464, 501
Kraigher, Sergej 520
Kralj, France 422
Kralj, Lado 515
Kralj, Tone 422–423
Kramař, Karel 365
Kranj 58, 82, 122, 125, 162–163, 203, 378, 421
Kranjc, Boris 487
*Kranjska čbelica*, literary almanac 293–294
Kregar, Stane 423, 501
Krek, Janez 431
Krek, Janez Evangelist 343, 360, 362–363, 367–368
Krek, Miha 431, 483
Krelj, Sebastjan 222
*Kresije* 255, 312
Kristan, Anton 366
Kristan, Etbin 366
Križaj, Franci 502
Krleža, Miroslav 500
Krpan, Martin 309

Krško 20, 163, 229, 243
Kržišnik, Zoran 501
Kučan, Milan 521, 532, 541, 549
Kulovec, Franc 431
Kumerdej, Blaž 259–260
Küzmič, Števan 233

La Tène period 28
Laibach 530
Lamberger, Caspar 169
Land Fund 476
Lassalle, Ferdinand 357
Latobici 55
Lauriacum 74
Lavant, Diocese of 147
Leben, Stanko 436
Lega Nazionale 352
Legionary camp 62–63
Levec, Vladimir 386
Levstik, Fran 295, 309, 324, 327–328, 334, 342
Liberal Democracy of Slovenia, LDS 540, 550
Liberal Party of Serbia 416
Liberalization 507, 522, 534
Liberals 328, 332, 337, 339–340, 350, 353, 357–358, 360–361, 363, 367–368, 371–372, 376, 406, 416, 418, 430, 459, 505, 507, 509, 516, 595
Liberals 398–399
Liberated territory 461, 464, 467, 470–471
Liberation Front, OF 454–458, 461, 463, 465, 470, 475, 501
– Executive Committee of, IOOF 459, 461, 465
– First Congress 475
Liburni 28
Licinius 76
Licinius Crassus, Publius 45
Lictorian Youth of Ljubljana, GILL 452

Limes 68
Linguistic equality 352, 379
Linhart, Anton Tomaž 262-263, 267-268, 571
Literature, Slovenian 169
Lithuania 175, 209
Littoral, Adriatic 465
Littoral, Slovenian 193
Liupram, Archbishop of Salzburg 111
Livy 41
Ljubelj, Pass 241
Ljubljana 125, 161-162, 165, 201, 208, 216, 224-225, 228, 246, 255, 261, 266-267, 273, 277, 281, 284, 287-290, 295, 299, 324, 326, 332, 335, 352, 359, 367, 377, 385, 398, 401, 407, 475, 520
– Congress of 288
– Cultural group 25
– Diocese of 148, 207, 232, 454, 483
– Emona 37, 48, 50-51, 56, 58, 61, 67, 72, 89
– Marshes 20-21, 58, 257
– Province of 450
Ljubljana Philharmonic Orchestra 422
Ljubljana Radio Orchestra 422
Ljubljana School of Graphics 502
*Ljubljanski zvon*, periodical 417, 436, 439
*Ljudska pravica*, periodical 437
Lloyd George, David 415
Log-boat 23
Lombards 61, 89, 106
– Friulian 89, 92
Lončar, Dragotin 364
Lothair, Frankish Emperor 108
Louis (Ljudevit Posavski), Prince of Lower Pannonia 104, 111
Louis II 110

Louis II of Teck, patriarch 153, 178, 207
Louis IV of Bavaria 174-175
Louis IV, emperor 140
Louis of Anjou 174
Louis the Child 116
Louis the German 110, 112, 115
Louis the Pious, Emperor 104, 110
Louis XIV 235
Lož 137
Lucius Metellus 49
Lucius Verus 67
Lunar, eclipse 65
Lupus, Duke of Friuli 92
Luther, Martin 215-216, 218
Lutherans 216, 223, 227, 233
Luxembourg dynasty 139
– Charles IV 175
– Sigismund 139, 171, 176, 181, 186, 190

Macedonia 45, 416
Macone, Stefano 189
Magnentius 76
Magnus Maximus 77
Magyar 229
Mahnič, Anton 355-356, 390
Maister, Rudolf 408
Majar Ziljski, Matija 297-298, 301-302
Mal, Josip 325
Maleš, Miha 423
Mamula, Branko 532
Mandelc, Janž 225
Mantua, Synod 108
Maraž, Adrian 511
Marčan, Jože 487
March Revolution 197, 309
Marcomannic Wars 60
Marcus Aurelius 61
Marenberk, monastery 169

Maribor  123, 163, 266, 290, 312, 345, 377, 422, 485
Maribor Program  329
*Mariborski Večernik*, periodical  436
Marinko, Miha  477, 501
Marković, Ante  542, 546, 549
Masaryk, Tomáš Garrigue  364, 423
Masonic lodges  280
Maxentius  75
Maximinus Thrax  72
May Declaration  402, 404
Mayr, Johann Baptist  239
Mecklenburg, Duchess of  40
Medžimurje  449
Megiser,Hieronimymus  224
Mehovo Castle  201
Mekinje, monastery  169
Melanchton, Philip  218, 224
Melik, Anton  417, 500
Mencinger, Jože  542, 544
Merania, Dukes.  *See* Andechs, Counts
Mercantile Province of the Littoral  255
Merian, Matthäus  242
Mertlic, Lenart  216
Meško, Kiar  511
Mesolithic period  15
Metallurgy  23
Metastasio, Pietro  267
Metelko, Franc Serafin  296
Methodius, missionary in Moravia and Pannonia  113
Metlika  163–164, 192
Metternich, Clemens  283
Metulum  52
Metzinger, Valentin  247
Michael III, Byzantine Emperor  110
Michelis, Gianii de  549
Mihailović, Draža  459
Mihelič, France  423, 471

Miklošič, Franc  268, 297–298, 301–302, 304
Miletić, Svetozar  333
Military Frontier  213, 223, 230, 234
Milošević, Slobodan  543, 548
Miniskirts  514
Ministerials  157
Minting  44, 53
Mithras cult  64, 70
*Mladina*, periodical  523, 530–532
*Mladinska revija*, periodical  511
Modernization  420, 503, 508, 512
Modestus, Bishop  94
Moesia Superior  73
Mohammed II the Conqueror  184, 192, 200, 211
Mohorjeva družba (St. Hermagoras Publishing House)  325–326, 382
Moline, William  286
Molzbichl, monastery  95
MoMA, New York  512
Montecuccoli, Raimondo  235
Montenegro  409, 416
Moratorium  549
Morgan Line  479
Munich Agreement  443
Municipal rights  163
Municipium  60
Murad II  184
Murad, sultan  183
Murn-Aleksandrov, Josip  383
Murnik Horak, Marija  390–391
Muršec Živkov, Jožef  298
Mušja Cave (Mušja jama)  27
Mussolini, Benito  427, 442, 444

Nanos, Mt. (Mt. Ocra)  37, 42
– Pass  42
*Naša sodobnost*, periodical  500
*Naši razgledi*, periodical  501
National Bank of Yugoslavia, NBJ  543

National Council 407, 409, 416,
  448, 461
National Guard 274, 300, 478
National Liberation 465
– Committee, NOO 467, 475–476
– National Committee of the
  Liberation of Slovenia,
  NKOS 466
– Slovenian National Liberation
  Committee, SNOO 456, 465–466
– Slovenian National Liberation
  Council, SNOS 466–467, 475, 477
National Party 357
National Progressive Party 357
National Radical Party 414, 442
National Socialist German Workers
  Party, NSDAP 452
National University Library 425
Natlačen, Marko 449
Nazi Punk Affair 524
– Fourth Reich 524
Nazis 450–451
Neanderthals
– Flute 17, 569
*Neodvisna slovenska revija*,
  periodical 436
Neolithic 18
Nero 65
Neue Slowenische Kunst, NSK 529
New Civil Movements, Coordination
  for 533
New Primitivism 524
– *Top Lista Nadrealista* 524
Nikopol, Battle of 140, 171, 175, 192
Niš, Declaration 402
Nonnosus, Deacon 90
Noreia 50
Norici 37
– Kingdom 37, 41
– Mines 60
– Noricum 43
– Tetradrachmas 53

Noricum
– Frankish occupation of 90
– province 90
Northern Atlantic Treaty
  Organization, NATO 536
*Nova revija*, periodical 522, 528,
  531
Novačan, Anton 440
Novak, Janež 371
*Novi kolektivizem* 530
Novo Mesto 30, 38, 61, 163–164,
  240, 459
Nuremberg 167

Oberstain, Paulus 209
Obii 66
Oblak, Jožefina 389
Octavian 49
Odilo, Bavarian Duke 94
Odovacar 81
Official Gazette 275, 352
OHO Group 512
Oman, Ivan 541
Ormož 164
Ortenburgs, Ortenburg 178
– Counts of 123, 127, 134, 138, 177
Ortenburg-Sternberg 141
*Osa*, periodical 328
Osimo Agreement 480
Osterc, Slavko 423
Ostrogoths 81
Oswald, Regional Bishop 113
Oswald, Stane 487
Otho 65
Otokar II Přemysl 131, 133,
  173, 186
Otokar II, Bohemian King 136
Otokar IV, Margrave and Duke of
  Styria 131, 158
Otokars. *See* Traungaus
Otto I, Emperor 121
Otto II, Emperor 121–122

Otto III, Emperor  117, 122
Ottoman
– Empire  183, 193, 250
– incursion  191

Pabo  110
Palacký, Fantišek  303
Palacký, František  331
Paleolithic period  15
Pannonia  43, 111, 114
– Province  90
– Superior  71
Pannonian basin  37
Pannonian, Plain  89
Pannonian–Dalmatian rebellion  54
Pan-Slavic Congress  304
Pan-Slovenian People's Party (Vseslovenska ljudska stranka)  363, 371
Papirius Carbo, Gnaeus  50
Paris, Peace Conference  408, 415
Parma, Viktor  384
Parthians  71
Partisan Army  458–460, 463–464, 468, 478
Party of Democratic Renewal, SDP  540
Pašić, Nikola  402–403, 405, 407, 409–410, 415
patrimonial
– Court  154, 157, 276
– Justice  281
Paul the Deacon  92, 100
Paulinus, Patriarch of Aquileia  104
Pavelić, Ante  407, 445
Pavia, Kingdom  109–110
Pavlica, Josip  361
Peasants  80, 125–126, 153–155, 166, 199–200, 267, 282, 300, 304, 312, 476
– Emancipation of  305, 345
– Peasants uprising  195, 197, 202, 222
Peca, Mt.  195
Pelzhoffer, Franz Albrecht  247
Pemmo, Friulian Duke  109
people's assembly  475
People's Committee  477
People's Front  438–439
People's Radical Party  413
People's Commisariat for Internal Affairs, NKVD  489
Peoples, Italian  35
Percl, Jože  487
Peregrine I, Patriarche of Aquileia  151
Perger, Bernard  208
Perseus  45
*Perspektive*, periodical  509–510, 522
Pertinax  70
Pesjakova, Luiza  342
Peter the Great  248, 250
Peterle, Lojze  541–542, 544, 549
Petkovšek, Jožef  384
Philharmonic Society  289
Philip V, Macedonian king  47
Philipp, Patriarch of Aquileia  133
Physiocratic
– Doctrine  257
– Principles  285
Piccolomini, Aeneas Sylvius  179, 207–208, 245
Pilon, Veno  422
Pinnes  54
Pippin  103–104, 106
Piran  161
Pirnat, Nikolaj  423, 471
Plague  68
Planinc, Štefan  512
Plečnik, Jože  423
Pleterje, Carthusian monastery  151, 178, 189
Pleteršnik, Maks  386

Pliberk 249
Pliny the Elder 40, 50
Plut, Dušan 541
Podbevšek, Anton 422
Poetovio (Ptuj) 37, 48, 58, 62–63, 65, 71–72, 77–78
Pogačnik, Jože 514
Pohlin, Marko 260–261
Poland 174, 209, 446, 459
Polič, Zoran 483
Pollini, Franc 269
Polybius 44
Pompeius Trogus 36
Poos, Jacques 549
Pope
– Adrian II 114
– Eugenius 153
– Eugenius IV 153
– Gregory I (Gregory the Great) 106
– John VIII 114
– Leo XIII 355–356, 362
– Nicholas I 113–114
– Pius VI 265
– Pius XI 434
– Zachary 94
Popit, France 532
Popovič, Janez Žiga Valentin 260
Popular Front 437
Poster affair 528–529
Postojna 15, 137
– Gate 48, 89
Požar, Miloš 501
Pozzo, Andrea 247
Praetorium Latobicorum (Trebnje) 62
Pragmatic Sanction 250
Prague Congress 304
Predjama Castle (Predjamski grad) 24
Predojević, Haasan Pasha 230
Prefecture, Bavarian 112

Pregelj, Marij 501
Prekmurje 25, 145, 148, 231–232, 237, 259, 310, 329, 331, 372, 408, 448, 450, 458, 466
Prelokar, Thomas 208
Prepeluh, Albin 364
Preprost, Brikcij 208
Pre-Romantic 266
Prešeren, France 292, 295, 297
Prešeren, Janez Krstnik 246
Pribićević, Svetozar 407, 409, 415
Pribina, Count of Pannonia 105, 110, 112, 124
Priscus of Panium 81
Privatization Act 553
*Problemi*, periodical 526
Protection of Human Rights, Council for the 535
Protestant grammar school 224
Protestantism 216, 218, 222, 227, 239, 245
Protić, Stojan 415
Provincial Diet 210, 317–319, 321, 336, 341, 404
– elections 319–320, 331, 338, 427
Prussia 252, 256, 266, 288, 333, 403
Prussian–Austrian War 334
Ptolemy 40
Ptuj 25, 80, 115, 121, 125, 162–163, 359
– Lords of 123
– Poetovio 37, 58, 89
Ptujska Gora 31, 189
Pučnik, Jože 540
Pufler, Janko 487
Pupienus 72
Pupilija Ferkeverk, experimental theater 515
Putrih, Karel 423

*Quadragesimo anno* 434
Quaglio, Giulio 247

Radgona 121
Radić, Stjepan 416
Radicals 409, 417–418, 430–431, 440
Radio Študent 513
Radovljica 209
Raič, Božidar 336
Raiffeisen cooperatives 343
Ramovš, Fran 452
Ranković, Aleksandar 484–485
Rapallo, Treaty of 408
Rastislav, Moravian Prince 113–114
Ratbod, Prefect of the Eastern March 111
Ratimir, Slavic Prince 111
Raubar, Christophorus 208–209, 216–217
Ravnihar, Vladimir 364
Reading societies 323
– Slavic Reading Society 323
– Styrian 324
Realism 423, 502
– Social 471, 497, 500–501
Rebirth (Preporod) movement 370, 399
Rein, Johann Frederick von 240
Reiner, Friedrich 465
Religious Affairs, Commission for 484
*Rerum novarum* 434
Resistance, anti-Fascist 430, 447, 450, 453–454, 456, 458, 465–466, 469, 471
*Revija 57*, periodical 501, 511, 522
Rialto 108
Ribnica 150, 248, 458
Ribnikar, Adolf 436
Richenburch 125
Richeri 110
Rijeka
– Tarsatica 67, 74
Rižana, Diet of 107, 109

Robba, Francesco 247
Robotti, Mario 458
Rock'n'roll 514
– Punk rock 522, 524
– Yu-rock 524
Rogerij of Ljubljana 240
Roma 411
Roman legions
– II and III Italica 67
– V Macedonica 63
– VIII Augusta 54
– VIIIth legion 57
– XIII Gemina 63, 68, 74
– XVth legion 57
Roman Republican weapons 40
Rome 47, 65
Romuald of Štandrež 228
Romulus 81
Romulus Augustulus 81
Roosevelt, Franklin D 457
Rösener, Erwin 462, 465, 483
Rosthorn brothers 284
Rostohar, Mihajlo 370
Rotar, France 511
Rovinj 161
Roxolani 74
Royal Court 418, 429
Rožanc, Marjan 514
Rožman, Gregorij 434, 463, 483
Rozman-Stane, Franc 464
Rugians 81
Rupel, Dimitrij 549
Rupnik, Leon 465, 470, 483
Ruše group 25
Russia 248, 281, 288
Russian Bolsheviks 405

Sachs, Jeffrey 544
Salacho, Frankish Count 111
Salzburg 111, 114
– Archbishop of 124, 163
– Archdiocese of 113, 122

– Metropolite 147
San Canziano d'Isonzo,
  Monastery 109
Santonino, Paolo 214
Santorio, Santorio 245, 248
Sarmatian, tribes 67
Sava, River
– Savus 25
Savaria 55
Savin, Risto 384
Savinja March 121, 124, 134
Savrič, Karol 487
Savus 60
Scarabantia 89
Schedel, Hartmann 179
Schell von Schellenburg, Jakob 238
Schmerling, Anton 317, 328
Schmidt (Kremserschmidt), Johann Martin 247
Schönbrunn, Treaty of 275
Schönleben, Janez Ludvik 239
school reform 308, 314, 336
Schwarzenberger, Felix 306
Scopolli, Giovani Antonio 268
Scordisci 47
Scythian incursion 31
Seckau, Diocese of 147
Seebach, Petrus 217
Segestani 49
Self-management 493–494, 508, 516, 535
Sempronius Tuditanus, Gaius 49
Sempt-Ebersberg 134
Septimius Severus 69
Serbia, Great 520
Serbian
– Army 403
– National defense 400
Serbian Radical Party 409, 415, 417

Serbs, Croats and Slovenes, Kingdom of, SHS 402, 405, 411, 413, 420
Settlements, hilltop 24, 79
Severan dynasty 71
Severi 70
Sexual revolution 514
Sibiu (Hermannstadt) 216
Siezenheim, Adam Sebastian 240
Sigismund Senior 209
Sigmund, King of Hungary and Bohemia and Holy Roman Emperor 140
Silesia 252
Silius Nerva, Publius 53
Simović, Dušan 446
Singidunum, present day Belgrade 47
Sirmium 55, 61, 72, 89
Sisak, Battle of 223, 230–231
Sisters of Scipio Nasica, Theater 529
Situla Art 28
Slatkonja, Jurij 209, 216
Slaves 80
Slavic Christian People's Association (Slovanska krščanska narodna zveza) 360
Slavonia 106, 116, 171, 176, 211, 368, 415
Slavs 84, 91, 108–109
– Alpine 92, 130
– Balkan 279
– Carantanian See Carantanians, Carantania
– resistance to Avars 91
– settlement 84, 89, 91, 563
– South 260, 267, 279, 296, 324, 327, 367, 369, 401, 407
Sloga
– Periodical 338
– Society 352

Slomšek, Anton Martin 291, 298, 309, 312, 325
*Slovenec*, periodical 328, 337, 355, 382
Slovenes
- Old 327, 332, 334, 338, 349
- Young 327–328, 332–334, 338, 349
*Slovenia, Fundamental Charter of* 537
Slovenia, Kingdom of 302
Slovenia, People's Republic of, LRS 477
- Constituent Assembly 477
- First Constitution of 477
Slovenia, United 290, 298, 300, 302–303, 307, 327, 330, 332, 340, 370, 454
Slovenia's Sovereignty, Declaration of 542
Slovenian Academy of Sciences and Arts 425, 498, 512
Slovenian Alliance 461
Slovenian Arts Society 385
Slovenian Catholic Convention 356
Slovenian Choral Societies, Association of 384
Slovenian Christian Democrats 541
Slovenian Christian Party, SKS 455
Slovenian Christian Social Association 363
Slovenian Democratic Alliance 536
Slovenian Music society 342, 384
Slovenian National Army 462, 468
Slovenian National Gallery 422
Slovenian national movement 281, 289, 292, 298, 300, 302, 308, 321, 323, 325, 337
Slovenian national program 303, 332, 528

Slovenian National Theater 422
Slovenian Parliament 542
Slovenian People's Party (SLS) 363, 369, 372, 376, 392, 414, 418, 429–433, 435, 438, 440, 442, 445, 455
Slovenian People's Party of Gorizia (Slovenska ljudska stranka za Goriško) 361
Slovenian Security Assembly 463
Slovenian society 434, 438
Slovenian Song Festival 499
Slovenian Women Teachers' Association (Društvo slovenskih učiteljic) 391
Slovenian Writers, Association of 530, 533, 535
Slovenian, Declaration 430
Slovenian-Croatian-Serbian-Bulgarian Republican Confederation 440
Slovenj Gradec 137, 163
*Slovenska beseda*, periodical 437
Slovenska Bistrica 163
Slovenska matica (Slovenian Literary Society) 323, 326, 338, 342, 361, 370, 385
Slovenska Matica (Slovenian Literary Society) 439, 500
*Slovenska zemlja*, periodical 437, 455
*Slovenski cerkveni časopis*, periodical 303
*Slovenski gospodar*, periodical 328, 337
*Slovenski narod*, periodical 328, 335, 356, 382
Smerdu, Frančišek 423
Smole, Dominik 514
Soča Front 397, 426
Soča, River 73, 82, 89, 123, 141–142, 150, 273
Social Democratic Party

- Austrian 357, 367
- Yugoslav 358, 364
Social Democrats 371, 392, 399, 405, 409
Social movements, alternative 522
Socialism 362, 364, 483, 493, 517
- Administrative 475
- Christian 433
- collapse of 9
Socialist Party 540
Socialists, Christian 433, 435, 438, 454–455, 463
*Sodobnost,* periodical 436, 439, 511
Solkan 125
Sonnenfels, Joseph von 267
Southern Railway 283, 312
Soviet Union 460, 490
Spanheims, Dukes of Carinthia 123, 131–132, 136, 143, 151, 172
- Bernhard von Spanheim 133
- Ulrich III 136, 151
Spittal an der Drau 90
Springtime of Nations 299, 389
St. Cyril and Methodius Society 352, 359, 380
St. Dismas, Society of 246
St. Gotthard, the Battle of 235
St. Jerome 58
St. John of Duino, monastery 150
St. Jurij beneath Rifnik 199
St. Mark, Republic of 109
St. Mary, societies 392
Stalin, Josip Visarionovič 460, 489
Stara Pravda, Faction of Slovenian Christian Party 455
Stare, Josip 325, 386
State Security Administration, UDBA 485
State Security Service, SDV 524, 530, 532, 539
State Treaty, Austrian 480
Steam engine 283, 286

Stefan, Jožef 386
Stephen II Kotromanić 174
Stepinac, Alojzije 484
Sternen, Matej 423
Stična 29–30, 85
- Cistercian monastery 151, 168
- Manuscript 168
Stojadinović, Milan 442
Stone-cutters 57
Strabo 40
*Straža v viharju,* periodical 440
Stritar, Josip 293, 323, 342
Struppi, Vincenc 257
Studenice, monastery 169
Stupica, Gabrijel 511
Styria 24, 115, 123, 130, 139, 151, 158, 169, 173, 182, 193, 210, 224, 323, 331, 351, 359, 372, 380, 467
Styrian Homeland League (Heimatbund) 452
subculture, youth 522
*Südslawische Zeitung,* periodical 334
Supilo, Fran 401
Svatopluk 114
Svetec, Luka 333
Svetokriški, Janez 240
*Svit,* periodical 511
Switzerland 246
Sylvester Patent 308
symbolism 383, 509
Šalamun, Tomaž 512
Šeligo, Rudi 515
Šinigoj, Dušan 542
Škocjan Caves (Škocjanske jame) 27
Škofja Loka 122, 127, 164, 228, 381
- monastery 169
štiftarstvo, štiftarji 215, 230
Štiglic, France 514
Štih, Bojan 502
Štrekelj, Karel 386
Štukelj, Leon 426
Šubašić, Ivan 466

Šubic, Janez 343, 384
Šubic, Jurij 343, 384
Šubic, Simon 386
Šuklje, Fran 372, 406
Šuštaršič, Marko 511
Šusteršič, Ivan 401, 440

tabor
- political gathering 324, 335–336
- stronghold 196, 203
Tacitus 57
Tasič, David 530
Tassilo III, Bavarian Duke 95, 103
Tattenbach, Hans Erasmus 235
Taufferer, Siegfried von 273
Taurisci 37
Tavčar, Franja 392
Tavčar, Ivan 342, 357, 391, 406
Taxes 107, 153–155, 199, 234, 252, 265, 274, 281, 287, 304, 319, 421
Territorial Defense 509, 531, 545, 547
Tetrarchy 75
Teurnia 89
Textor, Urbanus 218, 227
Theoderic, Regional bishop 95
Theodosius 75
Theodosius, Emperor 92
Theotmar, Archbishop of Salzburg 115
Theresian Reforms 252, 254, 259
Thirty Years' War 234
Thurn, Karl 235
Tiberius 54
Tiberius Pandusa 50
TIGR 427
Tihec, Slavko 511
Timavus 55
Times, The 405
Tisza, River 91

Tito, Josip Broz 459–460, 466, 483–484, 489, 493–494, 501, 518
Tito–Šubašić Agreement 467
Toleration, Edict of 264
Tolmin 31, 150, 198, 200, 234
Toman, Lovro 332, 334
Tomec, Ernest 434
Tonkli, Josip 356
Tonovcov grad 82
Tools 27
Towns
- Centers of craft 161
- Coastal 161
- Inland 162
Trajan 63
Transport 56, 132, 251, 278, 285, 346, 446, 482, 495, 510
Traungaus, Margraves of Styria 124
Treaty of Campo Formio 274
Trebonianus Gallus 73
Trieste 18, 24, 43, 108, 137, 139, 144, 161–162, 175, 179, 194, 200, 207, 210, 215–219, 225, 227–228, 236–238, 245, 251–252, 255–256, 261–262, 264, 266, 273, 275, 277–278, 281–283, 285–288, 290, 298, 303, 307, 310, 312–313, 318, 323, 328, 336–340, 344–346, 352, 358, 361, 364, 366, 369, 372–374, 376–377, 380–382, 384, 386, 391, 397, 408, 426–427, 467
- Dioces of 148
- Tergeste 37, 42
Trieste–Rijeka, road 89
Triglav 546
- Periodical 328
- Student Society 455
tripartite concept 368, 371
Tripartite Pact 446–447
Triumvirate 52
Trpinc, Fidelis 286
Tršar, Drago 511

Trstenjak, Davorin 297
Trubar, Primož 191, 210, 217–219, 221–223, 227, 232, 245
Trumbić, Ante 402–403, 407, 409
Tübingen 218, 223
Tudjman, Franjo 549
Tuma, Henrik 361
Turjak, Castle 462
Turkey 54, 235, 250–251, 265–266, 349, 415
Turkish raids 154, 160, 165
Turks, Turkish Empire 12, 155, 181, 186, 192–193, 195, 197, 200, 213, 221, 235–236, 245
Tvrtko II Kotromanić 177
Tyrol 91
Tyrš, Miroslav 455

*Ubi arcano Dei consilio* 434
*Učiteljski tovariš*, periodical 308
Ude, Lojze 463
Ulrichsberg 93
Umag 161
Umvölkung 451
Ungnad, Ivan 227
United Nations, UN 490
– membership 550
United Workers Association 438
University of Ljubljana 169, 420, 422, 484
Upper Carniola 55, 58, 122–123, 134–135, 138, 150, 198–199, 344–346, 421, 450, 452, 455, 467
Urach 221, 227
Urbanization 60–61, 374, 377–378, 503
Urnfield culture 24, 30
Uskoks 212–213
Ustaše 445, 447

Val Canale 92
Valentinian I 77

Valerianus 73
Valerius Maximianus, Marcus 63
Valerius Valerianus, Lucius 70
Valvasor, Janez Vajkard 241–242, 245, 248
Varna, Battle of 181
Vatican 484–485
– Index of Forbidden Works 221
Veber, France 440
Vega, Jurij 257, 270
Velesovo, Monastery 169
Velleius Paterculus 54
Veneti 28, 33
Venetian Republic 129, 137, 143–144, 150, 153, 163, 165, 189, 259, 280
Venetian Slovenia (Beneška Slovenija) 274, 302, 329–330, 341–342, 352, 368, 380, 382
– Venetian Slovenes 207
Venice 11, 106–108, 114, 138, 143, 162, 178, 200, 235, 274
Vergerius, Jr., Petrus Paulus 218, 221
Vespasian 60
Vetrinje 463
– Refugee camp 482
Victoriatii 50
Victorinus of Poetovio 64
Vidmar, Josip 421, 435–436, 477, 512
Vidmar, Josipina 392
Vidmar, Milan 387
Vienna 21, 89, 163, 186, 213, 236, 259, 267, 270, 279, 285, 290, 299, 302, 304, 343, 346, 364, 376, 383, 386, 402, 423
– Congress of 281
– Parliament 304–307, 331, 338, 372
– Reichstag 431
– Treaty of 233

– Universitiy of 208, 218
Vilhar, Miroslav 327
Villach 163, 202, 216, 231, 255, 310, 346
Vinča 19
Vinica 40
Vipsanius Agrippa, Marcus 54
Virgilius, Bishop of Salzburg 94–95
Virunum 89
Vischer, Georg Matthäus 243
Visegrád 181
Visigoths 79
Vitellius 65
Vitovec, Jan 180, 190
Vlah 90
Voccio 37
Vodnik, France 455
Vodnik, Valentin 261–262, 278–279, 292, 295, 310
Vogelsang, Karel 362
Völkermarkt 132
Volusianus 73
Vošnjak, Josip 323, 334, 356, 384
Vovk, Anton 484
Vraz, Stanko 297
Vrhnika 547
– Nauportus 37, 56
Vučedol 20

War Council 546
Weapons 27
Weichselburg, Counts of 124, 135
Weimar-Orlamünde 123
Wermannschaft, paramilitary formations 452
White Guard 459, 461
Wiener Neustadt 184, 190, 236, 448
Wiener, Pavel 216
Wilhelm II 134
William I, German Emperor 334
William II, Margrave of the Savinja March 124

William III, Prussian Emperor 283
Wilson, Woodrow 416
Windische Mark. *See* Prekmurje
Windischgrätz, Alfred 358
Winkler, Andrej 318, 351
Witigowo, Carantanian Count 110
Wittelbachs 177–178
Wittenberg 219, 224
Wiz von Wizenstein, Fran 240
Władysław II Jagiełło 175–176
Wolkenstein, Oswald von 170
Wooden cart 23
Working brigades 495
Working People, Socialist Alliance of the, SZDL 438, 501, 533, 535
Working People, Socialist Alliance of thw, SZDL 536
World War I 233, 282, 312, 325, 337–338, 352, 369, 371–372, 374, 376–377, 381, 386, 398, 419, 422, 429
World War II 127, 326, 328, 413, 416, 421, 423–424, 436–437, 439, 446–447, 469
Württemberg 202, 219, 221, 229

Yugoslav
– Government 407, 413, 418, 465, 484–485, 520
– Government 446
– Idea 337, 368
– Nation 406, 409
– State 405, 411, 413, 437, 441
Yugoslav Committee 401–403, 407
Yugoslav Democratic Party, JDS 413–414, 416
Yugoslav Muslim Organization, JMO 414, 442
Yugoslav National Army, JNA 531, 539, 545, 547, 549
Yugoslav Parliamentary Club 404
Yugoslav People's Army, JNA 519

Yugoslavia  411, 416–419, 426, 431, 440–441, 448, 465, 514, 541, 546
Yugoslavia, Democratic Federative  475–476
Yugoslavia, Federative People's Republic of, FNRJ  477, 494
Yugoslavia, Kingdom of  415, 468
Yugoslavia, Socialist Federative Republic of, SFRJ  504, 549
Yugoslavism  370–371, 418, 421, 436, 509

Zagreb, Diocese of  145
Zarnik, Valentin  334, 336
Zavrl, Franci  530
*Zdravljica*, national anthem  539
Zelenaši  409
Zeno  82
*Zgodnja danica*, periodical  308
Zimska pomoč  471
Zlobec, Ciril  541
Znojilšek, Janž  232
Zois, Karl  345
Zois, Michelangelo  238
Zois, Žiga  268, 279
Zollfeld  93
Zrínyi, Miklós  235
Zrínyi, Péter  236
Zupan, Vitomil  514
Zupančič, Beno  514
Zwitter, Fran  467
Železnikar, Franc  357
Žerjal, Gregor  364
Žiče, Carthusian monastery  151, 168, 264
Žitnik, Ignacij  360, 362
Živadinov, Dragan  529
Žovnek, Lords of. *See* Celje: Counts of
Žumberk  212
Župa  125
Župančič, Oton  383, 435, 470

www.ingramcontent.com/pod-product-compliance
Ingram Content Group UK Ltd.
Pitfield, Milton Keynes, MK11 3LW, UK
UKHW041324050325
4871UKWH00002B/12